HANDBOOK OF MANUFACTURING INDUSTRIES IN THE WORLD ECONOMY

Handbook of Manufacturing Industries in the World Economy

Edited by

John R. Bryson

Professor of Enterprise and Competitiveness, Birmingham Business School, University of Birmingham, UK

Jennifer Clark

Associate Professor, School of Public Policy, Georgia Institute of Technology, USA

Vida Vanchan

Associate Professor, Geography and Planning Department, State University of New York (SUNY) Buffalo State, USA

EE Edward Elgar
PUBLISHING

Cheltenham, UK • Northampton, MA, USA

Published by
Edward Elgar Publishing Limited
The Lypiatts
15 Lansdown Road
Cheltenham
Glos GL50 2JA
UK

Edward Elgar Publishing, Inc.
William Pratt House
9 Dewey Court
Northampton
Massachusetts 01060
USA

A catalogue record for this book
is available from the British Library

Library of Congress Control Number: 2014950966

This book is available electronically in the **Elgar**online
Business subject collection
DOI 10.4337/9781781003930

MIX
Paper from responsible sources
FSC www.fsc.org FSC® C013056

ISBN 978 1 78100 392 3 (cased)
ISBN 978 1 78100 393 0 (eBook)

Typeset by Servis Filmsetting Ltd, Stockport, Cheshire
Printed and bound in Great Britain by T.J. International Ltd, Padstow

Contents

PART IV POLICY NARRATIVES IN MANUFACTURING

PART V CONCLUSION

Contributors

Marilyn A. Brown is an Endowed Professor of energy policy in the Georgia Institute of Technology's School of Public Policy, USA and a distinguished visiting scientist at Oak Ridge National Laboratory, USA. The author of two books and more than 250 publications, she is a co-recipient of the 2007 Nobel Peace Prize, for contributing to the report on *Mitigation of Climate Change*. Brown has served on six committees of the National Academies of Sciences, including the Committee on America's Climate Choices and the Board of Energy and Environmental Systems. She co-founded the Southeast Energy Efficiency Alliance and is a member of the Editorial Board of the *Journal of Energy Efficiency*. In 2010, she was sworn onto the board of directors of the Tennessee Valley Authority, the USA's largest public power provider, following her nomination by President Barack Obama.

John R. Bryson is Professor of Enterprise and Competitiveness at the Birmingham Business School, University of Birmingham, UK. His research initially focused on the rise and role of business and professional services (BPS) but since 2005 has explored the changing economic geography of manufacturing in developed market economies. This research has explored the interactions between BPS and manufacturing and, in particular, the role design plays in the competitiveness of manufacturing firms. His books include *Service Worlds: People, Organisations, Technologies* (Routledge); *Hybrid Manufacturing Systems and Hybrid Products* (IMA/ZLW & IfU); *Design Economies and the Changing World Economy* (Routledge) and *Industrial Design, Competition and Globalization* (Palgrave).

Susan Christopherson is a Professor in the Department of City and Regional Planning at Cornell University, USA. An economic geographer with a commitment to comparative research, her central concern is the social and economic dimensions of urban and regional development.

Jennifer Clark is an Associate Professor in the School of Public Policy and the Director of the Center for Urban Innovation at the Georgia Institute of Technology, USA. Her research focuses on regional economic development, manufacturing, industrial districts and innovation. Her first book, *Remaking Regional Economies: Power, Labor, and Firm Strategies in the Knowledge Economy* (with Susan Christopherson; Routledge) won the Best Book Award from the Regional Studies Association in 2009. Her newest book, *Working Regions: Reconnecting Innovation and Production in the Knowledge Economy* (Routledge, 2013), focuses on policy models aimed at rebuilding the links between innovation and manufacturing in the US.

Margaret Cowell, PhD, is an Assistant Professor of Urban Affairs and Planning at Virginia Tech, USA. She is a member of the John D. and Catherine T. MacArthur Foundation-funded research project, 'Building Resilient Regions' and is also part of a team of researchers assessing the potential of the homeland security economy for community economic development at the St. Elizabeth's Hospital site in Southeast Washington, DC. Dr Cowell's research has been funded by the MacArthur Foundation, National Association of Counties and the United States Economic Development Administration.

Marc Doussard is an Assistant Professor in the Department of Urban and Regional Planning at the University of Illinois at Urbana-Champaign, USA and the author of *Degraded Work* (University of Minnesota Press). His research focuses on low-wage labour markets and the generation of living-wage jobs.

Dawn M. Drake is an Assistant Professor in the Department of History and Geography at Missouri Western State University, USA. Her research focuses on various geographic issues of competitiveness for the farm machinery industry in the US and abroad.

Colin G. Drury is Distinguished Professor Emeritus of Industrial and Systems Engineering, University at Buffalo, State University of New York, USA. He has over 200 publications in the field of aviation human factors and was awarded the Federal Aviation Agency's 'Excellence in Aviation Research' award in 2005.

Andrew Dugenske is the Manager of Research Services and a Principal Research Engineer at the Georgia Tech Manufacturing Institute, USA.

Maryann Feldman is the Heninger Distinguished Professor in the Department of Public Policy at the University of North Carolina, USA. Her research and teaching interests focus on the areas of innovation, the commercialization of academic research, and the factors that promote technological change and economic growth.

Paul L. Forrester is currently Senior Teaching Fellow in Enterprise and Management at Keele Management School, Keele University, UK. His main field of interest is in the design and organization of production and operations. His early research focused on production systems design, quality, flexibility, just-in-time and agility. The current focus of his research is on the management and organization of projects, process analysis and mapping, enterprise resources management and the management of international operations.

Chris Gibson is Professor of Geography and Director of the Global Challenges Research Program at the University of Wollongong, Australia. His research interests are in cultural economy and regional transformation. His recent books include *Creativity in Peripheral Places: Redefining the Creative Industries* (Routledge) and (with Andrew Warren) *Surfing Places, Surfboard Makers* (University of Hawaii Press, 2014).

Peter V. Hall is an Associate Professor of Urban Studies at Simon Fraser University in Vancouver, Canada. His research examines port cities, seaports and logistics, as well as local economic development. His publications include *Cities, Regions and Flow* (Routledge, 2013).

Teis Hansen is Associate Senior Lecturer at the Department of Human Geography, Lund University and Researcher at the Centre for Innovation, Research and Competence in the Learning Economy (CIRCLE), Lund University, Sweden. In addition to the development of low-tech industries, he has been working on collaborative innovation projects in the cleantech industry and cross-border innovation. Current research is on the geography of sustainability transitions and technology transfer in the field of environmental technologies.

Catherine Harris is a Postdoctoral Research Fellow at the University of Sheffield, UK. Her research interests are in the fields of entrepreneurship and enterprise, ethnic entrepreneurship, social difference, EU enlargement and EU migration. Her recent research includes the first large-scale study of the experiences of Polish migrant entrepreneurs in the West Midlands region of the UK.

Patrice Jalette is Professor at the School of Industrial Relations of Université de Montréal, Canada. Co-researcher at the Inter-University Research Centre on Globalization and Work, Canada, he researches questions relating to employment and labour relations in multinational companies and restructuring firms.

Ronald V. Kalafsky is an Associate Professor in the Department of Geography at the University of Tennessee, USA. Research interests include the geographies of manufacturing and trade, especially location-based competitive issues for firms in mature industrial regions, including Canada and Japan. Research has been published in journals such as the *Asia Pacific Business Review* and *Journal of Geography*.

William C. Kessler, before retirement, was the Director of Executive Programs in the Tennenbaum Institute and a Professor of Practice in the School of Industrial and Systems Engineering, Georgia Institute of Technology, USA. His specific focus was on aligning interdisciplinary and trans-disciplinary research with real-world enterprise system challenges such as capabilities for manufacturing in global complex (socio-technical-policy) enterprises. Kessler is co-editor of the book (published IOS Press, 2012) *Enterprise Transformation – Manufacturing in a Global Enterprise*.

Gyungwon Kim is a PhD student in the School of Public Policy of the Georgia Institute of Technology, USA. She started her PhD in 2008, with an interest in clean energy policy and associated economic development impacts. She also has a Master's in City and Regional Planning from Seoul National University and a BS in Urban Engineering from the University of Seoul in South Korea. Her current work is focused on policy to promote energy efficiency, and estimating its macroeconomic impact using the National Energy Modeling System and Input–Output modeling. She is a Graduate Research Assistant in the Climate and Energy Policy Laboratory.

Dieter F. Kogler is a Lecturer in Economic Geography at University College Dublin, Ireland. His research focus is on the geography of innovation and evolutionary economic geography, with particular emphasis on knowledge production and diffusion, and processes related to technological change and innovation.

Lauren Lanahan is a PhD candidate in the Public Policy Department at the University of North Carolina at Chapel Hill, USA. Her research interests include science and technology policy, economic development, entrepreneurship and policy evaluation.

Finbarr Livesey is currently Lecturer in Public Policy in the Department of Politics and International Studies, University of Cambridge, UK and academic visitor at the Institute for Manufacturing, Department of Engineering, University of Cambridge. He spent five years as a technology developer and consultant with Olivetti Research Labs and Cambridge Consultants Ltd. From 2004 to 2011 he ran the Centre for Industry and Government at the Institute for Manufacturing, University of Cambridge. His research focused on the role of government in supporting growth based on the emergence of new industries, and how policy making is changing due to the emergence of new technologies.

Nichola J. Lowe is Associate Professor of City and Regional Planning at the University of North Carolina at Chapel Hill, USA. Her research focuses primarily on regional economic and labour market adjustment in the North American context. Her work has been published in economic development and planning journals, including *Environment and Planning A*, *Economic Geography*, *Economic Development Quarterly* and *World Development*.

Lynn McCormick is an Associate Professor in the Department of Urban Affairs and Planning at Hunter College/City University of New York, USA. Her work focuses on community and economic development. Recent publications have investigated business associations' roles as workforce development intermediaries and the economic development goals of big cities' climate change plans.

Rachel Mulhall is a Research Fellow at the Birmingham Business School, University of Birmingham, UK. She is a co-author of a 2013 paper on 'The Energy Hot Potato and Governance of Value Chains: Power, Risk and Organizational Adjustment in Intermediate Manufacturing Firms', *Economic Geography*, 89. Her research interests are in firm and supply chain competitiveness, the regional economy, and the influence of energy, commodity markets and procurement in shaping competitiveness.

Sam Ock Park is Professor Emeritus of Seoul National University and Chair Professor at Gachon University, South Korea. He has published 20 books and 150 papers in English and Korean, focusing on Modern Economic Geography and Globalization and the Asia Pacific Region. His current research interests are the dynamics of economic spaces and global knowledge networks.

Petr Pavlínek is a Professor of Geography at the University of Nebraska at Omaha, USA and at Charles University in Prague, Czech Republic. His recent articles have appeared in *Economic Geography*, the *Journal of Economic Geography*, *Cambridge Journal of Regions, Society and Economy* and *European Urban and Regional Studies*. His research focuses on the regional development effects of foreign direct investment and the automotive industry in East-Central Europe.

John Provo, PhD, is Director of Virginia Tech's Department of Economic Development and an Adjunct Professor in the Urban and Affairs and Planning Program at Virginia Tech, USA. A 'pracademic' with more than 15 years of economic development experience in higher education and government, he builds partnerships that link university, community, and industry resources. He has researched and designed recent projects with manufacturers producing transportation equipment, composite materials and biopolymers. His work has been supported by various local, state and national funders, including the United States Economic Development Administration, United States Department of Labor and the Ford Foundation.

Megan Ronayne is an Economic Geographer based at the Birmingham Business School, University of Birmingham, UK. Her research interests focus on understanding the evolving economic geography of manufacturing in developed market economies with a particular focus on identifying and understanding survival strategies and competitiveness and emergent industries.

Greg Schrock is an Assistant Professor in the Nohad A. Toulan School of Urban Studies and Planning at Portland State University, USA. His research focuses on workforce development policies and their role in promoting industrial revitalization and social equity.

Carey Durkin Treado is an Economist whose research focuses on industrial organization in the context of international markets, including a four-year study of the steel industry value chain at the University of Pittsburgh, USA. In addition to academia, she has worked with the public sector and is the co-founder of a high-tech company. She is currently working as a consultant to the Competitive Industries Global Practice of the World Bank.

Vida Vanchan is an Associate Professor of Geography and Planning at SUNY-Buffalo State, USA. She is the author and co-author of a book and several journal articles on the competitive characteristics and dynamics of industrial design firms. Her research interests include industrial competitiveness and competitive characteristics of firms; development; international trade, foreign direct investment; multicultural management and negotiation; Southeast Asia; and emerging economies.

Susan Walcott is a Professor of Geography at the University of North Carolina at Greensboro, USA. She is the author and co-author of papers on China and the furniture industry, including a 2012 paper titled 'Developing China's West: Producer Services in Metropolitan Xi'an'(with Chen Ying), Special Issue *Regional Science Policy and Practice*; and 'The Furniture Foothills and the Spatial Fix: Globalization in the Furniture Industry' (2011) *Southeastern Geographer*. Her current research interests focus on industry changes relating to global competitive restructuring of the furniture industry, and Chinese outward foreign direct investment.

Ben Wang is Executive Director of Georgia Tech Manufacturing Institute, USA. He holds the Gwaltney Chair in Manufacturing Systems in the School of Industrial and Systems Engineering and is a Professor in the School of Materials Science and Engineering.

Michael Ward is Professor of Micro Electro Mechanical Systems (MEMS) Technology at Birmingham City University, UK. He initially trained as a physicist. His research focus is on various aspects of microelectronics and micro engineering. He worked at the RSRE (now Qinetiq), Malvern UK. At the RSRE he worked on the development of narrow band gap infrared detectors and radiation hard electronics, and became interested in MEMS. In 1999 he moved to the University of Birmingham, where he established a research activity on micro sensors. He is now based at Birmingham City University and remains interested in micro sensors and their applications, but has a special interest in energy harvesting technology, including micro nuclear power sources and additive manufacture.

Andrew Warren is Lecturer in Economic Geography at the University of Wollongong, Australia. His research interests include cultural and political economic theory, manufacturing, labour geographies and regional development. Andrew is a lifelong surfer. His recent book with Chris Gibson is titled *Surfing Places, Surfboard Makers* (University of Hawaii Press, 2014).

Sally Weller is an Australian Research Council Future Fellow at Monash University, Victoria, Australia. Her current research is focused on the restructuring of regional agricultural and energy sector labour markets as a consequence of shifts in government policy. Her 2001 book, *Re-fashioning the Rag Trade* (University of New South Wales Press), co-authored with Michael Webber, traces the decline of the Australian clothing industry after trade liberalisation.

Lars Winther is Associate Professor at the Department of Geosciences and Natural Resource Management, University of Copenhagen, Denmark and head of the research group Transformation of Cities and Landscapes. His research area is within economic geography, with special focus on industrial location and change, urban and regional development, and the geography of the knowledge economy.

Laura Wolf-Powers is an Assistant Professor of City and Regional Planning at the University of Pennsylvania, USA. Her research involves metropolitan labour markets and employment policies as well as the political economy of urban economic and community development.

Foreword

Wherever I go in the world people are talking about manufacturing. Some countries worry that they have too little while others are concerned that they have too much! There are many reasons for the rapidly rising interest in the manufacturing industries but perhaps the most significant is a renewed appreciation of the role that manufacturing plays in economies of all sizes.

Towards the end of the 20th century and the beginning of the 21st, there was much talk of post-industrial societies and the trend towards services. There were some justifications for these views. Wealthier societies were indeed consuming more services and advanced economies were finding it increasingly difficult to compete in labour-intensive manufacturing. What many commentators at the time failed to appreciate, however, was that manufacturing is not simply about changing the shape of materials; rather, it can be seen as a whole cycle of activities, from understanding markets through to design, production, distribution and service. This modern 'value chain' view of manufacturing meant that most economies want at least some part of the action. Indeed, losing capability in the physical part of manufacturing might make it more difficult to access other, potentially more lucrative parts of the cycle. So around the world there is a reappraisal of what is meant by manufacturing and how it might impact on local prosperity, growth and international engagement.

This book represents a major contribution to our thinking about modern manufacturing industries – and is not just timely, it is long overdue! The authors have done an outstanding job in bringing to bear a range of multi-disciplinary perspectives on a domain which all too often suffers from rather narrow disciplinary analyses. Ranging from engineering to social science and drawing on examples from the US, Europe and Asia, the book provides not only a wealth of fact and illustration but also a rich landscape to inform those charged with industrial policy and manufacturing strategies.

The book will, I am sure, be warmly welcomed by scholars and students as well as practitioners. It provides essential reading for those involved in the renaissance of manufacturing and will enable them to take a broad and informed view of this vital and fascinating activity.

Professor Sir Mike Gregory
Head, Institute for Manufacturing
University of Cambridge

Preface

Much of the literature and debate in the social sciences and amongst policy-makers over the last decade has been on understanding service-based economies, neglecting manufacturing and its effects on the economy. The financial crisis that commenced in 2008 has been associated with political debates that revolve around the 'real' economy and a call for rebalancing regional and national economies. The use of the term 'real' highlights the production of tangible goods rather than intangible services or even various forms of fictitious and financial capital. Over the last decade or so, manufacturing has been transformed but academic understanding has not kept pace with these alterations. Many of these alterations revolve around the development of new forms of expertise-driven manufacturing, new geographies of production, jobless growth, the development of hybrid products and hybridized production services that combine services and manufacturing tasks in novel ways, and new forms of financialization applied to manufacturing. These ongoing transformations suggest that it is timely to explore the development of a new geography of manufacturing and even to unravel the evolving geographies of manufacturing that have emerged since the 'crisis of Fordism' in the 1970s.

This interdisciplinary volume provides a critical and multi-disciplinary review of current manufacturing processes, practices and policies, and broadens our understanding of production and innovation in the world economy. Chapters highlight how firms and industries modify existing processes to produce for established and emerging markets through dynamic and design-driven strategies. This approach allows readers to view transformations in production systems and processes across sectors, technologies and industries. Contributors include scholars ranging from engineering to policy to economic geography. This collection demonstrates that manufacturing continues to matter in developed market economies as in emerging economies. It is important to appreciate the existence of varieties of capitalism or variegated capitalism but that manufacturing is also variegated in place and across space. In this context, it is important to appreciate place-based manufacturing differences as well as the role that trans-local or 'global' production relationships play in the production of manufactured products and hybrid products.

This book highlights contributions that critically engage in understanding manufacturing and production. It brings together chapters that explore the dynamics of manufacturing theoretically and empirically to advance debates and understanding of manufacturing from pre- to post-industrial worlds. The book also develops conversations between different but often unrelated debates in economic geography, for example manufacturing and financialization, manufacturing and creative work, manufacturing and logistics, and manufacturing and consumption. Chapters examine topics including theory and methods as well as theoretically grounded empirical analysis of production processes, pre and post production (from raw materials to marketing, design and consumption, including embedded services and supporting services), factors of production, financialization, and the growing policy debates about how to shape and sustain manufacturing in the 21st century world economy.

Particularly in light of the renewed policy interest in manufacturing as a core component of national and regional export strategies, there are targeted policy efforts underway to support manufacturing in advanced industrialized economies. A stated goal of these new policy initiatives is to 'push innovation down the supply chain' to create localized competitive advantage by combining production and innovation capacities. In a global economy often defined by the complexity or extent of its geographically distributed supply networks, such a policy priority requires a precise and nuanced understanding of the current state of manufacturing and its shifting dynamics. As this volume illustrates, that understanding is embedded in an interdisciplinary empirical approach to manufacturing processes that spans the social sciences, engineering and policy disciplines. It is our goal for this volume both to inform readers about the realities of manufacturing in the world economy and to make the case for the ongoing support of such critical, interdisciplinary research.

Acknowledgements

A book project like this represents (and requires) the work of many participants. We particularly thank the authors who contributed to the volume: Paul Forrester, Sally Weller, Susan Christopherson, Peter V. Hall, Margaret Cowell, John Provo, Patrice Jalette, Michael Ward, Ben Wang, William C. Kessler, Andrew Dugenske, Marilyn A. Brown, Gyungwon Kim, Dieter Kogler, Megan Ronayne, Susan Walcott, Ronald V. Kalafsky, Dawn M. Drake, Carey Durkin Treado, Rachel Mulhall, Colin G. Drury, Andrew Warren, Chris Gibson, Catherine Harris, Finbarr Livesey, Petr Pavlínek, Sam Ock Park, Marc Doussard, Greg Schrock, Laura Wolf-Powers, Lynn McCormick, Teis Hansen, Lars Winther, Maryann Feldman, Lauren Lanahan and Nichola J. Lowe. We appreciate their time, patience, expertise and commitment to the broader project of understanding how manufacturing works in the global economy and, perhaps more normatively, how manufacturing can work better to produce equitable and sustainable growth.

We also thank the colleagues, institutions and individuals who supported the project, including the Georgia Institute of Technology, the Geography and Planning Department and School of Natural and Social Sciences at Buffalo State – The State University of New York, and The University of Birmingham, UK. We appreciate the support of the Center for Urban Innovation, the Ivan Allen College of Liberal Arts and the Office of the Executive Vice President for Research, all at Georgia Tech. We also appreciate the support provided by the Institute for Advanced Studies (IAS) at the University of Birmingham and, in particular, Professor Malcolm Press, Sue Gilligan and Sarah Myring. The IAS-funded project on 'Regeneration Economies: Transforming People, Place and Production' provided the funding to bring social scientists and engineers together to explore old and new forms of manufacturing.

Producing an edited collection is a partnership between many different individuals – the editors, chapter authors and their families, but also those manufacturing firms and employees who were willing to allow academics access and provided the empirical evidence that made many of the chapters in this collection possible. We should also not forget all the policy-makers who have given their time to discuss manufacturing matters. Finally, we appreciate the work of our editor at Edward Elgar, Matthew Pitman for encouraging us and seeing the project through to completion.

PART I

INTRODUCTION

Introduction: manufacturing matters: space, place, time and production

Vida Vanchan, John R. Bryson and Jennifer Clark

Manufacturing has undeniably transformed the world; from new inventions and technical improvements in machinery and manufactured products to the development of new production systems (Best, 2001; Owen, 2000). Manufacturing has played a critical role in creating the evolving global or international economy and in economic development. The ongoing formation of a global economy would be impossible without products produced by manufacturing production systems. The development of global commodity chains required innovations in logistics and the introduction of containers and container ports. While manufacturing dominated the 19th and early to mid 20th centuries as a significant component of developed market economies, it is sometimes considered to be of less importance from the 1970s. From the 1970s, developed market economies underwent a process of deindustrialization which involved a relative decline in industrial employment and a rise in service employment (Bazen and Thirlwall, 1991; Bluestone and Harrison, 1982).

Many accounts of developed market economies assume that manufacturing no longer matters. In a recent British guide for parents, written by an award-winning teacher, it is claimed that: "Britain's manufacturing industry is spent. We are in an age where our chief export is intellectual property. To design fantastic things that people will want to buy, we need a nation of people who can dream things up, not simply regurgitate and obey" (Beadle, 2007: 10). This claim misrepresents the role that manufacturing continues to play in developed market economies. Knowledge-based employment has become important, but making rather than just designing products is still important for developed market economies. Employment in manufacturing has declined with productivity improvements and the development of capital- and knowledge-intensive production, but manufacturing still forms a major element of exports from developed market economies and in the creation of gross domestic product (Bryson *et al.*, 2013; Cohen and Zysman, 1987; Fingleton, 1999; Moretti, 2013).

Accounts of the decline of manufacturing in the US and UK (Bluestone and Harrison, 1982) and the transfer of production to lower-cost production locations have tended to focus too much on the decline in employment. Much of the academic debate has emphasized the emergence of new global commodity chains and the continued fragmentation of production (Dicken *et al.*, 2001). Deindustrialization was explained by differentials in factor inputs with a focus on labor costs and the transfer of production from high- to low-cost labor locations. Nevertheless, labor is only one production input and not all manufacturing is labor intensive (Loch *et al.*, 2007).

While 1966 is etched in the United Kingdom's historical consciousness due to exploits with a football, it is also significant as the highpoint of manufacturing employment in the country. In 1966 manufacturing accounted for 35 percent of Gross Domestic Product

(GDP), employed 8.7 million workers (Select Committee on Trade and Industry, 2002) and was at the center of the country's post-war economic boom. Manufacturing industries are in a continual state of change as new technologies, products, processes and forms of competition emerge (Christopherson and Clark, 2009). In this sense, manufacturing sectors emerge, grow, restructure and decline in different ways and at different times. During the early years of the 20th century there was a debate regarding the decline of the British gun industry. According to two anonymous Birmingham gun manufacturers:

> The evidence of decline comprises a lessening output at the British centre of the industry [Birmingham]; increased outputs at foreign centres of production; British exports of fire-arms are decreasing in number and value; British imports are increasing; the number of fire-arms manufacturers or manufactories is less than it was twenty-five years ago, but the number of retailers and merchants has increased; certain branches of the trade have disappeared; the industry is less prominent than it was; it is not so profitable as it used to be. (Artifex and Opifex, 1907: 31)

This account of British manufacturing identifies many different ways of measuring the decline of manufacturing – from a decrease in exports to a decline in profitability. This analysis was concerned with the increase in imports and a decline in exports, but also with the erosion of Birmingham's gun cluster and a decline in skills. The decline in the gun cluster was measured by an increase in demand for parts made in other countries and it was noted that "[n]ot many years ago everything requisite to the production of a small arm of any known type could be found or made in Birmingham; now, so many parts of various guns are imported from Belgium, or elsewhere" (Artifex and Opifex, 1907: 25).

This decline in the Birmingham gun cluster reflects a particular period in the economic history of Britain's fire-arms industry; this period of decline was followed by growth related to the onset of World War I. This account of the fire-arms industry highlights the need for detailed studies of industrial sectors and company case studies (Christopherson and Clark, 2009).

Since the 1970s the focus of academic and political debate has shifted towards charting the decline of manufacturing in developed market economies (Bazen and Thirlwall, 1991) and the rise of service employment (Bryson *et al.*, 2004; Gallouj, 2010). From 1970 manufacturing in developed market economies experienced a series of recessions and also the onset of competition from emerging economies. This is a complex history that includes the rise of Japan and then Taiwan, followed by Indonesia and China as important centers for labor-intensive manufacturing (Midler, 2011). The production of products has been switched between locations as the cost base of older "emerging economies" was undermined by newly emerging economies. The design and fabrication of the Barbie doll is a classic example of the changing economic geography of American manufacturing (Bryson and Rusten, 2011: 117–122). Barbie is designed and managed from the USA by Mattel, but has never been manufactured there. She would be too expensive to produce in the USA. Initially she was manufactured in Japan, but then production shifted to Taiwan in 1967 and Indonesia and China in 1987. The opening of the Taiwanese factory was an important innovation in global outsourcing. It is production decisions like this that led to the transformation of eastern Asia into the new "workshop of the world." Barbie's design team has always been located in America and has always been separated from production. This represents an early application of an international division of labor. Taiwan's economy developed from 1967 and this led to better working conditions and higher salaries as Taiwanese manufacturing shifted from producing low-value products to high-value electronics – laptops, personal organizers and MP3 players. As wages rose, Mattel shifted the manufacture of Barbie to mainland China. It is interesting to note that the "daughters of Taiwan's first Mattel employees are [employed in making] computers and [in] public relations" (Spencer, 2005: 15). Barbie's Taiwanese legacy is a Barbie museum that was opened in Taishan and on display are bridal Barbies, combat Barbies and Barbies in Chinese dresses.

Since the 1970s, manufacturing has been transformed. Manufacturing companies that design, market but outsource fabrication to anonymous contract manufacturers have become more like service firms. There are still many companies that design and manufacture but also provide products with embedded services. The dividing boundaries between manufacturing and service activities have become increasingly blurred (Bryson, 1997; Bryson, 2009; Bryson 2010; Bryson and Daniels, 2010; Daniels and Bryson, 2002). Manufacturing firms have been transformed into creative companies in which workers from across the supply chain work together to develop products and services. This transformation reflects new forms of financialization (Dore, 2008; Epstein, 2005), geographies of production, employment, logistics and consumption. However, academic understanding of manufacturing has not kept pace with these transformations. Much of the academic debate and literature has revolved around understanding the new knowledge-based economy and the service sector in general, and there has been a much more limited focus on manufacturing (Bryson *et al.*, 2004). In developed market economies there have been difficulties with the public perception of the manufacturing sector. The manufacturing stigma is not easily obliterated from the minds of those who consider manufacturing employment as low waged, labor intensive, repetitive, outdated and temporary; making it less attractive to young workers and entrepreneurs than service employment.

Since the early 1970s, British industrial policy has not had a specific manufacturing focus (Livesey, 2012). Successive governments have focused on the deregulation of financial services, privatization, attracting foreign direct investment, encouraging the growth of clusters, promoting research and development, and entrepreneurship (Beath, 2002; BDO, 2010; Livesey, 2012). British policy emphasized the importance of financial and knowledge-intensive services, and manufacturing was often considered to be an activity that was best undertaken in countries with low labor costs. The credit crisis of 2008–2009 that was centered around failures in the financial system led to a new political debate that called for a rebalancing of the British economy towards manufacturing and away from financial services (HM Treasury/Department of Business, Innovation and Skills, 2010). In March 2011 the Chancellor of the Exchequer concluded the Budget statement by arguing manufacturing would play a central role in the UK's economic recovery and that: "We want the words: 'Made in Britain', 'Created in Britain', 'Designed in Britain' and 'Invented in Britain' to drive our nation forward – a Britain carried aloft by the march of the makers. That is how we will create jobs and support families" (House of Commons, 2011). This concern with rebalancing can be traced back to 2008 with the publication of a new industrial strategy that argued the "UK's future lies in a mixed and balanced economy, where manufacturing activities complement services to deliver the widest possible range of economic benefits across all regions, and create skilled jobs that span the entire value chain, from research through to fabrication, branding and sales" (Department for Business, Enterprise and Regulatory Reform (BERR), 2008:5).

This edited book brings together scholars working on manufacturing from many different countries and academic disciplines. The aim is to develop a state-of-the-art account of manufacturing research by exploring theory, method, empirical research and policy. Our intention has been to create an inter-disciplinary collection that includes academics from across the social sciences and also engineering.

STRUCTURE OF THE BOOK

In both an international and multi-disciplinary context, this *Handbook* aims to provide a critical review of current manufacturing processes, practices and policies to broaden understanding of the dynamics of manufacturing, both theoretically and empirically. It aims to engage with and reflect upon the current state of manufacturing as well as the politics and policy of how we think about and practice manufacturing policy (Clark, 2013). The chapters in this handbook seek to enhance understanding of the ongoing debates about related aspects of manufacturing which include production processes, labor, supply chains, logistics, financialization, design and innovation; regional strategies and policies; manufacturing and management; and manufacturing and labor relations. Furthermore, they include insights into advanced manufacturing, energy costs (Mulhall and Bryson, 2013, 2014), raw materials, and industry and firm case studies of manufacturing from around the world.

This handbook is organized into five parts. Part I is the Introduction. Part II comprises 11 chapters serving as building blocks in understanding factor inputs and product organization. Part III comprises 12 chapters that provide theoretically grounded empirical analysis of manufacturing sectors and firms. Part IV explores the policy context that supports manufacturing and includes seven chapters, followed by Part V, Conclusion.

Following the Introduction, the first chapter begins by exploring the emergence of manufacturing management theory and practice. The main principles and developments underlying the management of the manufacturing function within business are examined. The first chapter has taken a historical perspective through the contributions of practitioners and academics alike to arrive at models and frameworks for manufacturing strategy formulation as practiced today. It demonstrates how a rich heritage of factory studies, often adopting socio-ethnography as the investigative style, enabled these discussions to evolve, right through to studies of Japanization. The development of lean, agile and mass customization principles, or what has been termed as Japanization, has influenced what is deemed as best practice manufacturing today. The chapter concludes by reviewing some of the frameworks that are employed to assist in the analysis and development of manufacturing strategies today.

Labor as one of the key factors of production is discussed in Chapter 2 and elsewhere throughout the book. In Chapter 2, the focus is on the internal organization of manufacturing work and the evolution of the labor process within manufacturing firms. The relationships between firms' internal forms of organization and the external labor markets from which they draw their labor supply; the impact of globalization on the distribution of manufacturing employment; and the changes that have emerged in the organization of manufacturing work as firms reorganize their operations in internationalized production networks are also discussed. In the conclusion, some key challenges facing manufacturing workers are identified.

The contextualization of capital, another key factor of production, is critical in understanding the evolution of manufacturing. Hence, the examination of "financialization" and its role in the manufacturing world is crucial. As manufacturing continues to matter not only because of the jobs it creates, but also because of its stimulus to innovation and its contribution to exports and the balance of trade, Chapter 3 engages the debate over the role of manufacturing in the US and UK by examining evidence for a new location calculus favoring the location of manufacturing in the two economies. It also examines how processes associated with the "financialization" of the US and UK economies affect investment in firms in areas critical to manufacturing product and process innovation, including human capital and research and development.

Product and place linkages enabled by a complex and modern transportation system are one of the greatest achievements of capitalism. With this, however, come challenges associated with logistics, particularly in identifying the most efficient and effective global production and management networks. Chapters 4, 5 and 6 deal specifically with these challenges and their implications by examining how new transport modes and routes, steps in the production process and the locations of manufacturing have changed in tandem with new communication technologies, the emergence of logistics as an industry and new forms of corporate organization; by shedding light on "reshoring" and historically contextualizing reshoring within broader global dynamics and discussing the term's recent rise to fame; and by exploring the social dynamics inherent in the relocation of production from one location to another.

Chapters 7 to 11 delve into the complexity of a product from its inception to sales and marketing, including a discussion on required engineering and production techniques (Chapter 8) and market protection in the form of patents and intellectual property rights. We have witnessed technological and scientific developments since the early 18th century. The coupling of technical innovations, market need and efficient manufacturing methodology has resulted in new and improved products. It is this methodology, argued in Chapter 7, that has transformed manufacturing. For instance, in the automotive industry it was Henry Ford's approach to mass manufacture that transformed car manufacture from a traditional cottage industry to a global business. In this chapter, the author elucidates the greatest improvement in manufacturing technology and methodology, which he terms "tool-less manufacture", which is going to change the world. It is tool-less because a product can be directly manufactured from a computer file, without the need for any intermediate tooling. It is also known as 3D printing, rapid prototyping and additive manufacture. The author prefers the term tool-less manufacture, as this name captures the essence and novelty of the technology as a new manufacturing methodology. The characteristics of current and future visions of tool-less manufacture are explored in detail. Suitable business models and intellectual property right protection to support this methodology are also examined.

Chapter 8 proposes an engineering-oriented model aimed at increasing the domestic production of manufactured goods using the guiding principle: "Discover here – accelerate Translation – Build here" (DTB). In order to accomplish this goal, the authors propose a new Accelerated Readiness Level (xRL) model that accelerates the speed at which new concepts become products. They argue that this xRL model increases profitability by mitigating the time it takes to move an innovation to market. It also emphasizes the role of design and customization in advanced and high-tech manufactured goods that requires higher skills and will likely lead to more high-tech jobs.

Chapter 9 provides a valuable background for examining the role of energy from two perspectives: 1) manufacturing in energy production; and 2) energy use in manufacturing. Energy is not only a necessary input in the manufacturing of goods and services, but is also an industry in its own right, beginning with resource extraction and including processing and conversion as well as the transport of energy commodities. As a result, this chapter discusses opportunities for cutting energy costs while exploring international policies in six countries. Two case studies – combined heat and power technology in the pulp and paper industry and solid-state technology in the lighting industry – are provided to illustrate the role of manufacturing in the creation of green energy innovations.

The new economic geographies of production reflect the complex transformations and interwoven relationships between manufacturing and service companies. Thus, Chapter 10 examines the relationship between design and manufacturing. Within the context of new economic geographies of production, it examines the global competitiveness of industrial design companies in the United States, Europe and China. While foreign export markets contribute to long-term growth and survival of companies in developed countries, increasing foreign competition, particularly from emerging markets, has affected the competitive advantage of companies in the design industry. Key determinants of firm success and long-term growth are discussed, which include service provision, human capital, market orientation and adaptability. In tandem with design as part of knowledge creation and diffusion, protecting intellectual property rights (IPRs) is crucial. IPRs are recognized as having established the conditions necessary for the development of a market system, in which the search for new knowledge of economic value is rewarded. Hence, the focus of Chapter 11 is patents, which are one form of IPRs widely used in highly innovative manufacturing industry sectors that offer insight into patterns of knowledge creation and diffusion.

From an empirical perspective, Chapters 12 to 23 provide insights into industries ranging from traditional textiles, farm machinery and furniture to consumer electronics, aerospace and non-traditional industries such as the surfboard industry. They also uncover the intricacy of supply chains, global production networks and competitiveness. Real-life case studies are also included to provide in-depth understanding of the sophistication of various types of manufacturing and its factor inputs.

Textile manufacturing, a traditional industry, has experienced great transformations – economically and geographically. It has seen a major decline in employment in developed countries, such as the United States and United Kingdom, and a significant rise in less developed countries such as Bangladesh, Sri Lanka, Vietnam and China. Chapter 12 explores the survival and evolution of UK textile manufacturers in this context of understanding the changing geographies of manufacturing in a global marketplace. The evolution of the UK textile industry, presented in the chapter, is based upon individual manufacturing firms' decision-making and the competitive strategies of innovation implemented, which have created a highly advanced and niche industry involved in the production of technical textiles. This industry is identified as an example of Susan

Christopherson's concept of a "Phoenix Industry". Thus, these manufacturers play an integral role in developed market economies.

In a similar vein, Chapter 13 examines the US furniture industry. This industry has experienced economic downturn with the movement of furniture manufacturing to Asia, and particularly China, since the final decade of the 20th century. The revitalizing strategies for this industry are very similar to those developed in the British textile industry and include process innovations, targeting niche markets and localization of companies in centers of regional production. As lower-cost manufacturers shift production to Southeast Asian sites, increases in transportation costs and wage rates in China, and demands for rapid delivery times provide opportunities for stabilizing the US furniture industry. Proceeding on a more technologically sophisticated industry, Chapter 14 provides a critical analysis of the new geographies of advanced manufacturing in the case of the machine tool industry. It examines changes in the global machine tool industry since 2000. Machine tools are of interest as they are critical to many manufacturing processes. Therefore, as argued in this chapter, expertise in this industry can serve as a gauge of a country's overall manufacturing activities. China was found to have garnered a large share of worldwide production and consumption, while the machine tool sectors of Germany and Japan have maintained their longtime leadership positions, plus, more importantly, in terms of exporting. A number of other international changes in this strategic industry are also discussed.

The next chapter explores a relatively unknown industry – farm machinery as part of manufacturing industry. Contrary to its limited public reputation, this industry is crucial as it provides the ability for nations to feed themselves. This chapter examines the major players in this industry and the challenges they face, with a focus on the US farm machinery industry. Another industry that is of interest is the optics and photonics industry. This industry sits at the crossroads between science policy and economic development, falling along the spectrum from high-tech to an established manufacturing sector. As a consequence, as argued in Chapter 16, photonics often fails to garner the state and federal investment directed at seemingly newer and more fashionable enabling technologies like nanotechnology and biotechnology. It also falls beneath the radar of policy-makers focused on new materials like lightweight composites or additive manufacturing. Thus, this chapter focuses on the regional clustering characteristics of the optics and photonics industry by providing an important example of the dynamics that underscore working regions in two key examples: Rochester, New York and Quebec, Canada.

Steel, an important contributing factor to both global production and consumption, warrants in-depth understanding. As such, Chapters 17 and 18 examine the restructuring of global steel supply chains in both traditional and emerging regions of the world; and the role of intermediate metal processing manufacturers in the West Midlands region of the United Kingdom. The steel industry has experienced significant transformation similar to other industries discussed in this handbook as a result of technological change and globalization. The ways in which the twin forces of technological change and global market expansion have contributed to the de-verticalization of the steel industry, the rise of emerging market players and the transformation of the role of traditional steel-making regions are discussed in Chapter 17. Chapter 18 offers insights into the intermediate metal processing industry, which comprises two principal activities: forging and casting of metal. This industry too has undergone a continued transition to higher-value-added products and services to remain competitive.

Cost pressure represents a common challenge across sectors and technological levels. As examined in Chapter 19, aircraft and space vehicle manufacturing, an industry at the top of the manufacturing echelon in both technology and cost, has also been transformed. This industry has a unique set of requirements, centering on public safety of the product as, though accidents are rare, each has high visibility so that the consequences of error are high for manufacturers. Cost pressures and environmental demands have led to costly products that must be operated for decades to recoup the purchase price. These have caused the industry to evolve into aircraft design and assembly companies supplied with many specialist components such as engines or landing gear. Current challenges of design and manufacturing outsourcing, logistics, materials and finance are considered in Chapter 19.

Before proceeding to explore specific case studies of companies in different parts of the world, Chapter 20 offers a unique case study of the surfboard industry. It is a recurring theme in this handbook to offer insights into a broad array of industries, including those that are relatively unknown yet are relevant and critical in providing a comprehensive analysis and understanding of manufacturing industry and its evolution. This chapter explores the trajectories of commercial surfboard manufacturing, which has transitioned from a craft-based activity to a large-scale industry paralleling surfing's own expanding popularity. However, the difficulties and dangers of up-scaling craft-based production are highlighted. Consolidation of surfboard manufacturing has occurred with a refocusing on customization for local surfers and breaks. The key to maintaining profitability, argued in this chapter, is to combine a diverse range of skills and rely on enduring demand for customized boards.

From a specific case study perspective, Chapters 21 to 23 examine the history, characteristics, competition, challenges and successful business strategies of an ethnic bakery business in the UK; Škoda, a Czech-based automobile manufacturer owned by the German company Volkswagen (VW); and Samsung Electronics (SEC). In the first case presented, a Polish bakery, Chapter 21 considers the significance of translocal exchanges and global value chains in the establishment and evolution of Polish manufacturing enterprises in the UK. In the case of Škoda, the coupling of its assets and capabilities with the strategic needs of VW and the successful integration of Škoda into global production and distribution networks are credited for its success as a Tier-Two lead firm. The government's decision to privatize Škoda and its privatization policy, including the conditions of privatization and the selection of VW as its original joint venture partner, are also explored. Nevertheless, foreign ownership is not unproblematic and does not automatically lead to success in the automotive industry. In the case of SEC, presented in Chapter 23, the company has been transformed from a subcontract manufacturer of products to an innovation flagship for producing world-first new products. Design innovation, brand marketing and networking are considered to be important for a firm's competitiveness, reflecting the need for complex processes and strategies. The SEC case study identifies the importance of the processes of path dependence, restructuring and new path creation.

Part IV follows with a policy emphasis, tracing manufacturing employment in the US since the 1980s; exploring and examining manufacturing redevelopment policies in different urban areas and in a global city such as New York city; and examining the role of innovation and technology industries in restructuring manufacturing; as well as addressing the gap in the labor market by accentuating the meaning and interpretation of "skills". As stated at the start of this Introduction, manufacturing employment has significantly shrunk in the developed world since the mid 20th century. In particular, in the US manufacturing industry shed more than 12 percent of its total workforce between 1979 and 1983 (see Chapter 24). This chapter argues that while contemporary scholarship on manufacturing has generated significant new ideas about innovation systems and global–local dynamics, it has yet to be reconciled with a systematic characterization of the shifting geography of the US manufacturing sector from deindustrialization. Thus, this chapter traces the industrial, occupational and spatial development of US manufacturing over the course of the 1980s, 1990s and 2000s business cycles. The findings are discussed; particularly the regional occupational endowments with manufacturing that have remained stable over time.

Recently, the term "advanced manufacturing" has been promoted by policy-makers and scholars to illustrate and emphasize a renewed interest and advancement in the manufacturing sector. The evolution of the term "advanced manufacturing" is explored in Chapter 25. This chapter then explores the re-emergence of manufacturing policy in the United Kingdom and the United States in relation to advanced manufacturing. The author argues that it is difficult to define the term "advanced manufacturing" in absolute terms. The meaning depends entirely on the motivations or objectives that influence the person or institution attempting to develop or apply a definition of advanced manufacturing. A number of terms have been used not as synonyms but in some fashion interchangeably by researchers, policy-makers and commentators over the past 10 to 20 years.

In Chapter 26, a new national manufacturing agenda in the United States is discussed as it has sparked hope of strengthening production networks in ways that draw on the specific assets of cities and their metropolitan regions. This chapter argues that the gap between scales of action – that is, the lack of powerful policy-implementing institutions at the regional level in US policy-making – makes the project of manufacturing redevelopment in cities problematic. To illustrate this argument, a local urban area, Newark, New Jersey in the US is explored, and policy recommendations are offered. In the case of a nearby city also known as a global city, New York, Chapter 27 explores the extent to which the decline in manufacturing has been inevitable, given the shift in the United States to a services economy, or consciously achieved to sustain the City's position as a premier global city. It discusses the city's efforts to upzone and redevelop industrial land and explores the apparel and furniture manufacturing sectors to understand how the City's bifurcated policies have affected them.

Emphases on the role industries play in the knowledge economy continue to be at the forefront of academic and policy debate. In the global competitive arena, the knowledge economy is where acquisition, creation and utilization of knowledge are the key factors in explaining the competitive advantage of firms. This type of competitive advantage is considered to have contributed to sustained economic growth for firms located in high- rather than low-wage economies. Because different industries exist in the knowledge economy, it is imperative to highlight both ends of the spectrum via low- and high-technology industries. Chapter 28 introduces this taxonomy. It then examines the role low-tech manufacturing industries play in the knowledge economy. It is argued that low-tech industries continue to be of large economic importance in high-wage countries, as they actively pursue strategies to maintain competitiveness and increase the value added of products and processes. This chapter questions the current focus of policy-makers on R&D investments and high-tech industries as a strong synthetic knowledge base while investments in machinery and human capital allow low-tech industries to resist the increasing global price competition.

From a regional perspective, Chapter 29 revives the discussion of mature regions by providing an exploratory case study of one key mature regional economy in the United States – the cities of Pittsburgh, Cleveland, Akron and Youngstown, which have banded together to define themselves as the "Tech Belt" – a continuous geographic region with similar industrial structure and the need to secure its economic future. As older industrial cities struggle to redefine themselves, this chapter considers the process of regional restructuring for this Tech Belt. Rather than evaluate a specific public program, this chapter considers how innovative firms draw upon the array of local and federal programs. The findings stress the need to coordinate efforts, as government policy interventions do not reinforce one another to form a cohesive policy infrastructure but instead appear to operate in silos.

Part IV concludes with Chapter 30, which underscores a key factor of production and a critical component of the manufacturing economy as well as of the entire economy – human capital. In response to policy initiatives to generate job growth, it is imperative to explore skills and manufacturing by first identifying what skills are, how to obtain them and who will utilize them and how. Chapter 30 explores what is at stake if the skills needs of industry are narrowly presented as an educational policy problem. It reframes a recurring industrial debate by presenting skill as an ambiguous and fluid concept, and therefore subject to interpretation. In response, labor market intermediaries do not just supply skilled workers to the industry, but they engage employers in a negotiated process over what is meant by skill, who is presumed to possess that skill and what structures are needed to develop and defend skill by using strategies of skill reinterpretation.

Part V concludes this handbook by re-emphasizing the main arguments developed in this collection: manufacturing matters, and continues to matter. Manufacturing has been evolving and these transformations merit further research from social scientists, engineers and policy-makers.

It is possible to argue that within developed market economies there has been a manufacturing renaissance. Unfortunately, academics and policy-makers have not kept up with all these alterations and it is also possible to argue that many policy-makers no longer understand manufacturing. Nevertheless, the 2008 recession altered the political landscape and led to a call for developed market economies to rebalance their economies and to take manufacturing seriously. The second decade of the 21st century has seen major policy interventions intended to encourage growth in manufacturing exports and in manufacturing more generally. Thus, today manufacturing, in political terms, has become a hot topic and one that requires further sustained academic research. This handbook has identified many of the avenues that require further research, but perhaps the most important areas for research include a new focus on factor inputs, including energy and skills, and on understanding the local and global aspects of complex commodity chains. Other important topics include understanding the adaptation and survival strategies developed by manufacturing companies located in high-cost locations and also the ongoing blurring of services and manufacturing.

It is becoming increasingly important that a research dialogue occurs between the social sciences and engineering. This is an under-developed area for inter-disciplinary research, but it is an important area as engineering departments are involved in developing new process and production technologies and techniques that have the potential to revolutionize manufacturing. These innovations will lead to new economic geographies and sociologies and will have important implications for social policy.

REFERENCES

Artifex and Opifex (1907). *The Causes of Decay in a British Industry*, Longmans, Green and Co.: London.
Bazen, S. and Thirlwall, T. (1991). *Deindustrialisation*, Heinemann: Oxford.
BDO (2010). *The Changing Shape of UK Manufacturing*, BDO: London.
Beadle, P. (2007). *Could Do Better! Help your Child Shine at School*, Doubleday: London.
Beath, J. (2002). "UK Industrial Policy: Old Tunes on New Instruments," *Oxford Review of Economic Policy*, 18, 2: 221–239.
BERR (2008). *Manufacturing: New Challenges, New Opportunities*, BERR: London.
Best, M.H. (2001). *The New Advantage of American Industry*, Oxford University Press: Oxford.
Bluestone, B. and Harrison, B. (1982). *The Deindustrialisation of America*, Basic Books: New York.
Bryson, J.R. (1997). "Business Service Firms, Service Space and the Management of Change," *Entrepreneurship and Regional Development*, 9: 93–111.
Bryson, J.R. (2009). *Hybrid Manufacturing Systems and Hybrid Products: Services, Production and Industrialisation*, University of Aachen: Aachen.
Bryson, J.R. (2010). "Service Innovation and Manufacturing Innovation: Bundling and Blending Services and Products in Hybrid Production Systems to Produce Hybrid Products," in Gallouj, F. (ed.), *Handbook on Innovation in Services*, Edward Elgar: Cheltenham and Northampton, MA: 679–700.
Bryson, J.R. and Daniels, P.W. (2010). "Service Worlds: The 'Services Duality' and the Rise of the 'Manuservice' Economy," in Spohrer, J., Maglio, P. and Kieliszewski, C. (eds), *Handbook of Service Science*, Springer: Berlin: 79–106.
Bryson, J.R. and Rusten, G. (2011). *Design Economies and the Changing World Economy: Innovation, Production and Competitiveness*, Routledge: London.
Bryson J.R., Clark, J. and Mulhall, R. (2013). *The Competitiveness and Evolving Geography of British Manufacturing: Where is Manufacturing Tied Locally and How Might This Change?* BIS: London.
Bryson, J.R., Daniels, P.W. and Warf, B. (2004). *Service Worlds: People, Organisations, Technologies*, Routledge: London.
Christopherson, S. and Clark, J. (2009). *Remaking Regional Economies: Power, Labor and Firm Strategies in the Knowledge Economy*, Routledge: Abingdon.
Clark, J. (2013). *Working Regions: Reconnection Innovation and Production in the Knowledge Economy*, Routledge: New York.
Cohen, S. and Zysman, J. (1987). *Manufacturing Matters: The Myth of the Post-Industrial Economy*, Basic Books: London.
Daniels, P.W. and Bryson, J.R. (2002). "Manufacturing Services and Servicing Manufacturing: Knowledge-Based Cities and Changing Forms of Production," *Urban Studies*, 39, 5–6, 977–991.
Dicken, P., Kelly, P., Olds, K. and Yeung, H.W.-C. (2001). "Chains and Networks, Territories and Scales: Towards a Relational Framework for Analyzing the Global Economy," *Global Networks*, 1, 2: 89–112.
Dore, R. (2008). "Financialization of the Global Economy," *Industrial and Corporate Change*, 17, 6: 1097–1112.
Epstein, G.A. (2005). *Financialization and the World Economy*, Edward Elgar: Cheltenham and Northampton, MA.
Fingleton, E. (1999). *In Praise of Hard Industries: Why Manufacturing, Not the New Economy, Is the Key to Future Prosperity*, Orion: London.
Gallouj, F. (ed.) (2010). *Handbook on Innovation in Services*, Edward Elgar: Cheltenham and Northampton, MA: 679–700.
HM Treasury/Department of Business, Innovation and Skills (2010). *The Path to Strong, Sustainable and Balanced Growth*, HM Treasury/BIS: London.
House of Commons (2011). Budget Statement, March 23, c966, available at http://www.publications.

parliament.uk/pa/cm201011/cmhansrd/cm110323/debtext/110323–0001.htm#11032368000001, accessed October 17, 2013.

Livesey, F. (2012). "Rationales for Industrial Policy Based on Industrial Maturity," *Journal of Industry, Competition and Trade*, 12, 3: 349–363.

Loch, C.H., Chick, S. and Huchzermeier, A. (2007). "Can European Manufacturing Companies Compete? Industrial Competitiveness, Employment and Growth in Europe," *European Management Journal*, 25, 4: 251–265.

Midler, P. (2011). *Poorly Made in China: An Insider's Account of the China Production Game*, John Wiley & Sons: Hoboken, NJ.

Moretti, E. (2013). *The New Geography of Jobs*, Mariner Books: Boston.

Mulhall, R.A. and Bryson, J.R. (2013). "The Energy Hot Potato and Governance of Value Chains: Power, Risk, and Organizational Adjustment in Intermediate Manufacturing Firms," *Economic Geography*, 89, 4: 395–419.

Mulhall, R. and Bryson, J.R. (2014). "Energy Price Risk and the Sustainability of Demand Side Supply Chains," *Applied Energy*, 123: 327–334.

Owen, G. (2000). *From Empire to Europe: The Decline and Revival of British Industry Since the Second World War*, HarperCollins: London.

Select Committee on Trade and Industry (2002). *The Competitiveness and Productivity of UK Manufacturing Industry*, The Stationery Office: London.

Spencer, R. (2005). "Barbie Sends a Shiver through Taiwan with a Lesson in Globalisation," *Daily Telegraph*, June 25: 15.

PART II

(PROCESSES) BUILDING BLOCKS: FACTOR INPUTS AND PRODUCTION ORGANIZATION

1. Manufacturing management in theory and practice

Paul L. Forrester

INTRODUCTION

Manufacturing management has a diverse and rich history. The evolution of management thought through the 20th century has been inextricably linked to the practices and theories of factory management. The need to organize, plan and control resources emerged with the development of the Industrial Revolution (Babbage, 1835), though initially the practice was for a very general form of enterprise management with little by way of a theory of production management. This changed with practices introduced by, and later inspired by, the work of Frederick Winslow Taylor (Taylor, 1911) and the emergence of the Scientific Management School (Urwick and Brech, 1945, 1953; Bryson, 2000). Criticized as Taylorism has been for many years, it was clearly the first attempt to derive a set of principles and practices for the management of production and thereby improve efficiency and effectiveness, to replace the general factory management of earlier years.

The emergence of new, often conflicting paradigms for the management of manufacturing occurred progressively throughout the 20th century (Figure 1.1). The Human Relations, or Behaviouralist, School emerged following the early work of Elton Mayo and colleagues, with their famous Hawthorn Studies of the 1920s and 1930s (see Roethlisberger and Dickson, 1939), followed by a set of studies exploring the impact and influence of 'human factors' on production, leading to theories of teamwork, motivation and leadership. These studies relied heavily on empirical research and factory studies, putting manufacturing management at the heart of the development of management theory (the factory studies are expanded upon in the next section). From the mid-point of the 20th century a number of further schools of thought emerged: the Operations Research[1] School with its complementary sets of tools and techniques for optimization of operations and projects (e.g. statistical stock control methods, logistics and transportation solutions, project planning and control); the Computer Science paradigm with manufacturing operations being to the fore in terms of the application of computer-based programmes for the planning and control of operations (forecasting, capacity planning and materials requirements planning systems); and the Service Revolution with its customer focus and lessons from running service operations related back to manufacturing.

However, two paradigms emerged as key influences and still shape our notions of 'best practice' in manufacturing today: 'Japanization and Lean Thinking' and 'Operations Strategy'. Interests in Japanese manufacturing techniques and philosophies gained momentum in the 1980s with the competitive success of Japanese businesses in export markets around the world and later influences of Japanese foreign direct investment. From this paradigm principles of Just-in-Time and Total Quality Management emerged and later evolved in both pure and adapted forms of what is now termed 'lean' and 'agile'

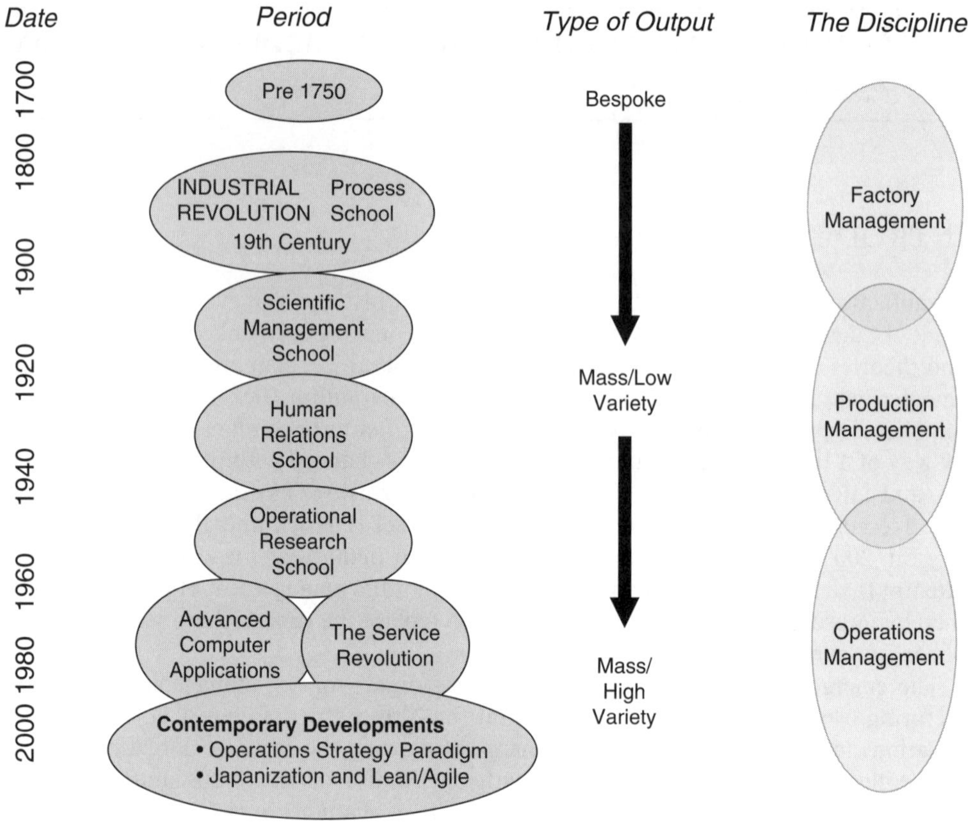

Figure 1.1 The evolution of manufacturing management

operations. The Operations Strategy paradigm in some ways went hand in hand with this emergent view of the importance of manufacturing operations and the Operations function as a critical success factor in business – and particularly the need to link corporate and marketing strategies and objectives to manufacturing and operations strategies and activities.

This chapter provides context and explanation of modern notions of manufacturing management. It does this by setting the historical context of classical factory studies which heavily guided early management thought and still provides a legacy for management thinking and practices today, generally and for manufacturing operations more specifically. The theme of Japanization is then explored and it is shown how the principles of 'lean' today provide a blueprint and, some would say, clear prescription over the management of manufacturing operations. Finally, exponents from the Operations Strategy School are introduced and it is shown how manufacturing businesses today seek to integrate strategies of production and supply with their broader business objectives.

CLASSICAL FACTORY STUDIES

There was a time when no management theory or, more correctly, what we now think of as management theory, existed. The momentum of the Industrial Revolution in advanced economies of the world, including the USA and much of Western Europe, was driven and primarily 'managed' by a combination of forward-thinking entrepreneurs and inventive engineers. A factory management style of control over resources was applied by the entrepreneurs, many using a hire and fire approach to exploit (literally) worker and productive resources for return on investment. Note the term 'factory management'; in the late 1800s this was yet to develop into coherent theories and practices of 'production management'. The pace of technological change was dictated and accelerated by inventors and innovators, mostly from the rising engineering fraternity. The emphasis was on the displacement of labour, especially skilled labour, and jobs of work were more often denigrated to menial low-skill tasks, with low pay and poor conditions. There were exceptions of course – in the UK George Cadbury's Quaker beliefs led to the development of Bournville Village in Birmingham (Bryson and Lowe, 2002). Other entrepreneurs who developed their factories with a social conscience include Sir Titus Salt and his mill operations at Saltaire in Yorkshire and, later, William Lever of Lever Brothers' Port Sunlight factory at the Wirral, near Liverpool (Crawford, 1995). All these paternalistic and enlightened developments featured modern factories, good conditions for workers and model village developments to house employees – a commitment to the workforce and some foresight into the later advocacy of the Human Relations School. However, these examples were the exception to the rule. Not only were most workplaces dire and devoid of social and human concerns but they also lacked methods and organization, relying instead on power and fear over the workforce to ensure they 'worked hard' (Braverman, 1974).

The landmark work of Frederick Winslow Taylor sought to place order in the chaotic situation he observed in his work as a mechanical engineer and work methods analyst (Taylor, 1911; Copley, 1923; Bryson, 2000). He was, in fact, one of the first of what we now call 'management consultants'. Taylor sought to develop a more planned, objective and scientific approach to methods and processes – hence the name he gave to his prescriptions: the 'Principles of Scientific Management'. Whole narratives have been produced on Taylor's ideas (often referred to as 'Taylorism'), most forcibly by Braverman (1974). Criticism centres on the tenor and impact of his work and the work of his later followers. It is without doubt that Taylor's ideas led to a dehumanizing effect in many factories (Braverman, 1974), but one can argue justifiably that he did attempt to study and improve manufacturing methods and organization from an informed and objective perspective, unlike the approach taken by many factory owners at his time of writing, around 1910. He effectively produced the first theory of production management in properly documented format. He was also keen to base his work on empirical factory study, leading to the development of work, time and motion study. However, the culmination of scientific management principles in practice was represented by Henry Ford's moving assembly lines and the emergence of Fordist techniques (Haber, 1964). The 'first commandment of mass production' was also formed – the economic batch size. Even at the time there were criticisms of these factory systems and forms of manufacturing management, most notably and entertainingly in the Charles Chaplin film *Modern Times* of 1936.

The period from the 1920s to the 1970s saw a continuation of factory-based empirical studies, but markedly critical and diverging in their views and perspective from Scientific Management. The work of Elton Mayo and his team, from the early Hawthorne Factory Experiments, is documented in minute detail in Roethlisberger and Dickson (1939) and formed the impetus for the Human Relations School (Gillespie, 1993; Bryson, 2000). This work showed clearly that people are not mere extensions of the machines they operate – that people can 'make or break' systems. Mayo and colleagues also pointed to the impact of teamwork and the general importance of considering the 'human factor' and people's needs. This later led to the emergence of motivation theory and insights into leadership styles in manufacturing businesses (Maslow, 1943; McGregor, 1960; Vroom and Yetton, 1973). Further studies adopting social-anthropological methodologies followed, including Elliot Jacques' (1951) *A Changing Culture of the Factory* and Wilfred Brown's (1962) study of the Glacier Metal Factory, *Piecework Abandoned*, – through to the socio-technical studies of the British coalmines (Trist and Bamforth, 1951), and the Volvo studies of the late 1960s/1970s documenting the emergence of autonomous work groups (Gyllenhammer, 1977).

These collections of studies impacted not only on the theory and practice of manufacturing management but also management theory in general. The period from the publication of Taylor's Principles in 1911 through the whole range of factory studies to the 1970s involved a debate on the interplay and prominence of the technical versus the social in manufacturing operations. A criticism one could make is that practitioners and academics alike, in their interest in work organizations, took their collective eye off the need for manufacturing to address business requirements and market needs. The 1970s saw a need for change, and the remainder of this chapter introduces new manufacturing paradigms that were developed from this time and still shape manufacturing management and production practices as they are today.

JAPANIZATION AND LEAN THINKING

The 1970s and 1980s signalled a sea change in the competitive environment for the majority of Western manufacturers, and particularly those in economies such as the USA and Britain. The years following the Second World War had seen an excess of demand in world markets over the capacity to supply goods, and so Western companies were able to continue to dominate world markets, selling pretty much anything they could produce, and often with scant regard to the need for reinvestment in capital resources and quality. The recovering economies of the world, most notably Japan and Germany, were rebuilding from a low base, but as a result were progressively developing their productive capacity and investing in new machinery and other technologies. At the same time and, in the case of Japan, with very limited access to indigenous natural resources and raw materials, these countries' manufacturers were devising new production methods enabling them, effectively, to capitalize on their manufacturing legacy and design and manufacture products of reasonable to good quality on, effectively, a shoestring. In such circumstances effective utilization of resources was paramount, and operations needed to be 'lean', not wasteful. A number of antecedents were set: Japanese and German goods in particular became increasingly competitive and attractive in world markets, including the West; world capacity expanded to the point that capacity exceeded demand; many traditional Western producers could not compete with the new competition and increasingly sophisticated customers; and attention also turned to trying to glean an understanding of the somewhat mysterious new Japanese production techniques.

A number of researchers turned their attention to discovering the 'secret' of Japanese manufacturing strategies, and production techniques in particular. Robert Hayes and Steven Wheelwright were prominent in their empirical investigations of 'Why Japanese factories work' (Hayes, 1981; Wheelwright, 1981) and accumulated their findings in a landmark book for manufacturing management practitioners entitled '*Restoring Our Competitive Edge: Competing Through Manufacturing*' (Hayes and Wheelwright, 1984), which also gave an insight into aspects of manufacturing in Germany and presented a framework for manufacturing strategy formulation. Hayes and Wheelwright therefore straggled the Japanization and Manufacturing Strategy paradigms, and presented very lucidly the reasons for the decline of the US manufacturing industry in particular, but also the means to redress this by learning from the Japanese and Germans. Their conclusion: there is no secret – just sound attention to resource management, a close eye on the market and customer needs, and timely investment and reinvestment in capital equipment and human resources. Many Western producers had just become too lazy and focused too much on internal concerns of cost and efficiency, ignoring the rising competition from re-emergent countries and, later, newly emergent economies such as Korea.

In the United Kingdom, Nick Oliver and Barry Wilkinson (1988) produced a seminal work entitled *The Japanization of British Industry*, with a second updated edition in 1993. This book struck a chord across a number of academic and practitioner disciplines in the field of management, including Operations Management, Human Resource Management and Industrial Relations. The work provided an insight into the causes of the decline of British manufacturing from the 1960s onwards, and mapped out the emergent influence of Japanese foreign direct investment on production management techniques, labour management and quality management. This work was timely. There had, for some years, been a fascination with tools and principles associated with Just-in-Time management (Schonberger, 1982; Voss, 1987) and Total Quality Management (Ishikawa, 1985), and the Oliver and Wilkinson publication put these Japanese-influenced approaches into their broader industrial, social and economic context. What is more, they differentiated between what they termed 'direct' Japanization, as found in Japanese transplant factories, 'mediated' forms (e.g. via Anglo-Japanese joint ventures, shared equities and collaborations) and 'emulated' Japanization, those indigenous firms which sought to adopt and adapt Japanese methods. Clearly, the spectrum from direct through to emulated Japanization represents a rough correlation from pure forms of Japanese manufacturing management through to highly modified forms.

Research and studies of Japanese companies in Japan and their transplant operations overseas, combined with context-bound adaptations of this alternative approach to manufacturing, progressively lifted the lid off the 'secrets' of Japanization and enabled new theories and philosophies to replace traditional manufacturing based upon Fordism, economies of scale and technology-push. What emerged was a more market-, customer- and design-oriented approach prescribed by writers such as Bennett and Forrester (1993) and Browne *et al.* (1996). The shift was more towards 'economies of scope', with flexible manufacturing solutions and market-pull-based enterprise resources planning systems. These pull-based principles reflected down to the level of the shop floor and system chains, with the adoption of kanban-pull systems, right-first-time attitudes to quality and service, and more intelligent management of inventory through the value stream. In short, a progression transition for the more successful manufacturers from mass production through to 'lean production' and 'mass customization'. In limited cases this represented direct adoptions and near copies of the Toyota Production System which, if truth be known, is at the heart of what many observers viewed as Japanization, but more often those manufacturers that continued to thrive managed to be selective and innovative in their choice of methods and the development of context-specific solutions.

LEAN, AGILE AND MASS CUSTOMIZATION IN MANUFACTURING

The Toyota Production System was considered in the highly influential work of Womack, Jones and Roos in their analysis of *The Machine that Changed the World* in 1990, alternatively known as the 'Lean Production Thesis', the machine being the 'car'. Womack *et al.* (1990) provided the thesis that the era of mass production was now over and pretty much defunct, and that a small number of successful Japanese producers had developed a new world of 'lean production'. The book was based on a large international research project investigating the worldwide automotive industry, both assembly and supply plants, run by the Massachusetts Institute of Technology, but with researchers located around the world. The book is in three parts: a historical perspective mapping the rise and emergence of the automotive industry, from its craft and coach-building origins through Fordism and ending with Toyotaism – incidentally and contentiously debunking socio-technical solutions such as Volvo's autonomous group working along the way as merely an anachronism, an unsustainable return to craft production. Though not the originators of the term 'lean' (this was first used by John Krafcik in 1988), this book certainly popularized the term, ensuring it entered the lexicon of manufacturing management practice centre-stage.

A further concept which is often associated with lean is the notion of 'agile' manufacturing. Agile manufacturing is often wrongly viewed as the latest updated version of lean – a further refinement. However, the use of lean and agile in conjunction can be a very powerful tool, and has enabled many of today's successful manufacturers to continue to flourish. In fact, it is highly likely that most of the products purchased today, by discerning customers demanding some choice and variants in product offerings of manufacturers, have been produced through a combination of leanness and agility. A very useful distinction between lean and agility was provided by Marshall Fisher (1997) and centred upon the distinction between functional products and innovative products. Functional products are those that serve a particular and continuing purpose, are not prone to constant design change and so can benefit from large-volume production. The example used was Campbell's Soups – only incremental changes in the decades since their introduction and so benefiting from volume production using lean production methods. Innovative products, however, are those that are constantly subject to design change, each design version having only a short life-cycle and meaning it is virtually impossible to establish standardized production lines. Agile operations focus on the need to deliver products with speed and flexibility, cost not being the prime driver. The example used was innovative IT products. But the differentiation between lean and agile is more subtle than this, and in the last 20 to 30 years some manufacturers have revolutionized manufacturing approaches in terms of process and product design, as will now be explained.

Systems of mass customization have now emerged which use lean and agile manufacturing principles in tandem. The notion is to strive to benefit from economies of scale and lean/low-cost production in upstream operations, back in the supply chain. Further downstream, and as close to the customer as possible, the idea is then to benefit from the speed and flexibility offered by agile operations. Agility implies the ability to customize and provide a wide range of product offerings to customer demand, but this is clearly more costly. The notion of mass customization is to ensure standardization of products and manufacturing operations as far downstream as possible, but then customize, or more accurately configure, downstream. Thus customization and individual configuration are offered, but are 'postponed' until as late in the process as possible (for more on this see Cousins *et al.*, 2008). The secret behind this lies in developing 'modular' product designs in the first place – developing standard platforms for products and components which can then be applied across a varied product range and/or configured for individual customers through a process of 'mass customization'.

One of the most poignant examples of mass customization was the emergence of Dell computers. In the early days of personal computing, having a customized or highly configured PC was an expensive undertaking. The product was configured at the factory at a considerable price premium, and had longer delivery lead-times and a higher potential for quality issues than the purchase of a standard PC. Then Dell entered the market – and customers could order a PC configured to their individual needs (by phone and then online), which was competitively priced and often less expensive than standard PCs provided by established companies, and which could be delivered within a day or two. Cheaper, faster, better quality – how did they do it? The answer is simple: the deployment of mass customization and postponement. Dell developed a series of PCs with standard platforms and modular designs. They could build the standard platform machines in volume, using lean techniques and economies of scale. These 'platform' machines would be shipped as core products to depots located close to customers. When a customer order for a configured machine was received, the local depot would add the necessary components; machine configuration or customization was undertaken close to the customer and the completed product delivered rapidly to the customer. This approach virtually wiped out the competition, including major computing companies. The concept was simple, but revolutionary, and has now been widely adopted. Consider the automotive industry and the ease with which a car can be individualized (e.g. a BMW Mini or Fiat 500) with some configuration processes undertaken by dealers. End-user configuration also occurs, for example with smartphones, with producers such as Apple, Samsung and Sony leaving the user to complete the configuration with the selection of Apps and aesthetics such as phone covers. These companies essentially sell standardized units which have been produced in high-volume lean manufacturing operations, and postpone the configuration to after the sale to the consumer. This approach also enables incremental redesign and the continuing development of the product, with individual component modifications and some final design configuration and detail postponed until as near to the product launch as possible.

This attention to leaning the process and ensuring fast response through downstream agility has also spurned a further set of developments, categorized under the term 'business process reengineering' (BPR). Taken from the work of Hammer and Champy (1993), this perspective argues that the operation of manufacturing firms (or any other operation, for that matter) is best viewed as a series of processes – upstream to downstream – which should be analysed for waste and inefficiencies and subsequently be removed. Any new processes should be designed with these principles, to ensure that only value-adding activities take place, and that non-value-adding and wasteful operations are removed. This approach has seen a resurgence of process mapping tools and techniques, derivatives of the tools developed by early practitioners of scientific management and systems analysis, and the term 'value stream mapping' has emerged as a popular phrase and process. It is clearly the case that efficient operations should be 'leaned' as much as possible to ensure the delivery of competitive products with speed and accuracy, and with the minimum of cost. But this has been a contentious area and critics have accused BPR practitioners of a return to an era of Taylorism, of dehumanizing work, of using BPR as a veil to downsizing and exploiting the workforce (Knights and Willmott, 2000). Nonetheless, mapping tools are now once more in wide usage and offer valid tools if used appropriately as manufacturing firms strive to implement lean principles in their operations.

MANAGING OPERATIONS STRATEGICALLY

Alongside Japanization and the convergent development of lean and agile operations, the period since 1970 has seen the emergence of the Manufacturing Strategy paradigm. This has resulted in a set of frameworks and methodologies to help manufacturing managers analyse their existing position and to develop effective manufacturing strategies. The starting point for these arguments goes back to the seminal article by Wickham Skinner (1969) in the *Harvard Business Review*, where he challenged the increasingly prevalent view at the time in corporate America that manufacturing operations were merely operational in nature and of very little or no strategic importance. Skinner accused most Western companies of viewing their factory operations as 'millstones around their corporate necks' – they constantly held the organization back in their strategic and market ambitions, using vast resources and constantly resisting change. The role of manufacturing managers was viewed as one of fire-fighting and sorting out problems often of their own making. In fact, should the firm not simply divest production operations altogether? Skinner challenged this, developing a stepwise plan to assist manufacturers to develop responsive and supportive operations processes and so compete effectively in world markets. The article's title indicates the main tenor of his argument: 'Manufacturing – Missing Link in Corporate Strategy'.

Skinner and later authors, including Hayes and Wheelwright (1984) and Terry Hill (1993), directly accused strategic managers and national policymakers in the USA and UK of having misconceptions about the role of the manufacturing function in businesses, one which saw operations as of little or no strategic importance. In fact, manufacturing managers often viewed their role as primarily reactive rather than proactive. Of course, compare this view with manufacturers such as Toyota and businesses such as Dell which used their supply chains, manufacturing operations and distribution systems to actively gain competitive edge over their rivals. These writers, amongst others, beat the drum hard to draw attention to the importance of manufacturing in business and to the economies of the USA and UK, to try to ensure that manufacturing was appreciated at all levels in business and government.

Hayes and Wheelwright (1984) built upon their observations of Japanese and German exemplars to present their four-stage 'Framework for Manufacturing Effectiveness' approach. The stages describe a progression, from the 'millstone'-type manufacturing operations defined by Skinner (1969) through to those manufacturing operations where the manufacturing strategy provides the company's core source of competitive advantage, the best example probably being Toyota. The stages are:

1. *Internally neutral: minimize manufacturing's negative potential.*
   The business views the manufacturing function as neutral, incapable of influencing competitiveness. In these circumstances the main concern is to minimize any negative influence that manufacturing may have upon the business. It is merely a 'millstone'.
2. *Externally neutral: achieve parity with competitors.*
   Manufacturing is perceived as important only in so much as it matches the effectiveness and efficiency of (i.e. it 'benchmarks') competitors in the same industry. The emphasis is on replicating industry practices and avoiding large step innovations in the design of products, processes or work organizations. Economies of scale and efficiency are the most important drivers in production.
3. *Internally supportive: provide credible support to the business strategy.*
   Manufacturing is expected to support and strengthen the competitive position. Manufacturing strategies and objectives derive from corporate and market objectives. Decisions made in manufacturing are consistent with corporate and marketing strategies and manufacturing managers are proactive in developing forward-thinking manufacturing strategies at the functional and factory levels.
4. *Externally supportive: pursue a manufacturing-based competitive advantage.*
   Manufacturing capabilities significantly influence competitiveness, and this is undisputed in the organization. Manufacturing strategy is not merely determined in a top-down cascade from higher strategies and marketing strategies, but manufacturing managers have a meaningful role contributing to the development of key corporate strategies as a whole. In some cases manufacturing is the key 'competitive weapon' for the business.

The notion is not that all manufacturers strive for stage 4. For some stage 2 or 3 is sufficient, given the relevance of manufacturing within the overall context of their business. But managers should be able to identify which stage their manufacturing operations are at in this framework and therefore identify changes that need to be made within the organization to progress, if need be, to the next stage. Hayes and Wheelwright reported, critically, that much of US industry at the time of writing in the early 1980s was based at stage 1. They argued that the best businesses would have manufacturing operations at stage 3 or 4. At stage 3 manufacturing managers and personnel are encouraged to come up with new product ideas and new methods, and suggest these to their customers, whilst stage 4 manufacturers redefine industry expectations. Consider the examples above, where Toyota was the original exponent of what is now termed lean production, and Dell completely reshaped the PC industry with the development of a mass customization approach.

Skinner's (1969) original approach to manufacturing strategy formulation and the Hayes and Wheelwright (1984) 'Effectiveness Framework' are both largely prescriptive. Though timely in the development of the Manufacturing Strategy paradigm, they are limited in the extent to which they provide a meaningful practical methodology for manufacturing strategy formulation. Developing Skinner, Hayes and Wheelwright's ideas, two notable contributions are now presented which bridge this gap: Hill's (1993) framework and the Platts and Gregory methodology (Department of Trade and Industry (DTI), 1988).

Hill summarizes his arguments and discussion within his 'framework for reflecting corporate and marketing strategy issues in corporate decisions'. Fundamental to the Hill framework is the concept that any manufacturing business has to have a clear understanding of 'how its products win orders in the market-place'. These could be on criteria related to price, quality, delivery, service, responsiveness to change, technical performance, and so on. The order-winning properties derive from both marketing activities as reflected in the marketing strategy, and from the manufacturing strategy in terms of the choice and operation of processes, supply chain and delivery systems plus the inherent manufacturing infrastructure, that is to say the planning and control systems. The Hill framework provides a means for determining how market and competitive decisions should be linked with decisions on manufacturing systems design. It also explores the potential for the proactive, bottom-up contribution of manufacturing reflecting, in effect, the third and fourth stage of the Hayes and Wheelwright framework. The framework can be used both to assess and to evaluate the effectiveness of manufacturing operations in relation to corporate objectives and product markets, but also, and more importantly, serves as a guide for developing innovative market-focused activities (Hill and Hill, 2011).

The Platts and Gregory Manufacturing Strategy Methodology explores how to determine market requirements, analyse competitor threats and measure the performance of existing systems, and then combine these into a comprehensible new strategy. The original methodology was creatively presented in 1988 as a workbook, enabling the user to work through a series of worksheets, performing an analysis of the market and competitive environment and then conducting an audit of existing operations facilities and competencies. The emphasis was to explore where manufacturing operations' performance may be improved to better align them to corporate and customer needs. The methodology involves three main phases:

1. *Stage 1: understanding the market position.*
   Gather data on all 'product families' in the manufacturing business and then indicate the relative importance of each family and its strength within the market. The key competitive criterion for each family is determined, whether price, delivery performance, and so on, and the current performance against these criteria is assessed. Potential areas of product profitability and vulnerability are also identified.

2. *Stage 2: understanding manufacturing operations.*
   Produce an assessment of the existing manufacturing strategy within the key areas of facilities, capacity, span of process, process design, human resources, quality, operations control, supplier management and new products.
3. *Stage 3: develop new strategy.*
   Taking each product family in turn (starting with the more 'important'), compare the current performance of manufacturing operations with that desired by the business strategy and market. Guidance is provided on integrating this into a coherent manufacturing strategy.

The Platts and Gregory methodology was a significant contribution to the manufacturing management literature and, in line with the sector where Platts and Gregory conducted their research, was relevant to, and applied by, engineering businesses in particular. Similar to Hill (1993) they suggest that the methodology can be applied both to audit existing manufacturing operations and as a framework for reviewing existing manufacturing strategies and for developing new strategies.

The contributions of Hill (1993) and Platts and Gregory (DTI, 1988) have been applied and used widely and have clearly been very significant in the development of the Manufacturing Strategy paradigm. It would appear bizarre today for any business which has not subcontracted or divested its manufacturing function to consider its manufacturing operations as unimportant and inconsequential. The approaches developed by Hill (1993) and Platts and Gregory (DTI, 1988) provide a useful insight into the process of operations strategy formulation and also provide practical guidance for the development of effective manufacturing strategies.

CONCLUSIONS

This chapter has reviewed the development and main principles underlying the management of the manufacturing function within business. It has taken a historical perspective through identifying some of the key contributions made by practitioners and academics alike to arrive at models and frameworks for manufacturing strategy formulation as practised today. Early contributions to understanding manufacturing and management were at polar opposites – from the scientific management principles of Taylor to the Human Relations School initiated by Elton Mayo and his colleagues.

This chapter has demonstrated how a rich heritage of factory studies, often adopting socio-ethnography as the investigative style, enabled these discussions and debates to develop, right through to studies of Japanization. Japanization, or more precisely the development of lean, agile and mass customization principles, has had an important influence on what is currently considered best manufacturing practice. The chapter ended by presenting some of the frameworks that are employed today to assist companies in the analysis and development of effective manufacturing strategies. Central to these debates has been a concern with understanding the efficient management and organization of manufacturing processes and operations.

NOTE

1. Generally termed 'Operations Research' in the USA, though in other places (e.g. the UK) the term 'Operational Research' is widely used.

REFERENCES

Babbage, Charles (1835), *On the Economy of Machinery and Manufactures*, London: Charles Knight.

Bennett, D.J. and Forrester, P.L. (1993), *Market Focused Production Systems: Design and Implementation*, Hemel Hempstead: Prentice Hall.

Braverman, H. (1974), *Labour and Monopoly Capital: The Degradation of Work in the Twentieth Century*, New York: Monthly Review Press.

Brown, W. (1962), *Piecework Abandoned – The Effect of Wage Incentive Systems on Managerial Authority*, London: Heinemann.

Browne, J., Harhen, J. and Shivnan, J. (1996), *Production Management Systems: An Integrated Perspective*, 2nd edition, Harlow: Addison-Wesley.

Bryson, J.R. (2000), 'Spreading the Message: Management Consultants and the Shaping of Economic Geographies in Time and Space', in Bryson, J.R., Daniels, P.W., Henry, N. and Pollard, J. (eds), *Knowledge Space Economy*, London: Routledge, 157–175.

Bryson, J.R. and Lowe, P.A. (2002), 'Story-Telling and History Construction: Rereading the Story of George Cadbury's Bournville model village', *Journal of Historical Geography*, **28** (1), 21–41.

Copley, F.B. (1923), *Fredrick W. Taylor, Father of Scientific Management*, 2 Vols, New York: Harper and Brothers.

Cousins, P., Lamming, R., Lawson, B. and Squire, B. (2008), *Strategic Supply Management: Principles, Theories and Practice*, Harlow: Prentice Hall.

Crawford, M. (1995), *Building the Workingman's Paradise: The Design of American Company Towns*, London: Verso.

Department of Trade and Industry (1988), *Competitive Manufacturing: A Practical Guide to the Development of a Manufacturing Strategy*, Bedford: IFS Publications.

Fisher, M.L. (1997), 'What is the Right Supply Chain for Your Product?', *Harvard Business Review*, **75** (2), 105–116.

Gillespie, R. (1993), *Manufacturing Knowledge: A History of the Hawthorne Experiments*, Cambridge: Cambridge University Press.

Gyllenhammer, P. (1977), *People at Work*, Reading, MA: Addison-Wesley.

Haber, S. (1964), *Efficiency and Uplift: Scientific Management in the Progressive Era 1890–1920*, Chicago: The University of Chicago Press.

Hammer, M. and Champy, J.A. (1993), *Reengineering the Corporation: A Manifesto for Business Revolution*, New York: Harper.

Hayes, R.H. (1981), 'Why Japanese Factories Work', *Harvard Business Review*, **59** (4), 56–67.

Hayes, R.H. and Wheelwright, S.C. (1984), *Restoring Our Competitive Edge: Competing Through Manufacturing*, New York: Wiley.

Hill, T.J. (1993), *Manufacturing Strategy: The Strategic Management of the Manufacturing Function*, 2nd edition, London: Macmillan.

Hill, A. and Hill, T.J. (2011), *Essential Operations Management*, London: Palgrave Macmillan.

Ishikawa, K. (1985), *What is Total Quality Control? The Japanese Way*, Englewood Cliffs, NJ: Prentice Hall.

Jacques, E. (1951), *The Changing Culture of a Factory*, London: Tavistock.

Knights, D. and Willmott, H. (eds) (2000), *The Reengineering Revolution? Critical Studies of Corporate Change*, London: Sage.

Krafcik, J.F. (1988), 'Triumph of the Lean Production System', *Sloan Management Review*, **30** (1), 41–52.

Maslow, A. (1943), 'A Theory of Human Motivation', *Psychological Review*, **50** (4), 370–396.

McGregor, D. (1960), *The Human Side of Enterprise*, New York: McGraw-Hill.

Oliver, N. and Wilkinson, B. (1988), *The Japanization of British Industry*, Oxford: Blackwell.

Oliver, N. and Wilkinson, B. (1993), *The Japanization of British Industry: New Developments in the 1990s*, 2nd edition, Oxford: Blackwell.

Roethlisberger, F.J. and Dickson, W.J. (1939), *Management and the Worker*, Boston: Harvard University Press.

Schonberger, R.J. (1982), *Japanese Manufacturing Techniques: Nine Hidden Lessons in Simplicity*, New York: Free Press.

Skinner, W. (1969), 'Manufacturing – Missing Link in Corporate Strategy', *Harvard Business Review*, **47** (3), 136–145.

Taylor, F.W. (1911), *The Principles of Scientific Management*, New York: Harper and Row.

Trist, E. and Bamforth, K.W. (1951), 'Some Social and Psychological Consequences of the Longwall Method of Coal-Getting', *Human Relations*, **4** (1), 3–38.

Urwick, L. and Brech, E.F.L. (1945), *The Making of Scientific Management, Vol 1: Thirteen Pioneers*, London: Management Publications Trust.

Urwick, L. and Brech, E.F.L. (1953), *The Making of Scientific Management, Vol 2: Management in British Industry*, London: Pitman and Sons.

Voss, C. (ed.) (1987), *Just in Time Manufacture*, Bedford: IFS Publications.

Vroom, V.H. and Yetton, P.W. (1973), *Leadership and Decision-Making*, Pittsburgh: University of Pittsburgh Press.

Wheelwright, S.C. (1981), 'Japan: Where Operations Really Are Strategic', *Harvard Business Review*, **59** (4), 67–74.

Womack, J.P., Jones, D.T. and Roos, D. (1990), *The Machine that Changed the World*, London: Harper.

2. Manufacturing and labor

Sally Weller

INTRODUCTION

Manufacturing is an important source of employment in both advanced and developing economies. In 2010, according to the International Labour Organization (ILO), manufacturing industries employed in excess of 670 million people worldwide and accounted directly for 26 percent of men's and 16 percent of women's jobs (ILO, 2012). In the United States, although manufacturing industries shed about four million jobs in the ten years 1998–2008, they still employed over 12 million people in 2012 and accounted for 9 percent of total non-farm employment (Bureau of Labor Statistics, 2012). In addition, many jobs found in the service industries rely on manufacturing activity. This chapter provides a critical overview of labor's role in and contribution to manufacturing. It restricts its scope to direct employment in manufacturing industries.[1]

The need to harness individual skills, abilities and effort to collective ends makes the efficient and productive organization of work tasks crucial to the competitiveness of manufacturing firms in capitalist economies. Although human labor is an essential input into any manufacturing process, it differs from other inputs in important ways. In the first place, it is not "produced" for the purpose of work and firms cannot control its rate of production (i.e. the population growth rate). Second, labor is produced by workers who have interests and responsibilities beyond their working lives. This conditions the amount of labor they are willing or able to contribute at different times of the day and across the life-course. Third, workers are living and sentient beings: physical capacities, feelings, emotions and perceptions of fairness all affect both their capacity to work and the amount of effort they expend in the workplace. These inextricably human characteristics mean that labor cannot be standardized. Consequently, matching workers with work tasks is a perennial challenge for manufacturing enterprises. It also means that labor's productivity is complexly related to the working arrangements and working conditions in workplaces. Fourth, while firms can move most other inputs from place to place at will, labor is comparatively immobile. As a result of differences in access to education, health services and cultures, the quantity and quality of labor vary from place to place. The spatial differentiation of local workforces has significant implications for the location choices of manufacturing firms.

The organization of work in manufacturing is a large topic with multiple dimensions. This chapter focuses on the internal organization of work in manufacturing firms, the relationships between firms and the external labor markets from which they draw labor, and changes in the organization of work among internationalized networks of firms. Throughout, labor is treated as a pseudo-commodity, work as an inherently social activity, and employment as a relationship shaped by the spatialized arrangement of firms, localities, regions and economies. The chapter focuses on matters specific to manufacturing employment, rather than more general issues like worker motivation, job satisfaction and well-being, education and skill development, or different ways of understanding the value of labor (see Castree *et al.*, 2004). It is organized in the following way. Following this introduction, the chapter describes the evolution of the labor process within manufacturing firms; how local labor markets match employers' labor needs with the characteristics of the local labor supply; and then the impact of globalization on the distribution of manufacturing employment. The chapter concludes by identifying some key challenges for manufacturing labor.

WORKPLACES AND THE LABOR PROCESS

The organization of work tasks and the task content of occupations constitute the *technical division of labor*. It is both socially constructed and dynamic, reorganizing incessantly in response to both technological change and changes in social, economic and political contexts. Within manufacturing workplaces, the evolution of the labor process, or the "process by which workers' capacity to labor is translated into actual work" (Edwards, 1986: 1), produces a variety of historically specific and contingent trajectories.

This account of the evolution of the manufacturing labor process begins with the industrial revolution in 19th-century England, when in newly established factories, craft-based production was reorganized into increasingly specialized sets of tasks. Adam Smith (1904), for example, noted the transformation of the trade of "pin-maker" from a one-person craft activity into a process with at least 18 sequential operations, each performed by a different worker. As workers became more specialized, productivity increased markedly, but so did the drudgery of more repetitive work demands (Braverman, 1974). Growing awareness of the relationship between workers' health and their capacity to work produced campaigns to improve workers' living conditions (Engels, 1958).

In the first half of the 20th century, firms sought to protect the unique advantages of firm-specific skills and processes by improving the wages and conditions of key workers, providing training and creating opportunities for internal career progression. The resulting "internal" labor markets were protected by limited points of entry and governed by rules and customs that were not readily transferable to other contexts. Positions and pay scales were set by administrative rules (e.g. seniority) rather than by reference to external market conditions. Some argue that internal labor markets also facilitated managerial control, using hierarchical distinctions between workers to prevent their organized mobilization in opposition to management's strategies (Edwards, 1986). This stable form of workplace organization flourished in larger and technologically advanced workplaces.[2]

By the 1970s, emerging manufacturers in Asian economies – which by virtue of low labor costs could produce manufactured goods at low cost – were penetrating higher income markets and challenging more expensive locally made manufactures. At the same time, as a growing fiscal crisis threatened the economies of the world's advanced welfare states and market-based policy approaches became prominent, states withdrew from active support of their manufacturing industries.[3] To survive in an increasingly competitive environment, manufacturing firms in the advanced economies concentrated on improving their productivity and competitiveness. Three central restructuring strategies were to save on labor costs, to increase labor flexibility and to reorganize workplaces.

An appetite for labor-saving encouraged automating labor-intensive tasks, minimizing materials-handling tasks, and contracting out non-core activities such as cleaning, bookkeeping and marketing. New technologies were designed to replace or limit the need for operators. This downsizing process reduced manufacturing employment considerably, with job losses concentrated in less skilled, labor-intensive occupations (Bluestone and Harrison, 1982; Massey and Meegan, 1982).[4] Technical innovations such as computerized controls, computerized design and inventory control made some specialized occupations redundant and transformed the work content of others. For example, secretaries (and their shorthand skills) were replaced by computer-literate personal assistants. This process was infused with power relations: tasks that had been the source of workplace power for trade unions – such as the crucial role of cutters in the clothing industry – were targeted for outsourcing, mechanization or disaggregation (Bluestone and Harrison, 1982). Emerging specialist and technical occupations were often established as services to manufacturing, rather than being integrated into manufacturing plants.[5]

As rapidly innovating and less predictable product markets produced increasingly volatile work flows, firms increasingly demanded numerically and functionally flexible workforces. With the introduction of "just-in-time" production scheduling, firms reorganized to operate with a core workforce of valued contributors (with more secure, full-time jobs) supported by a secondary workforce comprising less secure and less well paid contract, casual and temporary workers who could be engaged as required at peak times (Doeringer and Piore, 1971; Pollert, 1988). Work intensity increased across workplaces as precariously employed workers competed with each other for work and for job security, as core workers were induced to work harder to maintain their positions, and as improved technologies of work flow monitoring eliminated work "down" time and accelerated production lines. Over time, as these systems were institutionalized and computerized, the dual "core–periphery" structure eroded as contract employment exposed core workers to greater risk. In effect, the competitive character of the external labor market was being internalized within firms.

Another organizational innovation was the adoption of team-based forms of shopfloor organization first developed in Japan as the *Kanban* or Toyota system (Ohno, 1978), but known in the West as "flexible specialization" (Piore and Sabel, 1984) or "lean" production (Rutherford and Gertler, 2002). By providing a more functionally flexible and productive workforce, team-oriented organizational innovations enabled firms to respond more quickly to changing conditions, to entrust quality control to manufacturing subcontractors and to identify innovations emerging from the practical "learning by doing" knowledge of production workers. The upshot, often, was reduced worker autonomy and further labor-saving; for example, by making it possible to eliminate supervisory roles and flatten workplace hierarchies (Graham, 2005; see also Burawoy, 1979). As emerging computerized production technologies made it possible to monitor individual performance in relatively unobtrusive ways, coercive workplace rules and direct surveillance of workers were replaced by a combination of internalized self-surveillance and rigid performance benchmarking (Deetz, 1992). These technology-oriented innovations made it possible for firms to individualize the employment relationship and tailor pay and incentives to the value of each worker's labor.

Together, these changes increase employer prerogative in the workplace and make it more difficult for unions to organize workers. As the structure of workplaces and the frameworks of regulation have changed, confrontational industrial action in technologically advanced workplaces has become both less possible and less effective (Cumbers and Atterton, 2000). Declining union membership across advanced economies can be understood as both a cause and an effect of these workplace changes. Governments, through policy settings that are inhospitable to organized labor, have encouraged the individualization of work. Nonetheless, labor movements have demonstrated a capacity to influence both the direction of organizational restructuring and the nature of changes in the labor process (Herod, 1997). As well as a return to grassroots of workplace activism (Rutherford and Gertler, 2002), worker resistance now extends beyond the confines of firms, to worker–community alliances (Herod, 2001) and global social movements (Wills, 1998).

So far this account has focused on advanced economies. The workplaces of manufacturing plants in emerging economies have a quite different experience. In their case, relatively sophisticated manufacturing systems have been established at new sites, without the evolutionary development characteristic of early industrializers. Under the guidance of managers and engineers (or "transnational knowledge workers") brought from or trained in advanced economies, and often with the added discipline of foreign direct investment and export contracts within competitive global commodity chains (Yeung, 2002), these workplaces typically operate with more authoritarian forms of organization and feature weaker union activity (Depner and Bathelt, 2005). Typically, the workforces recruited to these plants are relatively new to manufacturing, with individual workers trained to perform specific tasks. Firms use spatial strategies – such as providing supervised dormitories adjacent to the workplace – as a means of disciplining workers, many of whom are migrants from poorer regions (Kelly, 2002; Fan *et al.*, 2009). Links between businesses and the private sphere, through family and place networks, also encourage patriarchal forms of organization (Chant, 1995; Olds and Yeung, 1999).

In combination, these workplace changes have increased the skill demands and social competences required of manufacturing workers. Workers with the capacity to undertake specialized tasks, monitor computerized controls and contribute to team goals are likely to need advanced technical skills, literacy and numeracy skills, critical faculties and the ability to communicate with their co-workers. As a consequence, the nature of labor demand in manufacturing has been transformed, and with it firms' relationships with external labor markets.

LABOR MARKETS

Labor markets are the institutions that govern the interaction between the demand for labor by firms and the supply of labor by citizens in communities. This interaction is shaped by local conditions and by localized regulatory frameworks. As well as matching work with workers, labor markets mediate interactions between many types of institution – principally between households and firms but also between employment intermediaries, training institutions, employer organizations, unions and various state bureaucracies. Just as labor is not a true commodity, labor markets are not true markets. In a true market, prices and quantities adjust until supply and demand come into balance. If labor markets were true commodity markets – like, say, the market for tomatoes – labor shortages would push up prices (wages) and bring more labor into the

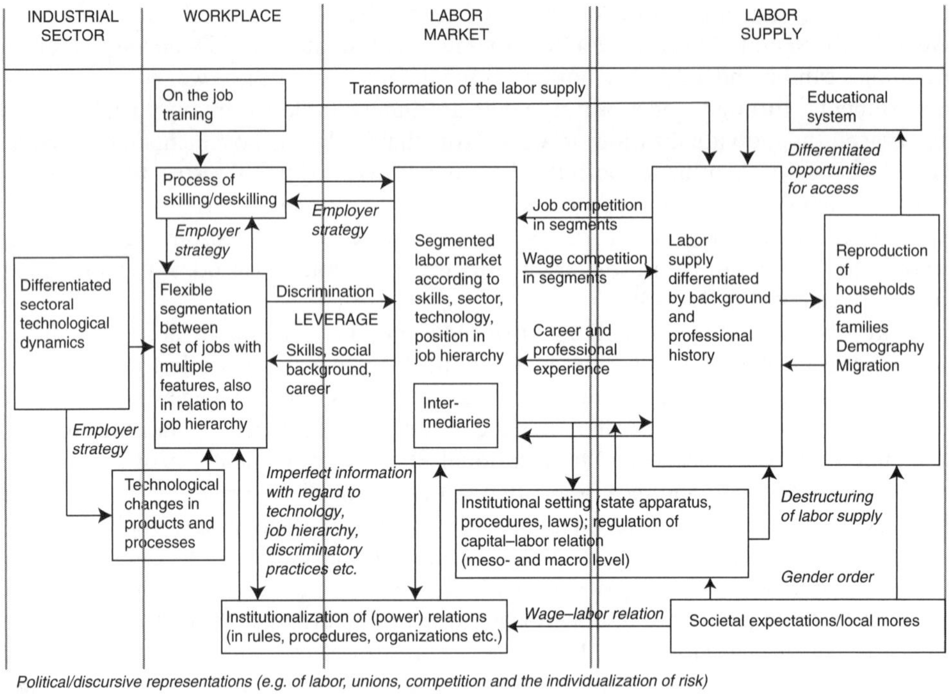

Source: Author's expansion of Moulaert 1987 and Storper and Walker 1983.

Figure 2.1 Institutional labor markets

market, while labor surpluses would have the reverse effect, suppressing prices (wages) and encouraging less competitive or more selective workers to leave the market. In reality, however, a lack of demand for labor does not eradicate an excess supply, and the price of labor cannot sustainably fall below the level required for workers to live, to continue to work and to support their families. It follows that viewing labor market processes as individualized market-based transactions between an employer and employee – as is the case in economics – is manifestly inadequate, even when supplemented by understandings of the role of worker skills or "human capital" (Becker, 1962), information (Stigler, 1962) or discrimination (Arrow, 1973).

From the institutional perspective that informs this chapter, on the other hand, the operation of the labor market is infinitely more complex. As depicted in Figure 2.1, labor markets link industrial sectors, workplaces, recruitment processes and the available labor supply in households. The nature of these linkages depends on local conditions and localized regulatory frameworks. Importantly, the relationships depicted in Figure 2.1 can be "read" from left to right – demand to supply – or from right to left – supply to demand: the "market" responds in both directions simultaneously.

Reading from left to right highlights that firms' orientations to the external labor market are shaped by industry conditions and their internal organizational forms.

Employers retain a high degree of discretion and prerogative over recruitment and staff deployment decisions (Rubery and Wilkinson, 1994). They can elect to meet their labor needs by reorganizing internal positions (e.g. offering overtime); they can provide training to alter the skill attributes of existing workers; they can adopt technologies that replace, alter or mechanize some job tasks; or they can recruit new workers from the external market. These strategies change with changing business environments and labor supply conditions (see Gough, 2005). If specialist skills cannot be found locally, firms might establish training programs, extend the scope of their search beyond the immediate locality or "head hunt" skills from their competitors (Dawley, 2007). Recruitment strategies reflect firms' broader strategic directions. Those seeking to maintain and reproduce existing internal workplace cultures and power relations will favor word-of-mouth or network-based recruitment strategies, drawing from the associates of their existing employees or from a known pool of temporary workers (such as contract or casual workers, interns and apprentices). Those seeking to actively reform a workplace culture, on the other hand, are likely to recruit from outside their existing networks.

Reading Figure 2.1 from right to left, on the other hand, highlights the challenges that face potential workers. In the first instance, their opportunities are influenced by their age, gender, education and social background (social class), as well as being circumscribed by their household circumstances and location. From a human capital perspective, the labor supply depends on individual choices – a trade-off between work and other activities – and depends principally on the price (wage) of job offers, which varies with skill and responsibility. One alternative view sees the labor market as a queue in which employers select from a pool of potential workers the applicant who best fits their task specifications and remuneration (Thurow, 1975). Others have emphasized that employers do not necessarily select the most technically skilled workers, but instead favor those who fit into the workplace socially or are more socially compliant (see Wright, 1999).

The effect of these recruitment processes is the "segmenting" of labor markets into multiple submarkets. Segmentation theories observe that labor market processes cluster together worker characteristics (age, gender, etc.), occupational tasks (groups of skills) and job characteristics (stability, career prospects, etc.). Thus, for example, men from middle-class homes move into well-paid executive roles, women are channeled into nursing and childcare jobs where pay scales are modest given the skill demands of the work, and less skilled migrants are condemned to even lower paid service jobs. The idea of the *gender* and *ethnic divisions of labor* recognizes that segmentation fragments local labor markets and allocates the poorest quality jobs to the most vulnerable groups of workers (Bauder, 2006; McDowell *et al.*, 2007). By viewing the employment relation as a social relation politically mediated through institutions and discourses, contemporary segmentation theories are able to explain why returns to training and skill are variable, why discrimination is endemic, and how relationships between households and workplaces shape the character of the labor supply.

Labor markets are also inherently spatial and localized (Peck, 1996). Their compositions reflect the demographic and skill characteristics of local communities, the effects of local regulations, social customs, and the history of relationships between labor (workers) and capital (firms). Over time, local labor markets develop unique characteristics or "labor control regimes" (Jonas, 1996) that reflect the historically produced idiosyncrasies of local economies and their regulatory frameworks. In practice, local labor markets are bounded geographically by realistic journey-to-work distances between workers' households and their workplaces, although these distances vary for different subgroups of workers and different occupations. Women with childcare responsibilities, for example, are often confined to work in areas closer to their homes (Hanson and Pratt, 1995), while higher paid workers with fewer domestic responsibilities can afford a longer commute. Segmentation processes imply that geographically bounded local labor markets actually incorporate multiple occupationally defined labor markets, each with a characteristic spatiality but superimposed over the same physical spaces simultaneously.

Local markets operate in the context of regional, national and sometimes international regulatory frameworks. Directly, states calibrate the supply of labor through rules about social security eligibility, migration, childcare and maternity leave. They fund work incentive schemes; build the quality of labor through interventions such as the provision of education, training services and health care services; and influence the demand for labor by calibrating interest rates, investment incentives, payroll taxes and company taxes. Governments stimulate the labor-matching process by funding recruitment, information and job search services. They also influence the labor market's operation indirectly. For example, corporation laws shape corporate management, investment and risk frameworks in ways that condition firms' approach to labor issues (Christopherson, 2002). In the United States, where corporate frameworks favor short-term returns on shareholders' investments, labor markets are more volatile and more individualized compared with those in parts of Europe, where regulation encourages firms to be more risk-averse. In Asia's Export-Processing Zones, it is not uncommon to find more stringent labor control rules than in other parts of the economy. The absence of a welfare system in some states, and the restructuring of welfare eligibility in others, actively "produces" vulnerable, low skill contingent workforces with little option but to accept low wages and poor working conditions (Peck, 2001).

LABOR, GLOBALIZATION AND RESTRUCTURING

Although labor market processes are localized in firms and communities, in aggregate the location decisions of manufacturing firms have produced an internationalized distribution of manufacturing work. The notion of the *spatial division of labor* refers to the reality that "employment not only takes place in space" but also the fact that businesses "use space as a strategic variable in the creation, destruction and re-creation of viable employment relations over time" (Walker, 1985: 182, see also Massey, 1984). In their decisions about industrial location, firms identify and exploit historically produced differences in the availability, talents and relative productivity of workers in different places. Firms can reduce their production costs by relocating to sites where needed skills can be purchased at minimum cost, or improve their capacity for innovation by moving to places rich with higher skilled or creative workers. These processes encourage places to compete for inward investment funds, both regionally and globally, on the basis of the qualities of their local labor forces. As processes of agglomeration reinforce existing strengths and weaknesses, labor markets in places are likely to become more specialized in the future.[6]

Forms of the internationalization of manufacturing production are historically specific and differ in pace and direction among industries and places. Since the 1970s, the increasing internationalization of manufacturing in global production networks and in large multi-plant transnational firms has transformed the distribution of manufacturing employment. More labor-intensive manufacturing production activities (such as clothing production) have relocated progressively to lower wage sites, spurred in part by firms' desire to escape from unionized labor in advanced economies (Fröbel *et al.*, 1980). The exit of uncompetitive firms and their replacement by new firms established at new "greenfield" locations, mainly in developing economy sites, reinforces this trend. These shifts have favored the concentration of manufacturing activity in Export-Processing Zones supported by government export-oriented industrialization policies; and in places that are the lower wage members of a regional trade group (such as manufacturing plants in Mexico's maquiladoras serving the US market).

International wage cost differences remain a major factor in the location of manufacturing jobs, but the relationship is complicated by wide variations in the productivity of labor and in the capital intensity of different manufacturing specializations. Countries like the United States, which retains a large manufacturing sector despite relatively high wages, tend to be specialized in industries like automotive and steelmaking that have high capital-to-labor ratios and high value added per worker.[7] The relative cost of manufacturing in different places changes with shifts in macro-economic factors and currency exchange rates. In the years 2003 to 2007, manufacturing labor costs were increasing, in relative terms, in high wage countries like Ireland, Canada and Australia but decreasing in the United States, Japan, Brazil, Taiwan and Mexico (ILO, 2012). In most parts of the world, manufacturing employment numbers have declined since the beginning of the 2007 global financial crisis, reflecting weaker consumer demand for manufactured commodities.

Countries vary in the extent to which manufacturing workers have been protected by labor law. These differences reflect more general differences in national "varieties of capitalism" (Wailes *et al.*, 2009). In general, in places with less workplace regulation firms have wider discretion over workers' benefits. Many countries do not meet the minimum standards of the ILO's (1998) Declaration on Fundamental Principles and Rights at Work, which sets out four minimum standards: freedom of association and the effective recognition of the right to collective bargaining; the elimination of all forms of forced or compulsory labor; the effective abolition of child labor; and the elimination of discrimination in respect of employment and occupation.

CONCLUDING REMARKS

This chapter has described the *social division of labor* – or the ways in which work is organized among enterprises, occupations, people and places – as a dynamic process that evolves with social, political and economic changes. The chapter concludes by outlining four areas where the frameworks of manufacturing labor are changing most rapidly.

First, with internationalized manufacturing and the development of increasingly specialized and integrated globalized production networks, conceptualizations of work relationships must now extend beyond discrete workplaces to take into account the dense interconnections and labor exchanges among firms, including both the dispersed branch plants of transnational firms and the distanciated but close relationships among linked firms in globalized production networks. As these connections reposition labor in complex socio-spatial relationships that span territories, places, scales and networks simultaneously (Jessop *et al.*, 2008; see also Coe and Hess, 2012), global networks might increasingly organize around the needs of highly mobile labor elites.

Second, at the same time as firms are becoming more globally integrated, the strategies of organized labor are internationalizing. Workers' strategies of resistance and activism readily transfuse from place to place, producing rapidly changing internationalized dynamics of activism (Wills, 1998 but see Burawoy, 2010). Activism now targets leading firms in production networks, targeting the reputations of their brands. These tactics are inducing firms to include labor standards in corporate social responsibility initiatives and institute voluntary labor codes of practice through their networked manufacturers. International solidarity alliances linking internationalized unions and human rights groups are becoming an increasingly important influence over the labor practices of transnational firms.

Third, contemporary manufacturing labor markets are being transformed by migration, which provides an inexpensive "fix" to the problem of coordinating geographically uneven patterns of labor supply and labor demand. This reflects revised government attitudes to training and skill formation that in effect shift responsibility for the costs of skill development from states to individuals (World Bank, 2009). In addition, the tendency of highly skilled workers to gravitate to "creative" places that offer rich vocational and social opportunities – and avoid places that do not offer these opportunities – appears to be reinforcing the physical separation of knowledge-based and manual work (see Florida, 2002). Despite the slowed growth associated with the global financial crisis, China's continuing development as a global manufacturing center has implications for manufacturing employment in both high wage countries and smaller low wage nations. Fourth, as policies designed to slow climate change increase the cost of energy and the cost of the energy-rich inputs used in manufacturing, manufacturing firms might be encouraged to locate closer to reliable power sources or to sites that reduce transport costs to markets. In the longer term, increasing transport costs might encourage a re-localization of manufacturing activities, bringing them closer to consumer markets.

Labor and labor markets are deeply embedded in economic, social and political systems. This means that the deployment of labor within manufacturing firms is both molded and limited by the social and cultural relations of the communities in which they operate.

NOTES

1. Manufacturing is defined by Groups 15–37 of the International Standard Industry Classification. This excludes home-based manufacturing and manufacturing activities undertaken in other industry sectors (e.g. agriculture).
2. This form of organization became known as 'Fordism' after its chief exponent, the Ford Motor Company. Critics argue that these arrangements were never generalized but existed only in larger workplaces and in sectors where workers had won better working conditions through industrial action.
3. After the collapse of the Bretton Woods agreement in 1971, Keynesian economic policies proved unable to curb inflation or to stem the growth of unemployment.
4. Many jobs lost from manufacturing were replaced by similar job functions in the services sector.

5. The once-clear division between manufacturing and services activities, based on the expectation that the former makes 'things' and the latter provides services, blurs with these shifts.
6. The notion of spatial divisions of labor continues to inform understandings of labor's sometimes oppressed and sometimes privileged positions in global production networks (Phelps and Fuller, 2000).
7. Higher wages reflect (1) higher skills, (2) higher productivity and (3) high rates of unionism, which increase the bargaining power of workers.

REFERENCES

Arrow, Kenneth (1973), "The theory of discrimination," in Orley Ashenfelter and Albert Rees (eds), *Discrimination in the Labour Market*, Princeton, NJ: Princeton University Press, pp. 3–42.

Bauder, Harald (2006), *Labor Movement: How Migration Regulates Labor Markets*, Oxford: Oxford University Press.

Becker, Gary (1962), "Investment in human capital: a theoretical analysis," *Journal of Political Economy*, **70**, 9–49.

Bluestone, Barry and Bennett Harrison (1982), *The De-industrialization of America*, New York: Basic Books.

Braverman, Harry (1974), *Labor and Monopoly Capital*, New York: Monthly Review Press.

Burawoy, Michael (2010), "From Polanyi to Pollyanna: the false optimism of global labor studies," *Global Labor Journal*, **1** (2), 301–313.

Burawoy, Michael (1979), *Manufacturing Consent: Changes in the Labor Process under Monopoly Capitalism*, Chicago: University of Chicago Press.

Bureau of Labor Statistics (2012), Economic News Release, Bureau of Labor Statistics, October 5, 2012. Table B1. United States Department of Labor. Available at http://www.bls.gov/news.release/empsit.t17.htm (last accessed October 10, 2012).

Castree, Noel, Neil Coe, Kevin Ward and Michael Samers (2004), *Spaces of Work: Global Capitalism and Geographies of Labor*, London: Sage.

Chant, Sylvia (1995), "Gender and export manufacturing in the Philippines: continuity or change in female employment? The case of the Mactan Export Processing Zone," *Gender, Place & Culture: A Journal of Feminist Geography*, **2** (2), 147–176.

Christopherson, Susan (2002), "Why do national labor market practices continue to diverge in the global economy? The 'missing link' of investment rules," *Economic Geography*, **78** (1), 1–20.

Coe, Neil and Martin Hess (2012), "Global production networks, labour and development", *Geoforum*. Available at http://dx.doi.org/10.1016/j.geoforum.2012.08.003.

Cumbers, Andrew and Jane Atterton (2000), "Globalisation and the contested process of international corporate restructuring: employment reorganisation and the issue of labor consent in the international oil industry," *Environment and Planning A*, **32**, 1529–1544.

Dawley, Stuart (2007), "Fluctuating rounds of inward investment in peripheral regions: semiconductors in the North East of England," *Economic Geography*, **83** (1), 51–73.

Deetz, Stanley A. (1992), "Disciplinary power in the modern corporation: discursive practice and conflict suppression," in Mats Alvesson and Hugh Willmott (eds), *Critical Management Studies*, London: Sage, pp. 21–45.

Depner, Heiner and Harald Bathelt (2005), "Exporting the German model: the establishment of a new automobile industry cluster in Shanghai," *Economic Geography*, **81** (1), 53–81.

Doeringer, Peter and Michael Piore (1971), *Internal Labour Markets and Manpower Analysis*, Lexington, MA: Heath.

Edwards, Paul (1986), *Conflict at Work: A Materialist Analysis of Workplace Relations*, Oxford: Blackwell.

Engels, Friedrich (1958), *The Condition of the Working Class in England*, first published 1845, Oxford: Blackwell.

Fan, Chuncui, Peter Hall and Geoffrey Wall (2009), "Migration, hukou status, and labor-market segmentation: the case of high-tech development in Dalian," *Environment and Planning A*, **41** (7), 1647–1666.

Florida, Richard (2002), "The economic geography of talent," *Annals of the Association of American Geographers*, **92**, 743–755.

Fröbel, Folker, Jürgen Heinrichs and Otto Kreye (1980), *The New International Division of Labour*, Cambridge: Cambridge University Press.

Gough, Jamie (2005), *Work, Locality and the Rhythms of Capital*, London: Abingdon.

Graham, Laurie (2005), "Manual workers: conflict and control," in Stephen Ackroyd, Rosemary Batt, Paul Thompson and Pamela Tolbert (eds), *The Oxford Handbook of Work and Organisation*, Oxford: Oxford University Press, pp. 338–355.

Hanson, Susan and Geraldine Pratt (1995), *Gender, Work and Space*, London: Routledge.
Herod, Andrew (2001), *Labor Geographies: Workers and the Landscapes of Capitalism*, New York: Guilford Press.
Herod, Andrew (1997), "From a geography of labor to a labor geography: labor's spatial fix and the geography of capitalism," *Antipode*, **29**, 1–31.
ILO (1998), *Declaration on Fundamental Principles and Rights at Work*, Geneva: International Labour Organization. Available at http://www.ilo.org/public/english/standards/decl/declaration/index.htm (accessed June 12, 2012).
ILO (2012), *Global Employment Trends 2012*, Geneva: International Labour Organization.
Jessop, Bob, Neil Brenner and Martin Jones (2008), "Theorizing socio-spatial relations," *Environment and Planning D: Society and Space*, **26** (3), 389–401.
Jonas, Andrew E.G. (1996), "Local labour control regimes," *Regional Studies*, **30**, 323–338.
Kelly, Philip (2002), "Spaces of labour control: comparative perspectives from Southeast Asia," *Transactions of the Institute of British Geographers*, **27** (4), 395–411.
Massey, Doreen (1984), *Spatial Divisions of Labor: Social Structures of the Geography of Production*, London: Macmillan.
Massey, Doreen and Richard Meegan (1982), *The Anatomy of Job Loss: The How, Why and Where of Employment Decline*, London: Methuen.
McDowell, Linda, Adina Batnitzky and Sarah Dyer (2007), "Division, segmentation, and interpellation: the embodied labors of migrant workers in a greater London hotel," *Economic Geography*, **83** (1), 1–25.
Moulaert, Frank (1987), "An institutional revisit to the Storper–Walker theory of labour," *International Journal of Urban and Regional Research*, **11** (3), 309–330.
Ohno, Taiichi (1978), *Toyota Production System: Beyond Large-Scale Production*, New York: Productivity Press.
Olds, Kris and Henry W.C. Yeung (1999), "(Re)shaping 'Chinese' business networks in a globalising era," *Environment and Planning D: Society and Space*, **17** (5), 535–555.
Peck, Jamie (2001), *Workfare States*, New York: Guilford Press.
Peck, Jamie (1996), *Work-place: The Social Regulation of Labor Markets*, New York: Guilford Press.
Phelps, Nick and Crispian Fuller (2000), "Multinationals, intracorporate competition, and regional development," *Economic Geography*, **76** (3), 224–243.
Piore, Michael and Charles Sabel (1984), *The Second Industrial Divide: Possibilities for Prosperity*, New York: Basic Books.
Pollert, Anna (1988), "Dismantling flexibility," *Capital & Class*, **34**, 42–75.
Rubery, Jill and Frank Wilkinson (eds) (1994), *Employer Strategy and the Labour Market*, Oxford: Oxford University Press.
Rutherford, Tod and Meric Gertler (2002), "Labor in 'lean' times: geography, scale and the national trajectories of workplace change," *Transactions of the Institute of British Geographers*, New Series, **27** (2), 195–212.
Smith, Adam (1904), *An Inquiry into the Nature and Causes of the Wealth of Nations*, first published 1776, London: Methuen and Co. Available at http://www.econlib.org/LIBRARY/Smith/smWN.html (accessed June 12, 2012).
Stigler, George (1962), "Information in the labor market," *Journal of Political Economy*, **70** (Supplement), 94–105.
Storper, Michael and Richard Walker (1983), "The theory of labor and the theory of location," *International Journal of Urban and Regional Research*, **7** (1), 1–43.
Thurow, Lester (1975), *Generating Inequality*, New York: Basic Books.
Wailes, Nick, Russell Lansbury and Anja Kirsch (2009), "Globalisation and varieties of employment relations: an international study of the automotive assembly industry," *Labour & Industry*, **20** (1), 89–106.
Walker, Richard (1985), "Class, division of labour and employment in space," in Derek Gregory and John Urry (eds), *Social Relations and Spatial Structures*, London: Macmillan, pp. 164–189.
Wills, Jane (1998), "Taking on the Cosmocorps: experiments in transnational labor organization," *Economic Geography*, **74**, 111–130.
World Bank (2009), *Economic Geography*, Washington, DC: World Bank.
Wright, Melissa (1999), "The politics of relocation: gender, nationality, and value in a Mexican maquiladora," *Environment and Planning A*, **31**, 1601–1617.
Yeung, Henry Wai-Chung (2002), "Producing 'the firm' in industrial geography III: industrial restructuring and labour markets," *Progress in Human Geography*, **26**, 366–378.

FURTHER READING

Bauder, Harald (2006), *Labor Movement: How Migration Regulates Labor Markets*, Oxford: Oxford University Press. (Presents detailed case studies of the impact of migration in labour markets.)

Bluestone, Barry and Bennett Harrison (1982), *The De-industrialization of America*, New York: Basic Books. (This important contribution to understanding the social consequences of industrial change describes the effects of 1970s manufacturing plant closures on local communities in the United States.)

Braverman, Harry (1974), *Labor and Monopoly Capital*, New York: Monthly Review Press. (A classic but much-criticized first-hand account of industrial workplaces, power, the labor process and technological change. Its central thesis – that labour was being deskilled by technological change – remains a cornerstone of debate on the labor process.)

Castree, Noel, Neil Coe, Kevin Ward and Michael Samers (2004), *Spaces of Work: Global Capitalism and Geographies of Labor*, London: Sage. (A geographical account that positions labor, both organized and disorganized, as a central agent in the production of households, regions and economies.)

Dicken, Peter (2010), *Global Shift: Mapping the Changing Contours of the World Economy*, Sixth Edition, London: Sage. (Explains the geographical shifts in the composition of economic activity, global production networks and transnational interconnectivities.)

Fine, Ben (1998), *Labour Market Theory: A Constructive Reassessment*, London: Routledge. (A critical review of labor market theories.)

Herod, Andrew (2001), *Labor Geographies: Workers and the Landscapes of Capitalism*, New York: Guilford Press. (This influential book documents the agency of collective labor, the spatial dimensions of union strategies and labor's capacity to influence the strategies of employers.)

Massey, Doreen (1984), *Spatial Divisions of Labour: Social Structures of the Geography of Production*, London: Macmillan. (Seminal work drawing on restructuring of manufacturing in the United Kingdom which develops the notion of the spatial division of labor.)

Peck, Jamie (1996), *Work-place: The Social Regulation of Labor Markets*, New York: Guilford Press. (Explains how inherently spatialized and localized labor markets are governed by a combination of formal laws and less formal practices and traditions, and that they evolve with related political institutions.)

Sayer, Andrew and Richard Walker (1992), *The New Social Economy: Reworking the Division of Labor*, Oxford: Blackwell. (Positions the division of labor as a key force driving the contemporary reorganization of households and corporations.)

3. How does financialization affect manufacturing investment? Preliminary evidence from the US and UK

Susan Christopherson

THE CONTEXT

In both the US and UK, policymakers have been excited to see a small but significant increase in manufacturing activity and exports in their national economies. Although growth in manufacturing jobs continues to be unpredictable, the creation of manufacturing jobs has been particularly noticeable in light of the major job losses since the early 2000s. The job numbers are not dramatic, however. Approximately 500,000 manufacturing jobs were created in the US between 2009 and 2012 but they represent a fraction of the over 4 million jobs lost in manufacturing since 2000 (Levinson, 2013a). In the UK a recent increase in manufacturing jobs has been slower but exports have been increasing from traditional manufacturing centers, such as the North East (Aldrick, 2013). While not a cause for celebration, these trends have re-awakened interest in manufacturing. There is recognition in both the US and UK that manufacturing matters, not only because of the jobs it creates but also because of its stimulus to innovation and its contribution to exports and the balance of trade. National and regional policymakers are taking a new look at manufacturing and attempting to lure more production firms with promises of tax incentives and workforce training grants.

The need for a strong manufacturing sector in advanced economies is not a new idea. Cohen and Zysman made an argument for maintaining a strong manufacturing sector in the 1980s in *Manufacturing Matters* (Cohen and Zysman, 1987). And, throughout the 30-year period of manufacturing decline, US and UK unions have maintained that the loss of manufacturing jobs was harming the economy. Recently, analysts of the sources of economic renewal have linked the slow pace of technological and organizational change in US industries to the loss of national manufacturing capacity (Kaushal et al., 2011; Pisano and Shih, 2009; Popkin and Kobe, 2010).

There are skeptics, however. Some simply want to leave manufacturing behind and focus economic policy on financial sectors and associated computer science and information industries. Others suggest that the manufacturing train has left the station. The draw of emerging global markets for manufactured products and the availability of global supply chains will prevent recapture of manufacturing in advanced economies. Still others understand the value of manufacturing but think that the conditions inhibiting manufacturing employment growth and firm establishment run deeper than those addressed by cost-based location decision-making options that can be ameliorated by state-subsidized incentive programs.

This chapter engages the debate over the role of manufacturing in the US and UK, examining the evidence for a new calculus favoring the re-location of manufacturing in the two economies. Two major questions are addressed. First, does it make economic sense to expect the growth of advanced manufacturing in what are considered "post-industrial" economies? And second, what conditions work against the establishment of robust export-oriented manufacturing in the US and UK? In particular, this chapter examines how processes associated with the "financialization" of the US and UK economies affect capital investment in manufacturing enterprises, particularly in the small and medium-sized companies that make up supply chains and provide access to the research and development resources that lead to product and process innovation in manufacturing.

For the purposes of this chapter, financialization is defined as a set of processes in which financial motives and instruments become central to capital accumulation. These processes have been more influential in economies such as those of the US and UK, the market governance institutions of which favor short-term returns on investment and emphasize the primary rights of one set of industrial stakeholders: capital investors.

A PIVOTAL MOMENT

This chapter illuminates a pivotal moment, one in which the recognition of the value of manufacturing in an advanced economy comes into conflict with the conditions that have been created by a market governance regime that favors short-term returns on investment. In the US and UK cases, incentives favoring short-term returns and continually increasing shareholder value encouraged disinvestment in manufacturing facilities, workforce and research and development, and the shift of capital resources to debt and higher short-term profit options. Disinvestment was achieved via cost cutting, facilities and program closure, and out-sourcing. This process is what is referred to as "financialization" in non-financial enterprises.

The conditions working against the development of robust manufacturing sectors in the US and UK have been interpreted as market or policy failures (Atkinson et al., 2012). This interpretation is predicated on a mechanical understanding of the economy, as evidenced by concepts such as "rebalancing." Within this frame, the responsibility for decreased investment by global manufacturing firms – in research that can spill over into the supply chain or in labor force skills – is framed within a territorial "competitiveness" discourse, and the responsibility for disinvestment shifted to the state. So, it is only through direct and indirect subsidies to global firms, through tax abatements, support for research and development and other mechanisms, that re-balancing and a higher proportion of manufacturing in the economy can be achieved.

Careful examination of how US and UK economic institutions and firm strategies have changed since the 1980s, however, leads to a more complex conclusion, in which global manufacturers based in the US and UK have responded to incentives to increase profits through a combination of disinvestment and financialization. The market governance "rules" that have shaped manufacturing firm strategies since the 1980s have significantly favored capital accumulation through financial markets and speculative investments over long-term investment in and capital accumulation through product manufacture.

As has already been noted, financialization can be understood as a continuum of processes in which capital accumulation strategies tied to financial products and services supersede those tied to the production of material commodities (Krippner, 2005). At the narrowest end of the definition, changing financial motives and products are connected with the rise of a global investor class and the growing activism of shareholders, especially among powerful institutional investors (Fligstein, 2001; Lazonick and O'Sullivan, 2000). As Krippner (2005) notes in her seminal piece, the description of tendencies in sectors and among consumers does not tell us about the effects of these tendencies on the overall structure of a national economy.[1] Her compelling evidence for examining the role of finance is based on relative profits in service, manufacturing and financial services, and the dramatic rise in financial service profits relative to the two other sectors of the economy, beginning in the mid 1980s.

An important dimension of research on financialization processes has explored how financial motives and incentives have altered behavior in non-financial firms, particularly those engaged in manufacturing. Krippner (2005) measures the impact of these processes in the importance of "portfolio" income (interest payments, dividends and realized capital gains on investments) in total earnings by comparison with earnings from productive activities (Figure 4, page 185), demonstrating a dramatic trend for the role of portfolio earnings beginning in the 1970s.

Epstein (2005) looks at the other side of this investment picture. They examine the negative impact of manufacturing company emphasis on financial market returns to investment in productive assets. Profits have been increased by cutting costs rather than through new productivity-enhancing investments and those profits have been reinvested to increase financial market returns and shareholder value. This has become an issue of considerable concern in the UK and US, where global manufacturing firms have retained capital through share buy-backs for example, rather than investing in the facilities, capital equipment or worker training that would strengthen process and product innovation capacity in the supply chain. At the same time, out-sourcing has enabled higher profits in non-financial firms and the reinvestment of those profits in share buybacks and dividend payments. The result is "a leakage from the system which reduces the dynamic gains from off-shoring by reducing reinvestment out of profits" (Millberg and Winkler, 2010: 290).

To set the stage for an examination of some ways in which financialization processes affect the potential for rebuilding manufacturing, I first examine the argument for why manufacturing continues to matter in advanced economies such as the UK and US, which have historically been dominated by service employment and are deeply engaged in processes of capital accumulation through financial markets and via financial instruments and products.

DOES MANUFACTURING MATTER IN "FINANCIALIZED", SERVICE-DOMINANT ECONOMIES?

For proponents of revitalized manufacturing, a narrow focus on job numbers ignores the broader potential offered by re-capturing manufacturing activity, including capital-intensive advanced manufacturing in the US and UK. They point to several reasons for wanting to rebuild manufacturing within what have long been service-oriented economies: the quality and multiplier effect of manufacturing jobs; the national and regional capacity for innovation; and the strength of the export base.

One reason why manufacturing is important is the quality of manufacturing jobs and their ripple effect throughout the economy. A significant proportion of US and UK manufacturing is now in medium-technology industries that provide good jobs for middle-skill workers. Although only 39 percent of the manufacturing workforce in the US is actually engaged in making things, production jobs remain important (Levinson, 2013a). These jobs require credentials and training beyond the secondary school level but, while it may require advanced technical skills, the work does not necessarily require an advanced degree. Workers who have some skills and can add to them through training are likely to benefit from manufacturing job growth in both the US and UK. In addition, because of the global market for manufactured products, more jobs (the other 61 percent) in manufacturing industries are in marketing, information, management, sales and other services. This trend is another demonstration of the financialization of manufacturing industries but it also reflects a closer integration of manufacturing and services, through financing and service contracts.[2]

Manufacturing jobs are critical to the US and UK economies because of their "multiplier" effect. Bivens (2003) estimates that manufacturing jobs have an employment multiplier of 2.91, compared with 1.63 in business services and 1.66 in transportation (meaning that one manufacturing job supports the creation of 2.91 other jobs in the economy). By extension, loss of manufacturing jobs in an economy also means loss of jobs in other sectors.

One example of the potential for supply chain capture and additional job development is being manifested in the US textile industry, which all would agree is a mature industry. A 2013 report on the US textile industry indicates that it is not only stable but "inching forward" (Reichard, 2013). The reason: the textile industry is now supplying carpeting for the revived construction industry and for the US-located automotive industry. The textile industry is now a capital-intensive industry, employing a much smaller workforce but one that can grow and, with increased product and process knowledge, diversify into specialized markets.

Even for those not concerned with employment, there is a compelling reason to pay attention to manufacturing production – its connection to innovation. The connection between making things and innovation is, in fact, demonstrable. A National Science Foundation Study released in 2008 found that at that time 67 percent of all private-sector research and development was conducted by manufacturing companies. From 2006 to 2008, 22 percent of US-based manufacturing companies reported a product or process innovation. By comparison, only 8 percent of non-manufacturing activities reported innovating in either domain. While these figures need to be evaluated within a context of increased out-sourcing of research and development and focus on financialized innovations in the form of intellectual property, they provide evidence that manufacturing is an important source of innovative capacity.[3]

Similar findings exist for the UK. According to a recent report (Haskel et al., 2011) manufacturing is responsible for 42 percent of innovation in UK industry. These innovations include manufacturing services that enhance production and transport efficiencies and provide customers with solutions to the complex problems associated with capital-intensive advanced manufacturing (Sisson, 2011).

In financialized economies, however, innovations associated with the capacity to make things may be less important than the development of new financial instruments, such as "collateralized debt obligations" or consumer-oriented applications, both of which require smaller investments and have shorter time horizons for the realization of profits.

One trend important to service-dominant economies such as the UK and US is the increasing inter-connection between manufacturing and services. There is an under-appreciation of the role of service provision in contemporary capital-intensive manufacturing and of the bridging of the services–manufacturing divide. There is new evidence that the highest paid services are tied to manufacturing (Sisson, 2011). A core problem in analyzing manufacturing trends, in fact, is that activities that were once carried out in vertically oriented manufacturing firms are now purchased on the market (Levinson, 2013a). So, economic activities deeply integrated with manufacturing, such as logistics services and customer relations, may appear as increases in professional services rather than within the classification of manufacturing. A key question in the current debates is whether original equipment manufacturers can provide these manufacturing-connected services without proximate access to the manufacturing processes with which they are combined and whose customers they serve.

There is also evidence that the problematic connection between making things and advanced business services is likely to expand, given new modes of manufacturing. For example, recent research by a group from the Massachusetts Institute of Technology (MIT) anticipates the new possibilities and challenges possible in "distributed production":

> Trends to offshoring and outsourcing have made manufacturing plants bigger and the distances goods traverse even longer. Tomorrow we can imagine technologies that would "destroy the tyranny of bulk" and distribute manufacturing, thus making it possible to manage capacity and demand flexibly through networks of small, localized manufacturers linked by Internet. (MIT, 2013: 19)

A third reason to support manufacturing is its contribution to the balance of trade. Because manufacturing industries are export oriented, manufacturing strengths translate into trade advantages. In the US, it is estimated that every $1 billion of additional exports will produce about 7000 additional jobs. According to the United States Department of Commerce (2012), more than one-third of manufacturing jobs in primary metals, computer and electronic products, textile mills and machinery are supported by exports. These jobs include those in the original equipment manufacturer but also in its supply chain. In the UK, the major organization representing manufacturing firms estimates that more than half of UK exports are in manufacturing (EEF, 2014). In the US, the contribution of manufactured goods to total exports is estimated at 70 percent. Small and medium-size enterprises (SMEs) play an important role in US manufacturing exports, constituting 28 percent of the total. They are also more likely to export directly to foreign customers than to foreign affiliates (United States Trade Representative, 2010).

The arguments in favor of supporting an enhanced manufacturing role in the economies of the US and UK have attracted considerable attention from policymakers as well as the media. However, those who support an expanded role for manufacturing in advanced economies and industries tend to focus on changing trends in manufacturing firm strategies – in organization of production, location and technology. They have neglected broad changes in the manufacturing environments in the economies that have lost manufacturing businesses and employment since the 1980s. An understanding of both trajectories and their dynamic interaction is necessary to calculate whether it will be possible to create a major role for manufacturing in the US and UK economies.

THE MANUFACTURING INDUSTRY TRAJECTORY: CHANGING INPUT COSTS AND NEW LOCATION CONSIDERATIONS

Manufacturing optimists tend to emphasize how the world of manufacturing is changing. Manufacturing is becoming more capital intensive and moving toward a customized, distributed model. Continuous product and process innovation will increase the knowledge content of many manufacturing industries and intellectual property considerations for manufacturing products and processes and associated services. Learning by doing and turning that reflective learning into new services, products and processes will be critical to the competitiveness of industrial enterprises. In some advanced economies, such as Germany, "labor" is considered as a partner in the innovation process rather than solely as a cost-driven input. Indeed, if one were to point to a single factor that is critical to the new location calculus it is the declining role of labor costs in total production costs, particularly in advanced manufacturing industries. One explanation for the decline in labor costs as a proportion of total production and distribution costs is the increased use of general-purpose technologies, particularly digital technologies (Brynjolfsson and McAfee, 2011; Levy and Murnane, 2004).

Apart from declining labor costs (in many industries), manufacturing optimists tend to look at changes in technology and its implications for manufacturing location. Their reasoning is framed by two trends in product and process technologies. First, technological breakthroughs that have led to the creation of a new set of industries based in key enabling technologies (KETs) (European Commission, 2011). These industries include photonics (laser optics and imaging), flexible electronics, nano-technologies and new materials. KETs fall broadly into the category of advanced manufacturing but are distinctive because they are typically combined in advanced final products, such as medical devices, transportation equipment and communications equipment. "ETs are knowledge and capital-intensive technologies associated with high research and development . . . intensity, rapid and integrated innovation cycles, high capital expenditure and highly-skilled employment" (European Commission, 2011: 10). These new industrial technologies have emerged from embedded knowledge and skill in what were once "end product" industries. So, for example, laser optics and imaging (photonics) emerged from knowledge in the photographic equipment industry.

KETs are typically developed in small manufacturing enterprises, linked together in regional, national or international supply chains. They are the advanced technology end of a trend toward what has been called "vertical specialization" (Herrigel, 2010), referring to global trade in inputs into finished products. Because KETs are legacy industry technologies (such as photonics from optics and imaging industries), they emerge in regional industrial agglomerations, including capacity for research and development in adjacent universities. They both benefit from regional industrial concentration and have global export potential (Christopherson and Clark, 2007).

The second trend is the application of digital information technologies to formerly routine manufacturing production processes, increasing labor productivity, the potential for customization and the need for manufacturing services (Brynjolfsson and McAfee, 2011; Sisson, 2011).

Related to both KETs and information technologies is an increase in the intellectual property content in manufacturing products, processes and services. Intellectual property embedded in industrial products and processes underpins the global competitiveness of many advanced manufacturing firms. It also re-enforces the need to recombine product production and research and development to achieve continuous product and process innovation and to create new products and industries utilizing key enabling technologies. By extension, increased embedded intellectual property in industrial products and the need for continuous product and process innovation proximate to end-user markets provide advantages to advanced economies with strong intellectual property laws and large end-user markets.

While these two trends construct the core for a new location calculus, they are accompanied by other influential factors that appear to be driving firms in some sectors to reassess their location alternatives. These include rising wages in China and increasing transportation costs.

The strength of the trend to reassess what these changes mean for location decisions has been documented in reports by major consulting firms, such as Booz Allen (Kaushal et al., 2011) and the Georgia Institute of Technology (2010), based on hundreds of interviews with firms expanding production in the US or repatriating production. These studies and what is now a wealth of anecdotal information indicate that many manufacturing firms are re-thinking their global production calculus (Reshoring Initiative, 2012).

This re-thinking has resulted in a trickle rather than a trend, however. As laid out in the next section, there are both structural and policy-based barriers to significant growth in manufacturing industries in the US and UK. Structural barriers, rooted in investment incentives, have limited investment in the capacities required for manufacturing expansion. At the same time, enhanced competition among nation states, particularly in corporate tax rates but also in the acceptable rate of profit taking, provides additional incentives to locate activities based on financial motives rather than those related to production capacity or market access (Devereux et al., 2008). Nation states are competing to enhance the financial assets of global firms, adding another dimension to the picture of how financial incentives are affecting location strategies in non-financial firms.

THE OPPOSING DYNAMIC: WHAT MAY INHIBIT ADVANCED MANUFACTURING LOCATION IN THE US AND UK?

Despite the optimism the manufacturing employment growth movement has engendered, there are skeptics. Their skepticism does not emerge from a disregard for the role of manufacturing in the economy but from an awareness that the conditions inhibiting manufacturing employment growth and firm establishment run deeper than those captured by a list of production location cost factors.

One set of skeptics sees barriers to the growth of manufacturing in the US and UK as a result of new forms of global competition, particularly over returns to capital investment (Atkinson et al., 2012). They advocate policies to compete with nation states, such as Ireland, that strategically compete for investment on the basis of tax policies that advantage multi-national firms (Sullivan, 2013). While enhanced competition over profit rates and corporate taxes would benefit multi-national firms, it does not address key problems in developing manufacturing capacity in the US and UK – that is, investment in small and medium-size companies, in workforce training and in product and process innovation.

From the perspective of analysts of processes of financialization, the reasons used to explain tepid growth in US manufacturing (presumably also applying to the UK) are symptomatic of underlying changes in economic incentives driving investors and companies to reduce their investment in the capacity required for the longer-term, more sustainable returns available from manufacturing and, instead, to emphasize the returns available through financial services and in international financial markets. In the next section, the dynamic underlying tepid investment in advanced manufacturing in the US and UK is viewed through the lens of financialization and the changing capital accumulation processes pursued by publicly traded global companies.

WHAT IS THE CONTEMPORARY PICTURE OF MANUFACTURING DISINVESTMENT?

In arguing that the trend toward a new production location calculus is necessary but not sufficient, Atkinson et al. (2012) point to the extent of manufacturing job losses in the US and UK in the 2000s, which were worse than those of the Great Depression of the 1930s. They also refute economists' contentions that the losses can be explained by productivity gains, pointing to declining output in 13 of 19 US manufacturing sectors in the 2000s and the absence of an increase in capital stock, conventionally associated with productivity-related gains in manufacturing. They contend that the patterns in US (and UK) manufacturing represent disinvestment rather than a movement to higher value-added work.

Critics of the potential in manufacturing "re-shoring" make a convincing argument that the problems facing economies with steep losses in manufacturing jobs and output are more serious and deeply rooted than is typically recognized either by economists or policymakers. Their causal explanation for these problems is not as effective, however. They argue that the loss of manufacturing jobs can be attributed to the declining "competitiveness" of the US and UK in the global economy. The implication is that the problem is one of taxes and business environment. Financial incentives and financialized economies are taken-for-granted as the ground on which competition for investment will occur.

In this competitive picture, however, the US and UK are frequently compared with Germany and Japan, both of which continue to have strong manufacturing sectors and exports. The underlying institutional reasons why these two economies have retained strong manufacturing sectors (investment in SMEs and in workforce skills, for example) are rarely examined. To do so would raise questions about the political interests that shape economic policy and encourage competition around financial and fiscal policies that enable super-profits and discourage investment in longer-term investment gains and manufacturing capacity.

From a politically inflected perspective, it is useful to examine the shape of those economies that are similar, not only those that are different. So, it is illuminating to look at the commonalities between the US and UK, particularly those that distinguish these Anglo-American economies from "coordinated" economies, such as that of Germany (Christopherson, 2002, 2011; Dixon, 2011; Peck and Theodor, 2007).[4]

A cursory look at the evidence supports a picture of the US and UK economies in which investment in manufacturing capacity has decreased in favor of short-term strategies to increase returns through financial market products and instruments. These strategies include incentives encouraging multi-national firms to: 1) leverage the equity based in manufacturing firm real property to move into riskier but potentially higher profit financial product markets and customer credit; 2) take advantage of international competition to increase profits among these firms in order to attract segments of the supply chain; and 3) encourage state assumption of necessary investments in research and development and workforce skills.

Examples of the processes that Krippner (2005) aptly describes as representing a move away from an activities-oriented economy to one that is accumulation oriented have expanded over time but primarily as case studies of individual iconic companies. They include the transformation of US Steel to service company USX and Ford Motor Company's emphasis on profit strategies through financing vehicles. One prominent example is that of General Electric (GE), which, like many major manufacturers, historically financed purchases of the appliances it sold. Over time GE became a conglomerate of unrelated businesses with a corporate brand that remained rooted in manufacturing (Froud et al., 2006). The conglomerate's major economic activity in the last third of the 20th century and into the 21st century was in commercial and consumer finance, including mortgage insurance. This activity was developed by leveraging the high credit rating associated with the lower growth GE industrial businesses to acquire riskier financial market assets and to borrow money at low rates to loan to both consumers and commercial borrowers. Recent analyses of the strategic actions of individual companies have re-enforced a picture in which financial motives have driven manufacturing firm strategies. So, from MIT (2013: 25):

> The great new American companies of the past 30 years like Dell, Cisco, Apple, and Qualcomm have little or no manufacturing in-house. Perhaps the single most compelling factor in the 1980s that led to shrinking the perimeter of the corporation and reorganizing it around "core competence" came from financial markets: higher stock market valuations of leaner, "asset-light" companies which had weeded out their less-profitable divisions and reduced their diversification. First among the business functions that companies started moving out of their own corporate walls was manufacturing – for that produced reductions in headcount and in capital costs that stock markets immediately rewarded. Advances in digitization and modularity in the 1990s made it possible to carry out this strategy and to outsource production to manufacturing subcontractors like Flextronics and Jabil and eventually to foreign suppliers and contractors like Taiwan Semiconductor Manufacturing Company, Quanta, and Foxconn.

While these case studies are illuminating and tell us much about how individual firms have transformed themselves into financial actors, they do not, as Krippner (2005) points out, tell us about the shape of the economy as a whole. To get this picture, even partially, we need to look at evidence concerning where firms in an economy with incentives to financialize are focusing their profit-seeking activities and how capital, infrastructural and human resources necessary to manufacturing are being deployed.

PROFIT SEEKING FROM FINANCIAL SERVICES NOT MANUFACTURING: THE IMPLICATIONS FOR CAPITAL INVESTMENT PATTERNS

The picture of the overall "shape" of the US economy that Krippner presented in 2005 before the great recession has been replicated by Atkinson et al. (2012). They examine manufacturing output as a portion of total GDP and profits in the economy attributable to manufacturing. In their analysis of the US economy, they note that:

> As share of the economy, manufacturing profits have been actually falling, especially since the late 1980s and again after 2000, in part as profits have flowed to the financial services industry. Any recent "recovery" in manufacturing profits is emerging from a very low base, and the profitability of manufacturing has been underperforming the corporate economy as a whole. (Atkinson et al., 2012: 58)

Their international comparison of manufacturing output to GDP growth shows that both the US and UK are performing poorly. The UK ranks 17th and the United States 16th of 19 countries in the change in the ratio of manufacturing real value added to real GDP.

The poor performance of the US and UK is a manifestation of disinvestment in manufacturing rather than a transformation of manufacturing into a generally higher value-added sector (as is the case in Germany, for example.) That this is the case is suggested by disparate trends across advanced economies. One notable aspect of those trends is that Japan and Germany basically maintained a significant manufacturing share over this period (Atkinson et al., 2012: 46).

The international differences in investment patterns raise questions about the contention that the decline in manufacturing investment in the US and UK is solely attributable to global competition and that lower corporate taxes or more entrepreneurial incentives would constitute solutions. Instead the better performance of Germany and Japan points to persistent institutional governance differences among advanced economies as responsible for influencing public and private investment strategies. These divergent investment paths support arguments about the institutional basis of advanced economies and how governance differences translate into the allocation of societal resources.

Atkinson et al. (2012: 47) note that over the decade of the 2000s, the amount of fixed capital investment (defined as investment in structures, equipment and software) by US manufacturers was at its lowest rate since World War II. An analysis by year shows that the annual rate generally declined in the 2000s, going under 1.5 percent for several years for the first time since 1950 (Atkinson et al., 2012: Figures 45 and 46). This decline represents the decreasing amounts invested, on average, in new manufacturing plants and equipment every year.

The argument that there has been disinvestment in manufacturing is supported in an assessment by the US Congressional Research Service:

> In . . . services, business start-ups and shutdowns account for a large proportion of job creation and job destruction. In manufacturing, by contrast, employment change appears to be driven largely by the expansion and contraction of existing firms, and entrepreneurship and failure play lesser roles. This may be due to obvious financial factors: the large amounts of capital needed for manufacturing equipment may serve as a deterrent to opening a factory, and the highly specialized nature of manufacturing capital may make it difficult for owners to recover their investment if an establishment shuts down entirely rather than reducing the scope of its production activities. (Levinson, 2013a: 10)

The trend toward disinvestment is indicated in a 58,000 decline in the total number of manufacturing establishments between 2000 and 2011 (US County Business Patterns, 2000–2011), indicating an erosion of supply chain capacity in what are now vertically disintegrated industries composed of SMEs.

It is also notable that a disproportionate amount of the capital investment in manufacturing industries that did occur was in the coal and petroleum industry (225 percent between 2000 and 2010) (Atkinson et al., 2012: 48). The oil and gas industry became a center of speculative financial investment during this period. By contrast, there were significant declines in investment in critical mature industries, including motor vehicles and furniture.

Again, if we look at the overall shape of the economy that emerged in the first decade of the 2000s, the lack of capital investment can arguably be connected to siphoning off capital that would have been invested in productive resources for alternative investments in financial products. One place where this capital went was into financial products and transactions, including mergers and acquisitions. The ability of large multi-national corporations to benefit from the profit possibilities presented by financialization spurred consolidation and the development of company forms such as that represented by GE, a multi-sectoral "engine for growth." The other venue for investment was in "innovation" and speculation in activities that were presented to the public and policymakers as alternatives to conventional manufacturing – in new firms and industries.

In the contemporary US and UK, however, innovation is rarely about building businesses that make things and employ people. Rather it is defined more narrowly, as a speculative process focused on extracting the financial value from ideas. The industries drawing investment capital during the go-go first decade of the 21st century, such as informatics and bioscience, were prized for their ability to produce intellectual property that could be sold on global markets and bring a financial return through sale of a start-up or intellectual property (Andersson et al., 2010).

OECD data indicate that research and development expenditures are targeted in very different ways across advanced economies, with US and UK research and development more concentrated in selected industries than spread across manufacturing sectors, and highly concentrated in intellectual property-producing (i.e. financialized) product development (Levinson, 2013b).

The favored industries were those, such as bio-tech and information and financial services, where innovations could be developed by small start-ups supported by venture capital expecting a rapid return on investment. These industries are capital and knowledge intensive and employ few people (scientists, engineers and technologists). Because they are speculative ventures rather than long-term businesses, they tend to locate in financial market centers, such as greater London, Boston or San Francisco, where they have access to venture capital and information on the global market for their innovative product or service. As a consequence, as is implied in the 2014 EEF report on the UK, financialization processes have had dramatic impacts on both the availability of capital to the SMEs that are critical to vertically disintegrated manufacturing and to the availability of investment capital in regions historically specialized in manufacturing.

At the same time, investment capital and policy interest dried up in the US for the incremental design and process innovations that could create new products and enhance efficiency across manufacturing industries. Major laboratory facilities, such as Bell Laboratories, focused on long-term product and process research, were closed in favor of short-term investment in applied research that would produce patent and license (i.e. financial) returns.

While this approach has provided economic opportunities for so-called "patent trolls," the hopes invested in innovative new industries as a job creation strategy were realized in only a few US regional economies (Feldman and Desrochers, 2003). According to an analysis by Mandel in *Business Week* (2009), the US Bureau of Labor Statistics predicted in the 1990s that new "innovative" industries would create 2.8 million jobs. Instead the jobs number was in the tens of thousands.

The focus on financial investments to capture short-term financial returns ultimately reduces the potential for socially and economically beneficial spillovers or externalities that can lead to new applications, products and industries (Jones and Williams, 1998; Mansfield, 1991). For example, many start-ups driven by financial motives are not intended to operate over time or to employ people. Their ultimate goal is not the establishment of a business but cash-out after "proof of concept" is established. Those few entrepreneurs who intend to create lasting businesses face daunting odds, rooted in competition for capital that migrates to faster returns based in financial markets. The lack of investment capital for companies which want to take a slower route to profit making has forced them to look internationally for commercialization capital. The aforementioned recent MIT study of choices made by start-up companies which wanted to establish manufacturing operations indicates this dilemma:

> many of them when they came to the stage of moving to full-scale commercialization, could not find finance in the U.S. As many of them made the transition from venture funding to high-volume manufacturing, they eventually had to look for foreign investors and often moved abroad to manufacture their products. (MIT, 2013: 17)

The problem of lack of investment in companies with longer-term goals or in existing manufacturing companies that needed to "re-tool" their operations in order to remain competitive has also been documented in the UK (EEF, 2014). A key manufacturing trade association, EEF, traces lack of financing for product and process innovation, basic industrial infrastructure and expansion to changes in the banking sector, including concentration. It points to differential regional access to investment capital and particularly to problems faced by small and medium-size companies in accessing capital as factors hampering manufacturing expansion in the UK.

In the US, a recent report by the National Institute on Standards and Technology (NIST, 2011) presents similar findings. The report reiterates that a central problem inhibiting manufacturing sector growth emanates from lack of capital for manufacturing start-ups, expansion or investment in research and development, human resources or access to global markets:

> a lack of available capital has restricted the ability of many small manufacturers to grow and compete in the current business environment. Surveys of manufacturing companies and numerous stakeholder interviews conducted as part of this assessment cited delayed capital investment as the biggest impact of the 2008 recession and financial crisis, and a constraint to future growth. (NIST, 2011: 1)

This report points to decisions on the part of increasingly concentrated and global financial institutions (and nation state policies) to favor larger over smaller investments and to raise requirements for loan approval. With concentration in the banking sector and emphasis on short-term returns, small manufacturing firms are finding it more difficult to access commercial credit to fill both their immediate and long-term needs for capital to expand their businesses.

There is also a critical regional dimension to the poor investment opportunities for the critical SME sector. It could be argued that the disinvestment in factor inputs that support manufacturing product and process innovation is most strongly felt in regions with the potential for the development of new technologies emerging from mature industries, most notably the KETs described earlier.

The locational pattern for financialized manufacturing start-ups in sectors such as bio-science and computer science is fundamentally different from that of manufacturing start-ups that emerge out of investment in legacy technologies. For example, because they are fundamentally financial products, start-up ventures in information and communication co-locate in financial centers. The spatial dynamic accompanying financialization encourages concentration of particular types of economic knowledge as well as its speculative investment activity in financial centers. The local knowledge that underpins longer-term capital investment in manufacturing firms, especially SMEs, in enabling technologies may be lost as capital and financial expertise are concentrated in global centers.

This pattern is more pronounced in the UK than in the US and is reinforced by the centralization of policy making affecting investment in productive resources. In the US, the ability and willingness of sub-national states to respond to interests rooted in manufacturing industries have mitigated the effects of disinvestment, replacing private with public capital and enabling some regions with manufacturing industries to benefit from the new location calculus.

Even in these instances, however, the kind of institutionally based, shared resources that underpin German or Japanese manufacturing productivity and international competitiveness are largely missing. As noted in the recent MIT study:

> in this company as in most of the others in this category that we interviewed in the U.S. all growth depended on their internal resources. They were not finding any complementary capabilities they could draw on in the industrial ecosystem as they tried to develop new components: no outside funding, no connections with community colleges, no trade associations, no research consortia (all regular fixtures, we would discover, on the landscape of German companies in the same industrial sector). As we wondered why the contributions to innovation of the Main Street manufacturers did not lead to greater profits and faster growth, the comparison with Germany was inevitable. (MIT, 2013: 20)

The combination of weak collective institutions to support the small and medium-size manufacturers that dominate manufacturing supply chains and the undermining of the financial expertise and long-term perspective required for effective investment in these SMEs is working against the potential created by the new location calculus.

ON BALANCE: WHAT ARE THE PROSPECTS FOR RECAPTURING MANUFACTURING IN FINANCIALIZED ECONOMIES?

There is accumulating evidence regarding the effects of financial motives and strategies in manufacturing firms, and their transition to financial firms with a manufacturing brand identity. In addition, because of the financial crisis, there is more understanding of the market-distorting role of financial firms that provide critical capital to the manufacturing sector for productivity-enhancing investment in facilities, equipment and human resources. However, the dominant economic narrative during the 30 years of declining manufacturing employment in the US and UK rationalized the loss as a necessary and inevitable transition to an economy dominated by financial products and services for a global market.

In the period since the onset of the recession, the importance of manufacturing to innovation and sustainable economic development has become more visible, and policymakers have responded with speeches about the need to "rebalance" the UK and US economies to capture more of the benefits of manufacturing. However, until the incentives that have driven disinvestment in manufacturing are carefully analyzed and altered, the prospects for a resurgence in UK and US manufacturing will only hint at the possibilities presented by a new production location calculus.

NOTES

1. The analysis in this chapter assumes that there are territorial economies, governed by political-economic institutions.
2. The higher portion of manufacturing jobs in services also suggests that increased manufacturing employment is unlikely to contribute toward alleviation of economic inequality.
3. The US lack of investment in manufacturing-based innovative activity does not hold in Germany, where incremental innovation in product and process by small and medium-size companies is the by-product of more coordinated industrial institutions (Christopherson, 2002, 2011).
4. In this chapter I am not addressing the question of China, the state-based economy and non-democratic government of which have undertaken mercantilist policies that have contributed to the financialization of the US and UK economies, both directly and indirectly.

REFERENCES

Aldrick, P. 2013. "UK Manufacturing Surge Bolsters Recovery Hopes." *The Telegraph*, September 2. Available at: http://www.telegraph.co.uk/finance/economics/10280177/UK-manufacturing-surge-bolsters-recovery-hopes.html (accessed October 10, 2014).

Andersson, T., P. Gleadie, C. Haslam and N. Tsitianis. 2010. "Bio-pharma: A Financialized Business Model." *Critical Perspectives on Accounting*, 21(7), October: 631–641.

Atkinson, R., L. Stewart, S. Andes and S. Ezell. 2012. *Worse Than the Great Depression: What Experts Are Missing About American Manufacturing Decline*. Washington, DC: The Information and Technology Foundation.

Bivens, Josh. 2003. "Updated Employment Multipliers for the U.S. Economy." Working paper, Economic Policy Institute. Available at: http://www.epi.org/page/-/old/workingpapers/epi_wp_268.pdf (accessed October 10, 2014).

Brynjolfsson, E. and A. McFee. 2011. *Race Against the Machine*. Lexington, MA: Digital Frontier Press.

Christopherson, S. 2002. "Why Do National Labor Market Practices Continue to Diverge in a Global Economy?" *Economic Geography*, 78, January: 1–20.

Christopherson, S. 2011. "The Geographies of Capitalism." In A. Leyshon, R. Lee, L. McDowell and P. Sunley (eds), *The Sage Handbook of Economic Geography*. London: Sage Publications, pp. 163–174.

Christopherson, S. and J. Clark. 2007. *Remaking the Region: Power, Labor, and Firm Strategies in the Knowledge Economy*. New York: Routledge Press.

Cohen, S. and J. Zysman. 1987. *Manufacturing Matters*. New York: Basic Books.

Devereux, M.P., Ben Lockwood and Michela Redoano. 2008. "Do Countries Compete over Corporate Tax Rates?" *Journal of Public Economics*, 92(5–6): 1210–1235.

Dixon, A. 2011. "Variegated Capitalism and the Geography of Finance: Towards a Common Agenda." *Progress in Human Geography*, 35, April: 193–210.

EEF. 2014. *Finance for Growth: Increasing Competition in SME Banking*. London: EEF. Available at http://www.eef.org.uk/resources-and-knowledge/research-and-intelligence/industry-reports/finance-for-growth-increasing-competition-in-sme-banking (accessed October 10, 2014).

Epstein, G. 2005. *Financialization and the World Economy*. Cheltenham and Northampton, MA: Edward Elgar.

European Commission. 2011. *Final Report, High Level Group on Key Enabling Technologies*. Brussels: European Commission.

Feldman, M. and P. Desrochers. 2003. "Research Universities and Local Economic Development: Lessons from the History of the Johns Hopkins University." *Industry & Innovation*, 10(1): 5–24.

Fligstein, N. 2001. *The Architecture of Markets: An Economic Sociology of 21st Century Capitalist Societies*. Princeton, NJ: Princeton University Press.

Froud, J., J. Sukhdev, A. Leaver and K. Williams (2006). *Financialization and Strategy: Narrative and Numbers*. New York: Taylor & Francis.

Georgia Institute of Technology (2010). *Enabling Manufacturers to Compete in the Global Economy*, Atlanta, GA: Georgia Tech Manufacturing Institute. Available at: http://stip.gatech.edu/wp-content/uploads/2010/10/GMS-2010.pdf (accessed October 10, 2014).

Herrigel, G. 2010. *Manufacturing Possibilities: Creative Action and Industrial Recomposition in the United States, Germany, and Japan*. New York: Oxford University Press.

Jones, C. and J. Williams. 1998. "Measuring the Social Return to R&D." *Quarterly Journal of Economics*, 113(4): 1119–1135.

Kaushal, M., T. Mayor and P. Riedel. 2011. "Manufacturing's Wake-Up Call." *Strategy + Business*, 64, Autumn.

Krippner, G. 2005. "The Financialization of the American Economy." *Socio-economic Review*, 3(2): 173–208.

Lazonick, W. and M. O'Sullivan. 2000. "Maximizing Shareholder Value: A New Ideology for Corporate Governance?" *Economy and Society*, 29: 13–35.

Levinson, M. 2013a. *Job Creation in the Manufacturing Revival* (7–5700), June 19. Washington, DC: Congressional Research Service.

Levinson, M. 2013b. *US Manufacturing in International Perspective* (7–5700), February 11. Washington, DC: Congressional Research Service.

Levy, F. and R. Murman. 2004. *The New Division of Labor: How Computers are Changing the New Job Market*. Princeton, NJ: Princeton University Press.

Mandel, M. 2009. "The Failed Promise of Innovation in the U.S." *BusinessWeek* cover story, June 15. Available at: http://www.businessweek.com/magazine/content/09_24/b4135000953288.htm (accessed October 10, 2014).

Mansfield, E. 1991. "Social Returns from R&D: Findings, Methods, and Limitations." *Research Technology Management*, 34(6): 24–27.

Millberg, W. and D. Winkler. 2010. "Financialisation and the Dynamics of Off-shoring in the US." *Cambridge Journal of Economics*, 34(2), March: 275–293.

MIT. 2013. *Report of the MIT Taskforce on Innovation and Production*, February 22. Available at: http://ilp.mit.edu/images/conferences/2013/pie/PIE_Preview.pdf (accessed October 10, 2014).

Haskel, J., P. Goodridge, A. Pesole, G. Awano, M. Franklin and Z. Kastrinaki (2011). *Driving Economic Growth: Innovation, Knowledge Spending and Productivity Growth in the UK*. London: National Endowment for Science, Technology and the Arts.

National Institute on Standards and Technology. 2011. Connecting Small Manufacturers with the Capital Needed to Grow, Compete, and Succeed: Small Manufacturers Capital Access Inventory and Needs Assessment Report. Gaithersburg, MD: NIST.

Peck, J. and N. Theodor. 2007. "Variegated Capitalism." *Progress in Human Geography*, 31, December: 731–772.

Pisano, G. and W. Shih. 2009. "Restoring American Competitiveness." *Harvard Business Review*, July–August: 114–125.

Popkin, J. and K. Kobe. 2010. *Manufacturing Resurgence: A Must for US Prosperity*. Washington, DC: The National Association of Manufacturers. Available at: www.nam.org/council (accessed October 10, 2014).

Reichard, R. 2013. "Textiles 2013: The Turnaround Continues: The U.S. Textile and Apparel Industry Remains on the Upswing, and the Prognosis Going Forward is Good." *Textile World News*, January/February, 2013. Available at: http://www.textileworld.com/Articles/2013/January/January_February_issue/Textiles_2013.html (accessed October 10, 2014).

Reshoring Initiative. 2012. Available at: http://www.reshorenow.org/ (accessed October 10, 2014).

Sisson, A. 2011. *More Than Making Things: A New Future for Manufacturing in a Service Economy*. Lancaster: The Work Foundation, Lancaster University, UK. Available at: http://www.theworkfoundation.com/DownloadPublication/Report/284_284_More%20than%20making%20things.pdf (accessed October 10, 2014).

Sullivan, M. 2013. "If Ireland Is Not a Tax Haven, What Is It?" *Forbes*, June 10. Available at: http://www.forbes.com/sites/taxanalysts/2013/11/06/if-ireland-is-not-a-tax-haven-what-is-it/ (accessed October 10, 2014).

United States. County Business Patterns, 2000–2011. U.S. Department of Commerce, U.S. Census Bureau. Available at: http://www.census.gov/econ/cbp/ (accessed October 10, 2014).

United States Department of Commerce (2012). "Jobs Supported by Exports 2012: An Update." Office of Competition and Economics Analysis (February 26). Available at: http://www.trade.gov/mas/ian/build/groups/public/@tg_ian/documents/webcontent/tg_ian_004021.pdf (accessed October 10, 2014).

United States Trade Representative. 2010. "Smaller Exporters Play Larger Role in Trade" (February 26). Available at: http://www.ustr.gov/about-us/press-office/press-releases/2010/november/smaller-exporters-play-large-role-trade (accessed October 10, 2014).

4. Manufacturing logistics

Peter V. Hall

INTRODUCTION

Manufacturing is now and has always been intimately bound up with transportation to move finished goods to market, to obtain raw materials and to integrate different parts of the production process. As Alfred Marshall (1890) and countless economic geographers who followed him have observed, manufacturing is more efficient and dynamic when it occurs at greater productive scale and in localized clusters. For precisely this reason, resources, raw materials, energy, parts and other inputs have to be brought to manufacturing locations, and products have to be shipped from there to widely dispersed points of final demand. What, if anything, is new about manufacturing logistics? The logistics industry encompasses a "wide set of activities dedicated to the transformation and circulation of goods, such as the material supply of production, the core distribution and transport functions, wholesale and retail and also the provision of households with consumer goods as well as related information flows" (Hesse and Rodrigue, 2004: 172). This suggests that logistics is not really all that new: it is merely the most recent attempt to distribute the inputs and outputs of manufacturing in more efficient ways. But the definition also draws attention to the idea that logistics entails the transformation of goods, and to the importance of information in supporting the ever more complex flow of goods. Bonacich and Wilson (2008: 3) provide a similar definition of logistics that usefully denotes the recent expansion of the term to encompass more activities, including notions of control: "[u]ntil recently, the term was generally limited to transportation and warehousing functions that any company needs to fulfill. But the idea of logistics has been expanded to refer to the management of the entire supply chain, encompassing design and ordering, production, transportation and warehousing, sales, redesign and reordering." They refer to this shift as the "logistics revolution." Logistics is different from transportation because it is also about new ways, or at least the possibility of new ways, of organizing and controlling the production process itself. It entails a redefinition of the boundaries of manufacturing; when manufacturers get into transportation, and when transporters get into manufacturing, something we call "manufacturing logistics" is the result. If manufacturing is the physical transformation of materials into products, and if transportation is the physical distribution of those material inputs and finished products, then manufacturing logistics is the increasing and systematic integration of transportation and manufacturing.

LOCATING MANUFACTURING, TRANSPORTING MANUFACTURES

To help untangle the complicated and shifting relationship between manufacturing and transportation, compare two broad schools of thought in economic geography. The first school, location theory, emphasizes that decisions about where to undertake manufacturing activity are strongly influenced by the relationships between places. Since transportation is one of the most important factors influencing those relationships, we can think of location theory as the idea that decisions about where (and how) to conduct manufacturing are patterned after transportation systems. Alfred Weber's (1909) original formulation of location theory placed transportation costs in such a central role. In trying to explain the punctuated distribution of manufacturing activity, he and his followers focused on identifying which spaces within transportation networks would be attractive to particular industries (Alonso, 1975). They predicted, for example, that manufacturing processes that increased the weight (or total transport costs) of the goods, such as brewing beer, were likely to locate close to market. In contrast, weight-losing industries such as lumber milling would be likely to locate close to the source of inputs. Points of break of bulk and transfer between transport modes too were understood to enjoy a cost advantage as sites for processing and recombining of goods, since if you have to change modes, you can save costs if you undertake other material changes at the same time and place.

A contrasting view of the transport–manufacturing relationship, localization theory, starts with the idea that the location of manufacturing is predominantly determined by what is scarce and hence particular to only some locations: a crucial raw material, the cost of labor, design and innovation capacity, or some combination of these and other factors. Once the optimal production location has emerged, or has been chosen, the job of transportation is to serve that location. Alfred Marshall (1890) fell more squarely within this second school. After establishing that the basis for the localization of manufacturing often lay in local knowledge exchanges, he asked how the location of manufacturing might change in response to any change in transportation costs and technologies. He identified both concentrating and dispersing effects:

> speaking generally we must say that a lowering of tariffs, or of freights for the transport of goods, tends to make each locality buy more largely from a distance what it requires; and thus tends to concentrate particular industries in special localities: but on the other hand everything that increases people's readiness to migrate from one place to another tends to bring skilled artisans to ply their crafts near to consumers who will purchase their wares. (1890: 227, IV, x, 4)

The first part of Marshall's observation is a clear statement of the manufacturing location as given, leading to the conclusion that cheaper transportation costs favor the most efficient producers who are now able to reach a wider market area. This is part of the same logic that has led observers to question whether the construction of continental transportation corridors will contribute to the development of peripheral regions and regional economic convergence (Vickerman et al., 1999). In the second part of his answer, however, Marshall builds on his emphasis on the importance of ideas embodied in labor to highlight one potential dynamic relationship between transportation and manufacturing.

The locations of manufacturing may change depending on the mobility of knowledge, people and technologies. Marshall did not take his analysis further than the first round of relocation of manufacturing to North America enabled by the mobility of English craftsmen; but he surely would not have been surprised to see manufacturing in North America displaced in turn.

The decline in transportation costs referenced by Marshall in 1890 continued, almost entirely uninterrupted, through the 20th century. At the same time, reliability and other dimensions of transport performance improved; and, as predicted, manufacturing activity shifted to lower cost locations and the flows related to manufacturing intensified. Glaeser and Kohlhase (2004) trace a decline in the average cost of moving a ton one mile within the United States from 18.5 cents (in 2001 dollars) in 1890 to 2.3 cents in 2001. Not only has transportation become much cheaper but the decline of canal and rail, and the rise of truck transportation, have also entailed a shift away from high fixed-cost systems. We have seen this pattern repeated again and again in the experience of globalization; the lowering of transportation costs and improvements in communication allow the distribution of goods from core locations, but also the relocation of production. However, even when manufacturers' location and other choices are chiefly driven by considerations other than transportation costs, physical distribution cannot be ignored. For example, with reduced trade barriers due to the 1994 North American Free Trade Agreement, manufacturers selling into the United States first looked to Mexico as a site for lower cost production. Fawcett (1990) wrote that one of the potential advantages of Mexican manufacturing, compared with that located in South East Asia, was simplified logistics. However, using a total logistics and total production cost approach, he found that "a significant portion of the production cost savings attained through production sharing are dissipated by the increased logistics costs required to support the maquiladora operations" (20). Sources of the higher costs included documentation, inventory, sourcing and transportation. It should not surprise us that these cost sources have subsequently received policy and regulatory attention, for example in the debate about how far into the United States to allow Mexican truckers to make deliveries, and in various cross-border infrastructure corridor initiatives.

A dynamic picture of the relationship between logistics and manufacturing emerges in this instance, one in which manufacturers are active agents in trying to reduce transportation and logistics costs. And the insights of both localization theory, which emphasizes the distinctive characteristics of places as potential sites of production, and location theory, which emphasizes the importance of transportation costs and systems, add to our understanding of the actual organization of manufacturing as well as the goods movements which sustain it.

FROM RIVER ROUGE TO ANCHORAGE

To the extent that a production process involves multiple steps, manufacturers have always had some choice about which activities to separate spatially by employing transportation services. The question about "when to make and when to move" is not new, but logistics enhances the range of options. Consider the transport choices of the archetype of large-scale mass-market manufacturing in the fordist era: Henry Ford's automobile assembly complex at River Rouge, Michigan. Located to the west of the city of Detroit, the River Rouge facility was special because it encompassed at a single site almost all of the steps involved in manufacturing an automobile. The mile by mile-and-a-half site contained an iron foundry and steel mill, a glass plant and other facilities that produced the components that fed directly into the automobile assembly line. The complex even had its own cement plant to build huge storage bins for raw materials. Over 100,000 people worked at the site in the 1930s.

Henry Ford's goal, to a great extent achieved, was to create a self-sufficient manufacturing complex. However, this is not the same thing as saying that the complex did not rely upon the careful choreography of distribution; within the site were miles of railway track, a bus service and miles of conveyor belt. We can think of these internal movements, and their co-ordination, as the internal logistics of large-scale manufacturing. And, the complex relied heavily on transportation connections to the rest of the world. Indeed, Ford chose the site at the confluence of the Rouge and Detroit Rivers precisely because it was such a convenient place to bring iron ore, coal and limestone for making steel, as well as other raw materials, by water transportation. Such large-volume movements of raw materials require specialized equipment, but their handling is not regarded as especially complicated. And since iron ore and coal do not degrade quickly, storage costs were low. River Rouge was also a convenient site from which to export vehicles by ship. Ford had begun exporting "completely knocked-down" (CKD) or kits, which contain all the parts for final assembly of a vehicle, to an assembly plant in Walkerville, Ontario, Canada in 1904. Sturgeon and Florida (2000) note that kits were desirable because they saved on transportation costs; eight kits cost as much to transport as one vehicle, and they were much less likely to get damaged in transit. By the late 1920s, when the first vehicles rolled off the line at River Rouge, Ford had CKD assembly plants in several European and South American countries, as well as South Africa, India and Malaysia. The other major automobile manufacturers followed suit, and CKD assembly remains an important feature of the global automotive industry to this day.

Logisticians today aspire to do this and more. While it is true that final assembly of the CKD is delayed until the product is close to market, the nature of the product itself is invariant once it leaves the integrated production site. The product is made and sent to market in response to a single information flow from the point of consumption to the point of production. What manufacturing logistics aims to do is to delay the final assembly of the finished product until the last possible moment, constantly adjusting production processes in response to new information about market demand.

What does this brief portrayal of Ford's River Rouge complex tell us about the relationship between transportation and manufacturing? Transportation was integral to the functioning of the complex from the very start. Raw materials, parts and energy moved into the complex, and kits and vehicles moved out. Furthermore, River Rouge had its own elaborate within-firm, on-site production logistics. Transportation, including on-site movements of components, was central to the successful operation of this single integrated production facility.

How has transportation changed within the automobile industry? Today the brand-name manufacturers still own and operate the automotive assembly plants, and some components are manufactured there. For example, body panel stamping often still occurs at the assembly plant (Sturgeon and Florida, 2000) because these components can be costly to transport. But most of the components arrive at the assembly line in pre-prepared modules or systems; the chassis, the drive train, the interior, the electronics, and so on. The work of assembling these modules is typically undertaken within a complex hierarchy of tiered suppliers. The rise of such modular production systems has contributed to relatively more transportation of intermediate goods; and modularity entails the potential for delayed and relocated assembly of interchangeable components, often by contracting out.

The transportation involved in moving all these components between the suppliers and up the hierarchy to the point of final assembly is enormous. In part, this is because each of these tiers of supply has its own geography. The last or highest tier suppliers are unlikely to locate more than a day's drive from the assembly line, but they in turn are aggregating widely sourced components. Similarly, while some lower tier suppliers may cluster around the higher tier suppliers to avoid being late with deliveries, for many components other cost factors influence where production takes place. Furthermore, because each lower tier supplier tries to avoid being held captive by having more than one higher tier customer, the pattern of movements quickly becomes multi-directional and complex. For example, automotive components, such as seat cushions or steering wheel covers, manufactured in low cost facilities in small towns in southern Ontario may cross the US–Canada border several times before finally becoming part of an automobile that rolls off the assembly line in either nation. Such internationally integrated supply chains place heavy demands on the infrastructure and control systems at border crossings (Anderson, 2012).

So, the task of manufacturing an automobile today involves more transportation, especially of parts and components, along a more complex set of routes. This is not just a change in the spatial organization of the production process; it is also about the broader structure of economic regulation within which that production is embedded (Dicken, 2011).

Today's manufacturing logistics occurs in a context of heightened global competition, shorter innovation and production cycles, extensive use of digital information, and fragmentation of consumer markets. With less competition, Ford's early models could afford to stockpile raw materials and parts for longer production runs. With intensified global competition, manufacturers, increasingly comfortable with outsourcing other parts of the production process, have sought help from logistics experts to achieve the desired speed-up. The semi-conductor division of the manufacturer, Motorola turned to the global logistics service provider, UPS when it sought a reduction in total shipment time to seven days from the time that an order was placed to when it was delivered to the customer (Bhatnagar and Viswanathan, 2000). Those seven days were to consist of four days in manufacturing lead time and 72 hours of distribution time. While Motorola was able to achieve the desired speed-up in manufacturing, after studying the issue with UPS they jointly concluded that they could not reduce the delivery time without re-engineering the supply chain. Average delivery times were 111 hours, and could only be pushed down to 90 hours under ideal conditions. The solution chosen was a new route that took all cargo from Asia through Anchorage, Alaska (the point of entry into the US) and Louisville, Kentucky (the UPS North American hub) to its final North American destination. In practice, this meant more reliance on air transport, bypassing Los Angeles, California as a point of entry and related intermediate warehousing facilities, and relying on UPS to co-ordinate all of the other transportation operators, such as trucking and warehousing. The new routing entailed more use of faster modes of transportation and more direct routes, the consolidation of routes into UPS's global hub-and-spoke distribution system, and the use of transfer facilities to consolidate/deconsolidate loads rather than warehouses as storage buffers. UPS achieved the desired reduction in distribution time, yet Motorola has since gone out of business: logisticians can improve a business's performance but they aren't necessarily its savior.

THE MANY FACES OF LOGISTICS

The Motorola case would seem to suggest that the world today more closely resembles the picture sketched by Marshall and the localization theorists: manufacturers decide what to make where, and ever more efficient transport providers have followed. However, this ignores the ways in which logistics blurs the dichotomy between here and there; as predicted by the location theorists, transfer points are used in manufacturing logistics to transform and add value to products. Hence, at the same time that logisticians have helped to reduce the cost of transportation, they have also made distribution more important to the smooth operation of production. With falling transportation costs, places such as Anchorage and Louisville that are well served by transportation have become vitally important points of transfer and transformation in the manufacturing process. The manufacturing logistics landscape is thus one that is more dispersed and shifting. In what follows, we will trace out how new transport modes and routes, steps in the production process and the locations of manufacturing have changed in tandem with new communication technologies and forms of internal and inter-firm corporate organization.

NEW MODES AND ROUTES

Contemporary logistics systems rely upon cheap, reliable means of transportation, exemplified by the modern intermodal shipping container. Containers allow the efficient integration of ocean- and land-based transportation systems; so we see bigger ships in ports but also more tractor-trailers on highways. Final manufactured products and intermediate components are shipped cheaply with reduced likelihood of damage (Levinson, 2006).

Increasingly, containers are also used to handle semi-processed raw materials such as cut lumber, paper pulp, high value agricultural products and even semi-processed metals. Beyond containers, ocean transport also handles other manufactures. For example, as already noted, Ford used CKDs to overcome the costs and losses associated with the shipment of fully assembled automobiles. However, in the 1960s Japanese automobile manufacturers worked with steamship lines to develop a new class of dedicated car carrier ships that could cost-effectively transport fully assembled automobiles (Hall and Olivier, 2005).

For high value-to-weight goods, the use of air cargo transportation by manufacturers has grown rapidly. Air cargo (freight per ton-kilometer) grew nine times faster than the world economy between 1970 and 2000, and even though maritime transport accounts for more than 90 percent of trade by volume, air cargo accounts for about 40 percent of world trade by value (Kasarda et al., 2004). Based on surveys of electronics firms in Singapore, Malaysia and the Philippines, Bowen and Leinbach (2003) found that it is not just the use of air transport per se that has increased, but also the use of advanced service providers which enable and control these movements. While the modes of transport of manufactured goods are changing, it is equally important to note that it is the organization and control of those flows that has changed with logistics.

NEW STEPS IN THE MANUFACTURING PROCESS

With new ways of moving goods come new ways of organizing the production process itself. One of the key targets of the logisticians has been the cost of holding goods between production and sale, as well as the related costs incurred when manufacturers produce too much or too little for consumers. In theory, these inventory costs can be reduced in a shift from push- to pull-based logistics. The ideal-type push model is one in which manufacturers produce goods in advance of sale according to their best estimate of profitable demand. The ideal-type pull model is one in which the product is only assembled once the sale has been contractually guaranteed, hence eliminating inventory costs. Neither ideal type exists in reality; no one can eliminate entirely the gaps between the time and place of ordering and producing a good, but logisticians do strive to move manufacturing away from a push-based towards a more pull-based system.

Postponement is one way to get closer to the pull model ideal. Van Hoek et al. (1998) trace use of the term "postponement" to the 1950s, and find its origins even further back, in the CKDs of the 1920s. Postponement has enjoyed widespread adoption:

> customization of products (form postponement) is delayed until the goods are ordered (time postponement) and have reached the international distribution chain, frequently followed by direct delivery to retailers or customers (place postponement). This system allows companies to separate the customization of products from the primary or basic manufacturing of standard products or generic modules. This separation frees primary manufacturing to focus more on large economical runs, while secondary or final manufacturing can be focused on responding to customer wishes. (33–34)

Note the explicit nexus between increased economies of scale and (at least some) elements of the pull model through delayed customization: hence the neat description, mass customization.

Another strategy of manufacturers is to externalize inventory costs. The idea behind just-in-time (JIT) production systems is that:

> every element needed during production is delivered only at the time when it is needed, which implies the effective sequencing and control of production processes, so as to anticipate the amounts to order and the right moment to place the order. JIT is appreciated among manufacturers as it enables drastic inventory reductions, which is particularly important as the average product life span keeps diminishing. (Lasserre, 2004: 76)

But off-loading inventory costs by requiring suppliers to deliver 'just-in-time' only works if the suppliers themselves are willing and able to comply.

NEW LOCATIONS

An important consideration in the ability to supply just-in-time is spatial proximity. Indeed, regional and local economic developers in developed countries battered by deindustrialization were initially somewhat optimistic that JIT production systems might relocalize manufacturing activity around lead firms in vertically integrated geographic production clusters. These might be expected to resemble "hub-and-spoke" industrial districts that are regarded as less vulnerable to plant closures and relocations than "satellite" or pure branch-plant districts (Markusen, 1996). However, lead firms can use logistics to engage in non-local sourcing, so undermining the potential for local and regional economic spin-offs. For example, van Egeraat and Jacobson (2005) found that global microcomputer assemblers in Scotland and Ireland imported most of their materials, leading them to conclude that "a strategy of building integrated vertical production clusters around subsidiaries of multinational enterprises is no longer suitable" (283). Contextual factors are important, but in general logistics implies a potentially important shift in the relationship between manufacturing and patterns of local and regional development because manufacturers can make use of logistics expertise to broaden their range of options for sourcing non-local inputs.

Logistics activities themselves have their own locational dynamics. The physical goods handling side of the logistics industry is often located in proximity to large transportation nodes such as ports and airports, at inland terminals and along corridors (Hesse and Rodrigue, 2004; Kasarda, 2010). Often these sites are found in the world's largest metropolitan regions (O'Connor, 2010), and at the metropolitan scale, logistics activities tend to cluster in suburban and even ex-urban locations (Bowen, 2008; Cidell, 2010; Hesse, 2008). Unlike the warehouses of old whose functions focused more on storage, the new logistics facilities are more oriented towards handling – including decomposition, consolidation and other means of load recombination – packaging, labeling, sorting and quality control, and related functions. However, some logistics activities, notably those that are most closely related to the business service sector, still cluster in core urban locations, where they are embedded in local relationships (Aoyama et al., 2006).

Information technology applications abound in the logistics industry, from bar-code scanners to complex terminal operating systems; but these have not replaced face-to-face relationships at key points of decision making and service innovation.

NEW TECHNOLOGIES OF COMMUNICATION

The story of manufacturing logistics is replete with new technologies and communication systems. In his account of the rise of the micro-computer firm, Dell, Fields (2003) places considerable emphasis on the Internet, both in creating new consumers for Dell's products and enabling the rise of web-based commerce that allowed Dell to move closer than ever before to a pull-based manufacturing system. The application of information and communication technologies – sometimes also referred to as electronic data interchange or e-commerce – has allowed more intense and speedy communications between firms, so allowing them increasingly to integrate their production processes. When a lead firm or assembler fills an order for a customer, this information is automatically transmitted to the lead supplier firm, and from there to its lower tier suppliers. This reduces the need for holding stock in hand, and is an integral part of JIT systems. At the same time, it accentuates the need for reliable, cheap and omnipresent transportation capacity. It is not true that Dell starts building a new computer from scratch every time an order comes in, but it gets as close to this situation as possible by relying on modular production, dispersed final assembly and air transportation. This is especially important because Dell is producing something that is relatively perishable, and on which a premium is placed on time to delivery. However, Fields (2003) is insistent that the new logistics-based manufacturing has geography. Dell demands that suppliers locate their warehouses in close proximity to its assembly facilities, and more than this, close control is exercised over them through the virtual integration of production, ordering and distribution. Hence when orders come in Dell can quickly build to order by mixing and matching modular components while externalizing the costs of maintaining inventory onto the suppliers.

NEW ACTORS, AND OLD ACTORS BEHAVING IN NEW WAYS

Coordinating all of these goods movements, and managing the information and communication technology, requires specialized expertise. Manufacturing logistics relies upon the rise of a new type of firm, the global logistics services provider (GLSP), which trades on its capacity to manage several fragmented components along the transportation chain as an integrated entity. These firms also go by the name "third-party logistics providers" (3PLs), which denotes their role in managing various transportation services on behalf of a cargo owner (Berglund et al., 1999). Skjott-Larsen et al. (2007) suggest that GLSPs succeed by combining physical and management services to create what they call "information-based logistics providers."

GLSPs have diverse origins: only some are new start-ups; others were enabled by deregulation and globalization to evolve from their original, more limited, functions (Dicken, 2011). Among the world's largest GLSPs are (former) divisions of ocean carriers (e.g. APL, Yusen or OOCL Logistics and Damco, a division of Maersk), railways (e.g. SNCF Geodis, BNSF Logistics, Union Pacific Distribution Services) and trucking firms (e.g. J.B. Hunt Transport Services; Ryder). GLSPs with these origins are sometimes known as asset-based logistics providers, having added inventory management, postponement and other functions to their traditional transportation and warehousing roles. Other GLSPs have origins as freight forwarders (e.g. Kuhne and Nagel, Expeditors International), as package delivery companies (e.g. Deutsche Post DHL, FedEx, UPS) or as general trading companies (e.g. Japanese sogo shosha). GLSPs with these origins are sometimes known as network logistics providers, having added track and trace, JIT and proof-of-delivery services to their traditional documentation and delivery functions. There is also a new generation of GLSPs that started as integrated logistics firms (e.g. Port Logistics Group, Menlo Worldwide Logistics). The 3PL industry has been dynamic, with multiple mergers (e.g. the rail-forwarder merger that created DB Shenker) and acquisitions, as well as the emergence of a hierarchical or tiered structure of lead and subcontracted logistics services providers. In their analysis of an annual survey of the use of 3PLs by the 500 largest manufacturers in the United States, Lieb and Bentz (2005) report that by 2004, 80 percent reported that they made use of 3PLs. This was up from 38 percent in the 1991 survey. American manufacturers tended to use 3PLs mainly to co-ordinate their transportation needs; over 40 percent of manufacturers reported using 3PLs for services such as freight payment, shipment consolidation, customs brokerage, warehouse management and freight forwarding, as well as direct transportation services. There was also a moderate level of involvement (25–42 percent) in the core postponement functions, such as tracking/tracing, order fulfillment, cross-docking and relabeling/repackaging. 3PLs did not appear to be playing a large role in manufacturing per se – the level of involvement in contract manufacturing, purchase of materials, assembly/installation and product testing ranged between 4 and 13 percent. Hence what this survey was picking up is the traditional role of transportation services in manufacturing, now increasingly contracted out to specialists in negotiating with multiple transportation providers; the growth in 'value added' services associated with postponement and mass customization; and the (hesitant) entry by logistics providers into core manufacturing functions.

A similar survey of the use of logistics outsourcing by Chinese manufacturers conducted in 2002 (Hong et al., 2004) revealed a slightly lower level of use of specialized logistics providers, 59.4 percent. However, the authors also noted the strong intention of Chinese manufacturers to increase their use of logistics providers. As with the American survey, respondents reported that they used logistics more for traditional transportation-related services (forwarding, transportation, warehousing); however, although only 11 and 14 percent respectively indicated that they outsourced for logistics system design and logistics information systems, 50 and 58.7 percent respectively indicated that they intended to make use of these in future.

NEW RELATIONSHIPS

It is important to appreciate the diversity and ongoing change in logistics systems. For example, a comparison of the import and distribution channels used by Japanese importers of fully assembled passenger automobiles to the United States revealed that while Honda assembled large volumes of relatively standardized vehicles in Japan and postponed limited customization to the point of sale, Toyota was more actively involved in managing imports of a diversity of models, integrating the distribution of these vehicles with North American-produced large-volume models (Hall, 2004). These differences reflect larger differences in corporate strategy, history and market embeddedness (see also Van Hoek et al., 1998). Cox et al. (2002) go further, arguing that supply chains are inherently unstable and hence diverse. Powerful supply chain actors, seeking to maintain flexibility, have no need to commit themselves to weaker "partners." Conversely, without definite commitments, weaker partners have every incentive to seek new customers and more favorable terms elsewhere. Paradoxically thus, supply chain actors both seek and abhor trust. These insights fit more generally into the value chain upgrading perspective developed by Gereffi et al. (2005). Once we accept that a firm's (or a region's; see Coe et al., 2004) place in a value chain is not fixed, this raises a dual-edged question: what is the role of logistics in assisting manufacturers to move up the value chain, and what steps might logistics firms take to move up the value chain? Hall and Olivier (2005) explored this idea with respect to automobile manufacturing and car carrier shipping lines. For example, Wallenius-Wilhelmsen Lines, a steamship line, has begun offering customs clearance, customization, landside transportation arrangement and other services to secure business for its asset-heavy ocean shipping operations. Bowen and Leinbach (2006), based on in-depth interviews with air freight forwarders in South East Asia (complementing their interviews with electronics manufacturers in the same locations), note that logistics intermediaries play an important role in allowing firms and regions to upgrade their position in value chains. They cite the example of airfreight forwarder, DHL in Penang, Malaysia, arguing that it is able to compensate for shortcomings in the local cluster: "DHL-Penang is integrated into a number of [global production networks], and these GPNs provide a means by which it can pull through the competitive advantages of other places and its own advantages can be harnessed" (161). Stock et al. (1999) make a similar argument about the bridging role of logistics firms, and the need for a fit between a manufacturing firm's broader strategy and structure, and the logistics services it chooses to employ. All of this implies that some manufacturers are increasingly reliant upon particular 3PLs (or GLSPs) for their business success. Lieb and Bentz (2005), in their survey of the use of 3PLs by American manufacturers, found indications that 3PLs were playing a co-ordinating role that extended beyond the individual manufacturing firm. The authors note that "increasingly, the 3PL users identified in our annual surveys report that their major vendors and customers are served by their primary 3PL service provider" (13). They also note that when manufacturers moved into new market areas, they pressured their 3PLs to move with them.

CONCLUSION: NEW SOURCES OF CONTROL, NEW SOURCES OF COMPETITION

These findings suggest that the adoption of 3PL services is something taken seriously by manufacturers; it entails a commitment, and it is consequential for the internal operation of the firm. Manufacturing logistics entails heightened attention to transportation within the production process, and heightened attention to production in the organization of logistics. Hence, the relationships between large global manufacturing firms and integrated logistics providers are understood to enhance strategic complementarities. For instance, Bhatnagar and Viswanathan (2000) concluded that "the UPS–Motorola alliance has therefore yielded strategic benefits to both parties – time postponement benefits (at an effective price) to Motorola and volume consolidation to UPS" (30). However, as several studies have made clear, large shippers, whether they are retailers or manufacturers, are able to exert considerable pressure on their supply chain partners (Fields, 2003; Gereffi, 1994). With increasing integration of logistics into the supply chains, transport providers are not immune to these pressures to reduce costs, especially labor costs (Bonacich and Wilson, 2008), and to externalize also the hidden costs of goods movement onto cities and communities (Hall and Hesse, 2013). There is thus every reason to expect that logistics providers will seek to develop the capacity to move up the value chain by offering additional "value adding" services.

The boundary between manufacturing and transportation is hence likely to continue to be highly dynamic. In one sense, this is an old story, as old as manufacturing and its punctuated, if shifting, geography. At the same time, logistics entails new configurations of the relationship between transportation and manufacturing, in which the boundary between them has become an arena of conscious corporate strategy; new modes of transportation and technologies of communication, and new routes, support new ways of organizing production, with profound consequences for regional development patterns. New actors, with their own histories, are contesting the terrain of logistics; they are forging new relationships as they seek to exert control over market space. Manufacturing logistics remains an unfinished project.

REFERENCES

Alonso, William (1975), "Location Theory," in John Friedmann and William Alonso (eds), *Regional Policy: Readings in Theory and Applications*. Cambridge: MIT Press.

Anderson, William P. (2012), *The Border and the Ontario Economy*. Windsor, Ontario: University of Windsor, Cross-Border Transportation Centre.

Aoyama, Y., Ratick, S. and Schwarz, G. (2006), "Organizational dynamics of the US Logistics Industry: an economic geography perspective," *The Professional Geographer*, 58(3), 327–340.

Berglund, M., van Laarhoven, P., Sharman, G. and Wandel, S. (1999), "Third-party logistics: is there a future?" *International Journal of Logistics Management*, 10(1), 59–70.

Bhatnagar, R. and Viswanathan, S. (2000), "Re-engineering global supply chains: alliances between manufacturing firms and global logistics services providers," *International Journal of Physical Distribution and Logistics Management*, 30(1), 13–34.

Bonacich, Edna and Wilson, Jake B. (2008), *Getting the Goods: Ports, Labor, and the Logistics Revolution*. Ithaca, NY: Cornell University Press.

Bowen, J.T. (2008), "Moving places: the geography of warehousing in the US," *Journal of Transport Geography*, 16, 379–387.

Bowen, J.T. and Leinbach, T.R. (2003), "Air cargo services in Asian industrialising economies: electronics manufacturers and the strategic use of advanced producer services," *Papers in Regional Science*, 82, 309–332.

Bowen, J.T. and Leinbach, T.R. (2006), "Competitive advantage in global production networks: air freight services and the electronics industry in Southeast Asia," *Economic Geography*, 82(2), 147–166.

Cidell, J. (2010), "Concentration and decentralization: the new geography of freight distribution in US metropolitan areas," *Journal of Transport Geography*, 18, 363–371.

Coe, N.M., Hess, M., Yeung, H.W., Dicken, P. and Henderson, J. (2004), "'Globalizing' regional development: a global production networks perspective," *Transactions of the Institute of British Geographers*, 29, 468–484.

Cox, Andrew, Ireland, Paul, Lonsdale, Chris, Sanderson, Joe and Watson, Glyn (2002), *Supply Chains, Markets and Power: Mapping Supplier and Buyer Power Regimes*. London: Routledge.

Dicken, Peter (2011), *Global Shift: Mapping the Changing Contours of the World Economy* (6th Edition). New York: Guildford.

Fawcett, S.E. (1990), "Logistics and manufacturing issues in maquiladora operations," *International Journal of Physical Distribution and Logistics Management*, 20(4), 13–21.

Fields, Gary (2003), *Territories of Profit: Communications, Capitalist Development, and the Innovative Enterprises of GF Swift and Dell Computer*. Stanford, CA: Stanford Business Books.

Gereffi, Gary (1994), "The organization of buyer-driven global commodity chains: how US retailers shape overseas production networks," in G. Gereffi, and M. Korzeniewicz (eds), *Commodity Chains and Global Capitalism*. Westport: Praeger Publishers.

Gereffi, G., Humphrey, J. and Sturgeon, T. (2005), "The governance of global value chains," *Review of International Political Economy*, 12(1), 78–104.

Glaeser, E.L. and Kohlhase, J.E. (2004), "Cities, regions and the decline of transport costs," *Papers in Regional Science*, 83, 197–228.

Hall, P.V. (2004), "Persistent variation: flexibility, organization and strategy in the logistics of importing automobiles to the United States, 1980–1999," *Environment and Planning A*, 36(3), 529–546.

Hall, Peter V. and Hesse, Markus (eds) (2013), *Cities, Regions and Flows*. Routledge: Abingdon.

Hall, P.V. and Olivier, D. (2005), "Inter-firm relationships and the evolution of transportation systems: the case of the car carriers and automobile importers to the United States," *Maritime Policy and Management*, 32(3), 279–295.

Hesse, Markus (2008), *The City as a Terminal: The Urban Context of Logistics and Freight Distribution*. Aldershot: Ashgate.

Hesse, Markus and Rodrigue, J.P. (2004), "The transport geography of logistics and freight distribution," *Journal of Transport Geography*, 12(3), 171–184.
Hong, J., Chin, A.T.H. and Liu, B. (2004), "Logistics outsourcing by manufacturers in China: a survey of the industry," *Transportation Journal*, 43(1), 17–25.
Kasarda, J. (ed.) (2010), *Global Airport Cities*. Twickenham, UK: Insight Media.
Kasarda, J.D., Green, J. and Sullivan, D. (2004), *Air Cargo: Engine for Economic Development*. Report commissioned by the International Air Cargo Association (TIACA) and presented at the TIACA Air Cargo Forum, Bilbao, Spain, September 15.
Lasserre, F. (2004), "Logistics and the Internet: transportation and location issues are crucial in the logistics chain," *Journal of Transport Geography*, 12(1), 73–84.
Levinson, Marc (2006), *The Box: How the Shipping Container Made the World Smaller and the World Economy Bigger*. Princeton, NJ: Princeton University Press.
Lieb, R. and Bentz, B.A. (2005), "The use of third-party logistics services by large American manufacturers: the 2004 survey," *Transportation Journal*, 44(2), 5–15.
Markusen, A. (1996), "Sticky places in slippery space: a typology of industrial districts," *Economic Geography*, 72(3), 291–314.
Marshall, Alfred (1890), *Principles of Economics*. London: Macmillan.
O'Connor, K. (2010), "Global city regions and the location of logistics activity," *Journal of Transport Geography*, 18(3), 354–362.
Skjøtt-Larsen, Tage, Schary, Philip B. and Mikkola, Juliana H. (2007), *Managing the Global Supply Chain*. Copenhagen: Copenhagen Business School Press.
Stock, G.N., Greis, N.P. and Kasarda, J.D. (1999), "Logistics, strategy and structure: a conceptual framework," *International Journal of Physical Distribution and Logistics Management*, 29(4), 224–239.
Sturgeon, Timothy and Florida, Richard (2000), *Globalization and Jobs in the Automotive Industry*. A study by the Carnegie Mellon University and the Massachusetts Institute of Technology for the Alfred P. Sloan Foundation, Cambridge, MA.
van Egeraat, C. and Jacobson, D. (2005), "Geography of production linkages in the Irish and Scottish microcomputer industry: the role of logistics," *Economic Geography*, 81(3), 283–303.
Van Hoek, Remko I. Commandeur, H.R. and Vos, B. (1998), "Reconfiguring the supply chain to implement postponed manufacturing," *International Journal of Logistics Management*, 9(1), 95–110.
Vickerman, R., Spiekermann, K. and Wegener, M. (1999), "Accessibility and economic development in Europe," *Regional Studies*, 33(1), 1–15.
Weber, Alfred (1909, 1929 translation), *Alfred Weber's Theory of the Location of Industries*. Chicago: University of Chicago Press.

FURTHER READING

Bonacich, Edna and Wilson, Jake B. (2008), *Getting the Goods: Ports, Labor, and the Logistics Revolution*. Ithaca, NY: Cornell University Press. (This book documents working conditions in the logistics industry, with a focus on the southern California port, trucking and warehousing industries.)
Dicken, Peter (2011), *Global Shift: Mapping the Changing Contours of the World Economy* (6th Edition). New York: Guildford. (Still the best single comprehensive text on globalization, with a separate chapter now devoted to logistics and distribution.)
Fields, Gary (2003), *Territories of Profit: Communications, Capitalist Development, and the Innovative Enterprises of GF Swift and Dell Computer*. Stanford, CA: Stanford Business Books. (*Territories of Profit* reveals how communications revolutions in different periods enabled businesses to innovate their operations, reorganize the structure of the firm, and reshape the geography of profit-making.)
Hall, Peter V. and Hesse, Markus (eds) (2013), *Cities, Regions and Flows*. Routledge: Abingdon. (Explores the role that freight plays in the development of cities and urban regions around the world.)
Levinson, Marc (2006), *The Box: How the Shipping Container Made the World Smaller and the World Economy Bigger*. Princeton, NJ: Princeton University Press. (Comprehensive and highly readable account of the rise of containerization in the context of economic globalization.)
Skjøtt-Larsen, Tage, Schary, Philip B. and Mikkola, Juliana H. (2007), *Managing the Global Supply Chain*. Copenhagen: Copenhagen Business School Press. (An important standard text on supply chain management with several case studies of manufacturing logistics.)

5. Reshoring and the "manufacturing moment"

Margaret Cowell and John Provo

INTRODUCTION

Outspoken politicians and academics have argued that we are on the verge of a "manufacturing moment," as manufacturing in the U.S. and other developed nations is poised to regain at least some of its competitive position. Though not yet robust, a continued growth of manufacturing activity in these nations seems increasingly plausible, at least in certain key industries. In the U.S., manufacturing employment has increased 4.4 percent since a recent low of 1.14 million jobs in January 2010 (U.S. Bureau of Labor Statistics 2013). Boston Consulting Group (BCG)[1] predicts that manufacturing growth could create between 2 and 3 million jobs in the next decade, with an estimated 600,000 to 1 million of them direct manufacturing jobs (Sirkin et al. 2012, p. 12). The Brookings Institution, a Washington, D.C.-based think tank, similarly predicts continued growth in high-wage manufacturing industries and those with high shipping costs, as well as modest growth in durable goods manufacturing, which is generally the middle-wage, more productive part of manufacturing (Helper et al. 2012). In the U.S., much of this predicted job growth is likely to come from the anticipated increase in exports that we can expect to accompany a burgeoning manufacturing sector. Some of this growth, however, will likely come from a less obvious source: the reshoring of manufacturing jobs as the pendulum creeps back ashore in certain key industry sectors. The BCG, the National League of Cities (NLC)[2] and others have predicted that sectors like transportation goods, computers and electronics, fabricated metals, machinery, plastics and rubber have reached a tipping point, and are likely to see increased production in the United States (Helper et al. 2012; McFarland and McConnell 2011). Many of the firms within these industries that have reshored did so because of growing concerns with quality, workforce competency, efficiency, speed to market and supply chain vulnerabilities abroad (PricewaterhouseCoopers 2012).

For many, the tale of reshoring is an appealing one; a firm that once left the United States for a lower-cost developing country is today welcomed back with open arms. For planners and geographers, the positive externalities associated with reshoring – including higher purchasing multipliers and the creation of good jobs – have generated increased interest in the literature (Giloth 2012; see also *Progressive Planning*'s Winter 2012, Special Issue on Manufacturing). It should come as no surprise, however, that the reshoring story is just a bit more complicated. Though still a fairly new concept, reshoring is increasingly being conceptualized as more than just the return of jobs to the U.S. and other developed nations from overseas locations (Reshoring Initiative 2013; Wu and Zhang 2013). Researchers have also begun to talk about reshoring as the expansion of existing facilities or firms within the United States, and to some extent in Europe, in industries that have been susceptible to offshoring, as well as new foreign direct investment (FDI). This more inclusive definition of reshoring reflects both the fuzziness of this concept and the nuanced nature of the emerging trend (Markusen 2003).

In the following, we shed light on this emerging concept by first historically contextualizing reshoring within broader global dynamics. We then discuss the more recent conceptualization of reshoring and its rapid rise, particularly in the popular press. The dynamics of manufacturing reshoring, including questions of why manufacturing firms reshore specifically, are then addressed. We conclude with an overview of the main industry sectors most likely to be affected by reshoring and a synopsis of relevant federal policies and programs within the U.S. context.

OUTSOURCING, OFFSHORING AND THEIR VARIANTS

A U.S. programmer was discovered in 2013 to have outsourced his job to China. He utilized the time to surf Reddit for cat videos, shop on Ebay and update his Facebook status (Chappell 2013). Reporters looking to go beyond the immediate humor of the situation stumbled on the term "Outsourced Life" to represent this case. That phrase was popularized by a former kick-boxer turned life-hacker who turned ideas about the global service economy popularized in Tom Friedman's *World is Flat* (2005) into a *Four Hours Work Week* (Ferris 2009). The 2012 U.S. presidential campaign between Barack Obama and Mitt Romney saw several news cycles devoted to debating the finer points of meaning between the terms "outsourcing" and "offshoring" (Ciliza 2012). The deeper meaning of these terms remains confused and contested territory as a dizzying array of variations across industry sectors continues to multiply at the speed of Internet memes (for a broader discussion of these terms, see Clark 2013).

However, the mobility of capital and labor and our struggle to understand it are hardly new phenomena. The rise of new manufacturing powers in Asia and elsewhere characterized the 1980s, stoking fears of deindustrialization in the U.S. (Bluestone and Harrison 1982; Magaziner and Reich 1983; Thurow 1980). By the 1990s offshoring had come to dominate the popular imagination. Companies followed the logic of free trade agreements and looked to maximize their opportunities to rationalize costs, especially in terms of labor. Maquiladoras along the Mexican border or massive factory campuses in China became the face of manufacturing for many. This reordering among nations was inexorable, and manufacturing became something developed economies could no longer retain and should no longer aspire towards (Gross 2012; Liveris 2011).

As the 21st century approached, reasons to reassess the dominant narrative began to enter the discussion. Shorter product life cycles made the ties between product development and production tighter and benefits to proximity more apparent. This meant that investments in research and development were clearly being dispersed along with manufacturing production. Service inputs like logistics and advertising were becoming increasingly important and redefining the meaning of manufacturing jobs. Further, following the September 11 attacks in the U.S., developed countries reacted to security concerns by tightening immigration, and an exodus of technical talent flowed to emerging nations like Brazil, India and China at the same time as opportunities opened up for them (Gross 2012; Liveris 2011; Manyika et al. 2012; National Governors' Association 2013).

At the same time, better communications technologies, oil price disruptions and changing labor costs were among factors leading investors to consider variations on outsourcing. In some sectors, often information technology and services, U.S. firms found logic in "farmshoring" to rural U.S. locations (Mayer and Provo 2007). To some extent, inward rural movement occurred within China as well. As labor markets in coastal cities heated up, many firms explored rural opportunities, though some sectors were burdened by weak infrastructure and governance challenges. In other sectors, labor arbitrage remained an issue and jobs were re-outsourced from China to markets with even lower labor costs such as Malaysia (Couto et al. 2006).

While our description mentions only a few of the terms that scholars have used to describe the global movement of production and capital over the last few decades, a host of others contribute to a growing lexicon. In Table 5.1, we provide an initial organization of these terms by their geographic implications. For example, "offshoring" always has explicitly international implications, with investments often crossing national borders. Other terms, such as "farmshoring," bring an explicitly domestic meaning. We also see terms with the suffix "-shoring," which have some geographic implications, as distinct from terms with the suffix "-sourcing," which are often ambiguous with respect to geography. This is in part because the usage of the "-shoring" terminology often has implications for investing capital within the firm, while "-sourcing" terms are generally applied to changes in business processes that may shift work patterns within the firm to a greater reliance on external partners, without necessarily moving the firm.

RESHORING: THE PENDULUM CREEPS BACK ASHORE

The most recent turn in the narrative describes reshoring, where new investment, often in manufacturing, is flowing into developed nations. This can be explained not only as the return of jobs from overseas but also as the expansion of existing facilities or firms in industries that have been susceptible to offshoring, as well as the attraction of new FDI in those sectors (Dholakia et al. 2012; Sirkin et al. 2011). There are several variations of this term in common usage, with "restoring" generally embracing both the return of offshored jobs and the relocation or expansion of foreign investment in sectors perceived to be susceptible to offshoring. As described in Table 5.2, other, less common variants of the term apply only to subsets of the drivers identified with reshoring. For example, the term "backshoring" – which has also found some currency, especially in Europe – typically refers solely to the return of jobs to an original host country from other foreign locations.

Estimates of the scale and pace of reshoring vary widely. The BCG forecast mentioned earlier, which identified 2 to 3 million jobs to be created in the U.S. over the next decade, reaches broadly to include reshored jobs, foreign investment and export-led growth in manufacturing (Sirkin et al. 2011, p. 12). A series of recent surveys paint a more complex picture. MIT's survey of firms with multinational manufacturing location found that one-third were considering bringing manufacturing to the U.S., and half of that number had definitive plans (Simichi-Levi 2012). A similar survey anticipated net-zero outsourcing in the near future (*Economist* 2013).

There is also certainly an array of anecdotal examples to consider, many involving large firms. For example, Ford and Apple have both announced plans to bring thousands of production jobs to the U.S. Other firms have announced specific investments,

Table 5.1 Select international and domestic variations on offshoring

Geography	Type	Definition	Example
International	Offshoring	Work is moved to another country	Furniture and textile manufacturing sectors in the U.S. following NAFTA
	R&D offshoring, innovation offshoring	Technically intensive work is moved to another country	Relocation of e-ink manufacturing industry from U.S. to Taiwan
	Nearshoring	Work is moved to another country close to the company location	Western European countries investing in Eastern Europe
Domestic	Farmshoring	Work is moved to a rural area in the same country	Placement of IT, call center and other back-office operations in the U.S.
	Homeshoring	Work is moved to individuals who connect virtually with their employer	Travel reservations in countries as diverse as the U.S. and China
International or domestic	Insourcing, co-sourcing	Work is provided on a company site by another entity under contract, or by a separate unit of the same firm	Private security companies used by the U.S. military during the wars in Afghanistan and Iraq
	Offsourcing, out-tasking	A portion of work is moved from a company site to another location internal to the company or under contract	Stanford University Medical Center shifted routine services to a contractor to focus on more strategic projects
	Crowdsourcing	Work inputs are solicited from large groups of individuals, often on-line	Hallmark Cards collaborated with an Open Innovation services provider to develop a new line of technology-enabled cards
	Outsourced life, personal outsourcing	Individuals sub-contract work out to others	Individuals use sites like Elance.com and Guru.com to identify project-based freelancers
	Outsourcing	Work is contracted by one company to another entity outside the firm	Sub-contracting non-core functions such as accounting, human resources or facilities management to specialist firms

Sources: Bloom et al. 2013; Brownell et al. 2006; Carmel and Abbott 2007; Erickson and Gratton 2007; Gramerian 2007; Liveris 2011; Mayer and Provo 2007; Robbins 2011.

including Caterpillar's new Texas plant and 15 General Electric manufacturing plants, which will bring lighting and appliances facilities to Kentucky, Alabama and Ohio (Simichi-Levi 2012).

However, U.S. manufacturing executives that were surveyed in 2012 did see movement

Table 5.2 Reshoring and select variations

Type	Definition	Example
Reshoring or insourcing	Return of jobs to developed nations, or new foreign investment in previously offshored sectors	Texas wins Caterpillar plant from Japan; Chinese firm Lenovo invests in North Carolina
Backshoring	Re-concentration of jobs to a host country from foreign locations	Apple's plans to return jobs to the U.S. from China
Inshoring	Location of jobs to supporting export trade	U.K.-based Rolls-Royce Aerospace in Virginia

Sources: Dholakia et al. 2012; *Economist* 2013; Fishman 2012; Gross 2012; Madsen et al. 2011; Simchi-Levi 2012; Sirkin et al. 2011.

away from China but anticipated firms would be more likely to nearshore to Mexico than reshore to the U.S. (Holden 2012). Further, a trade group representing privately held U.S. manufacturing firms cited continued growth in market share for manufactured imports as reason to remain skeptical of forecasts for growth (Tonelson 2013).

In Europe, evidence of reshoring is also mixed. Limited surveys of U.K. and German firms that had offshored production showed a slightly smaller movement than in the U.S. (Leibl et al. 2011). Continued economic uncertainty in Europe may be in part to blame. Europe also simply offshored less production in the first place than did the U.S. Often firms were likely to nearshore production to Eastern Europe or other low-cost locations on the periphery, which may reduce savings related to reshoring (*Economist* 2013).

Factors limiting the growth or slowing the pace of reshoring may include the capacity of the U.S. workforce to provide a technically skilled manufacturing production workforce. More than 75 percent of manufacturing managers surveyed in 2011 cited such shortages as hindering their ability to grow (*Economist* 2013). In Europe, rigid labor markets, with the possible exception of the U.K., may slow growth by negating the impact of labor cost savings (Associate Parliamentary Manufacturing Group 2013). With all of this in mind we focus on the U.S. for the remainder of the chapter; first, however, we focus on the main reasons why firms may choose to reshore.

DYNAMICS OF MANUFACTURING RESHORING

The main factors driving location decisions for most firms tend to change over time. In the case of offshoring during the 1990s and 2000s, most manufacturing firms were motivated to offshore because of lower costs in developing countries, namely due to lower wages abroad and an appreciating U.S. dollar. However, recent changes in traditional offshoring locations, including China and India, have brought about gradually increasing wages, as well as rising problems with legal ambiguities and quality control. These challenges, along with an ongoing recognition that the U.S. economy is still the most robust on a per capita income basis, have contributed to a growing interest in reshoring for many firms.

Recent research by PricewaterhouseCoopers (2012) identified a number of reasons why firms explore the reshoring option, including: increases in transportation and energy costs; U.S. market demand; currency fluctuations; U.S. talent; availability of capital; tax and regulatory climate; and realization of absolute labor costs in the U.S. Related research from BCG adds that an increasingly flexible and productive workforce, along with a comparatively low total cost of production, have made the U.S. and other developed countries more and more competitive for companies looking to reshore (Sirkin et al. 2011). BCG analysts predict that, within five years, the total cost of production for manufacturing in coastal Chinese cities will only be about 10 to 15 percent lower than it will be in many U.S. locations (Sirkin et al. 2011, p. 11). Fluctuations in shipping and transportation costs, as well as costs related to legal uncertainties and quality control issues, could eventually minimize or negate any perceived savings due to offshoring. Some American companies, including Apple and Caterpillar, have also cited the recent resurgence in consumer preferences for products that are "Made in America" as an important reason for their decision to reshore all or parts of their production process (Caterpillar 2010; Hardy 2012). In general, the main reasons why firms consider reshoring broadly fall into one of three categories: increasing costs abroad; supply chain vulnerabilities; and quality control and intellectual property concerns.

Increasing Costs Abroad

Gradual increases in wages across Southeast Asia have coincided with increased economic and urban development, making coastal export zones in these countries increasingly more expensive for both producers and wage workers (Sirkin et al. 2012). Though some firms choose to combat the rising costs by moving farther inland to interior areas where wages and land are comparatively lower, businesses that locate inland for cost savings face extended exporting operations and the additional risks that come with a longer supply chain. Meanwhile, as labor costs rise in countries like China and India, the cost of labor in the U.S., especially for manufacturers, has decreased in some cases or, at the very least, has increased comparatively more slowly. This shift has meant that the differential labor cost between the United States and many of the traditional offshoring locations like India and China, though still substantial, is shrinking (White House 2012). The cost differential between domestic production and production in places like China has also narrowed with the general depreciation of the U.S. dollar (Sirkin et al. 2012). Projected increases in transportation and energy costs are also of great concern to manufacturing firms which rely heavily on both inputs (Chen 2011). Certain key manufacturing processes are particularly susceptible to increases in transportation and energy expenses, including machinery and fabricated metals, and are often mentioned as prime targets for reshoring (Sirkin et al. 2012).

Supply Chain Vulnerabilities

Stretching their global tentacles across the globe has meant longer and more complicated supply chains for manufacturers which offshore. Though supply chain logistics have certainly improved since offshoring began, supply chains remain a significant challenge for manufacturers which continue moving towards leaner and more efficient forms of production. Just-in-time deliveries, significant customization and rapid responses from supply chain partners have become the norm, but can prove difficult when partners are far away. Recent disruptions due to natural disasters such as the Japanese earthquake of 2011, political upheavals like the Arab Spring of 2010/11, or economic crises, most prominently the Greek debt crisis of 2010, have reminded us that shorter, or less complex, supply chains can sometimes mean the difference between seamless production and lengthy, costly disruptions. Along with the smaller inventory levels and reduced lead times associated with a shorter or simpler supply chain, the potential to decrease supply chain vulnerabilities through reshoring may prove tempting for many manufacturers.

Quality Control and Intellectual Property Concerns

Just as longer and more complex supply chains have meant increased vulnerabilities in terms of the timing and reliability of deliveries, overseas production of goods has also led to growing concerns over the quality of the goods being produced and with intellectual property violations abroad (Clark 2013). Quality control remains a vexing challenge for many firms given differences in culture, language, laws and workforce training (McCormack 2011). Institutional advantages in education and experience are two key components that give the United States a competitive edge over developing countries abroad. Workforce training has also been frequently mentioned as a driver in manufacturer capital spending decisions, meaning that manufacturers are keen to locate where there are vocational programs already in place to cultivate a pipeline of skilled workers like machinists, craft workers and operators who can ensure quality production (PricewaterhouseCoopers 2012).[3] With substantial differences in contract law between sending and receiving countries, legal issues have also arisen in the area of intellectual property protections. Although some low-skill and labor-intensive manufacturing is likely to continue offshore operations, other high-skill manufacturing and service providers have long remained in the U.S. due to its strong intellectual property protection (Simchi-Levi 2012). Firms which have chosen to stay in the U.S. or to reshore in recent years have frequently cited these stronger protections and higher standards of production as motivations for their decision (Simchi-Levi 2012).

SPECIFIC MANUFACTURING INDUSTRIES

In recent years several research groups have analyzed the reshoring trend to provide further detail about the specific sectors most likely to experience this pendulum shift. According to a report prepared by the NLC, FDI in the U.S. has increased dramatically in recent years, especially in key industry areas such as transportation equipment, industrial machinery and equipment, chemicals, electronics, information and communications technology, and electrical equipment (McFarland and McConnell 2011). For companies in these industry sectors, domestic production has become more appealing in recent years as the cost–savings gap of manufacturing abroad shrinks. BCG similarly concludes that these and a few other key industry areas may have reached a "tipping point," or a point of indifference between producing abroad and producing in the U.S. (Sirkin et al. 2012). As Table 5.3 shows, there are many similarities between what the NLC identifies as key FDI sectors and what BCG considers "tipping point" sectors. Below we consider the implications of reshoring in the five sectors in which there is an overlap.

Table 5.3 Comparing the changing dynamics of industries abroad with industries and sectors investing in the U.S.

	BCG's "tipping point" industries	NLC's sectors investing in the U.S.
Computer/electronics	✓	✓
Machinery	✓	✓
Transportation	✓	✓
Fabricated metals	✓	✓
Plastics and rubber	✓	
Food and beverage		✓
Chemicals		✓
Electrical equipment	✓	✓
Furniture	✓	
Business services		✓

Sources: McFarland and McConnell 2011; Sirkin et al. 2012.

Computer/Electronics

BCG estimates that the U.S. purchases about 26 percent, or $122 billion worth, of the electronics it consumes from China. These items mainly include computers, wireless phones and televisions. In 2010, U.S. consumers purchased $467 billion worth of computers and electronics (Sirkin et al. 2012, p. 9). Over the last decade, computer and electronics imports have increasingly penetrated the U.S. market. While domestic manufacturers are likely to continue sending some low-value production overseas, IBISWorld (a market research firm specializing in long-range forecasting) predicts that this trend will slow substantially in coming years because most of the easiest to implement outsourcing and offshoring activities have already taken place (IBISWorld 2013a). Some portions of the computer and electronics sector may even expand within the U.S. in coming years as firms shift their focus to high-quality, high-value items.

Machinery

Items in this category include air conditioners, heaters, pumping equipment, office machinery, power tools, optical products, photocopiers and farm equipment. U.S. consumers imported approximately $16 billion worth of machinery from China in 2010 but the U.S. still manufactures 61 percent of machinery sold domestically, or $251 billion worth in 2010 (Sirkin et al. 2012, p. 4). As both the BCG and NLC note, logistical issues, including time to market, shipping costs and proximity of research and development to production lines, remain relatively important and may explain the fairly strong performance of the machinery sector. IBISWorld also anticipates steady continued expansion in key machinery subsectors because of overseas demand for U.S.-made machinery, domestic demand from recovering construction activity, and emerging trends in food and beverage consumption (IBISWorld 2013b).

Transportation

Both BCG and NLC identified the production of transportation and related goods as tipping point industries with growth potential. According to BCG, the U.S. manufactures 70 percent of transportation goods sold domestically (Sirkin et al. 2012, p.4). Though China has become a major source of car and truck components, motorbikes, bicycles and aircraft parts, U.S. imports from China only totaled $6 billion in 2010 (Sirkin et al. 2012, p.8). Aircraft engine and parts manufacturers in the United States are expected to see continued growth despite projected reduction in federal funding for defense (IBISWorld 2012a). Domestic passenger travel and a large backlog of orders for commercial carriers as well as unmanned aerial vehicles (i.e. drones) should drive a growing demand from domestic and foreign airlines over the next decade (IBISWorld 2012a).

Fabricated Metals

U.S. imports from China of fabricated metal goods including plumbing fixtures, hardware, hand tools, cutlery, and pots and pans totaled $10 billion in 2010 (IBISWorld 2012b, p.4). IBISWorld analysts note that fabricated metals industries rely heavily on demand from other sectors of the economy, especially manufacturing and construction, which saw substantial losses during the 2008 recession. As both consumer confidence and spending continue to improve, demand for cars and buildings will rise, which is expected to trigger increased demand for iron and steel foundry products within the U.S. and overseas. Major international markets are also expected to support the industry's recovery and growth within the U.S. (IBISWorld 2012b).

(Appliances and) Electrical Equipment

Each year, U.S. consumers purchase from China nearly $4.5 billion in lighting products and $6 billion in small appliances, including fans, vacuum cleaners and microwave ovens. U.S. imports from China of larger appliances, including refrigerators, freezers and dishwashers, totaled $25 billion (Sirkin et al. 2012, p.4). While smaller appliances with lower shipping costs will likely continue to be manufactured in lower-wage countries, IBISWorld predicts that a rebound in the U.S. housing market, a rise in exports and new energy-efficient products will contribute to a growing demand for larger appliances and electrical equipment like refrigerators, water heaters and trash compactors (IBISWorld 2013c). FDI is also expected to increase as large foreign multinational firms work to establish a manufacturing presence in the United States and gain access to engineering talent and new technologies (IBISWorld 2013c).

POLICIES IN SUPPORT OF RESHORING

Though not explicitly focused on these five "tipping point" industries, a series of U.S. government initiatives, mainly at the federal level, have been designed to encourage reshoring and manufacturing in the U.S. While some policies are industry based and designed to encourage firms to invest in America, others are focused on workforce development and preparing workers for technology-driven manufacturing firms, which increasingly seek highly skilled employees.

In 2010, the White House announced the "Skills for the Future Initiative," which is an industry-led initiative designed to improve partnerships between industry and community colleges in order to facilitate a nation-wide network to improve workforce development strategies, job training programs and job placement (White House 2011). In tandem with the "Skills for the Future Initiative," a private sector-driven collaboration with community colleges and trade schools is also emerging through the Manufacturing Skills Certification System. Developed and administered by the Manufacturing Institute, which is a nonprofit think tank affiliated with the National Association of Manufacturers (NAM), the certification system supports President Obama's goal of credentialing 500,000 community college students with advanced manufacturing certifications by 2016 (White House 2011). The NAM currently offers programs in areas ranging from basic foundation skills to machining and metalworking, and is also developing certification pathways for aviation and aerospace, and bioscience (Manufacturing Institute 2013). Bolstering the domestic workforce in these and other advanced manufacturing sectors is in line with broader supply-side efforts to encourage reshoring.

The National Institute of Standards and Technology's Hollings Manufacturing Extension Partnership (NIST MEP), a national network with over 1300 field staff in 50 states, is also working to strengthen American manufacturing and to encourage reshoring. The MEP works with partners across the United States to help firms assess whether it makes sense to reshore, and provides more general assistance in areas of workforce, technology adoption, environment and energy, quality and more (NIST MEP 2013). MEP efforts include SelectUSA (an information portal for businesses operating in the U.S.) and the Total Cost of Ownership Estimator (TCO). The TCO is a tool used to examine aggregate costs for firms and is available to any who are considering offshoring or reshoring decisions. It was developed by Harry Moser of the Reshoring Initiative, which is another industry-led effort to bring manufacturing jobs back to the U.S. and encourage manufacturers to utilize local sourcing and production.

Both the workforce development efforts and the information provision are in line with a broader political push to encourage reshoring at the national level. In the 2012 U.S. presidential campaign both candidates focused on manufacturing and the importance of advanced manufacturing (Vavra 2012). President Barack Obama's platform built upon his earlier National Export Initiative, which was a government-wide export-promotion strategy focused on doubling exports between 2010 and 2015 (U.S. Department of Commerce 2010). In March 2012 President Obama announced his plans for a National Network for Manufacturing Innovation, which would support regional hubs – public–private partnerships known as Institutes for Manufacturing Innovation – to create a robust manufacturing and research infrastructure (National Science and Technology Council 2013). In August 2012 an initial federal investment of $30 million helped launch the first of 16 Institutes, the National Additive Manufacturing Innovation Institute. In his 2013 State of the Union Address, President Obama highlighted both the continued importance of advanced manufacturing and education in Science, Technology, Engineering and Math in schools, colleges and universities, as well as making "America a magnet for new jobs and manufacturing" (Obama 2013). President Obama remains committed to the idea that manufacturing matters, most recently with the announcement of the "Make it in America Challenge." Partnering with the U.S. Department of Commerce's Economic Development Administration, NIST MEP, the U.S. Department of Labor's Employment and Training Administration, and the Delta Regional Authority, the challenge provides funding to foreign and domestic businesses to build and/or expand their operations in the United States. Initial funding for the program was announced in March 2013 for up to 15 awards to foster reshoring of productive activity by U.S. firms, increase FDI, encourage U.S. companies to keep or expand their businesses in the United States, and train local workers to meet the needs of those businesses (NIST MEP 2013).

Together, these initiatives represent a broad-based and multi-faceted effort at the federal level to encourage the pendulum to move back ashore. The combination of industry-based investments and targeted workforce development efforts may not be enough to encourage every firm to invest in the U.S., but it may prove successful in some key sectors. Though it is still too early to assess whether these initiatives will garner success, their mere existence suggests that the federal government has embraced reshoring and has begun to seize the manufacturing moment.

NOTES

1. BCG is a global management consulting firm. A recent series, *Made in America, Again* focused on the changing global economics that are starting to favor manufacturing in the U.S. For more information, see https://www.bcgperspectives.com/content/articles/manufacturing_supply_chain_management_made_in_america_again/ (accessed September 7, 2013).
2. The NLC is a U.S. advocacy organization representing 19,000 cities, towns and villages, and encompassing 49 state municipal leagues. It provides training for public officials and lobbies, and provides assistance to U.S. cities. In April 2011, the NLC held a "Leadership Academy on Local Economic Competitiveness in a Global Era," for local elected officials, staff and key stakeholders to discuss foreign direct investment and international trade. Promising practices identified during that event were highlighted in *Strategies for Globally Competitive Cities: Local Roles in Foreign Direct Investment*, available at: http://www.nlc.org/find-city-solutions/city-solutions-and-applied-research/economic-development/strategies-for-globally-competitive-cities (accessed October 23, 2013).
3. Despite the current accessible talent base, other research groups stress that the United States does not have adequate resources to supply a high-skilled workforce long term. For instance, Sirkin et al. (2011) estimates that the U.S. is short some 80,000 to 100,000 highly skilled manufacturing workers.

REFERENCES

Associate Parliamentary Manufacturing Group (2013), "Reshoring: Bringing making back?" *APMG Term Paper*, available at http://www.policyconnect.org.uk/apmg/events/reshoring-bringing-making-back (accessed April 17, 2013).

Bloom, Nicholas, James Liang, John Roberts and Zhichun Jenny Ying (2013), "Does working from home work? Evidence from a Chinese experiment," *White Paper*, Stanford University, available at http://www.stanford.edu/~nbloom/WFH.pdf (accessed April 17, 2013).

Bluestone, Barry and Bennett Harrison (1982), *The Deindustrialization of America*, New York: Basic Books.

Brownell, Bill, David Jegen and Karthik Krishnamurthy (2006), "Strategic out-tasking: A new model for outsourcing," Cisco Systems Internet Business Solutions Group, available at http://www.cisco.com/web/about/ac79/docs/wp/Out-Tasking_WP_FINAL_0309.pdf (accessed April 17, 2013).

Carmel, E. and P. Abbott (2007), "Why nearshore means that distance matters," *Communications of the Association for Computing Machinery*, **50**(10), 40–46.

Caterpillar (2010), "Caterpillar to expand manufacturing and increase employment in the United States with

new hydraulic excavator facility in Victoria, Texas" (Press Release), available at http://tinyurl.com/7bxjskn (accessed April 17, 2013).
Chappell, Bill (2013), "Outsourced: Employee sends own job to China; surfs web," available at http://www.npr.org/blogs/the two way/2013/01/16/169528579/outsourced-employee-sends-own-job-to-china-surfs-web (accessed April 17, 2013).
Chen, Jess (2011), *Renewable Energy Supply Chains: A Guide for Economic Developers*, Washington, DC: International Economic Development Council, available at http://iedconline.com/Downloads/Powering_Up/IEDC_Supply_Chain_Issues.pdf (accessed April 17, 2013).
Ciliza, Chris (2012), "Romney response to Bain story: Outsourcing vs offshoring," *Washington Post*, June 22, available at http://www.washingtonpost.com/blogs/the-fix/post/romney-response-to-bain-story-outsourcing-vs-offshoring/2012/06/22/gJQAMWVXvV_blog.html (accessed April 17, 2013).
Clark, Jennifer (2013), *Working Regions: Reconnecting Innovation and Production in the Knowledge Economy*, New York: Routledge.
Couto, Vinay, Mahadeva Mani, Arie Y. Lewin and Carine Peeters (2006), *The Globalization of White Collar Work: The Facts and Fallout of Next-Generation Offshoring*, Duke: Offshoring Research Network, available at https://offshoring.fuqua.duke.edu/pdfs/gowc_v4.pdf (accessed April 17, 2013).
Dholakia, Nikhilesh, Rama K. Kompella and Douglas N. Hales (2012), "The dynamics of insourcing," in *Proceedings of the Knowledge Globalization Conference, Pune, India*, Boston: Knowledge Globalization Institute, pp. 88–95.
Economist (2013), "Here there and everywhere," *Special Report on Outsourcing and Offshoring*, January 19.
Erickson, Tamara and Lynda Gratton (2007), "What it means to work here," *Harvard Business Review*, pp. 1–10.
Ferris, Tim (2009), "Outsourcing life," available at http://www.fourhourworkweek.com/blog/outsourcing-life/ (accessed April 17, 2013).
Fishman, Charles (2012), "The insourcing boom," *Atlantic Magazine*, December, pp. 45–52.
Friedman, Thomas (2005), *The World is Flat: A Brief History of the Twenty-First Century*, New York: Picador.
Gamerian, E. (2007), "Outsourcing your life," *Wall Street Journal*, June 2, available at http://online.wsj.com/article/SB118073815238422013.html (accessed April 11, 2013).
Giloth, R. (2012), "Planners and manufacturers: An uneasy alliance," *Progressive Planning*, 190, 9–12.
Gross, Daniel (2012), *Better, Stronger, Faster: The Myth of American Decline . . . and the Rise of a New Economy*, New York: Free Press.
Hardy, Q. (2012), "Why Apple got a 'Made in U.S.A.' bug," *New York Times Online*, December 7, available at http://bits.blogs.nytimes.com/2012/12/07/why-apple-got-a-made-in-u-s-a-bug/?_php=true&_type=blogs&_r=0 (accessed May 10, 2013).
Helper, Susan, Timothy Krueger and Howard Wial (2012), *Why Does Manufacturing Matter? Which Manufacturing Matters? A Policy Framework*, Washington, DC: Brookings Metropolitan Policy Program.
Holden, G. (2012), "U.S. manufacturers coming home?" *Research Technology Management*, **55**(5), 4–6.
IBISWorld (2012a), *Industry at a Glance – Aircraft, Engine & Parts Manufacturing in the US*, Los Angeles: ibisworld.com, IBISWorld Industry Report 33641a.
IBISWorld (2012b), *Industry at a Glance – Ferrous Metal Foundry Products in the US*, Los Angeles: ibisworld.com, IBISWorld Industry Report 331151.
IBISWorld (2013a), *Business Environment Report: Import Penetration into the Manufacturing Sector*, Los Angeles: ibisworld.com, IBISWorld Industry Report.
IBISWorld (2013b), *Industry at a Glance – Plastics & Rubber Machinery Manufacturing in the US*, Los Angeles: ibisworld.com, IBISWorld Industry Report 33322.
IBISWorld (2013c), *Industry at a Glance – Major Household Appliance Manufacturing in the US*, Los Angeles: ibisworld.com, IBISWorld Industry Report 33522.
Leibl, Peter, Roger Morefield and Rolf Pfeiffer (2011), "A study of effects of back-shoring in the EU," *Journal of Business and Behavioural Sciences*, **23**(2), Summer, 72–79.
Liveris, Andrew (2011), *Make It In America*, Hoboken, NJ: Wiley.
Madsen, Erik Skov and Dmitrij Slepniov (2011), "The path from outsourcing to backsourcing: Debating its logic and considerations," *Proceedings: Det Danske Ledelsesakademis Konference*, København: Det Danske Ledelsesakademi, available at http://www.det-danske-ledelsesakademi.dk/2011/Papers-2011/Madsen_Slepinov.pdf (accessed May 1, 2013).
Magaziner, Ira and Robert Reich (1983), *Minding America's Business: The Decline and Rise of the American Economy*, New York: Harcourt Brace.
Manufacturing Institute (2013), "NAM-endorsed certifications," available at http://www.themanufacturinginstitute.org/Skills-Certification/Certifications/NAM-Endorsed-Certifications.aspx (accessed April 10, 2013).
Manyika, James, Jeff Sinclair and Richard Dobbs (2012), "Manufacturing the future: The next era of global growth and innovation," McKinsey Global, available at http://www.mckinsey.com/insights/manufacturing/the_future_of_manufacturing (accessed April 12, 2013).

Markusen, A. (2003), "Fuzzy concepts, scanty evidence, policy distance: The case for rigour and policy relevance in critical regional studies," *Regional Studies*, **37**(6–7), 701–717.

Mayer, Heike and John Provo (2007), *Farmshoring in Virginia: Domestic Outsourcing Strategies for Linking Urban and Rural Economies in the Commonwealth of Virginia*, Blacksburg, VA: The Economic Development Studio @ Virginia Tech.

McCormack, Richard (2011), "Accenture: Offshore outsourcing has not worked," *Manufacturing & Technology News* **18**(7), available at http://www.manufacturingnews.com/news/11/0415/Fellowes.html (accessed April 17, 2013).

McFarland, Christiana K. and J. Katie McConnell (2011), *Strategies for Globally Competitive Cities: Local Roles in Foreign Direct Investment*, Washington, DC: National League of Cities, Center for Research and Innovation.

National Governors' Association (2013), *Making Our Future: What States Are Doing to Encourage Growth in Manufacturing through Innovation, Entrepreneurship, and Investment*. A National Governor's Association Policy Academy Report. Washington, DC: National Governor's Association.

National Institute of Standards and Technology's Hollings Manufacturing Extension Partnership (2013), "Make it in America," available at http://www.nist.gov/mep/america.cfm (accessed April 17, 2013).

National Science and Technology Council (2013), "National Network for Manufacturing Innovation: A preliminary design," available at http://manufacturing.gov/docs/NNMI_prelim_design.pdf (accessed August 7, 2013).

Obama, Barack (2013), The White House, Office of the Press Secretary. *Remarks by the President in State of Union Address*, available at http://www.whitehouse.gov/state-of-the-union-2013 (accessed April 17, 2013).

PricewaterhouseCoopers (2012), "A homecoming for US manufacturing? Why a resurgence in US manufacturing may be the next big bet," PricewaterhouseCoopers Publications, available at https://www.pwc.com/en_US/us/industrial-products/publications/assets/pwc-us-manufacturing-resurgence.pdf (accessed April 1, 2013).

Reshoring Initiative (2013), "Why reshore?" available at http://reshorenow.org/why_reshore/ (accessed March 19, 2013).

Robbins, Jason (2011), "Open innovation: Manufacturers' idea factories," Crowdsouring.org., available at http://www.crowdsourcing.org/document/open-innovation-manufacturers-idea-factories/7290 (accessed April 17, 2013).

Simchi-Levi, David (2012), *U.S. Re-shoring: A Turning Point*, MIT Forum for Supply Chain Innovation, Cambridge, MA: MIT.

Sirkin, Harold L., Michael Zinser and Douglas Hohner (2011), *Made in America, Again: Why Manufacturing Will Return to the US*, Chicago: Boston Consulting Group.

Sirkin, Harold L., Michael Zinser, Douglas Hohner and Justin Rose (2012), *U.S. Manufacturing Nears the Tipping Point: Which Industries, Why, and How Much?*, Chicago: Boston Consulting Group.

Thurow, Lester (1980), *The Zero-Sum Society*, New York: Basic Books.

Tonelson, Alan (2013), *Import Penetration Rises Again In 2011: Challenges Manufacturing Renaissance, Insourcing Claims*, Washington, DC: US Business and Industry Council.

U.S. Bureau of Labor Statistics (2013), *Current Employment Statistics. Employment, Hours, and Earnings from the Current Employment Statistics Survey*, available at http://www.bls.gov/ces/ (accessed April 1, 2013).

U.S. Department of Commerce (2010), "Commerce Secretary Gary Locke unveils details of the National Export Initiative," (Press Release) available at http://www.commerce.gov/news/press-releases/2010/02/04/commerce-secretary-gary-locke-unveils-details-national-export-initiat (accessed April 17, 2013).

Vavra, Bob (2012), "The 2012 election: Where do Romney, Obama stand on manufacturing issues?" *Plant Engineering*, available at http://www.plantengineering.com/single-article/the-2012-election-where-do-romney-obama-stand-on-manufacturing-issues/b6b83f6547d39cb1f9f5afd1f9bafe50.html (accessed April 17, 2013).

White House (2011), "President Obama and Skills for America's Future Partners announce initiatives critical to improving manufacturing workforce," press release, available at http://www.whitehouse.gov/the-press-office/2011/06/08/president-obama-and-skills-americas-future-partners-announce-initiatives (accessed April 17, 2013).

White House (2012), "Investing in America: Building an economy that lasts," available at http://www.whitehouse.gov/sites/default/files/investing_in_america_report_final.pdf (accessed April 17, 2013).

Wu, Xiaole and Fuqiang Zhang (2013), "Home or overseas? An analysis of sourcing strategies under competition," SSRN, available at http://ssrn.com/abstract=1871425 (accessed April 17, 2013).

6. Relocation of production activities and underlying social dynamics: an analytical framework based on a Canadian perspective

Patrice Jalette

The relocation of activities related to the production of goods and services profoundly disrupts industrial relations dynamics. In recent years, the practice of transferring or outsourcing production activities to other locations in Canada or abroad has been facilitated by factors such as trade liberalization and capital flows, technological development, lower transportation costs and access to skilled labour in low-wage countries (Berger, 2006; Murray, 2010). This globalized context is clearly favourable to employers since it adds to the arsenal of strategies and tactics at their disposal to pressure workers and their trade unions to accept concessions on working conditions (Jalette, 2011). The third actor in industrial relations – the state – must also deal with this context of transnationalization of production which poses considerable challenges to national regulations and public policies, highlighting the vulnerability of both the state itself and local communities in the face of employer options to relocate production activities (Murray, 2012). An in-depth knowledge of business decisions regarding the transfer of production activities is thus needed in order to understand the social dynamics that such decisions bring to bear in labour–management relations and beyond. In the current context, it is generally held that workers, the state and local communities can only stand by and watch the flow of production location transfers, being utterly powerless and lacking the capabilities or resources to stop them. But is this indeed always the case?

This chapter aims to broaden our understanding of production relocation.[1] First, we will situate the transfer of production activities within a two-way dynamic, thus going beyond the classic conception of production relocation options centred only on transfers away from the firm. Second, we will present an analytical framework which examines relocation processes from a broader perspective. The dynamics at play among the actors involved will be illustrated throughout the chapter, with emphasis on cases in the Canadian province of Quebec.

1. PRODUCTION RELOCATION DYNAMICS[2]

The transfer of production activities from one site to another, within the same firm or outside it, in the home country or abroad, can be described using many different expressions: offshoring, inshoring, outsourcing, insourcing, domestic outsourcing, international outsourcing, consolidation of production activities, repatriation of production activities, and so on. All of these transfers involve moving the production of goods and services across geographical and/or organizational boundaries. They highlight the importance of business decisions regarding the specific physical and organizational location of the production of goods or services. The choice of production location is among the key decisions made by business executives and, as pointed out by Berger (2006, p.57), it is these decisions, combined, which have led to the current globalization: 'Globalization is not something that just "happened" but the product of millions of choices: choices made by corporations about what to do within the four walls of a company, choices about what to buy from others, and choices about where to locate all these activities.' Seeking the most advantageous global location and mode of production has become a best business management practice. However, since decisions regarding the location of production activities are constantly reviewed as the context evolves, these decisions are far from definitive (Berger, 2006).

The various production location and relocation options open to firms can be categorized based on two fundamental decisions. The first involves ownership or organizational mode of production: will the production activity be carried out by the firm itself at its own sites or be outsourced to another firm, beyond its organizational boundaries? The second involves the geographical location, another important consideration: will the production activity be carried out in the domestic country or be located in or transferred to a foreign location, outside the geographical boundaries of this country?

Figure 6.1 illustrates the way the management and economics literature usually represents the options open to firms regarding these two dimensions of production relocation decisions. The first option (Box 1) is to integrate the various steps of the production process within the firm itself, within the same national territory. This option is often viewed as the starting point in the process of moving production away from a given site. However, as in other jurisdictions around the world (Gereffi et al., 2005; Berger, 2006), the number of firms in Quebec that do not outsource any production activities at all is very small, as shown by a survey indicating that 95 per cent of Quebec manufacturing firms outsourced

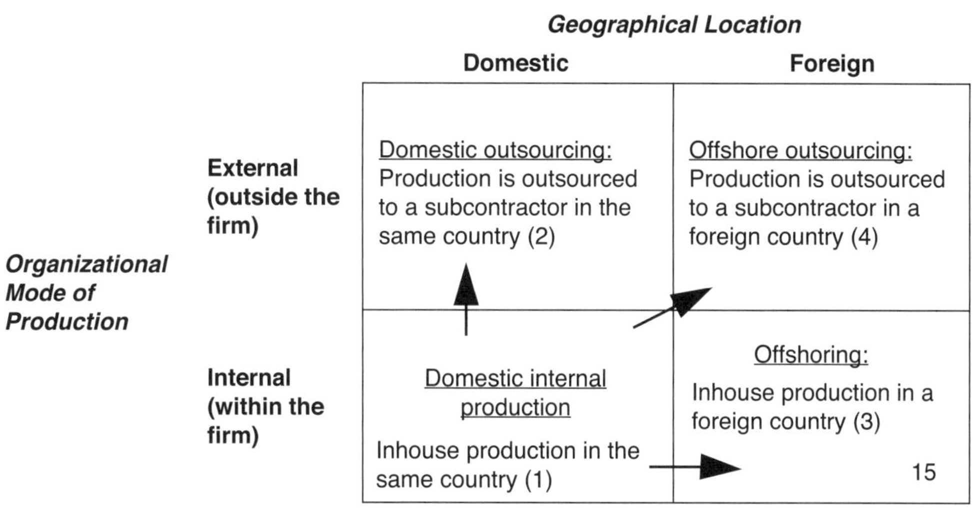

Source: Adapted from Kirkegaard 2007, p. 42.

Figure 6.1 Standard production relocation options matrix

at least one production activity (Jalette, 2004). The second option is domestic outsourcing (Box 2), that is, outsourcing the production activity to a specialized subcontractor within the same national territory. The third option (Box 3) is offshoring, wherein the production activity is transferred to another production location within the same firm but in a foreign country. The fourth and last option provided for in this matrix (Box 4) is offshore outsourcing, that is, outsourcing to a subcontractor in a foreign country.[3]

The problem with this analytical framework is that it considers production relocation solely in terms of outward relocation, with production seen as being inevitably drawn away from the actual site to a more favourable context: 'Implicitly, this matrix frames companies' production relocation choices in a globalizing world economy as a one-way movement – typically from a high-wage location to a low-wage destination' (Kirkegaard, 2007, p.4). This creates a serious distortion in our understanding of production relocation dynamics. There is, however, a more accurate way to portray the real options available to firms.

In addition to the options outlined above, Figure 6.2 presents new configurations of the geographical and organizational boundaries involved from a Canadian perspective, allowing for the consideration of all possible production relocation options open to a site in Quebec. In this respect, this matrix is socially and geographically embedded since, as has been shown by the new economic sociology (Bair, 2008) and economic geography (Rainnie et al., 2007), economic behaviour must be understood in the context of social interactions between actors embedded in a given territory.

First, the geographical dimension included in this matrix is more specific, reflecting the various industrial relations systems which co-exist in the different provinces of Canada and the federal jurisdiction, and which differ from one another with respect to labour legislation in particular they all have their own labour code. For example, the terms and specific application of provisions aimed at protecting trade union rights in the case of the sale or transfer of a business ('successor rights') vary across Canada and do not always apply in cases where production activities are transferred between jurisdictions. It is thus important to consider inter-provincial and inter-jurisdictional transfers of production activity in Canada with this context in mind.

Second, this matrix includes an option which was not considered in the first matrix; namely, the transfer of the production of a good or service to another production location within the same firm and within the same territory, in this case, Quebec (Box 1). Sometimes, this strategy is used to rationalize operations by closing one site and transferring production to another site, as was the case for the residential furniture manufacturer Shermag which, in 2006, closed some distribution centres across Quebec in order to concentrate its activities in Montreal. This strategy can also be used to relocate operations to regions that are favourable for various reasons (e.g. lower wages, government incentives, regional development). For example, in 2001 Desjardins decided to move all administrative tasks related to managing student loans, which had previously been carried out in branches across the province, to a single location in Gaspé. Intra-provincial and inter-provincial relocation options have thus been added to the domestic and international options included in Kirkegaard's initial complete relocation matrix.

Third and lastly, this second matrix can be said to be dynamic insofar as it shows that production relocation is not merely a one-way phenomenon (Kirkegaard, 2007). In addition to transfers away from the site (outward relocation), it also provides for transfers towards the site (inward relocation). Thus, inward relocation (Boxes 7 to 12) refers

Geographical location					
Quebec (movement within Quebec)		**Canada** (movement to/ from the rest of Canada)		**Foreign Country** (movement to/ from other countries)	
Within-firm	Outside the firm	Within-firm	Outside the firm	Within-firm	Outside the firm
Outward Relocation (production activity transfers away from the site) →					
Outward inhouse relocation within Quebec: To other inhouse production locations in Quebec (1)	**Outsourcing within Quebec:** To production locations of subcontractors in Quebec (2)	**Outward inhouse relocation within Canada:** To other inhouse production locations in Canada (3)	**Outsourcing within Canada:** To production locations of subcontractors in Canada (4)	**Offshoring:** To inhouse production locations in a foreign country (5)	**Offshore outsourcing:** To production locations of subcontractors in a foreign country (6)
Inward Relocation (production activitiy transfers towards the site) ←					
Inward inhouse relocation within Quebec: From other inhouse production locations in Quebec (7)	**Insourcing within Quebec:** From production locations of subcontractors in Quebec (8)	**Inward inhouse relocation within Canada:** From other inhouse production locations in Canada (9)	**Insourcing within Canada:** From production locations of subcontractors in Canada (10)	**Inshoring:** From inhouse production locations in a foreign country (11)	**Offshore insourcing:** From production locations of subcontractors in a foreign country (12)

Source: Adapted from Kirkegaard 2007, p. 43.

Figure 6.2 Comprehensive dynamic production relocation options matrix

to the case where a given site obtains or wins back a production activity that was being carried out at a different production location (within the firm or outside it, within the same territory or abroad). The inclusion of these options in the matrix allows for a more complete picture of production relocation dynamics and shows that, in addition to losing production activities, a site can also gain such activities. For example, AFT in Lennoxville in 2008 and AMI in Boucherville in 2007 both repatriated manufacturing activities which had previously been carried out in the United States (Box 11), while Xstrata closed its copper smelter in Timmins, Ontario and transferred production to Rouyn-Noranda and Montreal in 2010 (Box 9). Such transfers, aimed at consolidating existing production activities at a single location or allocating new activities to an existing location can be explained by the dynamics of attractiveness described in the foreign investment literature (Bognanno et al., 2005). The advantages of a particular production location (skilled workforce, technology, proximity to the market or natural resources, government assistance, etc.) can be promoted by local actors and often lead business executives to choose one location over another (Murray, 2012).

The notions of inward relocation, insourcing or repatriation recognize that the production of a good or service can be brought back from where it was initially sent out. For example, in 2009, dissatisfied with the results of offshore outsourcing, the communication company BCE repatriated technical assistance services which it had previously transferred to call centres in India (Box 12). A similar phenomenon was also observed in the municipal sector in Quebec, when trade unions succeeded, through various strategies, in winning back for their members some activities which had previously been outsourced (Box 8) (Cyr-Racine and Jalette, 2007).

What, however, explains these 'boomerang' transfers wherein some organizations arrive at the conclusion that it would be better to bring back production activities which had previously been outsourced and/or offshored? In fact, since the context is continually evolving, the parameters of the analysis of coordination costs, which originally led to the decision to relocate outward, can change sufficiently that a re-examination of the situation will lead to the decision to bring production activities back to the original site (Gereffi et al., 2005). Moreover, not all outward relocation is successful, for many well-documented reasons: insufficient cost savings, decline in the quality of the outsourced work, and so on (Barthélemy, 2003; Bergeron, 2008; *The Economist*, 2009).

The dynamics portrayed in this matrix show, first, that production relocation involves a 'two-way street' and, second, that production relocation transfer decisions can be revised based on new circumstances and social interactions, meaning that they are not set in stone. This broader conception of production relocation dynamics thus calls for an analysis of the underlying social dynamics which goes beyond economic and managerial paradigms.

2. ANALYTICAL FRAMEWORK

In this section and the next, we will present an analytical framework which examines the transfer of production activities from a broader perspective and sheds light on the dynamics among the actors involved in these processes. First, rather than considering globalization as an exogenous event, a *fait accompli* and an irreversible force, we consider it to be an incomplete and indeterminate process which can be strengthened or countered through a range of strategies deployed by the actors involved (Giles, 2000). The social dynamics to which these transfers give rise are expressed through the relations between these actors. For example, the power relations between the employer and the union come into play in these processes, as shown by a Quebec survey indicating that approximately 63 per cent of unions surveyed in manufacturing had been subject to a threat by their employer to transfer production activities elsewhere within the firm during the period studied, a threat which came to pass in two-thirds of cases (Jalette, 2011). As will be seen, the network of relations that develops in response to the possible relocation of production activities also involves the state and the local community; hence the need for a broader analytical framework.

Figure 6.3 presents the factors to be taken into account when examining cases of

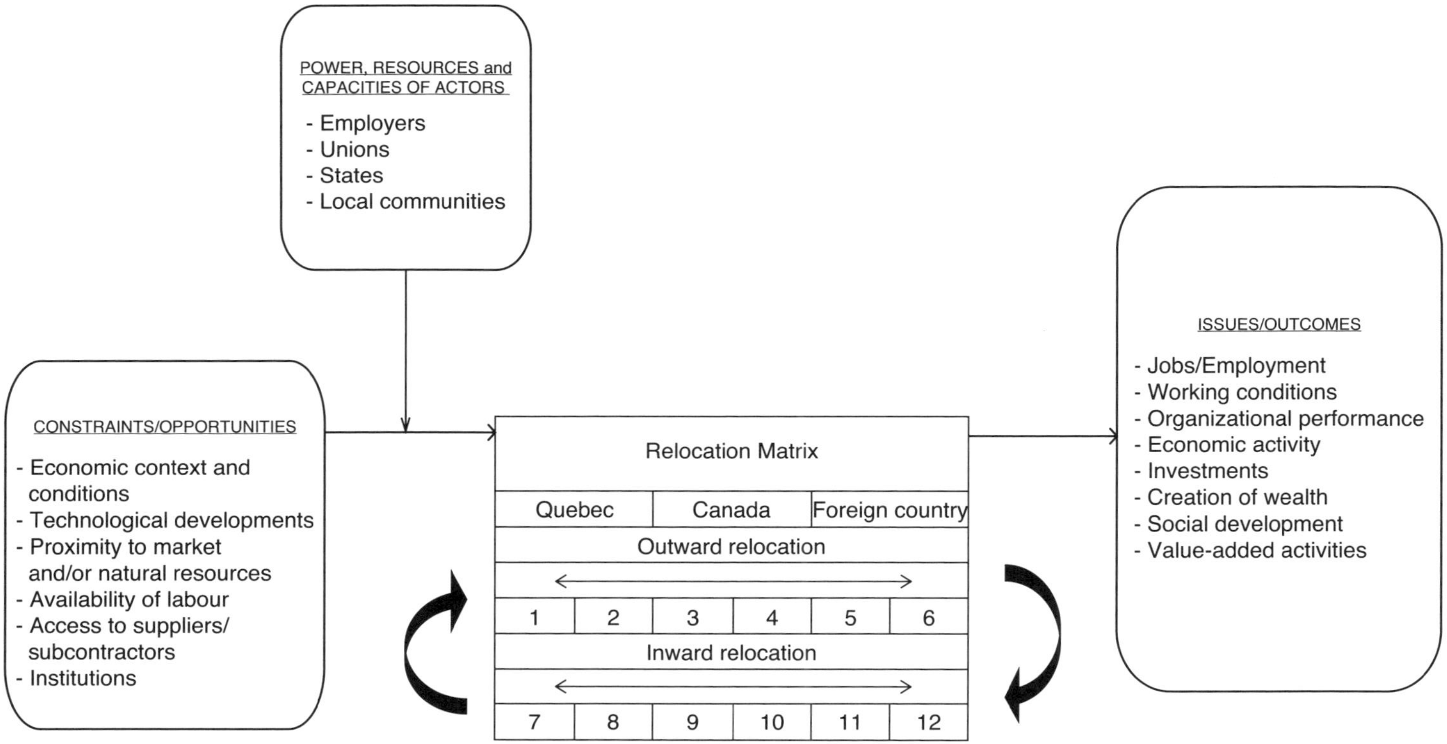

Figure 6.3 Analytical framework of production activity relocations

production relocation. This analytical framework takes for granted that production activity transfers are not only influenced by contextual factors such as economic conditions and technological developments, but also by the actors involved.

2.1 Contextual Factors

Different contextual factors can act as constraints or opportunities for the transfer of production activities. It should be noted that these contextual factors can evolve over time such that those that may currently be favourable to relocation will not necessarily always remain so.

The current economic context, characterized by strong competition for investors and consumers and by developments in information technology, communications, production and transportation, has provided ample motives and opportunities to review the previous production location decisions made by firms (Berger, 2006; Reich, 2008; Murray, 2010). Economic conditions may therefore have constituted an incentive for production relocation in recent years. In Canada, for example, the rise in the value of the Canadian dollar since 2002 may have led some firms operating in Canada to decide to transfer their production activities outside the country (Lamoureux, 2010). On the other hand, the economic context does not always lead to offshoring or outsourcing. For example, the OECD (2009) concluded that the recent economic crisis threatened global production networks not only because of a decline in international trade, but also because of the financial difficulties of subcontractors that were likely to go bankrupt. Thus, some firms may have been prompted to repatriate production activities in order to protect themselves against possible disruptions in their production network. Another example of an economic factor which, following a change in economic conditions, could go from presenting an opportunity to acting as a constraint for production relocation is the price of petroleum. In his book, *Why Your World Is About to Get a Whole Lot Smaller*, Rubin (2009, p.144) shows that 'the global market is . . . an artifact of the era of cheap energy' and a major change in the price of a barrel of oil – by raising the cost of transportation – could greatly diminish the advantages of offshoring and lead to an increase in regional trade. The introduction of a carbon tax to combat climate change could have a similar effect (Bergeron, 2008; Murray, 2010). The influence of the economic context can thus play out in both directions.

The high mobility of capital does not free firms from the constraint of having to choose a particular production location in which to set up operations (Rainnie et al., 2007). The choice of location is based on various logics. First, the logic of reducing operating costs is perhaps the most talked about in the globalization literature wherein the transfer of production activities from developed (high-wage) countries to developing (low-wage) countries is generally taken for granted. However, as seen above, the choice of production location is also based on other logics (natural resources, labour, etc.) which can counterbalance the attraction of low wages. For example, Bognanno et al.'s study (2005) on the production location decisions of U.S. multinational corporations over a period of 10 years showed that host country market size was by far the main determinant of such decisions, being even more important than wages and/or industrial relations context. Moreover, an increase in the size of the global labour pool, brought about by the inclusion of economies such as those of China and India (Murray, 2010), has given firms access to a low-wage skilled workforce which is obviously favourable to the transfer of production to these countries. The same is true for the presence, in a given territory, of suppliers or subcontractors that can be integrated into the firm's value chain.

Lastly, when firms set up operations in a given territory, they must fit these operations into the institutional context of the country concerned (laws, public policies, business and knowledge networks, industrial relations system, etc.). Business executives inevitably take these factors into account when making production location decisions. Moreover, it appears not only to be a question of the presence or absence of institutions establishing the rules of the game and restricting or supporting local action, but also the way in which local actors use their capacities to mobilize these institutional resources to improve their strategic position (Kristensen and Zeitlin, 2005; Morgan and Kristensen, 2006). It should be noted, furthermore, that while production relocation represents an opportunity for the firm to choose the most favourable location for its operations, it can also be a way to escape a constraining environment, whether in terms of labour legislation or union presence (Murray, 2012). Indeed, despite the possible constraints on such transfers within a given country, firms are completely free to proceed with such production transfers outside national boundaries since they are not subject to any regulation in this regard at the international level.

2.2 Actors and Issues

The proposed analytical framework goes beyond the traditional view of production activity transfers whereby the employer is the only actor that can influence these processes. According to this view, the workers, unions, governments and local communities are confined to the role of passive victims embedded in a territory while confronted with the mobility of capital. Our analytical framework, on the other hand, takes for granted that production relocation processes involve several actors, each having resources that can be mobilized to address their own interests and concerns.

2.2.1 Employers

As seen above, the inward or outward relocation of production activities is the result of choices made by business executives. By deciding where and when to relocate production activities, these actors control two major dimensions of relocation decisions (location and timing) which affect the capacity of other actors to react to this type of restructuring (AGIRE, 2008). The ability to initiate production activity transfers is thus a major source of power for employers, and reinforces the asymmetry of power between the actors involved. As described above, the current context has also contributed in various ways to strengthening the capacity of employers to relocate production activities. Moreover, as shown by Jalette (2011), the capacity of managers to relocate production outward is also determined by some organizational characteristics, such as the organization of the value chain or access to plants in countries offering low wages and/or making similar products. Lastly, the main issue for the employer with regard to production relocation is organizational performance, which can be obtained by reducing operating costs, gaining proximity to markets or natural resources, meeting the performance requirements of investors, or any other considerations which the employer considers to be strategic.

2.2.2 Trade unions and workers

Considering the asymmetry of power in favour of employers, workers and the union organizations representing them inevitably find themselves in a reactive position when faced with relocation decisions made by the employer (Pedersini, 2006; Jalette, 2011). Faced with a *fait accompli*, the union will seek to lessen the consequences for its affected members by negotiating for severance pay or measures to help them relocate within the firm, upgrade their skills or look for other employment. On the other hand, when a space for negotiation with the employer opens up – for example, in the case of a threat to relocate or the announcement of a relocation plan which is not yet definitive – the union has a margin of manoeuvre to change the outcome. This can involve negotiating concessions aimed at reducing labour costs (lower wages and benefits, changes in work organization, early retirements, targeted layoffs, etc.). A more proactive position involving, for example, a strategy to intervene on the reasons for the relocation can also be adopted by the union. As in the case of concessions, this strategy aims to reduce the incentive for the relocation and can include, without being limited to, searching for alternatives, drawing up comparisons between the source location and the possible destination locations, or negotiating a partnership whereby, in exchange for concessions or commitments to improve the firm's performance, the union can obtain guarantees regarding job security or investments in technology and training. This non-mutually exclusive repertoire of union actions is also applicable when a site wishes to win back previously relocated production activities or attract new investments. Lastly, while a union strategy of opposition based on power relations and pressure tactics remains conceivable, its effectiveness can be limited given the employer's capacity to thwart it simply by transferring production outside the zone of union influence, which is usually in Canada and restricted to the site itself. However, Bain and Taylor (2008) identify a repertoire of actions available to unions opposing offshore outsourcing, ranging from raising the awareness of clients and political lobbying to strong-arm pressure tactics (refusing to do overtime, strike action) and including global campaigns aimed at putting pressure on the firm or supporting the unionization of workers in the foreign location.

One of the resources available to unions opposing production relocation is the collective agreement, which can, for example, include clauses prohibiting layoffs resulting from the subcontracting of production activities or providing for union involvement in the decision-making process (Jalette and Laroche, 2010). Union power resources (internal solidarity; network embeddedness; narrative and organizational resources) can also be mobilized (Lévesque and Murray, 2010). For example, as shown by Jalette (2009), a union that is involved in decision-making and whose members display a high degree of cohesion and participate in union life (internal solidarity) is more likely to succeed in reducing the negative consequences of subcontracting. The main issues for unions regarding production relocation are jobs, working conditions, the expertise and skills of their members, retaining production activities within their zone of influence, and even, in some cases, the very survival of the union and the site.

2.2.3 The state

While the state is traditionally viewed in industrial relations as the omnipotent arbitrator standing above the fray, helping to resolve conflicts and establishing the rules of the game, the transfer of production activities reduces the state's capacity to intervene.

This is especially true for transnational transfers since the state can always exercise its power to regulate transfers within its national territory or jurisdiction, such as pursuant to section 45 of the Quebec Labour Code (R.S.Q., chapter C-27), making intra-national transfers less problematic for politicians at the national level. Moreover, while the state is generally unable to prohibit international transfers, it nevertheless enjoys a significant capacity to intervene in the institutional context involved in such transfers. The state has at its disposal various measures by which to increase or limit the choices and power of the actors involved: guaranteeing workers access to information and the right to participate in the process, increasing the costs of relocation for the employer (e.g. the obligation to reimburse government loans or assume the costs related to the reclassification of workers), improving the social safety net for workers (e.g. employment insurance), investing in training and innovation, and so on. Given that employment, value-added activities and the creation of wealth are its main concerns with respect to such transfers, the state plays a crucial role in attracting and retaining foreign and domestic investment within its national territory. In this respect, by offering government grants, tax incentives or loan guarantees, the state can negotiate with firms and encourage them to invest within the national territory in order to create new productive capacities or retain and improve those already in place (Murray 2012), as exemplified by the closing and reopening of the Kenworth (Paccar) truck plant in Sainte-Thérèse, Quebec (see Pichette and Hamelin, 1997).

2.2.4 Local communities

Production relocation also raises issues for local communities in terms of employment and economic activity, but also social development and sometimes even the long-term viability of the community. Community actors include citizens, politicians, business people, suppliers and all the different groups attached to a given sub-national territory (municipalities, regions), more restricted in scope than the national territory. These actors are concerned by transnational production activity transfers, certainly, but also by intra-national transfers, such as, for instance, those between regions within the same country. Although their power is more limited than that of the state, local communities nevertheless have the capacity to influence relocation decisions. Community leaders can also enter into the game of attracting and retaining investment in their area through various strategies adapted to their level of intervention (e.g. offering property tax exemptions and land grants, building infrastructure, etc.). Like workers, local community actors are virtually immobile in the face of the mobility of capital because of the economic (e.g. property) and social (e.g. family) ties that bind them to their community, meaning that they often have no choice but to engage in 'local boosterism' in order to attract mobile capital (Rainnie et al., 2007). Lastly, it should be noted that 'nomadic' firms end up setting down roots in communities that welcome them 'with open pockets',[4] thus compensating for the significant costs involved in setting up operations in a new environment: developing relations with suppliers, training workers, and so on (Rainnie et al., 2007). Private investments thus constitute an important determinant of the economic activity and social development of local communities, at home and abroad, and are seen by these communities as an opportunity to reduce unemployment, poverty and social inequality (Kaplinsky and Morris, 2002).

3. THE DYNAMICS INVOLVED

Our analytical framework takes for granted that production transfers contribute to the development of production networks – whether globally or more locally, within the firm or between firms – stemming from the fragmentation and dispersal of production activities in different geographical locations and the need to coordinate them through various means (Gereffi et al., 2005). Taking these networks or value chains into account brings out the interconnectedness of the actors and production activities across geographical space (Bair, 2008). Because of the economic, social and political issues underlying them, these networks constitute spaces for opposition and accommodation, domination and subordination, competition and collaboration/solidarity between the many different actors, each with their own interests, power and agenda (Rainnie et al., 2007; Levy, 2008). It thus appears that production relocation processes, by contributing to the development of such production networks, give rise to the expression of different dynamics and relations between the actors, as illustrated in Figure 6.4.

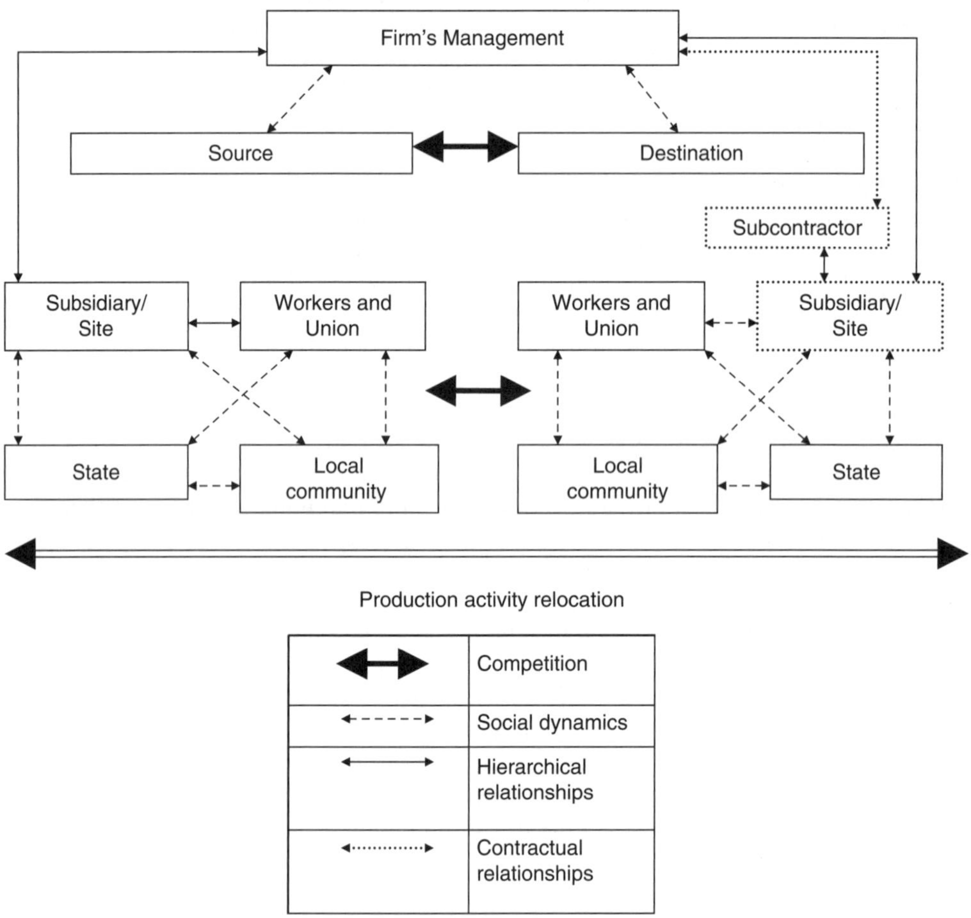

Figure 6.4 Social dynamics involved in the relocation of production activities

Figure 6.4 highlights the first principle of production relocation dynamics described above: every production transfer involves a source location and a destination location. The implication of this principle with regard to the proposed analytical framework is that, in both of these locations, there are socially and geographically embedded actors who are greatly concerned. A broader analysis of production relocation processes thus tends to increase the number of actors considered by including not only those at the source location, but also those at the destination location. Nevertheless, the main actor of production relocation remains the firm's management, for which the transfer constitutes a source of power that can be asserted in its relations with the other actors. As the initiator in charge of the entire process, the firm has the capacity to bring the source and destination locations and the other actors involved into competition with one another. This capacity constitutes the driving force behind relocation processes. However, in addition to this dynamic of competition, these processes also lead to other dynamics between the actors: cooperation, collaboration, solidarity, accommodation, negotiation, coercion, domination, and so on.

Figure 6.4 shows the dynamics that come into play within the firm's production network. First, depending on whether it is a case of subcontracting or within-firm relocation, contractual or hierarchical relationships will be established, and the implications of each will be different. In the case of subcontracting, the firm that wishes to transfer production activities and the firm to which these activities are contracted out will be tied by a contract that will likely specify the terms and conditions regarding cost, quality standards, division of labour, controls, and so on. These inter-firm relations involve an asymmetry of power between the client firm and the subcontractor, the extent of which will be determined by various factors such as the degree to which the products and processes involved are standardized, the added value of the activities undertaken by each of the parties in the value chain, and the level of mutual dependence between them (Gereffi et al., 2005). A significant power asymmetry usually leads to the domination of one firm by the other, with the dominant firm being in a position to impose the conditions of the business relationship on the other. For example, in 2007 BCE threatened not to renew its contracts with Expertech, a network infrastructure service provider, if the latter did not reduce its costs by 25 per cent. Since these contracts represented 90 per cent of Expertech's business income, this firm asked its employees for wage concessions corresponding to this demand. However, other configurations of the value chain can lead to more egalitarian inter-firm relationships (Gereffi et al., 2005).

In the case of within-firm relocation, a dynamic of competition and coercive comparison can develop between the sites considered. Head office will thus seek to play the different production locations off against one another, forcing them to show high levels of performance in the hope that they will be favoured by investment decisions. The management of the subsidiary plays an important role in these decisions by promoting the advantages of the site and its location to head office (Morgan and Kristensen, 2006; Murray, 2012). Thus, beyond its image as a highly rational entity, the firm – especially multinational firms – must be seen as an arena wherein internal rivalries and political games play out for the allocation of resources and responsibilities (Kristensen and Zeitlin, 2005).

Firms make the most of the opportunities provided by the current context to establish a similar dynamic of competition between the workers, states and local communities in the source and destination locations considered. Given the importance of relocation issues for the local actors who are embedded in both the source and possible destination locations, these actors will strive to promote the advantages of their location so as to attract, retain or win back the production activities of firms which, having the upper hand, can force them to compete with one another. This competition between possible production locations gives rise to a dynamic of 'regime shopping', 'whereby firms search for the locations that they believe can guarantee the most convenient regulatory regime' (Pedersini, 2006).

As for the workers, this competition is expressed at the level of the work sites, which are compared among themselves on the basis of various indicators (labour costs, productivity, etc.) (Morgan and Kristensen, 2006) or to subcontractors. These comparisons may then be used by the employer to ask the union to make concessions on working conditions in order to make their work site more advantageous. With these concessions, the union hopes to minimize the damage caused by a balance of power that is fundamentally unfavourable to it (Lévesque and Murray, 2002). My study in the manufacturing sector in Quebec confirmed this conclusion, showing that union concessions reduce the probability that a threat to relocate will actually be carried out (Jalette, 2011). With every threat, however, the union faces a serious dilemma: is it better to accept poorer working conditions or deal with the consequences of relocation? Other strategies can also be used by unions in their negotiations with employers. In a study on offshore outsourcing, Bain and Taylor (2008) concluded that the unions that had been better able to arrive at satisfactory agreements to protect jobs were those that had managed to develop a credible strike threat and thus make the firm step back from its plans. Moreover, Jalette and Hebdon (2012) show that a union strategy based on proposing alternatives to the transfer of activities to subcontractors reduced the probability of such plans being carried out in some Canadian municipalities, which likely obtained what they were seeking in the union proposals put on the table.

During location negotiations with the firm, the union can ally itself with other actors sharing the same concerns. In some cases, a collaborative relationship between the union and the local management can influence the decisions made by head office (Kristensen and Zeitlin, 2005) and lower probability of relocation (Jalette, 2011). Other, more natural, allies of the local union – and sharing many concerns with it – are the local community, the state and other unions. Alliances with these actors, based on an exchange of information and experiences and shared action, can be part of a wider strategy of promoting the advantages of the site. According to Lévesque and Murray (2010), integrating unions into these networks allows them to better influence the change processes and obtain their goals. However, local actors, engaged in promoting the advantages of their community to the firm, can also intensify the pressure on workers to accept the employer's conditions. Indeed, local actors often fear that the union's militancy and demands will discourage the firm from choosing their community as an investment destination (Rainnie et al., 2007). Lastly, the inter-site competition created by the employer and the alliances with local management to attract investment do not facilitate the development of solidarity links between the unions at different sites within the same firm, which helps to divide the workers (Lévesque and Murray, 2002; Grenier, 2006). The same is true with regard to solidarity links between workers of the subcontracted firm and those of the client firm, as well as between workers in different countries. Such solidarity is not always selfless either, since helping workers at another site to improve their working conditions can help employees secure their own situation by reducing the competitiveness of that other site (Rainnie et al., 2007).

CONCLUSION

This chapter presents an analytical framework for understanding production relocation by highlighting the need to consider the issues, contexts and actors involved, at both the source and destination locations, as well as the social dynamics underlying such transfers. Without dismissing the effect of context on production activity transfers, this framework considers the strategic decisions of the actors involved and the relationships between them as significant factors in explaining these transfers. Taking the social dynamics into account leads to a richer understanding of these events and the repertoires of action available to the actors involved. While it cannot consider all possible cases, this analytical framework brings out the points that the different production relocation options have in common and their implications for the actors involved.

Future studies in this field could help validate and enrich this analytical framework. Studying the relations between the source location and destination location, especially when they are located in countries of the North and South, respectively, would certainly be a promising avenue of research. Nevertheless, the transfer of production activities across organizational and geographical boundaries remains a complex subject when the trouble is taken to go beyond the usual preconceptions associated with it. The challenge for researchers will be to produce research that will be useful to the stakeholders involved, shedding light on a phenomenon with which they are likely to have to deal in the foreseeable future.

NOTES

1. Drawing on Kirkegaard (2007), the term 'production relocation' is used here in its broadest sense to refer to all possible types of production location transfer, including those that are kept within the firm and those that involve outsourcing production to a subcontractor, either domestically or abroad.
2. This section is based in part on an article which appeared previously in the journal *Effectif* (Jalette and Chevance, 2008). The business cases referred to in this chapter were documented in the media in Quebec. References are available from the author upon request.
3. Jalette and Chevance (2008) illustrate these various options by referring to the different types of production activity transfer undertaken by planes manufacturer Bombardier Inc. over the years.
4. In December 2010, Electrolux, a Swedish-based company, announced that it would be closing its l'Assomption plant in 2012 and transferring its applicances manufacturing activities to a new plant in Memphis, where the authorities literally offered them a golden bridge. The agreement between the parties can be viewed at http://stimulus.memphistn.gov/Portals/0/pdf_forms/ElectroluxAgreement.pdf (accessed on 29 March 2013).

REFERENCES

AGIRE (Anticipating for an Innovative Management of Restructuring in Europe) (2008), *Anticipation of Restructuring in the European Union*. Final Report, May.

Bain, Peter and Phil Taylor (2008), 'No Passage to India? Initial Responses of UK Trade Unions to Call Centre Offshoring', *Industrial Relations Journal*, **39**(1), 5–23.
Bair, Jennifer (2008), 'Analysing Economic Organization: Embedded Networks and Global Value Chains Compared', *Economy and Society*, **37**(3), 339–364.
Barthélemy, Jerome (2003), 'The Seven Deadly Sins of Outsourcing', *Academy of Management Executive*, **17**, 87–100.
Berger, Suzanne (2006), *How We Compete: What Companies Around the World Are Doing to Make It in the Global Economy*, New York: Doubleday.
Bergeron, Ulysse (2008), 'Délocalisation: le monde à l'envers', *Commerce*, November, 19–26.
Bognanno, Mario, Michael P. Keane and Yang Donghoon (2005), 'The Influence of Wages and Industrial Relations Environments on the Production Location Decisions of U.S. Multinational Corporations', *Industrial and Labor Relations Review*, **58**(2), 171–200.
Cyr-Racine, Cathy S. and Patrice Jalette (2007), 'What Have Unions Got To Do With Reverse Privatisation?', *Journal of Collective Negotiations*, **31**(4), 302–317.
Gereffi, Gary, John Humphrey and Timothy Sturgeon (2005), 'The Governance of Global Value Chains', *Review of International Political Economy*, **12**(1), 78–104.
Giles, Anthony (2000), 'Globalisation and Industrial Relations Theory', *Journal of Industrial Relations*, **42**(2), 173–194.
Grenier, Jean-Noël (2006), 'Local Unions and the Restructuring of Work within the Multinational Company: Internal Solidarity and Local Context', *Labor Studies Journal*, **31**(3), 65–84.
Jalette, Patrice (2004), 'Subcontracting in the Manufacturing Sector: A Quebec–Ontario Comparison', *Workplace Gazette*, **6**(4), Winter, 73–86.
Jalette, Patrice (2009), 'Workplace-Level Evidence of Outsourcing Consequences in Unionized Canadian Manufacturing', in R. Blanpain (ed.), *The Modernization of Labour Law and Industrial Relations in a Comparative Perspective*, Netherlands: Kluwer Law International, Bulletin of Comparative Labour Relations Series, 133–156.
Jalette, Patrice (2011), 'Relocation Threats and Actual Relocations in Canadian Manufacturing: The Role of Firm Capacity and Union Concessions', *American Behavioral Scientist*, **55**(7), 843–867.
Jalette, Patrice and Stéphanie Chevance (2008), 'De la délocalisation à la relocalisation: carte routière et implications pour les relations industrielles', *Effectif*, **11**(4), 50–53.
Jalette, Patrice and Robert Hebdon (2012), 'Unions and Privatization: Opening the "Black Box"?', *Industrial and Labour Relations Review*, **65**(1), 15–33.
Jalette, Patrice and Mélanie Laroche (2010), 'Organisation de la production et du travail', in P. Jalette and G. Trudeau (eds), *La convention collective au Québec*, 2nd edition, Montreal: Chenelière Éducation, 187–218.
Kaplinsky, Raphael and Mike Morris (2002), *A Handbook for Value Chain Research*, Institute of Development Studies, available at http://www.prism.uct.ac.za/Papers/VchNov01.pdf (accessed 5 July 2013).
Kirkegaard, Jacob F. (2007), *Offshoring, Outsourcing and Production Relocation – Labor-Market Effects in the OECD Countries and Developing Asia*, Washington: Peterson Institute for International Economics. Working Paper Series.
Kristensen, Peer Hull and Jonathan Zeitlin (2005), *Local Players in Global Games: The Strategic Constitution of a Multinational Corporation*, Oxford: Oxford University Press.
Lamoureux, Jean-François (2010), *Canada's Trade Performance: An Examination of Eight Indicators*, Ottawa: Export Development Canada, December.
Lévesque, Christian and Gregor Murray (2002), 'Local versus Global: Activating Local Union Power in the Global Economy', *Labor Studies Journal*, **27**(3), 39–65.
Lévesque, Christian and Gregor Murray (2010), 'Comprendre le pouvoir syndical: ressources et aptitudes stratégiques pour renouveler l'action syndicale', *La Revue de l'IRES*, **65**(2), 41–65.
Levy, David (2008), 'Political Contestation in Global Production Networks', *Academy of Management Review*, **33**(4), 943–963.
Morgan, Glenn and Peer Hull Kristensen (2006), 'The Contested Space of Multinationals: Varieties of Institutionalism, Varieties of Capitalism', *Human Relations*, **59**(11), 1467–1490.
Murray, Gregor (2010), 'La transnationalisation des modes de production et ses consequences', in P. Verge (ed.), *Droit international du travail: perspectives canadiennes*, Cowansville: Éditions Yvon Blais, 5–32.
Murray, Gregor (2012), 'Can Multiple Weak Ties Reverse the Social Regulation Deficit? Multinational Companies and Labor Regulation', *Comparative Labor Law & Policy Journal*, **33**(4), 715–747.
OECD (Organisation for Economic Co-operation and Development) (2009), *Responding to the Economic Crisis: Fostering Industrial Restructuring and Renewal*, Paris: OECD, Industry and Innovation, July.
Pedersini, Roberto (2006), *Relocation of Production and Industrial Relations*, EIRO On-Line, available at http://www.eurofound.europa.eu/eiro/2005/11/study/tn0511101s.htm (accessed 24 February 2013).
Pichette, Hélène and Louise Hamelin (1997), *Troc: Made in Québec*, Société Radio-Canada video document.

Rainnie, Al, Andrew Herod and Susan McGrath Champ (2007), 'Spatialising Industrial Relations', *Industrial Relations Journal*, **38**, 102–118.
Reich, Robert (2008), *Supercapitalism: The Transformation of Business, Democracy, and Everyday Life*, New York: Vintage.
Rubin, Jeff (2009), *Why Your World Is About to Get a Whole Lot Smaller*, Toronto: Random House Canada.
The Economist (2009), 'Big is Back', *The Economist*, August–September, 9.

7. Tool-less manufacture: digital fabrication, 3D printing and the third industrial revolution

Michael Ward

ON THE THRESHOLDS OF CHANGE: TECHNOLOGY DEVELOPMENTS

The current era is one of great technological and scientific developments in which new products appear to be released on an almost daily basis. Nevertheless, nearly all new product development occurs through a process of gentle refinement and evolution rather than via major disruptive technological innovations. Major manufacturing and technological developments, or technology disruptors, occur much less frequently. Over the last two hundred years of industrial development it is perhaps possible to identify just a handful of key technology disruptors: the development of an efficient steam engine to deliver almost limitless power, the construction of mass transportation systems to effectively deliver goods and raw materials, and the invention of the transistor that has so transformed communications, enabling the distribution of 'soft' content-based goods. However, it is not these technological innovations alone that have led to the explosion of the availability of cheaper and ever better quality products. Rather it is the coupling of these technical innovations with a market need and an efficient manufacturing methodology or process. It was Henry Ford's approach to mass manufacture, not the invention of the car, that so transformed car manufacture from a cottage industry into a global business (Chandler, 1964; Bryson and Henry, 2001).

While there have been many technological innovations and developments in consumer micro electronics, so aptly demonstrated by developments in smart phone products, the manufacturing approach remains essentially the same as that used to fabricate the original Model T Ford. Manufacturing is still based around the centralized mass manufacture of standardized products leading to cost minimization and improved product quality. However, a major potentially disruptive change has emerged in manufacturing technology and methodology. This new technology is so significant that it might be equal to the innovations in manufacturing processes introduced by Henry Ford. We may be about to witness a paradigm shift in our approach to manufacturing that will generate a change in the world order. This new world order is driven by a shift towards the production of products that contain digital elements or a shift that centres around digital fabrication (Bryson *et al.*, 2013). As good example of such a shift is reflected in the recent history of the Kodak company. This firm was once a global giant in the development and manufacture of cameras, film and film processing. During the late 1990s Kodak suffered from a decline in the sale of photographic film and the company was too slow in shifting its activities to reflect developments in digital photography. Kodak was nearly destroyed by innovations in digital photography that acted as a disruptive innovation destroying the market for photographic film and cameras. The application of digital technologies to manufacturing and fabrications is at the heart of the next manufacturing revolution. This 'third industrial revolution' reflects a shift towards 'tool-less manufacture' based around additive rather than subtractive production processes. In this chapter, the shift towards digital fabrication is explored as an innovation that has the potential to transform manufacturing processes and products. This chapter explores the engineering behind the emergence of 3D printing and examines 3D printing as an alternative to existing manufacturing processes. The analysis focuses on the engineering aspects of 3D printing, and this reflects my disciplinary background in physics and engineering.

TOOL-LESS MANUFACTURE: AN INTRODUCTION TO 3D PRINTING

The transformative technology that sits at the heart of the manufacturing revolution may best be described as 'tool-less manufacture'. 'Tool-less' since digital fabrication involves a process by which a product can be directly manufactured from a computer file, without the need for any intermediate tooling. It is also known as 3D printing, rapid prototyping and additive manufacture (Bryson *et al.*, 2013). Here we prefer the term, tool-less manufacture, as this name captures the essence and novelty of the technology as a new manufacturing methodology. It is worth noting that 3D printing is part of a much broader set of digital manufacturing tools and processes that include digital laser cutters and digital presses. In this chapter, the focus is on 3D printing.

In essence 3D printing technology is very simple, being based upon ink jet printing. The ink jet printer, that has had such an impact upon Kodak, works by using a computer to drive a small print head that moves across a page, depositing small droplets of ink to generate a photographic-quality image on the page. Not only is this a quick and effective way of manufacturing an image but it is incredibly flexible, as it requires no mask or litho plate to be produced prior to the production of the actual photograph. In this printing process, a computer file is created and used to manufacture a product without the need to produce any intermediate tooling. This leads to many benefits, including efficient use of resources, flexible just-in-time manufacture and reduced manufacturing costs.

Now imagine that instead of dispensing droplets of ink, that the print head deposits a small quantity of a liquid polymer that, once dispensed, is able to set to form a small, solid droplet. By repeatedly passing the print head over the page, or in this case a substrate, a three-dimensional object is created directly from a computer CAD file. As in the case of the ink jet-produced photograph, the object is created directly from a computer file without the need to produce any specialist intermediate tooling. A 3D printer produces an object by a process of addition rather than using tooling to remove or subtract material. The attractive features of tool-less manufacture can thus be summarized:

1. The component is made in one process step and there is no assembly of subcomponents. All the components of a gear box can be 3D printed simultaneously and the gear box can be printed ready assembled.
2. Complex interlocking and inter-penetrating components can be made with 3D printing that simply cannot be made by conventional subtractive or milling processes.

3. Only the amount of material required to produce the final component is used and there is no waste.
4. The 3D printing process is extremely flexible. A simple edit of the computer file containing the original design is all that is required to make every product unique and customized to meet individual customer needs.
5. Since each part of the structure is printed individually, the composition of the components can be graded from point to point within the structure, so minimizing cost and maximizing structural integrity. This is impossible with conventional manufacturing processes.

This technology is now a reality and commercial 3D printers are readily available. Currently the surface finish and available material properties limit the application and utility of the tool-less manufactured object. However, if we compare the development of 3D printers to ink jet printers it is surely only a matter of time before the quality of the printed product improves and eventually exceeds that of the conventional product.

One of the main applications of the technology at this time is to rapidly produce prototype structures of components that can be used by engineers and designers to assess the suitability of a new component in real-world applications. For this reason the technology is often referred to as 'rapid prototyping' technology. However, this term is rather limiting, implying as it does only a limited role for tool-less manufacture. When we look beyond the simple ink jet analogue we find that tool-less manufacture can also exploit metallic powders and high power laser beams to fuse and directly write a structure into the metallic powder. Indeed metallic components made this way can achieve 99.99 per cent of the strength of conventionally made components. Turbine blades, combustors and even complete jet engines have been produced using 3D printers. The technology is also extremely useful in the entertainment industry, where 3D printing has been used to produce the 'original' Bond car (Warman, 2012).

Tool-less manufacturing is not limited to mechanical components. Recent developments in organic electronics mean that in certain application areas viable alternatives to conventional silicon-based electronics are now available. These organic electronic materials are amenable to ink jet printing technologies. There is a very real possibility that tool-less or printed electronics manufacturing may develop. Therefore, in the future it will be possible to manufacture complex products with little more than a glorified ink jet printer and a suitable CAD file. This may sound over-optimistic, but if we look at the rapid development of ink jet printer technology that has taken place over the past twenty years, along with the price reduction, there is no reason to suppose that similar advances in tool-less manufacturing machines will not also take place. There appear to be no fundamental physics that would limit the development of this technology; however, there may well be other constraints that will limit its uptake and development.

CHARACTERISTICS OF TOOL-LESS MANUFACTURE: THE CAPABILITIES OF CURRENT 3D PRINTERS

A recent report by the Technology Strategy Board (TSB, 2012) in the UK identified seven 'tool-less' processes: powder bead, material feed, material and binder jetting, filament and resin feed and sheet lamination. These processes were able to produce components made from a variety of materials, including metals, ceramics, polymers and wax.

For metal and ceramic components the process usually follows a simple three-step process. A powder of the ceramic or metal particles, typically a few micro metres in size, is dispensed in a thin layer on to a flat substrate. Then the particles within the powder are selectively bonded together. For metallic powders this bonding is usually achieved by laser sintering of the powder, while for ceramic powders it is achieved by dispensing a small quantity of glue or binding agent to stick the particles together. Once the structure has been completed, unbound powder may be shaken from the structure and the structure itself subjected to further heat treatments to fully bind the powder particles together.

Polymer systems tend to use just two deposition methods. In the first case, a liquid thermo plastic is dispensed from a small nozzle, just like the ink jet process, but here the ink is replaced by the liquid thermo plastic. Following deposition the thermo plastic cools and sets. Alternatively, a liquid polymer is dispensed and an ultra violet light is used to cure the liquid polymer to form a solid. In both cases, a second material may also be deposited. This second material is referred to as a 'scaffold'. As the name implies, this scaffolding material is used to support the structural material during the course of deposition. At the end of fabrication it can be removed just as the unused powder is removed in the ceramic and metallic process. In both cases care must be taken during the design of the component to ensure that unused powder and scaffolding material can be easily removed at the end of the process. This imposes limits on the smallness and interconnectivity of voids that can be included in components if all the non-structural materials are to be removed from the completed object. A third 3D technology uses one or two laser beams to selectively cure a photo-sensitive polymer or resin. The resin is contained in a tank, which is slowly topped up as a laser scans the surface, selectively curing the resin. An alternative is to use two laser beams that selectively cure at the intersection of the two beams. Using this technology, micro-sized components have been produced with micron feature sizes.

Generally the polymer technologies are capable of delivering minimum resolutions of a few tens of micro metres, while the powder-based technologies are somewhat poorer. The limited resolution of power-based technologies limits the quality of the products that may be produced. These limitations produce poor surface finishes. Additionally the technology does not yet provide access to the full range of engineering grade materials that conventional manufacturing technology allows. This limits the range of applications for 3D printing. While the strength of metallic components is almost identical to that of conventionally manufactured components there are concerns about the fatigue properties of 3D printed components. However, these issues are currently being addressed.

A CURRENT AND FUTURE VISION OF MANUFACTURING

The vision that is proposed here is radically different from the current situation regarding manufacturing. A typical product development process for a mass produced product might be as follows. Currently a product may be conceived of, designed and engineered anywhere in the world. This part of the product development cycle requires intelligent human capital but comparatively little in the way of hardware and infrastructure. Once a market and design have been completed the design may be sent electronically to specialist global manufacturers which can transform the design into reality, package it and ship it back to the designer or directly to the customer. These last two stages, manufacture and shipping, both require considerable capital investment in infrastructure, both for the manufacturing facility and the global shipping capability. Both of these are very expensive processes and their cost can only be justified by amortizing their cost over very large numbers of products. This has several implications; only large markets can be dealt with in this way and so minorities cannot have cost-effective access to specialist volume manufacturing facilities and their associated products. This means that certain lower volume but high value markets cannot be satisfied. For example, there is a global market for the iPhone and so the very latest technology is included in the iPhone, while other equally important market areas such as security and medical markets are unable to access the latest technologies as their markets do not support large numbers of products. This highlights the production benefits that come from the ability to scale up a process given the size of the market.

There is considerable concern that the next generation of global manufacturing facilities may become too expensive for many companies or countries to build. For example, electronic foundries and nuclear power stations are now becoming so expensive that the amortization periods are too long to attract investors. For some products, global centralized manufacture may have reached the end of the road.

So what might the future manufacturing system look like and what would be the advantages and disadvantages over today's system? In developing this view of the future I assume that the performance of tool-less manufacturing systems will continue to improve, much in the way that ink jet performance has improved over recent decades. This improved performance will address the surface finish problem and material properties that limit 3D technology to rapid prototyping. In the future the design and marketing operation will remain much as they are today. However, since tool-less manufacture is much more flexible than conventional centralized manufacture, designers will be able to address much smaller markets and offer the option for product customization to meet the demands of an individual customer. For example, when ordering a pair of designer glass frames or a fountain pen the customer will be able to take a 3D image of their head or their hand which can be interpreted by the designer's software to tailor the design to the customer. Once a design has been completed and customized it can be sent to a suitable tool-less manufacturing machine. Note we have not said a tool-less manufacturing facility or factory but a tool-less manufacturing machine. The implication of this statement is that the tool-less manufacturing machine may be sitting on the customer's own desk, much like a super ink jet printer today, or a more specialist machine perhaps may be located at a local hardware store or at some regional level where a more sophisticated machine is justified. Thus, we envisage a distributed manufacturing facility with a hierarchy ranging from the desk top to some national facility. Once fabricated, locally, to the customer's needs then the customer is able to take immediate possession of the product.

For designers this approach has several advantages. They may choose the best tool-less manufacturing facility available to support their product and similarly the customer may choose the tool-less manufacturing machine that best suits his or her interests, based upon cost, location and response time. In many cases the customer will be able to order and receive the production straight away as there is no need to store and transport the product from a global manufacturing centre to the customer – manufacturing will take place at or near the customer's location. We can therefore see that the distribution costs are greatly reduced. For now, it is only the last leg of the journey from designer to customer that requires physical transportation, namely from tool-less manufacturing machine to customer. In many cases, this may be unnecessary as the manufacturing machine may be in the customer's own home. This will, of course, have a major impact upon the traditional procurement process. We should also note that retail shopping, warehousing and physical transportation do not form any part of the procurement cycle.

Since the interaction between the designer and the manufacturing process now takes place via the web and no specialist tooling is required, individuals will be able to access manufacturing capabilities that would never have been possible with the conventional manufacturing model. This should allow more people to engage in the creative process and stimulate economic regeneration.

WHAT BUSINESS MODELS MIGHT SUPPORT THIS BRAVE NEW WORLD?

There is already a working model for this sort of distributed manufacturing and marketing activity. Recently Apps for the iPhone and Android smart phones have stimulated a huge new market for advanced software applications. In this market small software companies, often individuals, can conceive of a market need for new iPod, iPhone or Android device functionality. Then, so long as the basic functionality exists within the phone (accelerometer, microphone, camera, etc.) and the App designer sticks to the phone design rules – that is, uses the phone software correctly – then the designer can easily produce an App to meet market needs. However, the designer cannot distribute the App directly to the customer. Indeed that would not be desirable from the customer's or the market's point of view. For unless the App is checked and in some way certified, poor quality or malicious Apps could be released on to the market. Rather the designer submits the App to a central App store that checks the software and confirms it is fit for purpose. The customer then purchases the App from the App store knowing that it is fit for purpose. Nothing physical moves, the functionality flows from the designer to the customer, while value flows from the customer through the App store to the designer.

The App store approach could easily be adopted as part of a future digital fabrication facilitated or enabled manufacturing system. A designer would identify both a product and a suitable tool-less manufacturing machine suited to producing a product. The manufacturer of the tool-less manufacturing machine would publish well-defined design rules for what its machine can produce. Working within these design rules the designer would create a suitable product design. The design would then be submitted to the machine manufacturer's 'App' store, where either at cost or for an agreed share of future business the App store would fabricate and verify the designer's product. Successful designs would then appear on the machine manufacturer's App store for customers to select. Note the close similarity to the Apple App store business model. Customers could then have the product built on their own tool-less manufacturing machine at home or, if necessary, on other more specialist machines located at commercial outlets. This is similar to deciding to print a photograph at home on a desk top ink jet or having it printed by a specialist photographic shop which is able to provide a more specialist machine.

While we have considered that part of the business model that is related to the value chain from designer through to customer, we must also consider the business model that will support and develop tool-less manufacturing machines. It may be argued that ownership of the technology used in these machines, along with the materials they use, may become strategically important and be compared with those economic regions that currently own large silicon chip manufacturing facilities.

It would seem most likely that there will be two or three global manufacturers of tool-less manufacturing machines satisfying different market sectors – mechanical, electronic, and so on. Each of these manufacturers will either develop and supply materials for its own machines or work in partnership with specialist third party suppliers which will develop a range of materials. These materials will almost certainly be produced in large volumes at specialist manufacturing facilities and then transported via some network to individual tool-less manufacturing machines. Clearly in such a scenario transportation will be required. However, since we will not be transporting individual packaged products, but rather generic bulk feed stock, we can reasonably expect that transportation costs and packaging cost and associated environmental impacts will be reduced. It is also likely that the tool-less manufacturing machines will be built in centralized global facilities, much as ink jet printers are today. However, there are some interesting developments taking place in the open source community that may influence this.

OPEN SOURCE OR DEMOCRATIC MANUFACTURING

There is now a strong open source software community that works on a not-for-profit basis that provides a good quality alternative set of operating systems and software applications to the standard commercially provided products. These so called open source products are freely available for all to use and modify. While they may not be as 'polished' as their commercial competitors, open source products can often match the functionality of the commercial products. This can make them extremely attractive to small startup companies and individuals who are unable to support the high costs of commercial software.

A similar movement is taking place in tool-less manufacture. Various groups are developing tool-less manufacturing machines, selling kits to make the machines and freely distributing the designs of the machines to actively encourage small companies and individuals to exploit tool-less manufacture. This provides a very low cost entry into tool-less manufacture and for the first time allows many individuals access to advanced manufacturing techniques that they would never have been able to access before. For this reason the technology is sometimes referred to as 'democratic manufacture'. One feature of particular note is that some of these open source machines are designed in such a way as to allow them to replicate themselves. So once a company has one machine, if the business is successful, it can use the first machine to build a second clone, or indeed an improved next generation machine, to satisfy the needs of the developing business. In this way the company has a relatively smooth route to expansion.

While tool-less manufacture may be democratic and offer cost saving and environmental benefits, there must, however, be concerns over intellectual property rights (IPR) protection and with the development of monopolies in the supply of feed stock material. Once a design is published it can easily be copied. This is readily seen today, where the music industry loses huge revenues from file sharing and inappropriate copying. Music stores like iTunes can protect the copyright owner by ensuring sufficient computer security on its site and on the associated smart phone or tablet. This approach tries to ensure that only the legitimate owner of the music can play it and that no illegitimate copies can be made. If a similar model is used for distributed tool-less manufacture, then we can expect copyright holders to receive the same degree of protection afforded by producers of Apps. For example, a design will include aspects of 'know how' that the designer will use to create a successful design. However, these designs will be placed into some sort of machine by the App store before being sent to a customer's tool-less manufacturing machine and so could not be accessed by the customer or competitor. Additionally, the App store would limit the number of copies of the product that could be produced on the customer's machine. Thus, it should be possible to protect the IPR of the designer.

A tight technological coupling between the tool-less manufacturing machine and the feed stock material (3D powders) could easily lead to the feed stock material manufacturers exercising undue control over the price of feed stock material. Again, there is a parallel with ink jet technology, where ink jet printers are sold at very reasonable prices while the ink cartridge refills are able to command much higher profit margins. It is interesting to note that third party or recycled ink jet cartridges are increasingly available, despite software and technological devices designed to force customers only to use cartridges produced by Original Equipment Manufacturers. It would be expected therefore that similar markets will develop for tool-less manufacture machine powder cartridges and so a vibrant market will eventually limit the control of the feed stock manufacturers.

CONCLUSION

This chapter has explored the implications of innovations in digital fabrication for the future of manufacturing. The focus has been on developments in 3D printing. At the moment, 3D printing technology represents a novel and potentially disruptive technology. The technology is still under development, but has the potential to play an important role in the emergence of a third industrial revolution. In this new industrial revolution 3D printing is initially facilitating the prototyping process, but has the potential to transform both the design and manufacture of products. The analysis has highlighted both the advantages and limitations of 3D printing. There are problems related to speed, quality and cost, but advantages based on the ability to produce objects that cannot be fabricated using more subtractive approaches. It may be that the third industrial revolution will blend the advantages of additive approaches to manufacturing with subtractive processes. It is important that academics continue to explore the implications of new technological developments on manufacturing processes. Incremental innovations have over time the potential to radically alter manufacturing, but the emergence of digital fabrication has the potential to disrupt existing approaches by providing manufacturers and consumers with a new approach to product development and production.

As a final thought regarding the paradigm-shifting nature of digital fabrication technology, it is interesting to consider 'Star Trek' technology. Most of the 'Star Trek' technology is now feasible, for example smart phones and communicators. Clearly, however, the warp drive and transporter technology are not, and are never likely to be, feasible. But maybe it will be possible to teleport products to customers in the future – after all, that is all a tool-less manufacturing machine is!

BIBLIOGRAPHY

Bluestone, B. and Harrison, B. (1982), *The Deindustrialisation of America*, Basic Books: New York.

Bryson, J.R. and Henry, N.D. (2001), 'The Global Production System: From Fordism to Post Fordism', in Daniels, P.W., Bradshaw, M., Shaw, D. and Sidaway, J (eds), *Human Geography: Issues for the 21st Century*, Prentice Hall: London, Chapter 12: 342–373.

Bryson, J.R. and Rusten, G. (2011), *Design Economies and the Changing World Economy: Innovation, Production and Competitiveness*, Routledge: London.

Bryson, J.R., Clark, J. and Mulhall, R. (2013), *The Competitiveness and Evolving Geography of British Manufacturing: Where is Manufacturing Tied Locally and How Might This Change?*, Foresight, Government Office for Science, BIS: London.

Bryson, J.R., Daniels, P.W. and Warf, B. (2004), *Service Worlds: People, Organisations, Technologies*, Routledge: London.

Chambers, J.D. (1961), *The Workshop of the World*, Oxford University Press: Oxford.

Chandler, A. (ed.) (1964), *Giant Enterprise: Ford, General Motors, and the Automobile Industry*, Harcourt Brace and World: New York.

Court, W.H.B. (1965), *A Concise Economic History of Britain*, Cambridge University Press: Cambridge.

Darley, G. (2003), *Factory*, Reaktion Books: London.

Deane, P. (1969), *The First Industrial Revolution*, Cambridge University Press: Cambridge.

Dicken, P. (2003), *Global Shift: Transforming the World Economy*, Sage: London.

Kumar, S. and Kruth, J.P. (2010), 'Composites by Rapid Prototyping Technology', *Materials and Design*, 31: 850–856.

Rifkin, J. (2011), *The Third Industrial Revolution: How Lateral Power is Transforming Energy, the Economy, and the World*, Palgrave: New York.

TSB (2012), *Shaping our National Competency in Additive Manufacturing*, a technology innovation needs analysis conducted by the Additive Manufacturing Special Interest Group for the Technology Strategy Board, Technology Strategy Board: London.

Warman, M. (2012), 'The Name's Printing: 3D Printing', *The Daily Telegraph*, 1 December.

Wohlers, T. (2001), *Wohlers Report: Rapid Prototyping & Tooling State of the Industry Annual Worldwide Progress Report*, Wohlers Associates: Fort Collins, Colorado.

Yan, X. and Gu, P. (1996), 'A Review of Rapid Prototyping Technologies and Systems', *Computer-Aided Design*, 28, 4: 307–331.

8. Engineering and manufacturing: concurrent maturation of xRL

Ben Wang, William C. Kessler and Andrew Dugenske

WHY MANUFACTURING MATTERS

There is abundant evidence that manufacturing is a critical sector of a nation's economy for building wealth. A number of studies from governments, corporations, policy institutes and academic institutions support this assertion and indicate that "making things" is an important way to improve a society's standard of living (Duesterberg, 2013). The United States recently highlighted four key benefits of a robust manufacturing sector (Report to the President, 2012):

1. 70 percent of exports consist of manufactured goods;
2. one manufacturing job produces an additional six related supply chain jobs and ten jobs in the general economy;
3. 66 percent of scientists and engineers are employed in manufacturing; and
4. more than 50 percent of national research and development expenditures are made in manufacturing.

A robust manufacturing sector is also critical to a country's national security because modern militaries rely on advanced weapon systems and communication platforms to maintain superiority. In effect, high-tech manufacturing capacities in terms of technical knowledge and the ability and resources to manufacture high-tech weaponry are essential for national defense. If a nation relies on a military supply chain that extends significantly beyond its border, the ability of the country to obtain critical components during a time of conflict becomes increasingly risky. Often, the innovation required to conceptualize new products takes place in an industrialized country such as the US, and the manufacture of the products is then sent to another country that has lower labor rates in an effort to reduce the total cost of the product. When military supply chains follow this model that dominates civilian manufacturing, there can be national security and national competitiveness consequences.

In this chapter, we propose an engineering-oriented model aimed at increasing the domestic production of manufactured goods using the guiding principle: "Discover here – accelerate Translation – Build here" (DTB). In order to accomplish this goal, we propose a new Accelerated Readiness Level (xRL) model that accelerates the speed at which new concepts become products. This xRL model increases profitability by mitigating the time it takes to move an innovation to market. It also emphasizes the role of design and customization in advanced and high-tech manufactured goods that requires higher skills and will likely lead to more high-tech jobs.

THE DTB CONCEPTUAL MODEL

This chapter presents and describes our DTB model and how it methodically reduces risk, time and the cost of accelerating the translation of research to high-value, competitive products in the global marketplace. The DTB model of "Discover here – accelerate Translation – Build here" focuses on the concurrent maturation of xRLs – TRL (technology readiness level), MRL (manufacturing readiness level), BcRL (business case readiness level) and ERL (ecosystem readiness level). This chapter: 1) describes the critical and enabling role that universities have in the translation process in the US case; and 2) explains how the DTB approach meets two major challenges for industrialized countries, including the United States.

At the national level, there are two significant challenges to moving innovation into production. The first is a question of effectiveness and efficiency of innovation. Many analysts have pointed to the need to increase the speed of innovation and specifically the time it takes to turn research results into innovative products. Today's research translation takes too long, costs too much and the results are too uncertain. The second challenge is a question of geography, that is, location of manufacturing. Policymakers are increasingly convinced that what is invented in a place should be produced in that place. The DTB model accepts the premise that innovating and producing in place is a vital component of economic and national security policy for industrialized countries.

The DTB model for achieving these goals requires that the institutional actors, including research and development (R&D) centers like universities, stretch well beyond a successful research role to establish a capacity for focusing interdisciplinary research and provide translational leadership for seamless and capable DTB. In the US, research-intensive universities like the Georgia Institute of Technology, the Massachusetts Institute of Technology, the University of Michigan and Stanford have moved towards addressing the challenge of translational research capacity with the establishment of refocused university-level interdisciplinary institutes. These research centers have refocused to become more outward-facing, forming industry and government partnerships, and focusing the translation of interdisciplinary research to achieve real-world economic benefits and societal impact.

The DTB model we describe in this chapter requires capturing synergies of manufacturing-related expertise, aligning the regional economic ecosystem with the time and geographic challenges underscored in the DTB model, and establishing and enabling industry–government partnerships to accelerate the translation of manufacturing-related research to innovative products. This approach also requires the design and deployment of an Operating System to effectively institutionalize this future, or "to-be," manufacturing innovation process.

Figure 8.1 identifies both the "as-is" and "to-be" characteristics for the DTB innovation, value-creation chain. Recent policy reports, including the White House Advanced Manufacturing Partnership Steering Committee's final report, have documented that although the "discovery of knowledge" and "product development" phases of the innovation chain have been, to a great extent, successful in the US innovation system, the "translation" phase has been characterized by unaddressed challenges known as the "valley of death" or the "missing middle" (Report to the

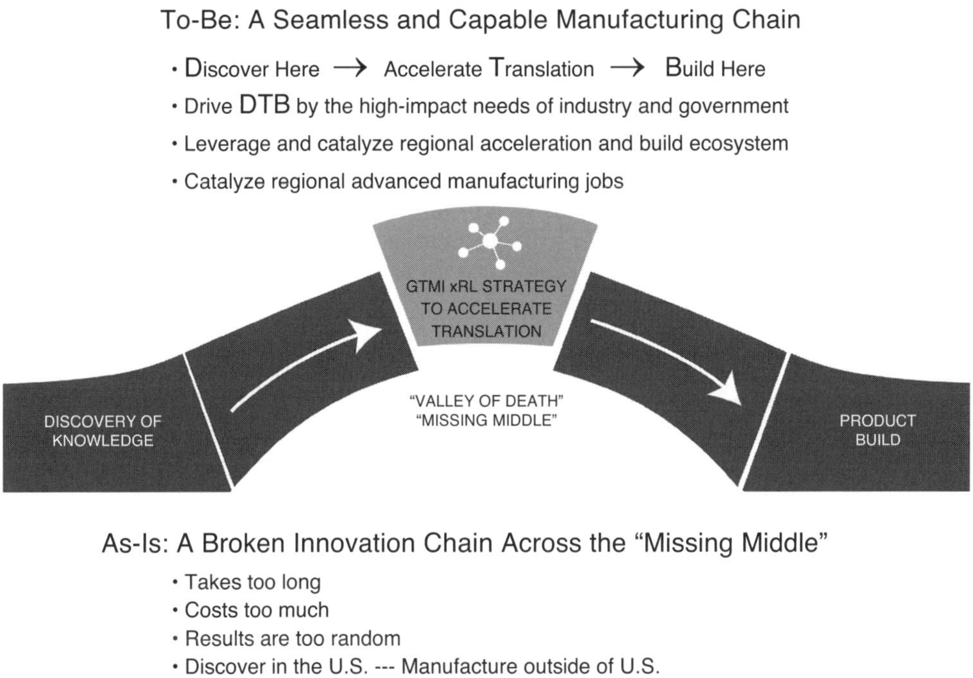

Figure 8.1 *Framing the DTB grand challenge: Discover Here – Accelerate Translation – Build Here*

President, 2012). The DTB model of concurrent maturation of xRL includes the entire innovation chain but emphasizes the problematic "to-be" translational capabilities and focuses on an integral view of technology, processes, methods, tools, infrastructure, policies and skills.

To address these gaps and better integrate the innovation system with an emphasis on commercialization and production, the DTB model carefully articles each stage in the innovation to production process: 1) discovery, 2) product development, and 3) translation:

- *Discovery*: This is predominantly the area of university-based researchers. Universities treat manufacturing-related research in a siloed manner, differentiated by disciplines (e.g., materials, processes, design, modeling and simulation, quality, supply chains, logistics, economics, finance, business, public policy and economic development). Disciplinary incentives are generally structured towards individual accomplishments and prioritize intellectual property strategies that are not conducive to collaborations between universities or across firms and industries. A seamless innovation model requires an interdisciplinary and holistic approach to translating fundamental discovery to technology development. To facilitate this, universities must develop a value proposition for engaging the "missing middle" as a viable "knowledge creation" workspace, and some university policies and practices will require modification.

- *Product development*: This is predominantly the area of emphasis for industry. In many industrialized countries, including the US, vertically integrated firm structures began to erode in the 1980s, leading to vertical disintegration and distributed supply chains. This has had an effect both on innovation systems and production networks. Well-known industrial research centers such as "Bell Labs" and other similar organizations have been disbanded due to corporate restructuring and budget cuts. Newer firms with a production-oriented vision are increasingly populating the ranks of those willing to pursue the basic knowledge required for making products. As a consequence, it is these firms driving translational research collaborations. Traditional US-based companies tend to reach into the "missing middle" for information but rarely reach all the way back, strategically and systematically, to collaborate with basic researchers or collaborate in the Discovery phase. Again, it is new firms that may reach back into the Discovery phase or may rely on "entrepreneurial researchers" for new products to bridge the gap between discovery and product development. Although industry may occasionally expend effort to address the "missing middle," a seamless and capable innovation chain does not universally exist.
- *Translation*: In developing the xRL operating framework, our innovation chain analysis indicated a number of missing capabilities in the current system. These include: 1) the absence of an integrated and concurrent technique to eliminate technological, manufacturing and business risks across the "missing middle"; 2) no clear method to accelerate maturation in a "realistic product use and manufacturing" environment; 3) the absence of "an Operating System" designed to accelerate research translation; and 4) no model for collaboration among relevant stakeholders. What our research found was that integrated and concurrent maturation across the "missing middle" requires measurement of integrated readiness of technology, both the manufacturing and the business cases, and the clear identification of gaps in readiness and identified actions to close the gaps. Our research also identified the need for a "realistic" environment, also known as a "relevant" or "representative" environment, for accelerating translation that can replicate and assess both the product operational condition (for instance, altitude, humidity, pressures, temperatures, etc.) and the manufacturing environment (proper validation in lab, scale up, prototype, pilot production, low-rate production and full-rate production) to be incorporated at the proper maturity readiness level. In other words, the availability of a scientific or technical prototype development, testing and evaluation environment is critical. Similarly, a business environment capable of absorbing innovations is essential. In other words, an awareness and understanding of the regional manufacturing innovation ecosystem is required to enable acceleration across the "missing middle" through the alignment (space and time) of the institutional infrastructure in the regional economy: technical colleges, workforce development intermediaries, capital sources and regional governance agencies.

Unfortunately, rebuilding such a successful corporate model as Bell Labs is likely to be far too expensive for most firms. However, replicating the techniques used by Bell Labs to commercialize research rapidly and effectively and in place could be recreated on a national or regional level through partnerships and consortium approaches like the DTB model shaped by the xRL operating framework. We have identified the need for four critical stakeholders to be involved to move the model forward: 1) universities possessing broad, interdisciplinary manufacturing-related research capabilities; 2) industrial and/or government entities that shape demand for new and high-impact products; 3) government and/or industrial entities that catalyze the translational process with sustained capital investments; and 4) a non-profit entity that can manage the overall process, act as a catalyst and unbiased broker, add value to the accelerated translation and assure that the appropriate space and environment are provided for the collaborators.

THE xRL APPROACH: OPERATIONALIZING THE DTB MODEL IN AN ENGINEERING RESEARCH CONTEXT

Our current research involves developing this framework to accelerate the commercialization of research conducted at the university level. The framework we have described in this chapter, xRL, focuses on establishing integrated technical, manufacturing, business and ecosystem maturity measures. The technical and manufacturing measures are based upon the TRL and MRL frameworks that have been developed primarily by the National Aeronautics and Space Administration (NASA) (Sadin, Povinelli and Rosen, 1988) and the Department of Defense (DoD) (Manufacturing Readiness, 2012; Rules and Regulations, 2011), respectively and are mature. The business and ecosystem readiness-level frameworks are in their preliminary development stages. In order to understand our proposed xRL model, it is important to understand the underlying characteristics of the TRL and MRL systems and the challenge of expanding and integrating these system models.

TRL

TRL is a method of classifying the maturity of a basic technology for use in a product or process. TRL defines a set of readiness levels from 1 to 9 that provide an efficient way of communicating a technology's current state along its development spectrum. A rating of 1 indicates that the technology is in a nascent state, just beyond the discovery phase. A rating of 9 indicates the technology is being used successfully in operations.

TRL was developed by NASA in the 1980s and is used by several large agencies throughout the world (Report to the President, 2012; Strategic Readiness, 2012; Department of Homeland, 2009; Technology Readiness, 2011). Fortunately, most implementations make use of the 1 to 9 rating scale, but the meaning of the ratings can vary based upon the primary mission of the user. NASA and the DoD are two large users of TRL. Table 8.1 shows the readiness levels used by NASA.

xRL makes use of TRL as one component of the overall readiness of a technology. Since TRL is well established and mature, xRL makes use of TRL without change. Currently, xRL makes use of the DoD readiness level definitions, but xRL may expand to use additional readiness level definitions to address various intents.

Table 8.1 NASA technology readiness levels

Level	Description
9	Actual system proven through successful mission operations
8	Actual system completed and qualified through test and demonstration
7	System prototype demonstration in an operational environment
6	System/subsystem model or prototype demonstration in a relevant environment
5	Component and/or breadboard validation in a relevant environment
4	Component and/or breadboard validation in laboratory environment
3	Analytical and experimental critical function and/or characteristic proof of concept
2	Technology concept and/or application formulated
1	Basic principles observed and reported

Source: Rules and regulations 2011.

MRL

MRL is another measure of overall readiness of a basic technology. As the name implies, it provides a way to communicate the readiness of a technology for use in a manufactured product or in a manufacturing process. It was initially developed by the DoD to improve the quality of procuring systems by the government. Specifics about MRL can be found in the "Manufacturing Readiness Level Deskbook" distributed by the DoD (Manufacturing Readiness, 2012).

Whereas TRL defines a single readiness level of a technology, MRL makes use of threads and sub-threads to provide a more extensive view of the readiness. Threads address nine manufacturing risk areas and consist of: Technology and the Industrial Base; Design; Cost and Funding; Materials; Process Capability and Control; Quality Management; Workforce; Facilities; and Management. Each thread is further divided into sub-threads that provide additional robustness and completeness to the analysis. MRL also defines a set of guiding/exit questions that are used to determine the readiness level of each sub-thread. The guiding questions are of great help to the users of MRL because they provide a systematic method of determining current state and provide semantics about the readiness definitions. xRL makes use of MRL without change to determine the manufacturing readiness of a technology within the xRL framework due to MRL's active user community and expanding use.

MRL provides a more robust view of readiness than TRL because MRL makes use of 22 topic areas versus a single TRL classification. This robustness requires additional analysis and is more challenging to communicate. Since xRL combines multiple readiness frameworks into an encompassing analysis, the challenges associated with MRL will be exacerbated in xRL's larger context. Therefore we propose that xRL be developed such that it can be applied without overtaxing resources and the results can be presented in a way that can be easily interpreted by the intended audience.

BcRL

No matter how innovative a product or process may be, if a firm cannot see the financial benefit then the new technology will likely be filed away. While engineers and researchers work within the realms of technology and manufacturing readiness – that is,TRL and MRL – corporate decision makers work within the realms of profits and earnings. To bridge the gap between innovation development and technology development, it is critical to incorporate BcRL and ERL into the process. Although very effective tools, TRL and MRL are not sufficient to guarantee successful and rapid commercialization of a new technology. Neither has gained such broad-based acceptance as Six Sigma – a doctrine that has gained tremendous popularity worldwide in manufacturing, services and the public sector.

Six Sigma is popular because it has a clear focus on achieving measurable and quantifiable financial returns by determining "product cost" – the metric of choice for firm decision makers. Executives can clearly see that when the quality level rises from Four Sigma to Five Sigma, the defects per million units produced drop from 6210 to 233. When the quality level reaches Six Sigma, another 230 defects per million units are eliminated. Decision makers prefer bottom line numbers, and Six Sigma readily offers these data where TRL and MRL do not. Although it is important from the new technology perspective to know maturity and manufacturing readiness, these metrics simply do not provide decision makers with the bottom line numbers they need to justify adoption of the technology.

This is where BcRL comes into play. As a companion measure to TRL and MRL, BcRL captures the "financial" or "business" reasoning for launching a new technology or manufacturing project. The intent of BcRL is methodically to build a business case and "market pull" as the technology matures, to shorten the time to market. It equips an integrated product and process design (IPPD) team with a disciplined maturation and evaluation process to bring the technology to market. Unfortunately, most technology projects ignore the importance of their business case until late in the development process. As a result, there is not enough market pull to justify the new technology insertion because associated benefits and risks have not been studied and articulated.

For this reason, it is critical to develop technology, manufacturing and business case readiness simultaneously. The IPPD effort was a significant step in the right direction. However, maturing technology and manufacturing concurrently without building a business case does not guarantee successful implementation of the new technology or manufacturing project. In other words, without the prospect of a solid financial return it is difficult to push a new technology into the marketplace, regardless of its level of innovation. A proper market pull – that is, sufficient business benefits – will ensure a smooth transition and insertion.

To build the business case, BcRL determines technology readiness for market transition (technology push), the targeted unmet needs (market pull), the product insertion timeline (technology road map), a market capture strategy and the financial benefits to the company. By incorporating BcRL into the process, a compelling business case is already in place when the technology reaches maturation.

BcRL is compatible with TRL and MRL because it is organized at nine readiness levels, as shown in Figure 8.2. The critical phases are BcRL 3–7, where the technology

Business Case Readiness Level (BcRL)		
PHASE	BcRL	READINESS LEVEL DEFINITIONS
Phase 3: Reaching the "Tipping point" and on to Full Scale Market Insertion	9	*Full Rate Production into National Markets – Future Product Improvements Planned*
	8	*Full Rate Production into Local Markets – Confirmation of Financial Metrics Estimate*
	7	*Product Insertion into one Target Market – Positive Market Focus Group Response*
Phase 2: Bridging the "Missing Middle"	6	*Market Ready Research Prototype Vetted to Outside Entity and Key Customers*
	5	*Financial Issues Defined – Return on Investment Required, Margin, Funding Source (Internal, External, or Both)*
	4	*Research Concept/Target Markets Presented to Industrial Partners – Fit to Strategic Plan Goals*
Phase 1: Technology/ Manufacturing for Market Readiness	3	*Research Concept Vetted to Outside Entity (Advanced Technology Development Center, Incubator Board, etc.) for Review*
	2	*University Team Review and Validation of Potential Research Concept Market Insertion*
	1	*Research Concept Proven in Laboratory – Principal Investigator Defines Usage of Potential Market Value*

Figure 8.2 Nine levels of business case readiness maturity

development reaches a tipping point, and firm executives are convinced of the potential business value of the new technology, allowing it to move forward.

BcRL is meant to evaluate technology starting at a TRL of 2 or 3 and ending at the tipping point, at TRL 6 or 7. This tipping point corresponds to BcRL 6 or 7, where the technical concept initially developed in the lab is transitioned to initial market insertion. A tipping point may be characterized by a commercial success during test market evaluation. The overarching objective of BcRL is to transition a technology from an academic or industry lab to market in a timely fashion so that product insertion immediately results in significant market success for the company. An additional benefit is when the owner of the technology receives a revenue stream through continuing royalty payments from the successful translation of its intellectual property. Ultimately, BcRL is a win–win solution for all partners. Combined with TRL and MRL, the triad addresses the first challenge of ensuring rapid innovation and seamless transition from lab to pilot to production.

ERL

The second challenge of "discover here, build here," however, requires a different set of capabilities. To ensure that a new technology or product can survive once it has reached maturity and entered the marketplace a supportive production environment or manufacturing ecosystem is required. Manufacturing ecosystems, also called industrial commons by Pisano and Shih (2012), provide a cluster of localized, interdependent businesses that offer design, production, distribution, workforce, infrastructure and investment capabilities to help a business thrive.

Take for example two recent companies that had the potential to have a significant impact on society but ultimately did not survive. These two firms were Solyndra and A123. Each company had received hundreds of millions of dollars in US government assistance, garnered much press and raised the hopes of many US-based manufacturers in cognate fields. Both had highly innovative technologies in the clean energy arena, rechargeable batteries for hybrid/electric cars (A123) and crystalline thin files (Solyndra).

Analysts speculate that these firms failed because the infrastructure to support these technologies had declined and departed the United States years ago. Although the United States was once a leader in battery design and production, for example, that industry largely followed the move of electronics manufacturing to Asia in the previous few decades, forming an industrial ecosystem in a new region. For A123, the "hollowed out" ecosystem for battery manufacturing that this migration created, ultimately led to the company's inability to find the localized group of suppliers and collaborators necessary to design a new product using a new technology for a new market.

Solyndra's story took a very similar route. The design and manufacturing skills and know-how needed to process ultrapure crystalline into wafers and apply thin films of silicon onto large glass sheets were lost decades ago. This was because the United States outsourced the seemingly mundane manufacturing of semiconductors, power supplies, controllers and similar components to lower-labor-cost economies. The center of knowledge and skill sets, or the locus of R&D and manufacturing, had moved to lower-cost locations many years before (Pisano and Shih, 2012).

This migration of ideas, skills and knowledge was a firm strategy employed by many firms as a simple solution to improve the bottom line for the short term primarily by reducing the labor cost. However, the net result of these business decisions by many US corporations over the past few decades has been, at the national level, lost competencies, lost jobs, and lost capacities for the future rounds of innovation in the United States. In other words, the "graze land" for battery-making, thin film-fabrication and many more foundational technologies has eroded in the United States. For A123 and Solyndra, there was no support system to provide assistance and nurturing. Developed in isolation, without a manufacturing ecosystem, they did not have a chance to grow and prosper.

It is important to note that, unlike other readiness-level tools, ERL does not remain constant once it reaches a certain level. A manufacturing ecosystem can ebb and flow as can any ecosystem. If certain pillars of sustainability begin to deteriorate, so does the ecosystem. That is why constant monitoring and upkeep is important in maintaining an existing ecosystem.

xRL: THE COMPILATION OF TRL, MRL, BCRL AND ERL

Some previous frameworks have been used to address specific areas of interest (i.e., technology or manufacturing); xRL addresses the meta-view of product readiness by providing a comprehensive view of product readiness status. It makes use of well-established sub-frameworks such as TRL and MRL and extends the overall readiness-level analysis by addressing BcRL and ERL readiness levels. Since BcRL and ERL are new concepts, these frameworks are being developed simultaneously with the overall xRL framework.

Our preliminary field tests of the xRL framework began with identifying the technology of interest and documenting exemplars (product emulators) chosen as tangible implementations of that generic technology. The exemplars were analyzed using the techniques defined in the various sub-frameworks. Each sub-framework defines a set of readiness levels that are used as ratings of maturity. Typically the levels range from 1 to 9 or 10 and are clearly defined so users of the frameworks know what is needed to achieve a particular level. Since the sub-frameworks address extensive areas of interest, the frameworks are divided into threads and sub-threads for a more granular analysis. Each of the threads and sub-threads is assigned a current readiness level so a more extensive view can be achieved. The current readiness level assigned to the threads and sub-threads is deduced from a set of exit or guiding questions.

Each question is associated with a readiness level and is answered true or false. The questions are applied from lowest to highest readiness level. If most of the questions relating to the readiness level are answered affirmatively, then the readiness level has been achieved. If most of the questions are answered negatively, then the readiness level has not been achieved. A comprehensive view of the readiness levels for the exemplars is determined by analyzing all the thread and sub-thread readiness levels of the frameworks. Currently, xRL makes use of radar charts to present the readiness findings. The xRL development team feels that determining more effective ways of presenting the xRL data is an excellent area for future development.

Once the overall readiness profile has been ascertained, the method to accelerate the technology used in products and processes is undertaken. Typically an inverse Pareto analysis of the readiness levels is used to determine the areas in most need of improvement. This is very similar to how a standard Pareto analysis is used to communicate, prioritize and address problems in a manufacturing facility. Identifying and documenting the comprehensive readiness view of a technology is one of the major strengths of xRL, for without knowing the complete deployment picture, knowledge of how best to deploy resources to accelerate technology adoption is hampered. Defining the specific actions required to accelerate technology adoption is also a major strength of xRL. Thus xRL provides a comprehensive view of maturity and the actions required to accelerate the change in maturity. Figure 8.3 contains a flowchart of the current xRL process.

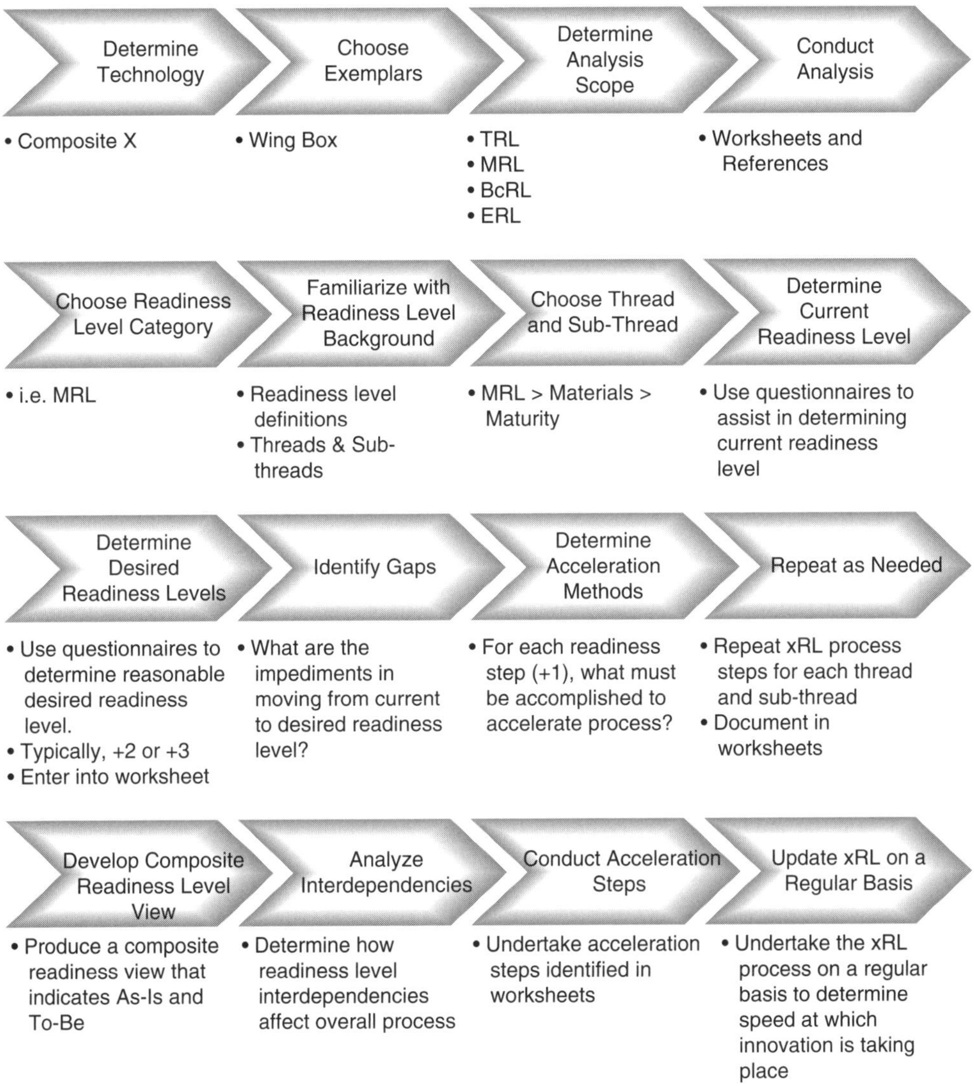

Figure 8.3 The xRL process

MOVING THE DTB MODEL FORWARD: THE FUTURE OF THE xRL FRAMEWORK

As the two firm cases demonstrate, it is critical for economic sustainability that innovation systems are not seen as distinct or divorced from a more nuanced, complex and dynamic manufacturing ecosystem. Although innovations have moved away from industrialized countries to emerging economies with lower costs, taking with them skills, ideas and talents, that trend is beginning to shift. As the recent 2008 recession has highlighted, industrialized economies have allowed the manufacturing base to dwindle, creating an economy based on services. The recession has reminded analysts and policymakers that regions that do not balance their industry mix do not stay in the lead and are extremely vulnerable to economic downturns.

Therefore, not only does the United States need to move its technologies to market rapidly, but it also needs to make sure that existing, localized production networks can absorb those technologies. Already, engineers have made crucial first steps by combining TRL and MRL. Now it is time to expand the model and incorporate the business and regional development elements into the framework used to evaluate, understand and invest in new technologies. Incorporating BcRL and ERL into the innovation chain can help move the process and the country in the right direction. The Discover here – accelerate Translation – Build here model begins to realign analytical processes to meet these persistent challenges and policy priorities.

BIBLIOGRAPHY

Department of Homeland Security Science and Technology Readiness Level Calculator (2009), "Final report and user's manual," Homeland Security Institute (Ver. 1.1), Arlington, VA.

Duesterberg, T.J. (2013), "The manufacturing resurgence: what it could mean for the U.S. economy, a forecast for 2025," The Aspen Institute, Washington, DC.

Dugenske, A., Kessler, W. and Wang, B. (2013), "GTMI xRL Workshop," unpublished workshop, Georgia Institute of Technology, Atlanta, GA.

"Global Manufacturing Competitiveness Index" (2013), McKinsey and Council on Competitiveness.

"Manufacturing Readiness Level Deskbook" (2012), OSD Manufacturing Technology Program; The Joint Service Industry MRL Working Group, vol. 2.2, available at http://www.dodmrl.com/MRL_Deskbook_V2_21.pdf.

Pisano, G.P. and Shih, W.S. (2012), *Producing Prosperity: Why America Needs a Manufacturing Renaissance*. Boston, MA: Harvard Business Review Press.

"Report to the President on capturing domestic competitive advantage in advanced manufacturing" (2012), The President's Council of Advisors on Science and Technology, Washington, DC.

"Rules and regulations" (2011), *Federal Register*, vol. 76 (223), available at http://www.gpo.gov/fdsys/pkg/FR-2011-11-18/pdf/2011-29894.pdf.

Sadin, S.R., Povinelli, F.P. and Rosen, R. (1988), "The NASA technology push towards future space mission systems," presented at the IAF, 39th International Astronautical Congress, Bangalore, India, available at http://ntrs.nasa.gov/search.jsp?R=19890030268&hterms=nasa+technology+push+towards+future+space+mission+systems&qs=Ntx%3Dmode%2520matchall%26t%3D0%26Ntk%3DTitle%26Ns%3DPublication-Date.

"Strategic Readiness Level – The ESA Science Technology Development Route" (2012), European Space Agency, Advanced Studies and Technology Preparation Division, available at http://sci.esa.int/jump.cfm?oid=37710.

"Technology Readiness Level" (2011), Public Works and Government Services Canada, Office of Small and Medium Enterprises, available at https://buyandsell.gc.ca/initiatives-and-programs/build-in-canada-innovation-program-bcip/program-specifics/technology-readiness-levels.

9. Energy and manufacturing: technology and policy transformations and challenges

Marilyn A. Brown and Gyungwon Kim

1. INTRODUCTION

Energy is both a resource used in the production of goods and services and an industry in its own right, with its own raw materials and finished products. Just as energy inputs in manufacturing are evolving, so are energy industries modernizing and shifting. These transformations are motivated by global competition, resource constraints, and environmental concerns, but they are also enabled and impacted by public policies and by advances in science and technology.

This chapter provides a unique and sweeping overview of energy as a manufactured or processed good, emphasizing past trends and future projections about global and US energy systems. This sets the stage for examining the role of energy as an input to manufacturing other goods and commodities, highlighting differences across industries, states, and countries. For example, opportunities to cut energy costs depend on the industrial process being targeted, the products being manufactured, the size of the manufacturing facility, and existing public incentives and regulations.

In addition to describing the role of energy in US manufacturing, we also summarize energy and manufacturing in six other countries: the United Kingdom, Denmark, the Netherlands, Sweden, India, and China, focusing in particular on their industrial energy efficiency policies. The strong policies implemented in these countries highlight the transformational potential of public/private partnerships, subsidies, and regulations. At the same time, they illustrate the interplay between industrial policies and the availability of energy resources, prices, and demand that uniquely define each country's energy economy.

Attention then turns to two technology case studies, focusing on US manufacturing. Our first case study examines the pulp and paper industry, which exemplifies all three facets of energy and manufacturing: energy is consumed in the production of pulp and paper products, electricity can be co-generated at pulp and paper mills, and biorefining can produce green chemicals and other high-value, petroleum-displacing co-products. The second case study examines the solid-state lighting industry. It illustrates the role of manufacturing in the creation of green energy innovations. The case study describes trends and directions in this key industry of the future. Finally, the chapter ends with a discussion of how energy and manufacturing relate to the economic, security, and environmental issues that shape industrial energy policies.

While we include several international case studies and discuss global trends, most of this chapter addresses energy and manufacturing in the US. With less than 5 percent of the world's population, the US is responsible for nearly one-quarter of its primary energy consumption and greenhouse gas (GHG) emissions. As is true of most industrialized

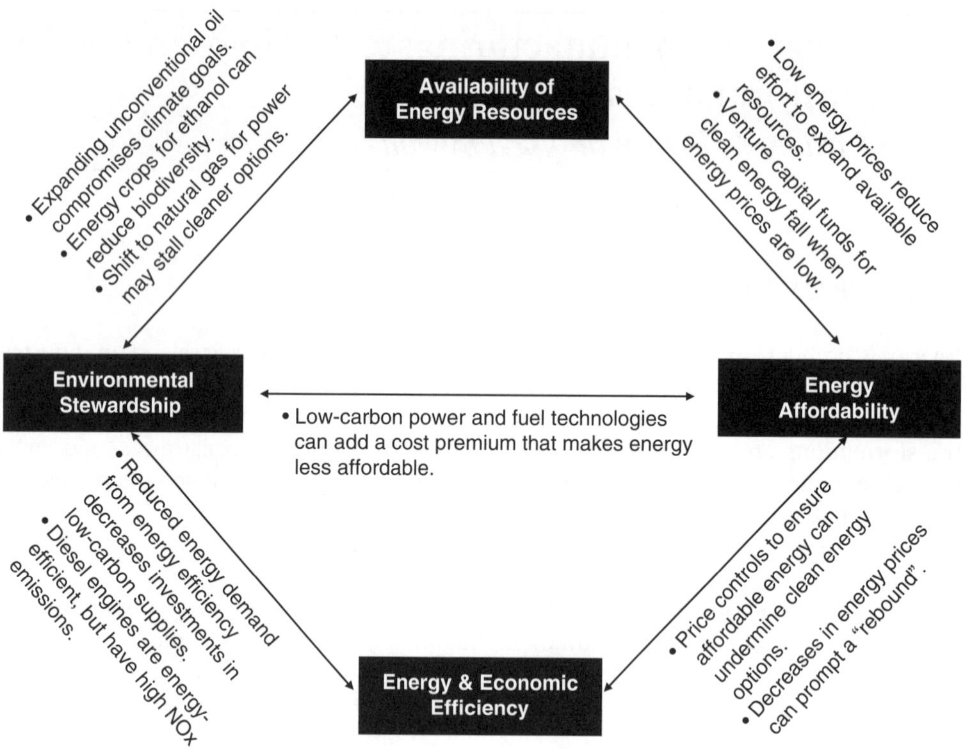

Figure 9.1 Alternative energy goals: conflicts and complementarities

nations, it relies on fossil fuels for both transportation and electricity generation. As a result, its strategic goals include energy supply security, energy affordability, energy efficiency, and environmental stewardship (Figure 9.1). Often these goals are in conflict, for example with the potential to develop domestic energy in the form of off-shore oil, Appalachian coal, and shale gas, competing with concerns for environmental protection and stewardship. Tradeoffs often occur between options that are effective along one dimension, but which adversely impact other aspects of security. Given the abundant illustrations of such conflicts, it would appear that most countries have overtly or inadvertently pursued technologies and policies that involve swapping progress in one energy domain with retreat in another. Without a fully articulated appreciation of these complexities, different strands of energy and climate policy run the risk of competing with each other or, worse, trading off so that the net result is continued emissions, higher prices, and threats to energy security. The US aspires to a low-carbon economy but its current energy system is carbon intensive; it is second only to China in total energy-related CO_2 emissions – at 5,610 million metric tons of CO_2 in 2010. These concerns provide a lens for describing the current link between energy and manufacturing, and for characterizing possible future trends.

Table 9.1 *US energy expenditures in 2012*

US Energy Expenditures	(billion US$)
Industry	$214
Commercial buildings	$173
Residential buildings	$234
Transportation	$755
Total expenditures	$1,424
Gross Domestic Product	$13,593

Source: Energy Information Administration 2014.

2. MANUFACTURING IN ENERGY PRODUCTION

2.1 Size and Evolving Nature of Energy Production and Trade

The value of energy products tends to be a large share of domestic economies across the world. In the US, for example, energy expenditures were 10.4 percent of GDP in 2012, or about $4,560 per capita (Energy Information Administration, 2014). Approximately half of these expenditures are spent on transportation fuels, which are about 95 percent petroleum based (Table 9.1).

While electricity is generated in nearly every country in the world, and many countries also refine their own petroleum products, the production of energy commodities is concentrated in a more limited number of countries that are well endowed with fossil energy resources. As Figure 9.2 shows, the world's known conventional oil reserves (1.2 trillion barrels) are concentrated in volatile regions where nationalized oil companies operate. The 12 largest oil companies are all state owned and they control roughly 80 percent of petroleum reserves. At the top of this list are the Saudi Arabian Oil Company and the National Iranian Oil Company. The distribution of other conventional energy resources, including coal, natural gas, and uranium, is equally consolidated. Eighty percent of the world's conventional natural gas is in 13 countries, and 80 percent of the world's coal is in six countries. Many of these same countries are among the six that control more than 80 percent of the world's uranium resources (Brown and Sovacool, 2011).

Some countries (e.g., Japan and Chile) have essentially no domestic conventional fossil fuels and thus are dependent on foreign supplies. At the same time, countries that are well endowed with one or two energy resources still rely on other energy imports. For example, Saudi Arabia is the largest exporter of crude oil but must import refined gasoline. Russia exports natural gas but must import uranium. The United States is a net exporter of coal but imports oil. This interdependence explains why the challenge of energy security must consider the interactions between countries as much as it considers the resources of individual countries.

The global energy map is changing, with potentially far-reaching consequences for energy markets and trade. It is being redrawn by the resurgence in oil and gas production in the US, by the global spread of unconventional gas production, and by the continued rapid growth of renewables – especially wind and solar technologies.

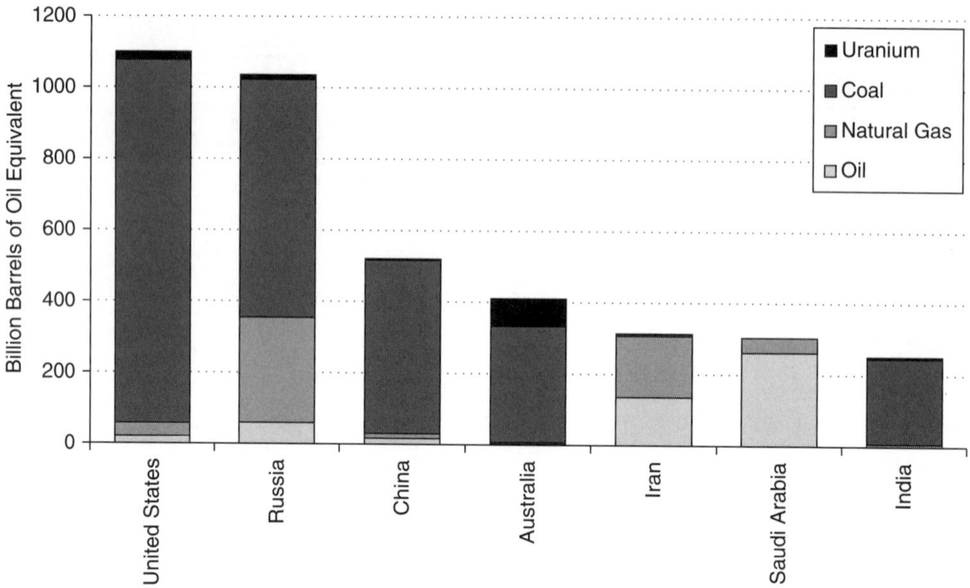

Source: Brown and Sovacool 2011.

Figure 9.2 Global distribution of conventional energy reserves

The resurgence of oil and natural gas production in the US may transform the national debate about energy policies (Figure 9.3). The recent rebound in US oil and gas production, driven by upstream technologies that are unlocking light tight oil and shale gas resources, is spurring economic activity – with less expensive gas and electricity prices giving industry a competitive edge – and steadily changing the role of North America in global energy trade. By around 2020, the *World Energy Outlook* (International Energy Agency (IEA), 2012, p. 23) projects that the US will become the largest global oil producer (overtaking Saudi Arabia). Low-cost natural gas is reducing coal use in the US, freeing up coal for export to Europe (where, in turn, it has displaced higher-priced gas). The development of unconventional gas resources may also transform energy markets worldwide, especially for the developing world. Trillions of dollars in resources lie buried in the backyards of many of the world's poorest citizens; if managed effectively and accountably, the economic development of these countries could surge. On the other hand, if the development of unconventional oil and gas continues, the climate change consequences of fossil fuel dependence may become more salient.

Globally, trade in energy commodities amounted to more than $3 trillion in 2011, including almost two-thirds of the oil produced in the world, and much of it was in natural gas, coal, and uranium (Brown and Sovacool, 2011). US energy prospects are now the envy of the world. The US is on a trajectory to becoming a major oil and gas exporter. At the same time, China, India, the EU, and Japan are all forecast to import 80–100 percent of their oil by the year 2035.

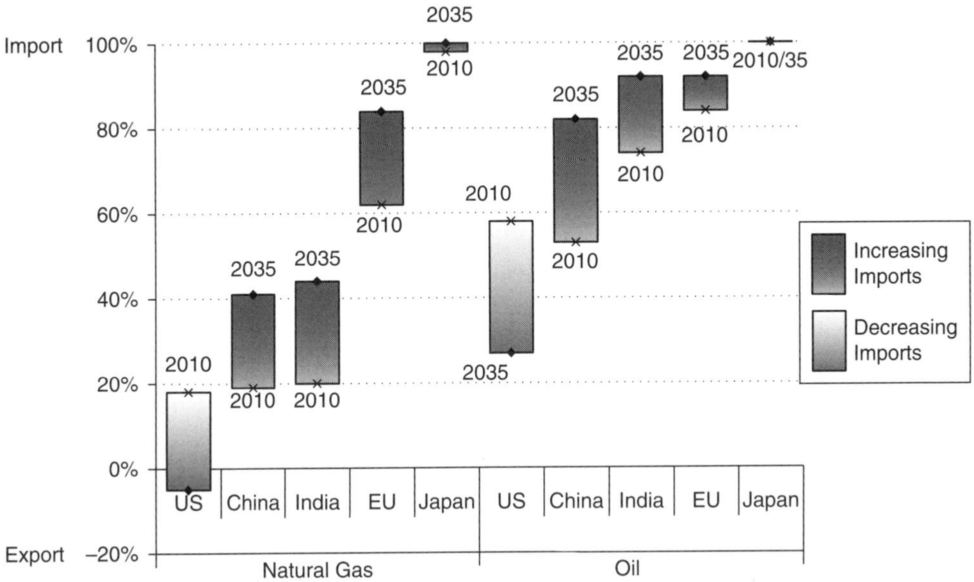

Source: International Energy Agency 2012.

Figure 9.3 Net oil and gas import dependency in selected countries

2.2 Extraction, Processing, Conversion and Transportation of Energy

Much of the cost of energy production takes place in the extraction phase – exploring and drilling for oil and natural gas, mining coal and uranium, growing biomass, building dams, and harvesting wind and solar energy for power production. The activity that is traditionally defined as manufacturing takes place in subsequent stages: refining oil, processing natural gas, cleaning coal, pelletizing and refining biomass, and transporting energy commodities (Table 9.2).

Manufactured products play a role in all of these extraction, processing, conversion, and transportation steps. For instance, the manufacture of drill bits, distillation columns, pipelines, and combustion turbines plays an essential role in the oil and gas industry. The manufacture of solar photovoltaics, wind and water turbine blades as well as transmission and distribution cabling, transformers, and other infrastructure assets is critical to the viability of renewable power and reliable electricity.

3. ENERGY CONSUMPTION IN MANUFACTURING

3.1 An Overview of Industrial Energy Use

Industry is the largest energy-consuming sector in most countries of the world, accounting for 37 percent of primary energy use worldwide (Intergovernmental Panel on

Table 9.2 The role of manufacturing in energy extraction, processing, conversion, and transportation

	Extraction	Processing and conversion	Transportation
Petroleum	Exploring, down-hole and horizontal drilling	Refining to multiple liquid fuels and chemical feedstocks; conversion to power with boilers and turbines	Pipelines, tankers, trucks
Natural gas	Vertical and horizontal drilling; capture from coal mines	Gas cleanup and liquefaction; conversion to power with boilers and turbines; conversion to chemicals	Pipelines, tankers, trucks
Coal	Mining	Coal cleanup; conversion to power with boilers and turbines; gasification; flue gas desulferization	Trains and barges
Uranium	Mining	Uranium enrichment (centrifuge and gaseous diffusion)	Trucking and shipping
Biomass	Agriculture, forestry, algae farming, landfill gas	Pelletizing, biorefining, conversion to electricity with combined heat and power	Pipelines, tankers, trucks
Hydro	Dams, run of river, tidal	Conversion to electricity with turbines	Transmission and distribution of electricity
Wind	Wind farms	Conversion to electricity with turbines	Transmission and distribution of electricity
Solar	Solar photovoltaics, solar thermal collection	DC/AC conversion with transformers	Transmission and distribution of electricity

Climate Change, 2007) and 32 percent (31 quads) of total US energy consumption. A large portion of this energy is used by four energy-intensive industries: petroleum refining, chemicals, pulp and paper, and primary metals (Figure 9.4). Less energy-intensive industries include the manufacture and assembly of automobiles, appliances, electronics, textiles, food, beverages, and other products. Since energy is a smaller portion of their overall costs, these industries have tended to pay less attention to finding ways to cut energy use. However, current evidence shows that this may be changing as the result of an increased focus on reducing carbon footprints (Prindle, 2010).

Large enterprises dominate most energy-intensive industries across the globe, especially in industrialized countries, and the US is no exception. Two-thirds of US industrial energy use is consumed by approximately 8,600 large firms – those with 250 or more employees. Their annual energy bills typically exceed $3 million. At the other extreme, more than 160,000 small firms with fewer than 50 employees account for only 7 percent of total industrial energy use in the US, and their annual energy bills tend to be less than $250,000 (Figure 9.5).

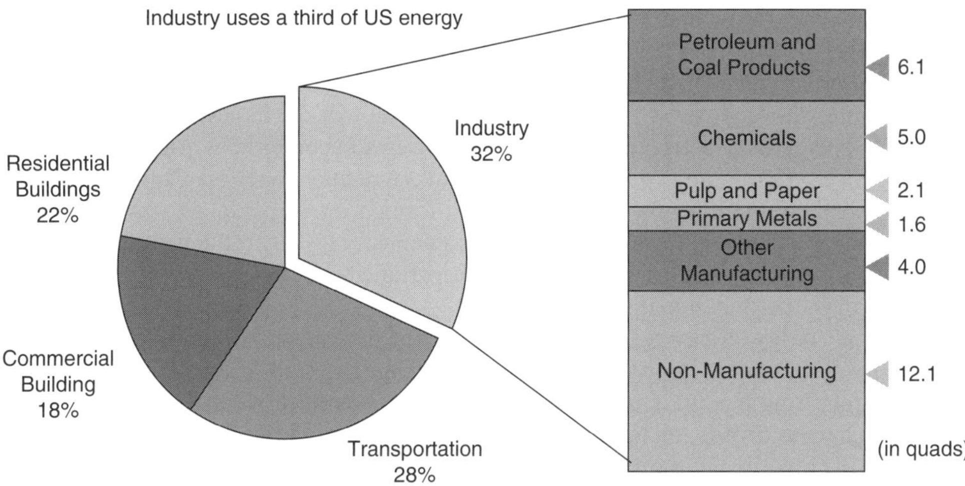

Note: Total US energy use in 2010 was 98.35 quads.

Figure 9.4 US energy consumption in 2010

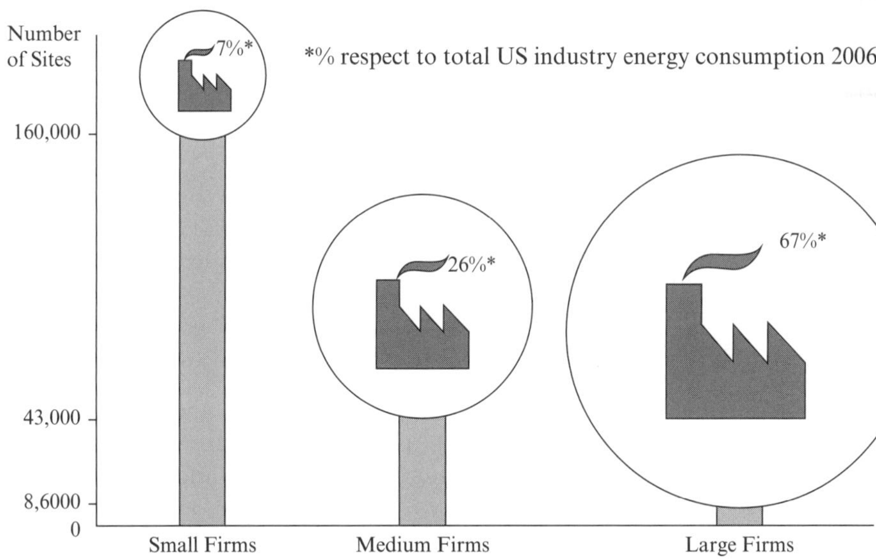

	Small Firms	Medium Firms	Large Firms
Number of Employees	5–49	50–249	250 or more
Annual Energy Consumption	$250,000 or lower	$250,000–$2.9 mill	Over $3 mill
% of Total	Lower than 70 BBtu	From 70 to 499 BBtu	More than 500 BBtu

Source: Brown et al. 2011, Figure 1.2.

Figure 9.5 US industrial consumption by size of firm

As populations grow and standards of living rise, the production of energy-intensive goods is likely to continue to increase worldwide. An increasing proportion of this production is moving to China, Korea, India, and other rapidly industrializing countries. For example, although the United States remains the world's largest producer of bulk chemicals and refined petroleum products, China has become the world's largest producer of steel, aluminum, and cement (Intergovernmental Panel on Climate Change, 2007). Global competition for export markets, foreign investments, and raw materials is intensifying, which will reward industries that can cut costs by reducing their resource requirements. In an increasingly resource-constrained world, improving the energy efficiency of industry is essential.

Industrial energy efficiency has improved over the past several decades in response to volatile fossil fuel prices, global and domestic competition, and technological advances. US manufacturing has undergone significant changes to improve market competitiveness and increase profits, including reductions in the energy intensity of manufacturing following the oil crises in the 1970s. Over the past several decades, however, the pace of investment in industrial efficiency has slowed, even though technological opportunities for clean energy transformations of industrial systems have increased. After a decade of stagnation, federal policymakers are finally considering options for accelerating clean-energy industrial transformation.

In 2009, the US National Academies, an organization chartered by US Congress and representing the nation's leading scientists, published a multiple-volume report on America's Energy Future. This report concludes that investments in available efficiency technologies could reduce US energy consumption in the industrial sector by 14 to 22 percent over the next decade (National Academies, 2009). At the current average rate of industrial CO_2 emissions per Btu of energy consumption, this would reduce emissions by 434 million metric tons of CO_2 annually by 2020.[1] Comparisons of the energy content of manufactured products across countries underscore the potential for US industry to reduce its energy intensity. Japan and Korea, for instance, have particularly low levels of industrial energy intensity.

3.2 Industrial Energy Efficiency – Some Technology Options

Advances in engineering, materials, thermodynamics, sensors and controls, and information technologies (among others) have the potential to transform industrial processes. There are numerous examples of existing technological innovations with the potential for industry-wide energy savings:

- In today's electricity system with large-scale centralized power plants and dispersed customer locations, there are large mismatches between thermal needs and waste heat streams. Tremendous overall energy savings could be achieved if systems were optimized so that wasted energy were recycled into productive uses. This can be done by cascading and recycling the waste heat and hot exhaust gases that are vented to the atmosphere, low-grade fuels that are typically flared off, and high-pressure steam and gas. Such combined heat and power (CHP) opportunities exist in many industries, including bulk chemicals, food processing, and pulp and paper production (Shipley et al., 2008).

- Opportunities to improve petroleum refining include high-temperature reactors, distillation columns for separating liquids, gas separation technologies, corrosion-resistant metal- and ceramic-lined reactors, sophisticated process-control hardware and software, pumps of all types and sizes, and more efficient steam generation (Lawrence Berkeley National Laboratory, 2005).
- Blending fly ash, steel slag, and other recycled materials with cement could cut energy consumption in the cement industry by 20 percent (Worrell and Galitsky, 2004). Further CO_2 emission reduction would result from replacing carbonates in the composition of cement.
- Motors, the largest single category of electricity end use in the US economy, offer considerable opportunities for saving electricity through technology upgrades and improvements in system efficiency achieved by selecting appropriately sized high-efficiency motors along with variable speed drives to match the motor use to the need of the system. Next-generation motor and drive improvements, including superconducting materials, are currently under development (National Academies, 2009).

A recent study by the McKinsey Group (a US-based global consulting research firm) and studies by many other organizations have documented the great potential for energy savings in energy-intensive US industries (Granade et al., 2009). The National Academies (2009) reviewed the literature on five energy-intensive industries to evaluate their potential for cost-effective improvements by 2020. Of the 34.3 quads of energy forecasted to be consumed by US industry in 2020 (Energy Information Administration, 2008), the National Academies concluded that 14 to 22 percent could be saved through the implementation of cost-effective energy-efficiency improvements (National Academies, 2009).

In the chemicals industry, potential cost-effective energy savings are estimated to reduce energy consumption by 3 to 18 percent by 2020. Larger potential savings could be made in the petroleum-refining industry, ranging from 5 to 23 percent of energy consumption in 2020. Estimates for the pulp and paper industry range from 6 to 37 percent reductions in 2020. While the broad range of these estimates highlights the lack of consensus about the magnitude of the opportunity, it is clear that the energy efficiency gap is great. Further, numerous studies show that sizeable energy savings in manufacturing can be made while providing positive cash-flows for investors (Brown et al., 2014).

3.3 The Efficiency Gap in Manufacturing

Numerous technical and market barriers have impeded the transformation of industrial energy systems (Brown et al., 2011, 2014). These barriers include:

- *Technical risks.* Uncertainties about the benefits and risks of new technologies are a major barrier in the current manufacturing environment, which often requires continuous operation. Reliability and operational risks are major concerns for industries adopting new technologies.
- *High costs.* New energy-efficient technologies often have higher up-front costs and longer payback periods than traditional equipment and they entail more financial

risk because of uncertain future energy prices. New technologies must compete for financial and technical resources against projects that achieve other company goals.

- *External benefits and costs.* External environmental benefits – including reductions in GHG emissions – are not usually considered by potential investors in energy efficiency.
- *Lack of specialized knowledge.* Industrial managers can be overwhelmed by the number of energy efficiency products and programs, which can be difficult to evaluate unless a company has in-house energy experts.
- *Incomplete and imperfect information.* Researching a new technology takes time and resources, and many industries prefer to spend their human and financial capital on other priorities. This is especially true of small firms that typically lack in-house expertise.
- *Market risks caused by uncertainty.* Uncertainties about future electricity and natural gas prices and long-term product demand can present a powerful barrier.

Existing policies can also stymie efforts to improve energy efficiency. For example, the Environmental Protection Agency's (EPA)'s New Source Review (NSR) Program can discourage improvements in energy efficiency at industrial facilities (EPA, 2002). As part of the 1977 Clean Air Act Amendments, Congress established the NSR Program and modified it in the 1990 Amendments, but old coal plants and industrial facilities were exempted from the New Source Performance Standards. These Standards are intended to promote the adoption of the best available air pollution control technologies, taking into account technology costs and other non-air quality, health, and environmental impacts and energy requirements. However, investing in an upgrade could trigger an NSR. Thus the threat of such a review has discouraged companies from investing in plant upgrades. NSR thus imposes pollution controls where they are least needed and artificially inflates the value of the dirtiest plants (Brown and Chandler, 2008).

The barriers described above have caused many firms to defer or reject investing in energy efficiency. Once an asset is installed, it becomes difficult to change, thus locking in a level of energy efficiency that will last for years or even decades. This is a reason to aggressively pursue "windows of opportunity" for putting energy-efficient technologies and systems in place.

At the same time, many energy-intensive industries have devoted considerable resources to increasing their energy efficiency. For many other industries, energy represents a small fraction of their costs and is not a priority. A variety of policy interventions have been implemented in countries around the world to improve the energy efficiency of manufacturing. The magnitude and design of these initiatives reflect each country's access to energy resources, energy prices, and other circumstances. The following case studies illustrate the interplay of energy and manufacturing in several countries.

3.4 International Case Studies: the United Kingdom, Denmark, the Netherlands, Sweden, India, and China

As an island nation, and home to the industrial revolution, *the UK* is no longer rich in on-shore coal and natural gas resources, but it has broad access to the North Sea oil reserve. The UK has managed to meet domestic demand by drilling oil in the North Sea, transitioning to an oil-exporting country in the 1980s. However, its off-shore oil production peaked in the year 2000 and has declined since then (Brown and Sovacool, 2011; UK Department of Energy and Climate Change, 2014).

The Great London Smog in 1952, which ranks as the worst fog in human history, triggered the domestic environmental movement and directly led to the Clean Air Act in 1956, which has been the basis of continuous regulatory effort to shrink SO_2 emissions. Liberalization of the UK energy market has been another driver of change. Under the Conservative Party in the 1980s and early 1990s, the government gradually lost direct control of energy markets as state-controlled energy companies became privatized (Ilie et al., 2007). Intended to build a competitive energy market and reduce energy prices, market liberalization, and subsequently induced investment and regulatory efforts, have stimulated the efficient transformation of the energy sector (Green and Pricef, 1995; Helm, 2002). The positive consequence of the combination of privatization and regulation was more diversified energy supply, which gradually led to carbon emission reductions of 20 percent from 1970 to 2007. Competition provides an opportunity for renewable energy and energy service companies to thrive. Utility companies are better driven to become "green" when environmentally aware customers have the option to switch to more progressive suppliers to fulfill their demand for green energy (Stanford, 1998).

A series of market-based incentives and regulations also contribute to the UK's accomplishments. Proposed in 1999 and introduced in 2001, the Climate Change Levy imposed a tax on delivered energy on energy users in business and the public sector, providing incentives for energy efficiency and carbon emission reduction. A set of domestic policies and programs reflect the UK's significant efforts targeting CO_2 emissions to comply with the Kyoto Protocol. In 2002, the Climate Change Programme was launched to facilitate the reduction of carbon emissions. In the next year, the energy white paper – *Our Energy Future* – was published to guide energy policymaking toward a low-carbon economy. These regulations and policies combined have motivated energy-efficiency and emission-reduction measures in the UK, particularly in manufacturing (Mitchell, 2008).

Denmark is the world's leader in installed CHP capacity, with approximately half of its total electricity generated by CHP technologies. This achievement is all the more impressive when considering that the country is roughly the same size and population as the state of Maryland.

The Danish CHP success story is based on a package of strategies that evolved after the First Heat Supply Law was introduced in 1979. This law included planning regulations and financial incentives that worked together to create desirable market conditions for CHP (IEA, 2009). In 1990, Denmark promoted environmentally friendly zoning to advance electricity investments for towns and villages outside major cities. Co-generation units were required to replace district heating units, and their previous use of oil, diesel, and coal was prohibited and replaced by natural gas. If the local market was not large enough to use co-generation, the district heating plants were required to utilize biomass. Concomitantly, all large utilities in the major cities were ordered to use biomass (especially straw), and required to obey mandatory energy efficiency regulations. This wave of environmentally friendly conversions and improvement of efficiency drove significant investment in the CHP market. Thus, co-generation now provides roughly

60 percent of all electricity and 80 percent of heat consumed in Denmark, making it a leader in Europe. The next closest country, the Netherlands, harnesses only 38 percent for electricity, Finland 36 percent, and every other country in the European Union less than 10 percent (Odgaard, 2007).

The social and environmental benefits from Denmark's energy taxes and support for energy efficiency, wind energy, and CHP have been impressive. As more renewable energy and co-generation units have come to replace less efficient and more polluting conventional fossil fuel units, total CO_2 emissions associated with fuel combustion and electricity generation have plummeted. In 1990, almost 1,000 grams of carbon dioxide were emitted for every kWh of electricity produced in Denmark. In 2005, less than 600 grams were emitted. Altogether, the carbon dioxide emission intensity – the amount of CO_2 emitted per unit of Gross Domestic Product – was 48 percent lower in 2004 than it was in 1980.

At the core of Denmark's successful approach has been sustained taxes on energy fuels, electricity, and carbon dioxide, which both create incentives for energy efficiency and provide government revenue for renewable energy research programs. These energy-related taxes were kept high and not lowered after fossil fuel prices dropped in the 1980s and 1990s so that the manufacturing industry could rely on stable fuel and electricity prices. The taxes furthermore sent price signals that encouraged energy efficiency measures in the Danish power market, and accrued government funds for R&D expenditures that were directed at wind power, biomass, and small-scale CHP units.

Primary energy consumption nationally grew just 4 percent from 1980 to 2004, even though the economy grew more than 64 percent in fixed prices. The carbon tax alone added about 1.3 eurocents per kWh of additional income for renewable power providers. Such taxes do mean that electricity prices are relatively high in Denmark: around 25 eurocents per kWh compared to 19 eurocents per kWh for Germany and 13 eurocents per kWh for the United Kingdom.

The Netherlands has specialized in Long Term Agreements with industry to promote energy efficiency. In the most recent phase, the most energy-intensive industries will be benchmarked to comparable industries worldwide. The targeted industries must be best in class in energy efficiency and, in return, the government will not implement additional stringent climate change policies. Analysis of the benchmarking mechanisms suggests that estimated energy savings will be smaller than under a continuation of the first phase of the Long Term Agreements (Phylipsen et al., 2002). This is due to a change in the policy from continued energy savings in the original Long Term Agreements to a benchmarking standard in the second phase. With the expiration of the second phase in 2012, it remains to be seen whether the initial increase in efficiency gains will be maintained over the entire period. Other industries remain covered under the goals of the Long Term Agreements.

Since the 1970s, *Sweden* has used fiscal policy to drive environmentally sound energy use, first through taxes on oil products (Ericsson et al., 2004; IEA, 2008a) and later differentiated across all energy sources and end uses. The energy tax's objective has evolved over time, shifting from reducing oil demand in the 1970s to its current focus of reducing both oil and electricity demand for heating. Carbon taxes were introduced in 1991, which resulted in increased biomass use for district heating systems since bioenergy and peat were tax exempt. A sulfur tax was implemented the same year, applicable to coal, peat, and oil (Swedish Energy Agency, 2009). These levies tilted the economics in favor of wood fuels and biofuels for electricity, which saw a greater than three-fold increase between 1990 and 2000 (Johansson, 2000).

Sweden has energy-intensive industries (in particular, paper and pulp, steel, and chemicals) for which energy represents a sizeable share of production costs (Thollander and Ottosson, 2010). Given that industry consumes 38 percent of final energy, energy efficiency instruments have been established, including the Program for Energy Efficiency Improvement in Energy-Intensive Industry, which refunds companies their electricity taxes if they implement in full the recommendations outlined in an energy audit (Swedish Energy Agency, 2009).

Both the Swedish government and the public are mindful of the country's contribution to climate change and have therefore enacted measures to curb their GHG emissions. Under the European Union's Climate and Energy Package, which entered into law in 2009, Sweden agreed to legally binding targets for a 17 percent reduction in GHG emissions over 2005 levels in sectors not covered by the EU Emissions Trading Scheme, and for renewables to supply 49 percent of final energy demand by 2020 (European Parliament, 2009). This was facilitated by opening the "Renewable Electricity with Green Certificates" Government Bill to international certificate markets in a measure considered in the Bill's 2012 review. At the same time, because GHG emissions were significantly slashed in the wake of the oil crises, less scope for further reductions exists, especially since hydropower and nuclear sources already generate 89 percent of total electricity (IEA, 2008b).

As a developing country, *India* does not have quite the same historically coordinated effort that the Europeans exemplify. Its industry makeup is also different, supporting more small and medium-sized companies (World Bank, 2008). The government has attempted to incentivize energy service companies to enter the industrial sector, but has had a difficult time doing so. Despite these difficulties, India is currently less energy intensive than the US, and aspires to match the efficiency of Japan (Lamont, 2009).

India's newest approach to the problem is innovative. It has introduced an energy efficiency trading program designed to reduce energy intensity by 5 percent a year through certificate trading. It is expected this market will be worth $15 billion and will cover nine sectors by 2015 (Lamont, 2009). This approach is very similar to other markets for efficiency credits, but India mandates the reductions and the program is designed like a cap-and-trade program. This is a unique approach for a developing country, with the expected outcome of more rapid deployment of efficient technologies throughout the Indian economy.

From 1980 through 2000, *China* experienced a reduction in national energy intensity of 65 percent (Zhang, 2003). These reductions were the result of process and technological changes, as well as structural shifts throughout Chinese industry. Rapidly developing countries typically see an increase in energy intensity; China was able to buck this trend through a series of policy reforms allocating capital toward energy efficiency and developing energy service conservation and energy management centers, which act similarly to energy service companies (Sinton et al., 1999; Wang 1999). China intended to continue this trend, with goals and mandates in the Energy Conservation Law of 1997 (ECL) and the 10th Five-Year Plan.

However, China has recently faced difficulties with these goals. The early 2000s saw energy consumption outpace GDP growth, and thus saw an increase in energy intensity for the first time in decades. Part of this increase was almost certainly driven by difficulties in implementation of the ECL itself, which required provincial energy plans that were slow to develop and difficult to enforce (Wang, 1999).

Noting the deteriorating conditions, the Chinese government announced a mandatory reduction in energy intensity of 20 percent by 2010 in late 2005. Initial responses were not sufficient to reverse the trend, inspiring new policies and strategies to meet the mandate (Lin et al., 2006). The ECL was revised, tax policy was modified for export products, tax credits for efficiency investments were granted, and numerous buildings and appliance policies came into effect, being adopted in the 11th Five-Year Plan. The Top-1000 Energy Consuming Enterprises program has promoted energy efficiency throughout large-sized industry.

It is anticipated that these top energy-consuming businesses will contribute 25 percent of the overall efficiency gains required by the 11th Five-Year Plan, and additional businesses are being added to the program. Even so, with energy-intensive industries including steel, cement, and so on, experiencing increasing demand for their products as the global economy recovers from the recent recession, continuing the progress may prove difficult, and increases in overall consumption are virtually guaranteed (Buijs, 2011).

Just before the December 2009 Copenhagen Summit began, China announced a commitment to reduce the carbon intensity of its economy to 40 percent below the 2005 level by 2020. This will require a 4 percent reduction in GHG emissions each year from projected emissions increases, at the same time that China's economy could grow at an annual rate of 8 percent or higher. Achieving such a goal may involve expanding the scope of major efficiency improvements to China's smaller industrial facilities in addition to potentially imposing new regulations and continuing to close inefficient plants (Friedman, 2009). Others have estimated that the 40 percent goal represents the business-as-usual case for China, and will be easier to meet with faster economic growth (Seligsohn and Levin, 2010).

4. MANUFACTURING WITH CHP: A CASE STUDY OF THE PULP AND PAPER INDUSTRY

4.1 Overview of CHP

Large mismatches abound between the thermal needs and waste heat streams of industrial facilities served by large-scale centralized power plants. If systems were optimized so wasted energy were recycled into productive uses, industrial energy consumption could be cut tremendously and low-carbon electricity could be a byproduct. The most important technology approach to achieving this is combined heat and power.

CHP is a suite of technologies that couple thermal systems with electricity production, thereby boosting overall system efficiencies. While the specific boost in energy efficiency depends on the manufacturing process, it generally leads to an improvement from the 35 to 50 percent range for plants that use boilers to generate heat and purchase power from electric utilities to the 70 to 80 percent range for plants that use an integrated system to meet both their thermal and power needs (Shipley et al., 2008).

Table 9.3 *CHP distribution by sector and top 3 applications in each sector*

Sector/application	Capacity (GW)	Number of facilities
Total	83.72	3,660
Industrial	72.15	1,409
Chemicals	24.11	286
Refining	15.13	105
Pulp and paper	12.09	256
Commercial	4.76	1,485
Colleges/universities	2.74	240
Hospitals/healthcare	0.76	189
Air transportation	0.21	17
Residential	0.12	203
Multi-family buildings	0.11	167
Private households	0.00	36
Other	6.70	563
District energy	2.32	41
Utilities	1.00	100
Agriculture	0.95	135

Source: ICF International 2011.

Co-generation is common in manufacturing industries, such as chemical plants (24 GW), refineries (15 GW), and pulp and paper mills (12 GW), which rely on both thermal and electrical energy (Table 9.3). Because of the concentration of such industries in particular US states, in conjunction with supportive state policies, Texas, California, Louisiana, and New York have particularly large levels of CHP capacity. In addition, California, Texas, New York, and Connecticut promote CHP deployment in small-sized industries as well as commercial buildings with their CHP policies, including interconnection standards, CHP-friendly standby rates, financial incentive programs, output-based environmental standards, net metering regulations, and CHP credit in statewide resource standards (Kim et al., 2013).

Numerous barriers to deploying CHP technologies include the use of input-based emissions standards that ignore the efficiency of the industrial system. Switching to an output-based emissions standard measured in emissions per unit of useful energy output would reflect the efficiency of the industrial system and enable the installation of more efficient technologies, without threatening the legality of the current environmental permits for the facility. Although the EPA supports the adoption of output-based emissions standards, only 17 states have done so.

A recent analysis using the Georgia Tech version of the National Energy Modeling System assessed the potential impact of a long-term investment tax credit (ITC) for CHP installations. Expanding current 10 percent ITC to 20 percent ITC for the next 20 years could increase industrial CHP electricity generation by 19 percent in 2035, compared with business as usual (Baer et al., 2013). This means that industries which are able to generate electricity on their site can be less dependent on central utility for their power supply and take additional benefits from enabling sales of excess power back to the grid, recycling waste resources and avoiding a risk of sudden power outage. In addition, enhancing industrial energy efficiency would lead to the potential of average electricity price reduction, which would benefit not only the industrial sector but also residential and commercial consumers. Because CHP systems are ultra-efficient, industrial energy consumption would shrink relative to the forecasted growth, as would CO_2 emissions and criteria pollutants.

4.2 The Pulp and Paper Industry: Manufacturer of High-Value Products and Low-Carbon Electricity

Using biomass-derived fuels and materials in place of alternatives derived from fossil fuels could substantially mitigate increases in atmospheric CO_2 (World Business Council for Sustainable Development (WBCSD), 2011, 2013). The US pulp and paper industry is a global leader in pulp, paper, and paperboard production: it accounted for 11 percent of total US manufacturing energy use while generating 5 percent of industrial GHG emissions in 2010 (Aden et al., 2013). In recent years, it has expanded its production of an array of products, and its co-production of low-carbon electricity is gaining in importance (Figure 9.6).

The US pulp and paper industry has increased the value of its shipments and the tonnage of its output without a commensurate increase in energy consumption, resulting in lower energy intensity over time. For example, the energy intensity of pulp and paper production between 1985 and 2004 decreased by 15 percent (Brown et al., 2010; U.S. Department of Energy, 2010). However, paper manufacturing still accounts for the second largest electricity intensity, behind primary metals (Aden et al., 2013). The principal products of the forest industry are lumber, wood panels, pulp, paper,

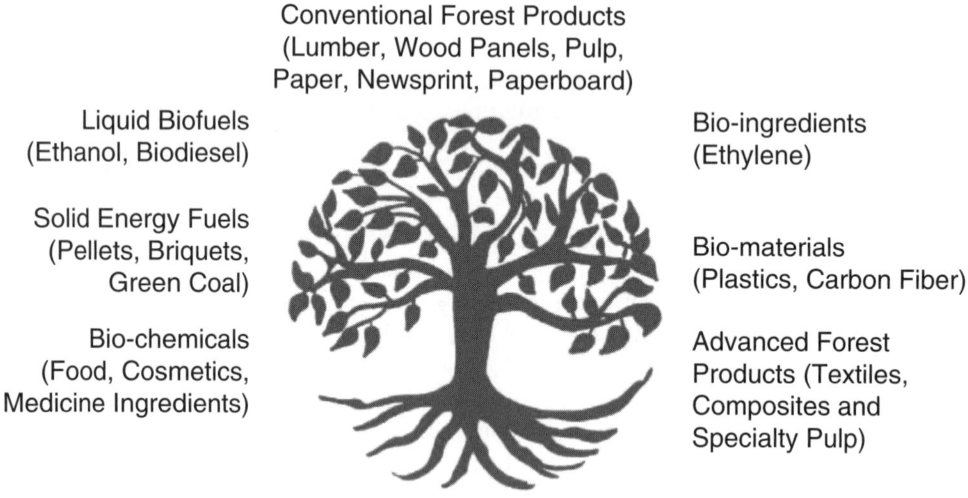

Source: Revised from Nilsson 2013.

Figure 9.6 Multiple biomass products

newsprint, and paperboard. The pulp and paper manufacturing process begins with wood preparation and chemical/mechnical pulping (e.g., grinding, cooking, screening, washing, and bleaching) and ends with papermaking (e.g., wet-end, pressing, drying, and calendering). Mills – including pulp mills, paper mills, and paperboard mills – consume 95 percent of energy use in this sector and require substantial amounts of heat. Mills for each of these products have shown important improvements in energy use. Jacobs and the Institute of Paper Science and Technology at Georgia Institute of Technology (2006) claim that the paper manufacturing process could save 28 percent of energy use by adopting best available energy efficiency technologies.

Within the pulp and paper industry, paper drying is the most energy-intensive step (Kramer et al., 2009). Several energy-efficient methods of drying have been developed, many of which are cost-effective today. One of these, a systems approach, involves using waste heat from heat-generating processes, including from power generation and ethanol production, as the energy source for evaporation (Thorp and Murdock-Thorp, 2008). These opportunities to recycle waste heat are only practical if the power production does not use condensing turbines – that is, if it is relatively inefficient – or if the ethanol distillation is conducted at relatively high temperature and pressures. Advanced water removal technologies, including the use of membrane and advanced filtration methods (Energetics Incorporated, 2005) can also substantially reduce energy use in drying and concentration processes. High-efficiency pulping technology that redirects green liquor to pre-treat pulp and reduce lime kiln load and digester energy intensity is another energy-saving method for this industry (Energetics Incorporated, 2005). Modern lime kilns are available with external dryer systems and modern internals, product coolers, and electrostatic precipitators (Energetics Incorporated, 2006). As high-efficiency papermaking technologies, extended nip press and condebelt drying have key technical potentials to reduce energy use and heat needs. The extended nip press, using a concave shoe instead of the traditional two rotating cylinders, increases drying efficiency by creating more efficient water extraction in the longer pressing time. Condebelt drying uses a hot steel band to evaporate the water continuously in the process. This method is different from conventional steam drying in a chamber and increases drying rates by 5 to 15 percent (Aden et al., 2013).

Recycling also conserves energy and represented roughly 40 percent of total paper production in the US in 2005 (Houser et al., 2008). This percentage of recycling is lower than the 69 percent waste paper pulp in the UK (Confederation of Paper Industries, 2009), which is close to the estimated limit of 81 percent in Europe (Edenhofer et al., 2014). US paper recycling has the capacity to double its current levels.[2]

In the papermaking industry, fiber optic and laser sensors can monitor water content, sheer strength, and the bending stiffness of paper, thus saving energy and improving paper quality. For example, a fiber optic sensor measures paper basis weight to improve wet-end control in papermaking and give paper a uniform basis weight and higher quality, as well as minimizing energy requirements. Another non-contacting laser sensor measures shear strength and bending stiffness. By measuring the rate of propagation of ultrasonic shock waves in the paper, this device could save the US paper industry approximately $200 million annually in energy costs.[3]

Kraft processing is a prominent way to produce wood pulp. In most Kraft mills today, the black liquor produced from de-lignifying wood chips is burned in a large recovery boiler. Because of its high water content, the combustion of black liquor is inefficient, and the possibility of electricity production from secondary steam production is limited by the steam's low pressures. Gasification of black liquor not only allows efficient combustion but also enables the use of a gas turbine or combined cycle process with a high electrical efficiency, thereby offering the potential for increasing the production of electricity within pulp mills. The surplus of energy from the pulp process also allows for the possible production of useful heat, fuels, and chemicals – that is, the operation of "bio-refineries" – a promising direction for the future. Numerous federal and state policies have been implemented to spur the adoption of biomass-based energy use. Renewable portfolio standards in 29 US states set goals for electricity generated from renewable resources by a target year and have given credits to biomass-based energy. The US renewable fuel standard, which requires bio-based fuel (e.g., corn-based or cellulosic ethanol) increases, could be a favorable instrument to the pulp and paper industry because forest-based biomass is a possible future resource for ethanol production (Brown and Baek, 2010). Unlike other renewable resources, biomass is significantly affected by environmental regulations because burning biomass might not meet GHG emissions requirements (see, for example, Searchinger et al., 2008). In 2010, the EPA adopted the Tailoring Rule to set thresholds for stationary sources under the Clean Air Act's New Source Review Prevention of Significant Deterioration (PSD) and Title V Operation Permit programs for large emitting facilities. In the initial rule, biomass was treated in the same manner as fossil fuels, disregarding the carbon-reduction benefits of using biomass for power production (Holtsmark, 2013). In 2011, the EPA deferred the application of the PSD and Title V permitting requirements on biogenic emissions in order to conduct a detailed scientific examination of biogenic carbon neutrality, which means that the EPA has authority to temporarily exempt large industrial sources burning biomass from GHG permitting requirements. However, with a lawsuit filed against the EPA by the Center for Biological Diversity, the final Tailoring Rule does not allow the EPA's deferment of biogenic emissions from GHG permitting requirements (EPA, 2013). The Clean Power Plan proposed by the Obama Administration in June 2014 also was noncommittal about the treatment of biomass for power generation, awaiting further analysis. The debate on the carbon benefits of forest bioenergy is ongoing.

5. MANUFACTURING GREEN ENERGY PRODUCTS: A CASE STUDY OF SOLID-STATE LIGHTING

5.1 Overview

Industry can be a powerful enabler of green energy innovations. It is not just a recipient of new technologies to meet production demands, but in fact is playing a key role in developing the next wave of green energy technologies. While many innovations are created at research hubs like top tier universities and national laboratories, manufacturing itself is often a source of technological innovation. Public–private partnerships with industry have been especially effective in this innovation process (Laitner and Brown, 2005).

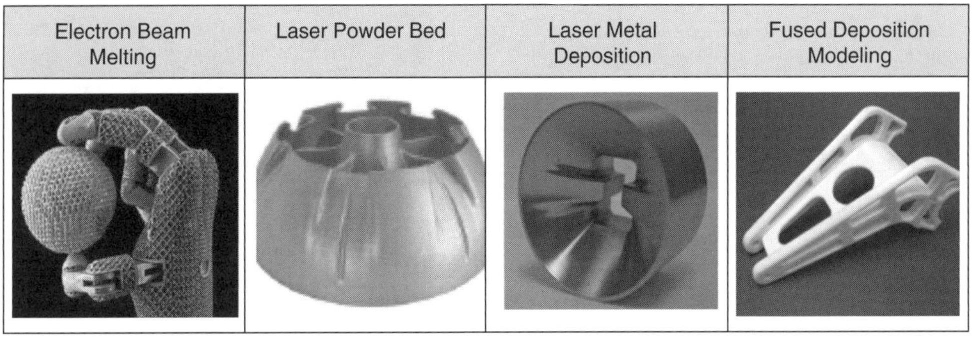

Source: Liby 2013.

Figure 9.7 Types of additive manufacturing

In the energy realm, many next-generation technologies promise to advance the performance and lower the cost of energy technologies. For example, lightweight materials such as carbon fiber composites can reduce the weight of vehicles, thereby improving fuel efficiency without compromising performance or safety. Smart grid technologies are evolving rapidly and are becoming enablers of distributed renewable electricity and efficient energy use (Brown and Zhou, 2013). The performance of green products will likely also be achieved by breakthroughs in additive manufacturing.

Figure 9.7 shows an array of additive manufacturing advances that could transform the production of green technologies. Precision melting of powder materials using electron beams allows the processing of complex geometries not possible through machining. Ultrasonic additive manufacturing is a solid-state process that allows embedding of optical fibers and sensors into equipment, which can then be used to optimize operation and reduce energy use. Site-specific material addition with laser metal deposition enables the repair of dies, punches, turbines, and other metal parts that have deteriorated through corrosion or wear, which means that the existing equipment can operate longer and the expense of wholesale replacement of major parts can be avoided. Finally, fused precision deposition of thermoplastic materials enables the development of high-strength composite materials for industrial applications (Liby, 2013; Oak Ridge National Laboratory (ORNL), 2013).

To illustrate the potential of advances in manufacturing that will enable new and improved green energy products, this section focuses on the manufacture of solid-state lighting. We couple this with an overview of the policy landscape facing the solid-state lighting industry.

5.2 Solid-State Lighting: Manufacturing Next-Generation Green Products

The past decade has seen a transformation of lighting technologies with the introduction and maturation of compact fluorescent bulbs and the continued improvement in linear fluorescent lighting (Sandahl et al., 2006). The market is now on the verge of being transformed by the introduction of solid-state lighting based on

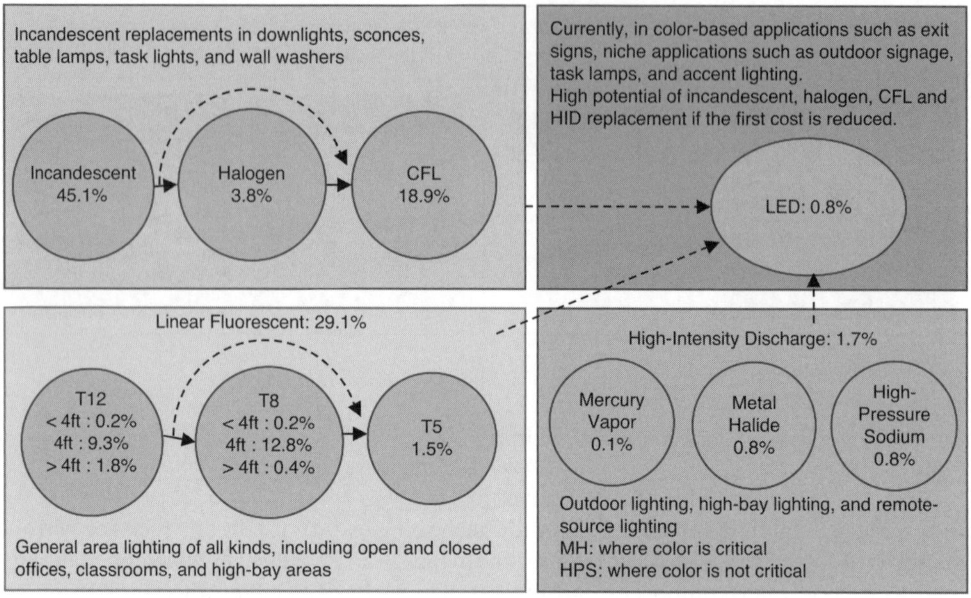

Notes: Percentage: Market share of lamps in 2010 (Navigant Consulting Inc, 2012a).

Source: Navigant Consulting, Inc. 2012a.

Figure 9.8 Lighting technology options and distribution of lamps (%) in 2010

light-emitting diode (LED) technology. Light sources can be grouped into four broad categories: incandescent, fluorescent, high-intensity discharge and LED (Figure 9.8). The first three of these sources is associated with a specific type of application: general area, task, and outdoor lighting. LEDs, in contrast, offer the potential to meet many different lighting needs, including indoor and outdoor lighting, as well as traffic and street lighting and flat screen displays.

A challenge with LED lighting implementation is consumer inertia. Successfully commercialized by Thomas Edison, the incandescent light bulb is well established in residential task lighting, accounting for 45 percent of market share. This inefficient lighting has already seen dramatic changes over the last decade with the emergence of compact fluorescent lamps (CFLs). This trend has been largely attributed to energy cost savings as well as federal regulations and a growing interest in energy conservation. Linear fluorescent bulbs are also common for general area lighting in commercial buildings. Markets have seen a migration from T12 linear fluorescent lamps to T8 and T5 lamps, reflecting improved lamp efficacy. LEDs represented only 1.3 percent penetration in the U.S. in 2013, primarily for LED exit sign installations. LEDs, however, have great potential to replace all lighting applications. In China, LEDs achieved an 8.9 percent penetration in 2013 (U.S. Department of Energy, 2014).

Transitioning to LED lighting offers the potential for significant energy savings. Lighting accounts for approximately 19 percent of electricity use globally (IEA, 2006), and in the US it accounts for 13 percent of electricity consumed in residential buildings and 21 percent of electricity consumed in commercial buildings (Energy Information Administration, 2013a). According to Navigant Consuting Inc. (2012b), LEDs in the US could reduce the energy required for lighting by 19 percent in 2020 and by 46 percent in 2030, saving $30 billion in energy costs in 2030 alone.

LEDs are semiconductor electronic devices with a p–n junction structure that produces light or lumens (lm) by applying an electrical bias. The light-producing mechanism is electroluminescence, which is the radiative recombination of electrons and holes in a semiconductor material. Their efficacy now exceeds 100 lm/W, which is five times greater than a typical incandescent bulb. Compared with a 65 W incandescent lamp that has an annual energy cost of about $85 when operated for two hours a day, an 8 W LED would have an annual energy cost of only $10. The problem is that the purchase price of the LED is more than $10 while the incandescent costs about 50 cents (Zhu and Humphreys, 2012). The economics of CFLs is in between these two extremes and currently has the lowest total cost of ownership. As a result, the challenge facing LED lighting is to reduce the cost of manufacturing in order to make first costs less of a barrier.

Manufacturing LEDs has little in common with traditional lamp production. The supply chain for LED semiconductors has many parts: wafer fabrication (involving materials deposition, etching, lithography, and doping), as well as assembly, packaging, and systems integration with sensors and lighting control technologies. Analysis by Sanderson et al. (2008) indicates that firms typically focus on a single part of the supply chain. Major lighting manufacturers, such as Philips, OSRAM, and GE, have responded to the opportunities in LED lighting by creating joint ventures with semiconductor firms to access preexisting expertise.

Because of the youth of the industry, quality remains variable across firms and is sometimes difficult to assess. Quality for an LED includes a long operating lifetime before substantial fading occurs, controls to ensure against defective chips, and consistency of operating characteristics, including color emission spectrum, light intensity, and viewing angle (Toniolo, 2006).

There are many kinds of LED, made from various materials. For example, green LEDs are made of zinc telluride materials and yellow LEDs are made of gallium phosphide materials. Recent white LEDs, the dominant type, have used GaN and InGaN materials, two alloys in the "III-nitride" family.[4]

The LED wafer fabrication process involves the crystal growth of III-nitride materials, GaN and InGaN. III-nitrides are grown in specially designed facilities that manipulate high-purity source gases and materials at high temperatures with various pressures for high-quality crystal growth. Three types of materials deposition process are used:

- *Metal organic chemical vapor deposition*. Involves the chemical reaction of ammonia and hydrogen gas with a metalorganic source of gallium or indium at over 1,000°C in a stainless steel chamber. It allows large-scale production by high growth rates and produces high-quality and abrupt interface growths without ultra-high vacuum conditions, resulting in lower maintenance costs. However, it lacks in-situ characterization and requires a post-growth doping process.
- *Molecular beam epitaxy*. Typically uses sublimed solid source materials and ionized nitrogen gas by a plasma generator to grow III-nitrides at about 700°C in ultra-high vacuum chambers. It offers the highest achievable purity in the films,

a highly controllable growth mechanism, and in-situ characterization, but it has high maintenance costs and low growth rates.

- *Hydride vapor phase epitaxy*. A simple growth technique with very high throughput. The process is carried out in a quartz furnace at 950–1,050°C. A disadvantage of this technique is the production of NCl_3 as a byproduct, which is highly explosive and corrosive. In addition, the high growth rate produces high levels of impurities (Manasreh, 2011; Graham, 2013; Lee, 2013).

The kinds of fabrication technique used to produce LED lighting are similar to those used to produce many other high-precision consumer products such as television displays as well as green energy products such as photovoltaic panels and fuel cell membranes. The full life-cycle energy requirements of these techniques have yet to be examined, and the proprietary nature of the processes would make such studies challenging. However, given the long-range benefits that they offer in terms of energy savings and low-carbon power production, it is likely that their net energy and environmental impacts are highly favorable.

Governmental policies support the development and deployment of energy-efficient lighting. In the US, federal energy efficiency standards and labels work together to push and pull the lighting market toward greater energy efficiency. The push for standards has come to the fore after the Energy Independence and Security Act (EISA) of 2007 that sets new energy efficiency standards beginning in 2012, playing a significant role in removing inefficient traditional products from the market. For example, the 100 W incandescent bulb is no longer manufactured or imported for sale because it does not meet the EISA 2007 standard.[5] Starting in 2011, new "lighting facts labels" pulled the market by allowing customers to compare the energy efficiency of products, including their estimated lumens, annual operation cost, and color rendition.

Besides these regulations, for more than a decade the US federal government has funded research and development of LED lighting consistent with a solid-state lighting market road map that sets performance targets for improving luminous efficacy while reducing costs. As a result, today's LEDs have better quality of color and brightness at lower cost,[6] and future LED lighting products will be increasingly attractive and cost competitive.

6. DISCUSSION AND CONCLUSIONS: ENERGY ECONOMICS, ENVIRONMENTAL STEWARDSHIP, AND SECURITY

As corporate sustainability has become better understood, industry has taken a much broader view of its energy and environmental responsibilities, extending its concerns to the sustainability of the products and services it offers, as well as the sustainability of its chain of suppliers. Many corporations have also taken voluntary action to reduce the GHG emissions of their operations. These efforts are not yet operating at the scale needed to address the challenges of climate change and clean energy (Southworth, 2009); however, they appear to be expanding as corporate commitments to sustainability grow, and as consumers and shareholders demand greater effort (Prindle, 2010).

Meeting the energy needs of future generations without overheating the planet is one of the most vexing challenges of our time. In an increasingly resource-constrained world, improving the energy and resource efficiency of industry must be part of the climate solution. Advanced industrial technologies and best practices in energy management are already working to improve energy efficiency and reduce GHG emissions. Energy efficiency is also delivering a return on investment that contributes to the profitability of enterprises. The manufacturing of a next generation of clean products, such as solid-state lighting, promises the opportunity to develop new markets while addressing environmental goals. By deploying CHP in industrial plants with significant thermal requirements, industry can also generate clean electricity from heat that would otherwise be wasted, producing a new source of low-carbon power and a new revenue stream.

The public's imagination is captivated by buildings that can generate more energy than they use and zero-emission cars that can provide back-up power to the grid. Similarly, manufacturing systems can restore ecosystems, contribute to human health, produce green products, generate green power, and promote energy self-sufficiency.

ACKNOWLEDGEMENT

The authors gratefully acknowledge the funding provided by Oak Ridge National Laboratory (Melissa Lapsa, Project Manager) and the Renewable Bioproducts Institute located at the Georgia Institute of Technology. Insights from Norman Marsolan, Director of the RBI, and Paul Baer, founder of EcoEquity, were particularly valuable in addressing our case study of the pulp and paper industry and combined heat and power. Samuel Graham of Georgia Tech and Kyoung-Keun Lee at Texas State University provided information on semiconductor electrical engineering, which strengthened the case study of solid-state lighting. Useful overall input and feedback on the entire manuscript was provided by Jennifer Clark of Georgia Tech's School of Public Policy. The views expressed in this chapter, and any errors, are attributable entirely to the authors.

NOTES

1. Author's calculation.
2. Paper recycling has some limits on recovery due to loss of quality when paper is used for permanent records, destroyed, or contaminated. To improve the quality, virgin wood fibers are needed to replace those that are damaged.
3. See http://www.physorg.com/news4221.html. Accessed November 1, 2013.
4. III-Nitrides include AlN: Aluminum Nitride, GaN: Gallium Nitride InN: Indium Nitride and InGaN: indium gallium nitride.
5. See the details of standards at the DOE's building technologies office website; http://www1.eere.energy.gov/buildings/, retrieved November 1, 2013.
6. Cree (www.creebulb.com, retrieved November 1, 2013)), a lighting and semiconductor manufacturing company, claims that its new 60 W-equivalent LED bulb cost only \$13–\$14 in 2013, and lasts 25 times longer and consumes 84 percent less energy than a traditional incandescent bulb.

REFERENCES

Aden, N., Bradbury, J., and Tompkins, F. (2013). Energy Efficiency in U.S. Manufacturing: The Case of Midwest Pulp and Paper Mills. *World Resources Institute*. World Resources Institute. Retrieved October 8, 2014 from http://www.wri.org/publication/energy-efficiency-us-manufacturing.

Baer, P., Brown, M.A., and Kim, G. (2013). The Job Generation Impacts of Expanding Industrial Cogeneration. Atlanta, GA. Retrieved from http://www.spp.gatech.edu/aboutus/workingpapers/job-generation-impacts-expanding-industrial-cogeneration. Forthcoming in *Ecological Economics*.

Brown, M.A. and Baek, Y. (2010). The Forest Products Industry at an Energy/Climate Crossroads. *Energy Policy*, *38*(12), 7665–7675. doi:10.1016/j.enpol.2010.07.057.

Brown, M.A., Baer, P., Cox, M., and Kim, Y.J. (2014). Evaluating the Risks of Alternative Energy Policies: A Case Study of Industrial Energy Efficiency. *Energy Efficiency*, *7*, 1–22.

Brown, M.A. and Chandler, S. (2008). Governing Confusion: How Statutes, Fiscal Policy, and Regulations Impede Clean Energy Technologies. *Stanford Law and Policy Review*, *19*(3), 472–509.

Brown, M.A., Cox, M., and Cortes, R. (2010). Transforming Industrial Energy Efficiency. *The Bridge*, (Fall), 22–30.

Brown, M.A., Jackson, R., Cox, M., Cortes, R., Deitchman, B., and Lapsa, M. (2011). Making Industry Part of the Climate Solution: Policy Options to Promote Energy Efficiency. Oak Ridge National Laboratory. Retrieved from http://info.ornl.gov/sites/publications/Files/Pub23821.pdf.

Brown, M.A. and Sovacool, B.K. (2011). *Climate Change and Global Energy Security: Technology and Policy Options*. Cambridge: MIT Press.

Brown, M.A. and Zhou, S. (2013). Smart-Grid Policies: An International Review. *Wiley Interdisciplinary Reviews (WIREs): Energy and Environment*, *2*(March/April), 121–139.

Buijs, B. (2011). Why China Matters. Chapter 15 in *Energy, Sustainability and the Environment*, ed. P. Fereidoon and P. Sioshansi, Elsevier Press. Retrieved from www.clingendaelenergy.com/inc/upload/files/Why_China_Matters_1.pdf.

Confederation of Paper Industries (2009). Recovery and Recycling of Paper and Board fact Sheets. Retrieved from http://www.paper.org.uk/information/factsheets/recovery_and_recycling.pdf.

Edenhofer, Ottmar, Ramon Pichs-Madruga, Youba Sokona, Susanne Kadner, Jan Minx, Steffen Brunner, Technical Summary of Working Group III: Mitigation of Climate Change (Geneva: IPCC, 2014).

Energetics Incorporated (2005). *Energy and Environmental Profile of the U.S. Pulp and Paper Industry*. Washington, D.C.: U.S. Department of Energy.

Energetics Incorporated (2006). *Energy Bandwidth for Petroleum Refining Processes*. Washington, D.C.: U.S. Department of Energy.

Energy Information Administration (2008). *International Energy Annual 2006*. Washington, D.C.: Energy Information Administration.

Energy Information Administration (2013a). *Annual Energy Outlook*. Washington, D.C.: Energy Information Administration.

Energy Information Administration (2013b). Manufacturing Energy Consumption Survey (MECS) Data. Retrieved from http://www.eia.gov/consumption/manufacturing/data/2010.

Energy Information Administration (2014). *Annual Energy Outlook*. Washington, D.C.: Energy Information Administration.

EPA (2013). Regulations and Standards.

Ericsson, K., Huttunen, S., Nilsson, L.J., and Svenningsson, P. (2004). Bioenergy Policy and Market Development in Finland and Sweden. *Energy Policy*, *32*, 1707–1721.

European Parliament (2009). *Directive 2009/28/EC of the European Parliament*. Brussels: European Parliament.

Friedman, L. (2009, November 30). China, U.S. Give Copenhagen Negotiators Some Targets. *The New York Times*. Retrieved from http://www.nytimes.com/cwire/2009/11/30/30climatewire-china-us-give-copenhagen-negotiators-some-ta-73618.html?pagewanted=1.

Graham, S. (2013). *Introduction to Solid State Lighting*. Georgia Institute of Technology, personal communication.

Granade, Choi, H., Creyts, J., Derkach, A., Farese, P., Nyquist, S., and Ostrowski, K. (2009). Unlocking Energy Efficiency in the U.S. Economy. McKinsey & Company. Retrieved from http://www.mckinsey.com/clientservice/electricpowernaturalgas/downloads/US_energy_efficiency_full_report.pdf.

Green, R. and Pricef, C. (1995). Liberalisation and Divestiture in the UK Energy Sector. *Fiscal Studies*, *16*, 75–89.

Helm, D. (2002). Energy Policy: Security of Supply, Sustainability and Competition. *Energy Policy*, *30*, 173–184.

Holtsmark, B. (2013). The Outcome is in the assumptions: Analyzing the Effects on Atmospheric CO_2 Levels of increased Use of Bioenergy from Forest Biomass. *GCB Bioenergy*, *5*(4), 467–473. doi:10.1111/gcbb.12015.

Houser, T., Bradley, R., Childs, B., Werksman, J., and Heilmayr, R. (2008). *Leveling the Carbon Playing Field*. Washington, D.C.: World Resources Institute.
ICF International (2011). Combined Heat and Power Installation Database. 2011. Retrieved March, 2011 from http://www.eea-inc.com/chpdata/.
IEA (2006). *Light's Labour's Lost: Policies for Energy-Efficient Lighting*. Paris: International Energy Agency.
IEA (2008a). *Sweden 2008 Review*. Paris: International Energy Agency.
IEA (2008b). Electricity for Sweden. Retrieved December 7, 2010, from http://www.iea.org/stats/electricity-data.asp?COUNTRY_CODE=SE.
IEA (2009). *IEA Scoreboard 2009*. Paris: International Energy Agency.
IEA (2012). *World Energy Outlook 2012*. Paris: International Energy Agency.
Ilie, L., Horobet, A., and Popescu, C. (2007). Liberalization and Regulation in the EU Energy Market. *Munich Personal RePEc Archive Paper*. Munich: Academy of Economic Studies, http://mpra.ub.uni-muenchen.de/6419/1/.
Intergovernmental Panel on Climate Change (2007). *Climate Change 2007: Mitigation of Climate Change*. New York, NY: Cambridge University Press.
Jacobs and the Institute of Paper Science and Technology at Georgia Institute of Technology (2006). The Pulp and Paper Industry: Energy Bandwidth Study. American Institute for Chemical Engineers. Retrieved from http://www1.eere.energy.gov/manufacturing/industries_technologies/forest/pdfs/doe_bandwidth.pdf.
Johansson, B. (2000). *The Carbon Tax in Sweden*. Workshop on Innovation and the Environment. Paris: OECD, pp. 85–94.
Kim, G., Baer, P., and Brown, M.A. (2013). The Statewide Job Generation Impacts of Expanding Industrial CHP, *Proceedings of the 2013 ACEEE Summer Sudy on Energy Efficiency in Industry*. Washington, D.C., pp. 2–1–2–16. Retrieved from http://www.aceee.org/files/proceedings/2013/data/index.htm.
Kramer, K.J., Masanet, E., Xu, T., and Worrell, E. (2009). Energy Efficiency Improvement and Cost Saving Opportunities for the Pulp and Paper Industry: An ENERGY STAR Guide for Energy and Plant Managers. Lawrence Berkley National Laboratory. Retrieved from http://www.energystar.gov/ia/business/industry/downloads/Pulp_and_Paper_Energy_Guide.pdf.
Laitner, J.A. and Brown, M.A. (2005). Emerging Industrial Innovations to Create New Energy Efficient Technologies. *Proceedings of the Summer Study on Energy Efficiency in Industry*. Washington, D.C., pp. 4–71–4–83.
Lamont, J. (2009). India to Launch Energy-Efficiency Trading. *Financial Times*, September 27.
Lawrence Berkeley National Laboratory (2005). Energy Efficiency Improvement and Cost Saving Opportunities for Petroleum Refineries, An ENERGY STAR Guide for Energy and Plant Managers. BNL-57260-Revision. Prepared by C. Galitsky, S. Chang, E. Worrell, and E. Masanet. Berkeley, CA: LBNL.
Lee, K.K. (2013). Texas State University, personal communications.
Liby, A. (2013). ORNL's Advanced Manufacturing Initiative. Presentation at the Georgia Institute of Technology, Core University Liaison Meeting. Atlanta, GA.
Lin, J., Zhou, N., Levine, M.D., and Fridley, D. (2006). Achieving China's Target for Energy Intensity Reduction in 2010: An Exploration of Recent Trends and Possible Future Scenarios. Berkeley, CA: Lawrence Berkeley National Laboratory.
Manasreh, O. (2011). *Introduction to Nanomaterials and Devices*. New York: Wiley.
Mitchell, C. (2008). *The Political Economy of Sustainable Energy*. Basingstoke: Palgrave Macmillan.
National Academies (2009). *Real Prospect for Energy Efficiency in the United States*. Washington, D.C.: The National Academies Press.
Navigant Consulting Inc. (2012a). *2010 U.S. Lighting Market Characterization*. Washington, D.C.: U.S. Department of Energy. Retrieved October 1, 2013 from http://apps1.eere.energy.gov/buildings/publications/pdfs/ssl/2010-lmc-final-jan-2012.pd.
Navigant Consulting, Inc. (2012b). Energy Savings Potential of Solid-State Lighting in General Illumination Applications. Washington, D.C.: U.S. Department of Energy. Retrieved October 1, 2013 from http://apps1.eere.energy.gov/buildings/publications/pdfs/ssl/ssl_energy-savings-report_jan-2012.pdf.
Nilsson, S. (2013). Forest Sector Transformations: Pathways to Success. IPST Workshop April 18–19. Atlanta, GA.
Odgaard, O. (2007). *Energy Policy in Denmark*. Copenhagen: Danish Energy Authority.
ORNL (2013). Realizing the Potential of Next-Generation Manufacturing. Retrieved May 26, 2013 from www.ornl.gov/manufacturing.
Phylipsen, D., Blok, K., Worrell, E., and Beer, J. de (2002). Benchmarking the Energy Efficiency of Dutch Industry: An Assessment of the Expected Effect on Energy Consumption and CO_2 Emissions. *Energy Policy*, *30*(8), 663–679.
Prindle, B. (2010). *From Shop Floor to Top Floor: Best Business Practices in Energy Efficiency*. Washington, D.C.: Pew Center.

Sandahl, L., Gilbride, T., Ledbetter, M., Steward, H., and Calwell, C. (2006). *Compact Fluorescent Lighting in America: Lessons Learned on the Way to Market*. Richmond, WA: Pacific Northwest National Laboratory.
Sanderson, S., Walsh, K., Simons, L., Walls, J.L., and Lai, Y.-Y. (2008). Unpublished manuscript prepared for a volume of the National Academies.
Searchinger, T., Heimlich, R., Houghton, R.A., Dong, F., Elobeid, A., Fabiosa, J., Tokgoz, S., et al. (2008). Use of U.S. Croplands for Biofuels Increases Greenhouse Gases Through Emissions from Land-Use Change. *Science*, *319*(February), 1238–1240.
Seligsohn, D. and Levin, K. (2010). *China's Carbon Intensity Goal: A Guide for the Perplexed*. Washington, D.C.: World Resources Institute.
Shipley, A., Hampson, A., Hedman, B., Garland, P., and Bautista, P. (2008). *Combined Heat and Power: Effective Energy Solutions for a Sustainable Future*. Oak Ridge, TN: Oak Ridge National Laboratory.
Sinton, J., Levine, M., Fridley, D., Yang, F., and Lin, J. (1999). *Status Report on Energy Efficiency Policy and Programs in China*. Berkeley, CA: Lawrence Berkeley National Laboratory.
Southworth, K. (2009). Corporate Voluntary Measures to Reduce Carbon Emissions: A Valuable but Incomplete Solution to Climate Change and Energy Security Challenges. *Policy and Society*, (27), 329–350.
Stanford, A. (1998). Liberalization of the UK Energy Market, an Opportunity for Green Energy. *Renewable Energy*, *15*, 215–217.
Swedish Energy Agency (2009). *Energy in Sweden 2009*. Eskilstuna: Swedish Energy Agency.
Thollander, P. and Ottosson, M. (2010). Energy Management Practices in Swedish Energy-Intensive Industries. *Journal of Cleaner Production*, *18*, 1125–1133.
Thorp, B.A. and Murdock-Thorp, L.D. 2008. Compelling Case for Integrated Biorefineries. Retrieved October 8, 2014 from http://www.epoverviews.com/oca/Compellingcaseforbiorefineries.pdf.
Toniolo, T. (2006). What Determines an LED's Quality? *Electronic Products*. Retrieved November 1, 2013 from http://www.electronicproducts.com/Optoelectronics/What_determines_an_LED_39_s_quality.aspx.
UK DECC (2014). UKCS Productions and Projections. Retrieved October 9, 2014 from https://www.gov.uk/government/uploads/system/uploads/attachment_data/file/287001/production_projections.pdf.
UK Department of Energy and Climate Change (2010). *UKCS Productions and Projections*.
U.S. Department of Energy (2010). U.S. Energy Intensity Indicators: Trend Data. Retrieved October 2010 from http://www1.eere.energy.gov/ba/pba/intensityindicators/trend_data.
U.S. Department of Energy (2014). *Multi-Year Program Plan Aug 14, Solid-State Lighting Research and Development*, prepared for the U.S. Department of Energy [Online]. Available: http://apps1.eere.energy.gov/buildings/publications/pdfs/ssl/ssl_mypp2014_web.pdf.
US EPA (2002). *New Source Review: Report to the President*. Retrieved October 1, 2013 from http://www.epa.gov/nsr/documents/nsr_report_to_president.pdf.
Wang, A. (1999). *A Comparative Analysis of the 1997 Energy Conservation Law of China and the Implementing Regulations of Shandong, Zhejiang, and Shanghai*. Washington, D.C.: Natural Resources Defense Council.
WBCSD (2011). The Sustainable Forest Products Industry, Carbon and Climate Change. SFPI Carbon System Action Team. Retrieved October 9, 2014 from https://mail.gatech.edu/service/home/~/SFPI-CarbonAndClimateChange.pdf?auth=co&loc=en_US&id=440408&part=2.
WBCSD (2013). *Issue Brief Biomass Carbon Neutrality*. Retrieved from http://www.wbcsd.org/Pages/EDocument/EDocumentDetails.aspx?ID=15347&NoSearchContextKey=true.
World Bank (2008). India: Energy Efficiency Lending Focuses on Industry Clusters. Washington. Retrieved from http://go.worldbank.org/RWBUAW0RM0.
Worrell, E. and Galitsky, C. (2004). *Energy Efficiency Improvement and Cost Saving Opportunities for Cement Making – An ENERGY STAR (R) Guide for Energy and Plant Managers*. Berkeley, CA: Lawrence Berkeley National Laboratory.
Zhang, Z. (2003). Why Did the Energy Intensity Fall in China's Industrial Sector in the 1990's? The Relative Importance of Structural Change and Intensity Change. *Energy Economics*.
Zhu, D. and Humphreys, C.J. (2012). Chapter 35 in Lighting/Appliances. In D. Ginley and D. Cahen (eds), *Materials for Energy and Environmental Sustainability* (pp. 474–490). Cambridge, UK: Cambridge University Press.

10. Design and manufacturing: the competitiveness of American, European and Chinese industrial design companies

Vida Vanchan and John R. Bryson

INTRODUCTION

Globalization, though not a new process, continues to intensify as new complex international divisions of labour emerge. The developing global economy is facilitated by place-based differences in factor inputs, including labour, energy, raw materials and skills. The varieties of capitalism literature has highlighted the variegated nature of capitalism with an emphasis on institutional differences (Hall and Soskice, 2001; Daniels *et al.*, 2011). Localized consumer cultures, however, continue to exist, reflecting different lifestyles, traditions, histories, cultures and even climate. The continued persistence and even intensification of local consumer cultures means that companies have to be increasingly sensitive to the variegated nature of consumer cultures. The continued existence of local consumer cultures represents an opportunity for companies that are able to differentiate or segment products by core markets.

The ongoing deepening of globalization continues to alter geographies of production and consumption and has also enhanced competition amongst companies operating internationally. These alterations reflect complex transformations and the development of new relationships between manufacturing and service functions and, in particular, between manufacturing and design. A tension exists between manufacturing and design. On the one hand, manufacturing companies want to produce standardized products but, on the other hand, some products have to be customized to meet the needs of a specific consumer culture. These are not conflicting processes as customized products can be produced using standardized components.

Developed market economies have been undergoing a sectoral transformation, with a shift in employment from manufacturing to services. Nevertheless, manufacturing continues to be a significant part of developed market economies, but has been transformed and now involves many more service tasks. Many manufacturing companies have been transformed into creative companies in which designers, production managers, engineers, sales teams and shopfloor workers from across the global supply chain work together to develop products and services. To date, a clear converging point between the academic and professional literatures on design is that design matters and will continue to matter (Bryson and Rusten, 2011). Industrial design companies provide critical intermediate inputs into the activities of manufacturing companies, including product and process design and redesign (Vanchan, 2007; Vanchan and MacPherson, 2008b; MacPherson and Vanchan, 2010b). Design companies have also been undergoing a process of transformation as specialist design firms have been forced to become more generalist providers of a range of specialist design services. Existing research on design has focused on understanding the economic geography of American and European design firms, but there has been limited research on design firms located in emerging markets (Bryson and Rusten, 2011).

This chapter examines the relationship between design and manufacturing; specifically, to evaluate the role design plays in the new economic geographies of production. The focus is on understanding the global competitiveness of design companies located in the US, Europe and China. Increasing global competition has intensified price- and design-based competitiveness (Rusten and Bryson, 2010; Bryson and Rusten, 2011). Developments in the division of labour have tended to separate the design function from manufacturing, and this is especially the case for products that are fabricated using an international division of labour or complex global supply chain. A product that is designed 'here' and manufactured 'there' has become the new norm. Complex and sophisticated products are produced using fragmented value chains. For many of these supply chains, design remains located in developed market economies. Thus, Apple's products are designed in California and many are manufactured by contract manufacturers located in Shenzhen, Guangdong, China.

The popular account of Chinese manufacturing is one that emphasizes that China is not a design-intensive or design-led economy, with many Chinese manufacturing firms relying on copying products that are designed elsewhere (Midler, 2011). Chinese companies are trying to move up the value chain by producing design-intensive and branded high value-added goods (Bryson and Rusten, 2011). The Chinese government is trying to enhance the knowledge-intensiveness of manufacturing and has realized that this requires the development of China's service economy. Producer services, and in particular knowledge-intensive services (KIS), have been included in China's development plans for long-term economic growth. During the 16th Meeting of the Communist Party of China, in late 2002, it was proposed that the development of 'modern services' should be part of an increase in the size of tertiary industries in the national economy. These 'modern' services included KIS and producer services. The development of producer services was explicitly stated in the *11th Five-Year Plan* (2006–2010). Indeed, in 2007 the State Council of China made a clear pledge to make producer services a strategic sector for upgrading the industrial structure and promoting economic transition. This commitment has been emphasized further and carried over into the *12th Five-Year Plan* (2011–2015), which is notable for an unambiguous intention to continue the transformation of the economy from traditional manufacturing towards emerging strategic industries and modern services (Daniels, 2013).

The emphasis that has been placed by the Chinese government on encouraging the development of high added-value manufacturing has important implications for the developing geography of manufacturing. Related research questions here include: 1) What are the competitive characteristics of companies in the industrial design sector? 2) What are the key determinants of success amongst these companies? 3) To what extent have Chinese companies eroded the competitiveness of companies located in developed market economies? and 4) What is the future outlook of industrial design firms and their position within global value chains? Before examining these questions, this chapter explores the relationship between design and manufacturing companies and the role design plays in the new economic geographies of production.

DESIGN AND MANUFACTURING IN THE NEW ECONOMIC GEOGRAPHIES OF PRODUCTION

New geographies of production, consumption and distribution are continuously being created while old ones are continuously being disrupted (Dicken, 2011). The increasingly widespread practice amongst manufacturers of durable goods is to subcontract routine production tasks to low-cost suppliers located in newly industrializing countries. This is often considered to be a search to reduce input costs (Dankbaar, 2007), but it may also be partly driven by a concern with gaining access to new emerging markets. Knowledge-intensive activities, including research and development (R&D), product design and engineering (Parker and Anderson, 2002; Lynn and Salzman, 2007), have also been outsourced by major industrial companies to lower-cost locations. Hiring external sources of technical expertise to support new product development has been a growing trend in recent years (Beesley and Rothwell, 1987; Chandra and MacPherson, 1994; Coffey and Drolet, 1996; Hagedoorn, 1996, 2002; O'Connor, 1996; MacPherson, 1997; Vanchan, 2007).

Specialized services are provided by suppliers that are often, but not always, based in high-wage nations, rather than by cheaper providers located in newly industrialized countries (Sturgeon *et al.*, 2006). It is often assumed that the core global cities are the sites for specialized advanced producer services (Bryson and Rusten, 2011), but this is not reflected in the geography of design services. Design services have concentrations in the global cities, but are also over-represented in places that specialize in particular manufacturing activities. Thus, the geography of industrial design has a distributed or dispersed geography with some urban concentrations, but also mirrors the geography of manufacturing more generally. Advanced technical functions have increasingly been outsourced to suppliers located in high-wage nations such as Japan, Italy and the United Kingdom (MacPherson and Vanchan, 2010b). These functions include an array of activities, from outsourcing of airframe component design by Boeing and Airbus, to custom manufacturers, design houses and large industrial corporations (Pritchard and MacPherson, 2007), such as Caterpillar the American manufacturer of earth-moving equipment, outsourcing almost all electronics-related design work to external design consultants (Costlow, 2002).

Design is the first step in the manufacturing process and it is where most of the important decisions are made that affect the final cost of a product (Boothroyd, 1994). Consultants in the industrial design sector, along with custom manufacturers, have become increasingly central to the product development efforts of US manufacturing companies across a growing range of sectors (Belecheanu *et al.*, 2006). Industrial design companies provide critical innovation inputs to their clients, which include a reduction in defect rates of new products, optimized ergonomics, aesthetic superiority, ease of manufacture and improved product performance (Vanchan and MacPherson, 2010b).

There is an acute shortage of design-related talent within the US manufacturing sector (Pritchard, 2001; Kalafsky, 2002). The strategic importance of the design sector has been underscored by studies from across the globe, including the United States (Vanchan and MacPherson, 2008a), Canada (Gertler and Vinodrai, 2004; Vinodrai, 2010), Norway (Bryson *et al.*, 2004), the United Kingdom (Design Council, 2006) and China (Balfour and Roberts, 2003). Although the design industry is still heavily concentrated in developed countries competition from firms located in developing countries such as China (Vanchan, 2013) and Brazil (Consoni and Quadros, 2006) has emerged, and this suggests that the geography of the design industry is altering to meet changes in demand. This chapter provides an overview of the developing design sector in the US, Europe and China. The following sections, therefore, explore the competitiveness of design companies in developed economies, including the United States and Europe, and that of an emerging economy, China. The aim is to explore the evolving economic geography and global competitiveness of industrial design firms and their relationship with manufacturing.

THE COMPETITIVENESS OF AMERICAN DESIGN FIRMS

There are approximately 1600+ industrial design companies in the United States, employing about 13,800 employees (US Census Bureau, 2013).[1] These companies are primarily engaged in creating and developing designs and specifications that optimize the use, value and appearance of products (US Census Bureau, 2013) for the mutual benefit of both users and manufacturers (Industrial Design Society of America (IDSA), 2013). In sum, industrial design is a synergistic fusion of art (aesthetically pleasing), engineering (easy to build, maintain or repair) and science (creating something that delivers superior performance within the confines of commercial feasibility). Most design companies are small, employing fewer than ten people. They serve clients across a spectrum of sectors, from household goods to automotive and aerospace. The diversity of services provided by design firms and the official classification of industrial design companies may explain a report published in 2010 that identified 40,000 design jobs and 1900 industrial design companies in the United States (IDSA, 2010).

Geographically, industrial design companies are unevenly distributed across the US, reflecting local agglomerations and the nature of service specialization (MacPherson and Vanchan, 2010a). In other words, design firms tend to cluster in major metropolitan areas in a few states (California and New York account for about 30 per cent of all firm establishments and about 40 per cent of total receipts in this industry (US Census Bureau, 2013)); and their sectoral focus reflects the nearby production structure (Vanchan and MacPherson, 2008a) and origins of the owner (Vanchan, 2013). The majority of the companies are young (<20 years) (Vanchan, 2009). Interestingly, despite relatively stable total employment and stability in the number of establishments over the last two decades, new companies continue to enter this market on a regular basis. What are the contributing factors to this trend? Which factors best differentiate successful from unsuccessful companies?

The continued trend of new firm entry and stable employment within the industrial design industry can be explained by the following factors:

1. Although the share of industrial production has been falling steadily since the 1950s, there are still plenty of buyers when it comes to design services (Vanchan and MacPherson, 2008b). MacPherson (2007) notes that inflation-adjusted spending on external design services increased by close to 75 per cent over the period 1994–2004. Furthermore, the strategic importance of design illustrates the increasing demand for and recognition of design services.

2. It is evident that graduates with professional degrees in industrial design rarely seek permanent employment inside the manufacturing sector (Vanchan and MacPherson, 2008b; Vanchan, 2013). Sectors that face cyclical demand are especially unattractive to young graduates with design skills (e.g. commercial aerospace, automotive products, machine tools). Many of these graduates join existing design consultancies or establish their own companies rather than seek long-term employment in the manufacturing sector. Although most of these graduates initially gain experience or visibility while working for original design manufacturers (ODMs), it would appear that many of these people eventually end up in the service sector (Vinodrai, 2006).
3. The size and dynamic nature of industrial design companies produce a quick turnover rate. On the one hand, many design firms employ few people and in many cases none on a full-time basis, or have just the owner/s as the employee/s. This means that overall operating costs are low, and many small design companies have the ability to keep their companies afloat for long periods of time compared with larger companies. On the other hand, this has also made it easier for firms to close or for new firm formation to occur.

Four key determinants of firm success can be identified from the extant literature on firm competitiveness within the industrial design sector (Vanchan and MacPherson, 2008a, b; Rusten and Bryson, 2010; Bryson and Rusten, 2011; Vanchan, 2013). They are *service provision*, *human capital*, *market orientation* and *adaptability*, and these will be examined in turn.

Service Provision

The most commercially buoyant companies are those that have diversified their service offerings beyond product or component design (Vanchan and MacPherson, 2008b). These companies have developed strategic competencies in fields such as contract research, prototype development, product testing, technological forecasting, market analysis and even advertising. Design firms need to respond to technological developments. The emergence of smart phones and web-based marketing, for example, extended the range of services that are provided by design firms. Design research has emerged as a competitive strength of US design companies (Vanchan and MacPherson, 2008b; Vanchan and Ren, 2013b). There is no need to provide a detailed discussion of the multiple interpretations of design research in this chapter (see Faste and Faste, 2012), but the definition of design research used in this analysis is the scientific exploration of alternative engineering options with respect to critical dimensions such as material selection, styling, ergonomics, product performance or enhancement of manufacturing efficiencies. The goal is to create radically improved design solutions that surpass contemporary standards (IDSA, 2005). It is broadly accepted by the professional literature that while most design companies can create alternative designs for any given product, only companies with research capability have the potential to create innovative solutions on a consistent basis. The extent of research activities performed by companies varies according to the markets they serve. In addition, not all design companies have research capability. While companies serving medical, automotive, machinery, household goods and electronics markets were found to be more research intensive than those that primarily serve textile/apparel markets (Vanchan and MacPherson, 2008a; Vanchan, 2009), most design companies are research oriented.

From a technological innovation front, Design Trade-Off Analysis (DTA) is an important component that allows clients to statistically evaluate potentially competing design objectives (e.g. product performance versus ease of manufacture). It has resulted in considerable cost savings for clients (Vanchan and MacPherson, 2008b). As indicated in Vanchan and MacPherson (2008b), not all design companies have DTA capability. Vanchan and MacPherson (2008a) suggest that the most commercially successful US design companies operate as 'total package providers', or TPPs. This also holds true amongst the more innovative design communities of the European Union (EU), notably in the UK, Italy and France (Chiesa *et al.*, 2004). Based on recent literature, three main factors explain the shift towards a TPP mode of service provision. They include, first, an increasing pursuit by client companies for 'one stop' product development solutions to avoid dissipating their external search or procurement efforts across too many channels (Laursen and Salter, 2006). This not only minimizes transactions costs, but it also creates logistical efficiencies in terms of input search and retrieval.

A second factor is the broader skill base provided by recent graduates from the US design schools, which enhances the depth and quality of the industrial design industry's human capital base. These graduates are more likely to have received professional training in design research than their peers from 10 or 20 years ago (Okudan and Zappe, 2006).

A third factor is that highly skilled young design specialists possessing both artistic and engineering talents rarely seek permanent jobs with manufacturing companies if they can help it (see Kalafsky, 2006). This is because of the cyclical demand within the US manufacturing sector.

TPPs, or 'Tier I companies', refers to the companies that deliver a wide range of services, including research, design and marketing (Vanchan and MacPherson, 2008a). Tier II refers to companies that offer only two types of design service, mainly design and marketing services; and Tier III refers to companies that offer only one type of design service – product/component design (Vanchan and MacPherson, 2008a). TPPs were found to lead the non-TPPs in employment creation, sales growth and profitability (Vanchan, 2009). They are export intensive and likely to invest in human capital and R&D. In addition, they have a core focus on design research, a complex network consisting of collaborators and specialists, and prior employment experience with ODMs.

Human Capital

Human capital is central to any business operation and the quality of human capital is especially important for successful business. Vanchan and MacPherson (2008a) found that the quality of human capital is one of the key differences between successful and unsuccessful design firms. TPPs and non-TPPs, according to Vanchan and MacPherson (2008a, 2008b), do not differ with regard to firm size, age, sector focus (main client groups), occupational structure, regional or metropolitan location, or export-involvement. TPPs were also found to be more likely to invest in new software technology, worker training, new service development, design research and new ways of doing business (e.g. combining research services with design and marketing services).

In a follow-up study involving in-depth interviews with 22 US industrial design companies in five US metropolitan areas that are home to more than 30 per cent of all design establishments, Vanchan (2013) found that TPPs employed workers with at least a Bachelor's degree; none of the TPPs in the sample employed workers with an Associate's degree or lower. The hiring decision was commonly based on an applicant's portfolio. This portfolio, according to Vanchan (2013), provides employers with insight into the applicant's capabilities and know-how. Although many companies do not consider hiring from top design schools as their priority, more TPPs than non-TPPs cited the importance of hiring workers from the top design schools.

One of the common traits amongst all design companies in the study when it comes to training, was the provision of on-the-job training that was tailored to a particular task/project (Vanchan, 2013). Training was also provided for new software updates or other technologically related issues. TPPs appeared more structured in their training programs and exhibited a greater propensity to invest in new software technology and worker training than their counterparts (Vanchan, 2013).

Trust-based relationships are an important contributor to business success (Vanchan, 2009). Most design companies survive on the basis of repeat business with longstanding customers. While websites and advertising are important, word-of-mouth referrals appear to be a more important method for obtaining new customers (Vanchan and MacPherson, 2008b). Vanchan (2013) found that revenues are mostly generated by referrals. These referrals come from friends, former co-workers and employees. Many owners of these companies have degrees in industrial design (at least an Associate's degree). Furthermore, they have at least five years' of prior work experience in design or manufacturing companies.

Market Orientation

Evidence from studies of US industrial design companies suggests that growth rates have been fastest amongst companies serving non-local markets (Vanchan, 2009; Vanchan and MacPherson, 2008a; Vanchan, 2013). It also indicates that TPPs are more likely to engage in export than are Tier II and III companies. The last section of this chapter discusses this further in comparison with Chinese design companies.

Adaptability

In a world of constant change, adaptability is critical. It is the new competitive advantage (Reeves and Deimler, 2011). Reeves and Deimler (2011) argue that sustainable competitive advantage no longer arises from positioning or resources; instead, it stems from four organizational capabilities that foster rapid adaption. They are: 1) the ability to read and act on signals of change; 2) the ability to experiment rapidly and frequently (i.e. business models, strategies, processes, products and services); 3) the ability to manage complex and interconnected systems of multiple stakeholders; and 4) the ability to motivate employees and partners. These four organizational capabilities can be observed amongst US TPPs. This, however, does not mean that successful non-TPPs are not capable of adaptation, but TPPs are more likely to possess the attributes of adaptability. For example, one of the design companies based in California that serves medical and consumer electronics markets experienced growth as it also functions as an original contract manufacturer with an office in Taiwan (Vanchan, 2013). A declining automotive design firm in Michigan failed to keep up with technology (use of CAD or new software) and had a limited marketing strategy (no website or means to attract new customers) (Vanchan, 2013).

Clearly, the above organizational capabilities that foster adaptation were embedded in the discussion on the first three key determinants of firm success. Design companies compete on the basis of quality, reputation and image (Vanchan, 2013). Here, it is important to emphasize the need for companies to adapt, particularly in this rapidly changing global economy. While adaptation may be difficult for large companies (Reeves and Deimler, 2011), it works to the industrial design companies' advantage because of their average size and dynamism. More importantly, while companies can choose different paths to adapt, only those that excel in service provision and human capital, with clear and practical market orientation and timely adaptive capability, are able to attain long-term growth and sustainability.

THE COMPETITIVENESS OF EUROPEAN DESIGN FIRMS

The European design industry consists of 410,000 designers who in 2006 had an annual turnover of 35 billion Euros. In 2002, designers accounted for 5.4 per cent of the added value produced by business and professional services throughout Europe (Bureau of European Design Associations (BEDA), 2006: 3). The European design industry includes a range of different design interventions, including styling, redesigning products and services to enhance production efficiencies, user–product interface, eco-design, market appeal, identity and branding.

The geography of European design reflects the geography of production. Thus, in France in 2001 there were between 11,000 and 13,000 designers (excluding fashion and web design), but 60 per cent were based in the Paris area (BEDA, 2006: 12). In Italy, in 2003, there were 5,000 companies specializing in product design employing around 14,800 designers. Over 700 foreign designers are located in Italy and are predominately based in the Milan area (BEDA, 2006: 12).

The European design industry is polarized between countries that have a design industry that focuses on supporting local clients and countries that have a design industry that is also involved in exporting design services. In 2007 an analysis of the European design industry revealed that the turnover of the United Kingdom's design industry was 16.7bn Euros. The UK design industry is the major European provider of design services, followed by Germany with a turnover of 6.9bn Euros, the Netherlands (2.6bn), France (2.4bn) and Switzerland (1.7bn) (BEDA, 2007: 1). It is interesting that countries that have international reputations for the quality of their design have relatively small design communities, for example Italy and Sweden, but this reflects differences in the size of national economies and also differences in industrial structure. The BEDA analysis may not include all in-house designers.

The UK design industry, in 2006, employed 185,500 designers. There were 12,450 design consultancy firms and 47,400 freelance or self-employed designers. This highlights the importance of small and also micro-firms in the design industry. In 2010 the

Design Council (UK) undertook a survey of designers. This estimated that there were 232,000 designers in the UK, and that 65,900 (28 per cent) were freelance designers while 83,600 worked within in-house client teams (Design Council, 2010: 1). There are around 6,500 design teams inside non-design businesses with more than 100 employees. The size of the UK design industry is reflected in the higher education system; in 2009 55,310 undergraduates were studying for degrees in design. The top 5 per cent of design firms had annual fee incomes of over £2m, but 58 per cent of freelance designers earned less than £50k in 2009 while only 1 per cent of freelancers had fee incomes between £250 and £500k. More research is required into the growth in freelance self-employed designers in the UK. These designers may be working as associates to larger firms or be part of networks of self-employed designers who work together to provide services to clients. They may also be providing in-house design teams with additional capacity. Since 2005 there has been a 10 per cent increase in the number of larger firms employing in-house design teams and this highlights the increasing importance firms are placing upon design as a key source of competitiveness. Just over 60 per cent of design firms employed four or fewer people and 27 per cent employed between five and nine staff.

The geography of UK design is concentrated on London and the South East; 23 per cent of designers are located in London and 17 per cent in the South East. The rest of the industry is relatively evenly distributed with, for example, 8 per cent located in the West Midlands and 7 per cent in the North East. The UK design industry is dominated by designers involved with various aspects of communication followed by digital and multimedia design. The UK design industry is polarized between firms that focus on providing services to firms located in the UK and designers that are also exporting services. International design competition has intensified, but 90 per cent of British design firms noted in the Design Council Survey that their main competitors were located within the UK (Design Council, 2010: 7). Countries like China, South Korea and Singapore are beginning to become important design centres and this has implications for the UK. British design firms need to ensure that they do not compete on price alone, but provide clients with specialist design expertise that is related to European markets. Design localization is important within the industry, with firms having to customize designs to meet the requirements of key markets. This means that British firms need a design presence in the US and Asia Pacific and also that design firms in places like Singapore and South Korea need to establish European design centres. Design localization can also be undertaken through the establishment of strategic relationships with design firms located in key foreign markets. More research is required to understand the complex evolving geographies of the developing international design industry.

THE COMPETITIVENESS OF CHINESE DESIGN FIRMS

China's rapidly growing economy has captivated the attention of economic geographers. Accomplishments that China has achieved in the last three decades have required half a century or more for many developing countries to achieve (Guthrie, 2012). The Chinese economy, from a total GDP (Purchasing Power Parity) perspective, ranks second in the world only to that of the European Union (Central Intelligence Agency (CIA), 2013). More importantly, China, known for its manufacturing activities, is the world leader in gross value for the production of industrial output (CIA, 2013). China's rapidly expanding industrial base has attracted foreign companies and also the growth of industrial design firms (Rocks, 2005; Toloken, 2010; Zhang *et al.*, 2011). More importantly, other critical factors contribute to China becoming a major force in industrial design. These include a growing popularity of design as a major amongst college students in China (Rocks, 2005), the international reputation of Chinese designers (Nussbaum, 2006; Zhang *et al.*, 2011) and strategies employed by Chinese firms (Rocks, 2005; Zhang *et al.*, 2011). These strategies involve Chinese firms building internal design firms or accessing external expertise to develop products for both local and global markets (Rocks, 2005). Rein (2006) points out that the success of Chinese companies like Haier and Lenovo is explained by successful investments in industrial design. Set against this backdrop, the following examines the competitive characteristics of Chinese industrial design firms and the extent to which they have eroded the competitiveness of firms located in developed countries such as the United States and United Kingdom.

In an exploratory study of Chinese industrial design firms, Vanchan and Ren (2013a) found that industrial design firms in China are relatively young (<5 years) and employ more people than American firms. The geographic distribution of Chinese firms is urban-centric, reflecting agglomeration economies and uneven geographic development in China, with a high concentration of firms located in the eastern part of the country. Similar to their US counterparts, Chinese firms serve clients in various sectors ranging from automotive to textiles and apparel. A notable difference is that the majority of Chinese industrial design firms' clients (over 60 per cent) were involved in the production of household goods and machinery, whereas these two sectors represented only about 30 per cent of US clients (Vanchan and Ren, 2013b).

From a service provision perspective, service diversity was found to be very strong amongst Chinese firms (Vanchan and Ren, 2013a). Applied research represented a growing trend amongst the service offering provided by both Chinese and US firms, while component and final product design services served as their principal service offering and source of revenues. Noticeably, the majority of Chinese firms claimed to provide a stronger level of assistance compared with American firms in various categories, including applied research, component design, final product design, improved product quality, improved product ergonomics and aesthetics/styling, improved client technological capability and increased client revenues (Vanchan and Ren, 2013b). Interestingly, this evidence suggested that US firms are more likely to assist clients by providing more sophisticated services including reducing product defect rates and increasing the ease of manufacturing procedures compared with Chinese design firms. This further suggests that US firms, in general, are more specialized and proficient in advanced technical services than Chinese firms. However, it is important to point out that the level of service sophistication varies amongst Chinese firms based on their service taxonomy.

From a service taxonomy perspective, Chinese Tier I (TPP) firms were found to offer a higher level of assistance to their clients in research performance, improved product quality, improved product ergonomics and improved client technological capability than did the non-TPPs (Tier II and III) (Vanchan and Ren, 2013a). No statistically significant differences were found between TPPs and non-TPPs in terms of age, size, sector focus, customer locations, client size, export involvement, technological capability and expected growth rates (Vanchan and Ren, 2013a). In addition, no statistically significant differences were found between the firms in employment, R&D spending, sales, profits and export growth. Besides the difference in service typology and intensity, the key areas with statistically significant differences amongst Chinese firms included human capital, investment in worker training and export propensity (Vanchan and Ren, 2013a).

From a human capital perspective, Chinese TPPs and non-TPPs exhibited statistically significant differences in workforce composition, skills, manufacturing experience and education. Chinese TPPs employed more workers with Associate's (three years of formal schooling) and Master's degrees compared with the non-TPPs (Vanchan and Ren, 2013a). Chinese firms exhibited a stronger association with manufacturing compared with the American firms, with an overwhelming majority of their employees (98 per cent) having manufacturing experience, of which half (51 per cent) had at least five years' manufacturing experience (Vanchan and Ren, 2013a, 2013b). Amongst the Chinese firms, TPPs reported more positive growth in investment in worker training over the study period compared with non-TPPs (Vanchan and Ren, 2013a). This aspect is also shared by the US TPPs (Vanchan, 2009).

From a market orientation perspective, Vanchan and Ren (2013b) found that Chinese companies primarily serve local and regional markets (75 per cent of companies serve local and provincial markets). However, about a third of Chinese companies indicated that they export (Vanchan, 2013) – about the same percentage as US companies involved with exporting their services (Vanchan, 2013). This underscores the increasing internationalization process of Chinese companies and China's role in the global economy (Dicken, 2011; Coe *et al.*, 2013). Regarding the propensity to export, TPPs in both the United States and China are more likely to export than non-TPPs. In addition, both American and Chinese exporters exhibited positive growth in sales, profits, job creation, R&D spending and investment in worker training (Vanchan and Ren, 2013b; Vanchan and MacPherson, 2008b).

The top five factors that contributed to the success of Chinese design firms included service quality, creativity, customer service, having specialist knowledge/abilities and existing relationships with clients (Vanchan and Ren, 2013a). This ranking is almost identical to the American design firms, with only a small change in the ranking of the last two factors, that is, existing relationships with clients ranked fourth and having specialist knowledge/abilities ranked fifth for the American firms (see Vanchan, 2009). The number of locations was the least important factor for Chinese firms (see Vanchan and Ren, 2013a) and this was also considered less significant for the American firms (see Vanchan, 2009). Interestingly, the key distinguishing factors that contributed to the success of Chinese TPPs and non-TPPs included service creativity, use of advanced or recently introduced technologies, having specialist knowledge/abilities, marketing strategy and existing relationships with clients (Vanchan and Ren, 2013a). Noticeably, marketing strategy was found to be the most important distinguishing factor (highest statistically significant difference) between the Chinese TPPs and their counterparts. This is a significant finding as 'marketing strategy' was not identified as important by many American firms (Vanchan, 2009).

Chinese companies were found to have higher growth rates compared with American companies (Vanchan and Ren, 2013b). This growth rate can be explained by the continuing growth of the Chinese economy and its ability to attract foreign investment, particularly in the manufacturing sector. More importantly, China is moving up the value chain by promoting quality design and products. Chinese contract manufacturers continue to build up their in-house design teams (Balfour and Roberts, 2003; Bryson and Rusten, 2011) and attract foreign designers (Toloken, 2010). These efforts by Chinese companies challenge the existing international division of labour between development and design and production (Bryson and Rusten, 2011) and the competitive advantage of industrial design companies located in developed market economies. They also show the adaptability of Chinese companies, including industrial design companies, to develop products to meet the requirements of export and local markets.

CONCLUSIONS

Companies compete in many ways: on the basis of price, process, product, innovation, patents, brands, fashion and design. These different forms of competitiveness are mirrored in Chandler's three pillars of corporate form – management, marketing and production. These represent generic business skills that are applicable to the majority of firms (Chandler, 1977: 419). Chandler's account of industrial production highlights the relationship that exists between production and consumption, but through the medium of marketing. Marketing plays an important role in corporate competitiveness, but it is a supporting role that is wrapped around the production of products and services. Marketing also involves market research in which firms try to learn from consumers. There is an important missing element in Chandler's account and this omission is design. Design plays an important role in corporate and national competitiveness. Chandler's three pillars of corporate form need to be extended to include a fourth – design. Design should be understood as a process that is located at the interface between industry, market technology and culture (Bonsiepe, 2007).

Design is related to culture and society as it is involved in incorporating place into product and product into place (Rusten and Bryson, 2007). Design not only concerns the shape and functionality of end products but is also implicated in the creation of sustainable or green products and in the development of inclusive products.

This chapter has explored the evolving design industries in the US, Europe and China. There are a number of important issues that require further research. First, the relationship between design and corporate competitiveness requires further attention from social scientists. Thus, what role does design play in enhancing the competitiveness of different economic sectors – from manufacturing to services and also social enterprises? Second, the competitiveness of design firms requires further research. In particular, research needs to explore the ways in which design firms are developing strategic relationships with firms located in other countries. Research needs to focus on understanding the approaches that are being developed to create design firms that are able to access competencies in most consumer markets – the US, Europe and Asia Pacific. This involves foreign direct investment and the development of strategic alliances. Third, the emergence of freelance designers as a significant element within the design industry needs to be explored. Thus, what role do freelance designers play in the provision of design services? Is this role about supporting and providing additional capacity for design firms or in-house design teams, or does this involve working with other freelance designers to assemble teams that are able to compete for larger client projects?

The emergence of China as an important provider of manufactured goods raises important questions regarding the relationship between Chinese design and manufacturing. Design is an emerging industry within China as most Chinese manufacturing firms are currently competing on price rather than design. This is beginning to change as labour rates rise and China's competitiveness, founded on low-cost labour, is eroded. Chinese companies have shown a commitment to move up the value chain by promoting quality design and products. Companies that provide well-designed and manufactured branded products will be more successful and profitable than anonymous contract manufacturers. Competition is intensifying as Chinese contract manufacturers establish in-house design teams (Balfour and Roberts, 2003; Bryson and Rusten, 2011), attract foreign designers (Toloken, 2010), and promote human capital and excellence in industrial design (Zhang *et al.*, 2011). These efforts by Chinese companies challenge the existing international division of labour between development and design and production (Bryson and Rusten, 2011), and illustrate the growing manufacturing base and market in China. Nevertheless, there are many small and medium-sized Chinese manufacturers that rely on copying designs produced by companies based in America and Europe. These firms have yet to develop an in-house design capability and compete on the basis of price rather than design.

The emergence of Chinese design has perhaps important implications for European and American designers. It is important to appreciate that different countries have different attitudes and expectations regarding products. This is especially the case in relation to colours and patterns. The implication is that American and European companies wanting to sell products to Chinese consumers must customize them to meet the expectations of Chinese consumers. This process also means that Chinese producers must design products that meet the expectations of consumers located in America and Europe. The emergence of Chinese brands and a Chinese design culture has implications for America and Europe as Chinese firms will have to establish design centres in these core markets or use design firms that have capabilities in these foreign markets. All large Chinese firms with aspirations to target the American and European markets will have to establish design capabilities in these markets and this will generate local employment. While TPPs seem to be the leader in the design industry across geographic space, the key to success and long-term growth lies in the ability of companies to adapt in a timely manner, to have clear and practical market orientation, and to excel in service provision and human capital.

The relationship between design and consumption operates at a number of different scales. Some products can be sold in all markets, but many products and services require alterations to meet local consumer demand (Bryson and Rusten, 2011). These alterations may be driven by local standards, but they may also be about local fashions and cultures. Research needs to be undertaken that focuses on unravelling the complex relationships between geography, design and consumption. These relationships involve all types of design – from packaging, product names and brands, advertising to style. This means that a product's appearance may be the same in key markets, but the supporting materials (brand, packaging, advertising and marketing) will be different. The global economy should be conceptualized as a mosaic of different consumption communities. For firms trading beyond their home market this complexity is negotiated by the activities of design companies. In this context, design must be conceptualized as a primary process that is fundamental to the competitiveness of all types of company.

NOTE

1. The analysis presented in this chapter draws upon datasets that use different definitions of 'design' and this will explain some of the differences in the size of the design sector. These differences are not critical for the argument developed in this chapter.

REFERENCES

Balfour, F. and Roberts, D. (2003), China's dream team: industrial designers are making a mark. *Business Week*, 1 September.

BEDA (2006), *European Design Report: The European Design Industry in Facts and Figures*. Bureau of European Design Associations: Brussels.

BEDA (2007), *European Design Report: Supplement on Design(ers) in Europe*. Bureau of European Design Associations: Brussels.

Beesley, N. and Rothwell, R. (1987), Small firm linkages in the United Kingdom. In Rothwell, R. and Bessant, J. (eds), *Innovation, Adaptation and Growth*. Elsevier: Amsterdam.

Belecheanu, R., Riedel, J. and Pawar, K.S. (2006), A conceptualisation of design context to explain design trade-offs in the automotive industry. *R&D Management*, 36(5): 517–529.

Bonsiepe, G. (2007), The uneasy relationship between design and design research. In Michel, R. (ed.), *Design Research Now*. Essay and Selected Projects, Board of International Research in Design, BIRD. Birkenhäuser Verlag, AB: Basel, Boston, Berlin.

Boothroyd, G. (1994), Product design for manufacture and assembly. *Computer-Aided Design*, 26(7): 505–520.

Bryson, J.R. and Rusten, G. (2011), *Design Economies and the Changing World Economy: Innovation, Production, and Competitiveness*. Routledge: London.

Bryson, J.R., Daniels, P.W. and Rusten, G. (2004), Design workshops of the world: the production and integration of industrial design expertise into the product development and manufacturing process in Norway and the United Kingdom. *Working Paper No. 53/04*, Design Norwegian Competitiveness, Institute for Research in Economics and Business Administration.

Chandler, A.D. (1977), *The Visible Hand: Managerial Revolution in American Business*. The Belknap Press of Harvard University Press: Cambridge, MA.

Chandra, B. and MacPherson, A. (1994), The characteristics of high-technology manufacturing firms in a declining industrial region: an empirical analysis from western New York. *Entrepreneurship and Regional Development*, 6: 145–160.

Chiesa, V., Manzini, R. and Pizzurno, E. (2004), The externalization of R&D activities and the growing market of product development services. *R&D Management*, 34: 65–75.

CIA (2013), *World Fact Book*, China, accessed online via www.cia.gov (2013).

Coe, N.M., Kelly, P.F. and Yeung, H.W.C. (2013), *Economic Geography: A Contemporary Introduction* (2nd Edition). John Wiley & Sons, Inc.: Oxford.

Coffey, W.J. and Drolet, R. (1996), Make or buy: internalization and externalization of producer service inputs in the Montreal Metropolitan Area. *Canadian Journal of Regional Science*, 19(1): 25–48.

Consoni, F. and Quadros, R. (2006), From adaptation to complete vehicle design: a case study on product development capabilities of multinational assemblers in Brazil. *International Journal of Technology Management*, 36(1–3): 91–107.

Costlow, T. (2002), Outsourcing takes hold in design. *Design News*, 57: 8–12.

Daniels, P.W. (2013), The transition to producer services in China: opportunities and obstacles. In Yeh, A.G.O. and Yang, F.F. (eds), *Producer Services in China: Economic and Urban Development*. Routledge: London and New York.

Daniels, P.W., Rubalcaba, L., Stare, M. and Bryson, J.R. (2011), How many Europes? Varieties of capitalism, divergence and convergence and the transformation of the European services landscape. *Tidschrift voor Economische en Sociale Geografie*, 102(2): 146–161.

Dankbaar, B. (2007), Global outsourcing and innovation: the consequences of losing both organizational and geographic proximity. *European Planning Studies*, 15: 53–61.

Design Council (2006), Design in Britain 2005–2006. Accessed via www.designcouncil.org.uk/factfinder (19 July 2007).

Design Council (2010), *Design Industry Insights: Comments and Conversations on the Business of Design in the UK*. Design Council: London.

Dicken, P. (2011), *Global Shift, Sixth Edition: Mapping the Changing Contours of the World Economy*. The Guilford Press: New York.
Faste, T. and Faste, H. (2012), Demystifying 'design research': design is not research, research is design. *IDSA Education Symposium 2012*. Boston, August.
Gertler, M. and Vinodrai, T. (2004), *Designing the Economy: A Profile of Ontario's Design Workforce*. Report prepared for the Design Industry Advisory Committee of the Ontario government by the University of Toronto, Department of Geography and Munk Center for International Studies (copy available from meric.gertler@utoronto.ca).
Guthrie, D. (2012), *China and Globalization: The Social, Economic and Political Transformation of Chinese Society*. Routledge: New York.
Hagedoorn, J. (1996), Trends and patterns in strategic technology partnering since the early seventies. *Review of Industrial Organization*, 11: 601–616.
Hagedoorn, J. (2002), Inter-firm R&D partnerships – an overview of patterns and trends since 1960. *Research Policy*, 31: 477–492.
Hall, P. and Soskice, D. (eds) (2001), *Varieties of Capitalism: The Institutional Foundations of Comparative Advantage*. Oxford University Press: Oxford.
Industrial Design Society of America (2005, 2010, 2013). Accessed via http://ww.idsa.org.
Kalafsky, R. (2002), The role of location in a mature manufacturing sector: an examination of the U.S. machine tool industry. PhD dissertation, Department of Geography, University at Buffalo, New York.
Kalafsky, R. (2006), Performance and practice: examining the machine tool industries of Japan and the United States, *Tijdschrift voor Economische en Sociale Geografie*, 97(2): 178–194.
Laursen, K. and Salter, A. (2006), Open for innovation: the role of openness in explaining innovation performance amongst UK manufacturing firms. *Strategic Management Journal*, 27(2): 131–150.
Lynn, L. and Salzman, H. (2007), 'Innovation shift' to the emerging economies: cases from IT and heavy industries. *Sloan Industry Studies Working Paper*. www.industry.sloan.org/industrystudies/workingchapters (accessed 18 July 2014).
MacPherson, A. (1997), The contribution of external service inputs to the product development efforts of small manufacturing firms. *R&D Management*, 27(2): 127–144.
MacPherson, A. (2007), New product development services and industrial innovation: panel data from a sample of New York State scientific instruments producers. Canada–US Trade Center, Occasional Paper 32, State University of New York at Buffalo.
MacPherson, A. and Vanchan, V. (2010a), Locational patterns and competitive characteristics of industrial design companies in the United States. In Rusten, G. and Bryson, J.R. (eds), *Industrial Design, Competition, and Globalization*, Palgrave Macmillan: Basingstoke, UK.
MacPherson, A. and Vanchan, V. (2010b), The outsourcing of industrial design by large U.S. manufacturing companies: an exploratory study. *International Regional Science Review*, 33(1): 3–30.
Midler, P. (2011), *Poorly Made in China: An Insider's Account of the China Production Game*. John Wiley & Sons: Hoboken, NJ.
Nussbaum, B. (2006), The best product design of 2006. *Business Week*, 29 June.
O'Connor, K. (1996), Industrial design as a producer service: a framework for analysis in regional science. *Chapters in Regional Science*, 3: 237–252.
Okudan, G.E. and Zappe, S.E. (2006), Teaching product design to non-engineers: a review of experience, opportunities and problems. *Technovation*, 26: 1287–1293.
Parker, G.G. and Anderson, E.G. (2002), From buyer to integrator: the transformation of the supply-chain manager in the vertically disintegrating firm. *Production and Operations Management*, 11: 75–91.
Pritchard, D. (2001), The globalization of commercial aircraft production: implications for U.S.-based manufacturing activity. *International Journal of Aerospace Management*, 1: 213–226.
Pritchard, D. and MacPherson, A. (2007), Strategic destruction of the North American and European commercial aircraft industry: implications of the systems integration business model. *Airframer: The Journal of Aircraft Manufacturing*, February, Issue 17:1–8.
Reeves, M. and Deimler, M. (2011), Adaptability: the new competitive advantage. *Harvard Business Review*, 135–141.
Rein, S. (2006), Chinese tech firms combine form and function to capture the local market. Seeking Alpha Web platform for investment research. Accessed via: http://seekingalpha.com/article/18412-chinese-tech-firms-combine-form-and-function-to-capture-the-local-market (accessed 7 January 2015).
Rocks, D. (2005), China design. *Business Week Online*, 13 October.
Rusten, G. and Bryson, J.R. (2007), The production and consumption of industrial design expertise by small and medium-sized firms: some evidence from Norway. *Geografiska Annaler*, 89(1): 75–87.
Rusten, G. and Bryson, J.R (eds) (2010), *Industrial Design, Competition, and Globalization*. Palgrave Macmillan: UK.

Sturgeon, T.J., Levy, F., Brown, C., Jensen, J.B. and Weil, D. (2006), *Services Offshoring Working Group Final Report*. Industrial Performance Center, MIT, http://web.mit.edu/ipc/www/IPC (accessed 11 July 2007).
Toloken, S. (2010), Foreign designers open shop in South China. *Plastics News*, 22(27): 6.
US Census Bureau (2013), Economic census. Accessed via www.census.gov.
Vanchan, V. (2007), Communication and relationships between industrial design companies and their customers. *Industrial Geographer*, 4(2): 28–46.
Vanchan, V. (2009), *The Competitive Characteristics of U.S. Industrial Design Companies*. VDM Publishing Company: Germany.
Vanchan, V. (2013), The nature of contracts, network cultivation, and relationships between industrial design companies and their clients. *Working Paper, Department of Geography and Planning, Buffalo State, The State University of New York*.
Vanchan, V. and MacPherson, A. (2008a), The recent growth performance of U.S. companies in the industrial design sector: an exploratory study. *Industry and Innovation*, 15(1): 1–17.
Vanchan, V. and MacPherson, A. (2008b), The competitive characteristics of US companies in the industrial design sector: empirical evidence from a national survey. *Competition and Change*, 12(3): 262–280.
Vanchan, V. and Ren, H. (2013a), Competitive dynamics of industrial design companies in China: exploratory study. *Working Paper, Department of Geography and Planning, Buffalo State, The State University of New York*.
Vanchan, V. and Ren, H. (2013b), Are Chinese design companies a threat? A comparative study between U.S. and Chinese companies. *Working Paper, Department of Geography and Planning, Buffalo State, The State University of New York*.
Vinodrai, T. (2006), Reproducing Toronto's design ecology: career paths, intermediaries, and local labor markets. *Economic Geography*, 82(3): 237–264.
Vinodrai, T. (2010), Designed here, made there? Project-based design work in Toronto, Canada. In Rusten, G. and Bryson, J.R. (eds), *Industrial Design, Competition, and Globalization*, Palgrave Macmillan: UK.
Zhang, D., Hu, P. and Kotabe, M. (2011), Marketing–industrial design integration in new product development: the case of China. *Journal of Product Innovation Management*, 28: 360–373.

FURTHER READING

Bryson, J.R. and Rusten, G. (2011), *Design Economies and the Changing World Economy: Innovation, Production, and Competitiveness*. Routledge: London.
Vanchan, V. (2007), Communication and relationships between industrial design companies and their customers. *Industrial Geographer*, 4(2): 28–46.
Vanchan, V. and MacPherson, A. (2008), The competitive characteristics of US companies in the industrial design sector: empirical evidence from a national survey. *Competition and Change*, 12(3): 262–280.
Vanchan, V. and MacPherson, A. (2008), The recent growth performance of U.S. companies in the industrial design sector: an exploratory study. *Industry and Innovation*, 15(1): 1–17.

11. Intellectual property and patents: knowledge creation and diffusion

Dieter F. Kogler

INTRODUCTION

The "profit motive" is considered the principal objective that guides businesses' drive for success. The underlying goal in this context is to generate profits, which in turn generate benefits for the entity itself, as well as its shareholders. While the prospect of simplifying the complex structures and processes that guide the activities of corporations into one argument is rather appealing, it is probably too narrow a view to explain the stark variations among ventures in innovative output, entrepreneurial engagement or performance measures, which certainly persist even after controlling for input factor conditions (Hollis and Nell, 1975). In essence, a diverse set of aspects more or less determine firm performance at various levels, and not necessarily at the same intensity over time (Dess and Robinson, 1984; Afuah and Utterback, 1997).

Changing the point of view from an individualistic to a general perspective, to be innovative seems another objective that is common to the vast majority of market-oriented businesses. The innovation machine in this framework is considered the engine of the economy that creates the growth miracle of capitalism (Baumol, 2002). More explicitly, free market processes, or variations thereof (Hall and Soskice, 2001), lead to the formation of companies, which subsequently compete with each other based on price, quality and by constantly generating novel products and processes, that is, innovations (Greenhalgh and Rogers, 2010). It is exactly this constant progression of creating, improving and replacing existing processes, methods of production and products that keeps the economy in motion. This was acknowledged by Schumpeter (1942), who termed this process "creative destruction," and further noted that this is endogenously generated. He considered this to be the engine of growth.

Intellectual property rights (IPRs), and especially patents, are recognized as having established the conditions necessary for the development of a market system where the search for new knowledge of economic value is rewarded (Scotchmer, 2004). IPRs create the incentive for entities to engage in a process that is driven by continuous innovation, which in turn will initiate technological change in the broader sense. Thus, inventions that hold potential economic value constitute the foundation of technological change and economic growth. Once an invention is successfully commercialized, finds adopters and subsequently starts to diffuse across the economy, it is usually referred to as an innovation (Rigby, 2000).

The origin of IPRs dates back to early advances in science in 15th-century Europe; that is, the Renaissance. It is considered that it was the institutional reforms and social changes that took place during this period which laid the foundation for the Industrial Revolution and the consequential economic success of the Western economies

(Rosenberg and Birdzell, 1986; Mokyr, 2009). While it was first monarchs and local authorities that rewarded either local inventors, or foreigners who agreed to import their new technologies, by providing them with patents that ensured protection for a limited term, it was later national governments that further refined the system of IPRs. Over the course of history, public entities, or the public sector in general, have sought ways to encourage and promote innovation by putting in place institutions that supply ideas and encourage investments in research and development (R&D) (Feldman and Kogler, 2008). It is the interaction between institutions and organizations that shapes the evolution of an economy. In this context we can consider institutions, or government in general, as the entities that define the rules of the game (North, 1991) that affects everything from the supply of ideas to the demand for final products, the legal and regulatory framework that determines the functioning of market, and the ultimate accountings of what are profits and how these will be allocated. Organizations and their entrepreneurs, on the other hand, are the players that engage through their drive for profit and their creative potential in a game where the primary objective is to take advantage of the defined opportunity set in order to generate returns (Geels, 2004).

IPRs are an integral part of this game, as they remove the fear of individuals who engage in innovation that other agents will appropriate ideas and usurp the gains that should accrue to the inventor. Of course, this largely depends on the predictable implementation and enforcement of laws and regulations that promote competition and innovation by penalizing opportunistic actions (Gould and Gruben, 1996). Thus, economic growth depends on society's ability to create institutions that are productive and stable, and which are able to adapt to changing political and economic situations (North, 2005). Overall, this suggests that institutions facilitate the efficient performance of human activity, creativity and inquiry, all of which are essential for economic development and subsequent prosperity.

The following section of this chapter provides a brief overview of different forms of IPR, and in particular their use in various sectors, and also a more detailed discussion of the patent system. Subsequently, the main focus will be directed at patent data, and especially patent statistics and databases as they are applied in an attempt to measure knowledge creation and diffusion. What follows is a case study that aims to demonstrate how patents can be used to generate more accurate patent metrics in terms of ranking places according to their innovative output. The final section of the chapter provides some general concluding remarks.

INTELLECTUAL PROPERTY AND MANUFACTURING SECTORS

What is the right way to protect our novel product, process or design? This is the principal question individuals and entities face when they are presented with an invention that has potential economic value (Hall et al., 2012).[1] Various possible avenues can be pursued in this context – copyrights, trademarks, trade secrets and patents are the most obvious ones that spring to everybody's mind. Which of these will be applied is subject to the type or kind of invention that should be protected, as well as the economic sector, as outlined in the brief overview below (Maskus, 2000).

Copyright

Literary and creative works usually depend on a copyright as their primary way of IP protection. In its most basic definition, the copyright gives the creator of an original work exclusive right of use for a specific length of time, which can vary by jurisdiction (Goldstein, 2001). In addition to being credited for the work, a copyright also ensures that the creator has a say in who may adapt or perform the work, and also who may financially benefit from it. This IP protection is most frequently used for creative works like books, publications, plays, performances, music, art work, and motion pictures and broadcasts.[2] Copyright is usually an unregistered right, although creators have the opportunity to register their copyright at the U.S. Copyright Office, and frequently is only enforced as a civil matter. With the advent of advanced information and telecommunications technologies, traditional copyright laws have come under pressure, and business ventures in particular have sought an extension of their exclusive rights to copy, reproduce, distribute and adapt their creations within the legal framework. In addition to individuals, the media and publishing industry sectors rely heavily on this form of IP protection (Corrigan and Rogers, 2005).

Trademarks

Trademarks, on the other hand, are used in a large variety of economic sectors; probably more so than any other kind of IP protection mechanism. In brief, a trademark is a recognizable sign, which can be a design, logo, picture, word, expression, color, smell or even a sound, which identifies and distinguishes a product or service of a particular entity – that is, an individual, firm, organization, public institution, and so on – from those of others. The idea is that a trademark conveys a certain standard of a product, process or design (Tadelis, 1999). It can be owned by individuals as well as public and private entities, and is usually visibly located on the product, in the software, packaging, and so on. Like most IPRs trademarks can also be owned or licensed; however, trademarks, unlike patents and copyrights, do not necessarily have a limited life-span as long as they are in continuous use by their respective owner. While many countries have common law rights that protect unregistered trademarks, it appears that most valuable trademarks have been registered, which in turn provides better protection in potential cases of trademark infringement. In general, trademarks aim to indicate and protect the source of a product or service rather than a new and useful invention; something that is the objective of trade secrets and patents, as outlined in turn.

Trade Secrets

While trade secrets are part of the wider intellectual property framework, they certainly occupy a special status. In brief, trade secrets are proprietary information about a particular production process. The most infamous of all trade secrets seems to be the recipe for Coca-Cola's Coke. However, notable examples also include composition formulae of well-known products, such as WD-40, McDonald's Big Mac special sauce recipe and the methodology that is used to generate the *New York Times*' Best-Seller List. Trade secrets are usually a valid form of protection, which allow appropriation of the economic benefits of an invention, where patents and copyrights seem unattractive means to do so. This is especially true if the innovator's aim is to disclose very little information about the novel product or process in order to avoid the leakage of business proprietary information (Cohen et al., 2000). Besides the advantage of non-disclosure, trade secrets are not limited in time, as they continue as long as the secret is not revealed to the public, and they do not incur any registration costs, especially when compared with patents. The obvious disadvantage of trade secrets is that a competitor can potentially discover the secret through either reverse engineering or research activities, which in turn would entitle this entity to make use of it (Rockman, 2004). Other alternative IP protection mechanisms, which follow close in line with trade secrets, are non-disclosure and non-compete agreements in employment contracts. It should be noted that non-competes are not enforced in all states, which in turn has an impact on inventor mobility (Marx et al., 2009).

Patents

While an invention is categorized by the development or discovery of a new, useful and non-obvious process or product, it is the commercial realization of economic value derived from it that makes it an innovation in the broader sense. Thus, inventions, unlike random ideas, constitute an initial, crucial step in the process of innovation.[3] In its most basic definition, a patent is a temporary property right to the commercial use of a device, characterized by a product or process, and is granted if the invention is nontrivial and has a potential commercial value. In other words, a patent is the commercialization of new knowledge constituting the essence of technological change.

The focus in the reminder of the chapter is on patents awarded by the United States Patent and Trademark Office (USPTO). There are some significant variations between U.S. patent right and that exercised by the European Patent Office (EPO) as well as other national patent offices (Rockett, 2010). The most prominent, and for researchers highly relevant, of these differences pertain to how the patent search is carried out – that is, the evaluation of an invention by a patent examiner – and how patent citations – that is, references to prior art – are derived (Michel and Bettels, 2001). However, the rationale of obtaining and granting a patent is common to all national patent offices.

Various types of patent are issued by the USPTO, including utility, design, plant and reissue patents. Nevertheless, because the vast majority of patents granted at the USPTO are utility patents, these are the primary focus in most available patent database and research efforts. A utility patent, according to the USPTO (2000), is:

> issued for the invention of a new and useful process, machine, manufacture, or composition of matter, or a new and useful improvement thereof; it generally permits its owner to exclude others from making, using, or selling the invention for a period of up to twenty years from the date of patent application filing, subject to the payment of maintenance fees. Approximately 90% of the patent documents issued by the PTO in recent years have been utility patents, also referred to as "patents for invention".[4]

The USPTO is a federal agency in the Department of Commerce. Since its foundation 200 years ago the basic role of the USPTO has remained the same: to promote the progress of science and the useful arts by securing for limited times to inventors, the exclusive right to their respective discoveries (Article 1, Section 8 of the United States Constitution). While the agency's mission has not changed since its inception, significant policy changes have been implemented throughout the decades. Most notable of these are the Invention Secrecy Act of 1951, which aims to prevent disclosure of new inventions and technologies that potentially present a possible threat to the national security of the United States, and the Bayh–Dole Act, adopted in 1980, which was the result of a long-lasting debate about the U.S. research system's failure to turn research and scientific achievements funded by the federal government into innovations. In essence, the latter entitled universities and other not-for-profit research organizations to take ownership of intellectual property rights from research funded by the federal government, including the opportunity to issue licenses. The implementation of the Bayh–Dole Act did not change the patent granting mechanism, but it certainly did change the economic landscape in terms of the additional entities that now pursued patent rights (Nelson, 2001). Until recently, the USPTO followed the principle of "first-to-invent," which aimed to ensure that the grant of a patent for an invention was awarded to the individual or entity that indeed first developed a novel product or process. However, following the America Invents Act (AIA), which came into effect in March 2013, the agency now follows a "first-inventor-to-file" system for patent applications. Some of the arguments for this change include that the implementation of the AIA will speed up patent review and cut down on the current backlog of applications, reduce patent litigation, and help to reduce the risk of infringement for start-up companies with fewer resources than their well-funded and established counterparts, all of which in turn should create jobs and secure the competitive position of the U.S. in the global technology market. Of course, there are also a number of counter arguments that claim that the AIA will make no difference, or in the worst case, will hurt small inventors with limited resources (Hurst, 2013).

Empirical studies have shown that industries vary significantly in the average number of patents generated by each dollar of R&D invested. In essence the propensity to patent is not the same in every industry (Scherer, 1983). The patent system does not cover all industries – for example, many of the service sectors rely on other mechanisms of intellectual property protection rather than patents, whereas others operate in progressive environments where the time lag between new products and processes is so short that it is not feasible to apply for patents. Finally, if an invention has the potential to generate revenues for a time period that is longer than 20 years (the USPTO currently guarantees intellectual property protection for up to 20 years), and the product or process is considered not to be replicable through reverse engineering or by other means, it is very likely that the inventor will choose "secrecy" as the best option to extract as much value as possible from the invention.[5] Examples of this would be the formula for Coca-Cola, or the process of generating crystal glass developed by Swarovski.[6] These examples clearly show that patent statistics only capture certain segments of innovation and technological change (Teece, 1986; Levin et al., 1987; Klevorick et al., 1995).

However, even if patenting is the primary form of IP protection in a specific industry sector, there are strong variations in the observed sectoral growth rates over time. There are several possibilities that might explain these variations. For example, the overall rate of patent generation, in a particular sector, may be the result of the maturity level of an industry, or a general trend towards or away from patenting as a primary form of IP protection. Another reason for uneven growth among industry sectors could be that some of them have become less important in the overall national economy, and thus inventiveness as such has declined over time, which in turn translates into smaller growth rates. Some of these issues, and in particular the propensity to patent among sectors, will be discussed in further detail in subsequent sections.

On that note it should be mentioned that many patents are never commercialized, but rather serve as an intermediary to other, more promising technologies (Ter Wal and Boschma, 2009). Sometime patents are also used solely for the purpose of building up a patent portfolio of a company. There are several reasons that could explain such a move, including the aim to exclude competitors or new enterprises from entering a specific technology space, or the goal to improve a firm's position in future business negotiations, or finally to improve the reputation and technological image of a firm (Blind et al., 2006). More recently there has been a surge in the number of applied patents at the USPTO (Kortum and Lerner, 1999). Kortum and Lerner (1998) list several reasons why this is not unique to the USPTO. These include evidence that the United States has not become a more attractive destination for patents over time, that the increase is among a wide spectrum of technologies, and that the growth of biotechnology and software patenting alone does not explain the increase. In addition, this study also suggests that the surge in patenting is not driven by increased research efforts, but rather by changes in the management of innovation.

Appropriability conditions, that is, the factors that govern an innovator's ability to capture profits generated by an innovation, strongly determine the type of IPR that is most suitable (Teece, 2010). Patents confer strong appropriability in sectors where knowledge is highly codified, for example in the biotechnology or mechanical sectors (Levin et al., 1987). In cases where innovation is embedded in processes, or where it relies on a high degree of tacit knowledge inputs (specialized knowledge acquired through practice), other forms of IP protection might be more feasible. Nevertheless, IPRs are essential in avoiding market failure that is likely to occur in the absence of IPRs due to the positive externalities generated by novel products and processes, and knowledge in general. In this context, government and public entities play a very important role, as outlined above (Feldman and Kogler, 2008). Essentially, without patent protection an inventor has little to no incentive to invest time and money into something that can easily be copied by competitors without incurring significant R&D costs. Patents therefore correct this market failure by providing a temporary monopoly to the inventor.

MEASURING INVENTION: PATENT STATISTICS

The creation of original knowledge and its transformation into novel products and processes, which are subsequently employed in the marketplace in the form of innovations, constitute the very essence of the process of technological change that leads to economic growth (Nelson and Phelps, 1966; Dosi et al., 1988). Patents and patent statistics encompass an incredible wealth of information with the potential to facilitate a multitude of approaches in the investigation of knowledge creation and diffusion processes (Jaffe and Trajtenberg, 2002). Jacob Schmookler (1966) is regarded as the pioneer in the use of patent statistics, first outlined in his seminal work "Patents, invention, and economic development"; in addition, Frederic Scherer (1965) should also be mentioned, as he was another key initiator of this approach. While the idea of using patent data as indicators of the economic sources of inventive activity goes as far back as Schmookler (1962, 1966) and Scherer (1965), it was later significantly expanded on by Scherer (1982) and Griliches (1984), and more recently has been pursued by a growing number of researchers in various disciplines. Academics and practitioners alike have expressed a strong interest in uncovering and explaining the sources of invention and innovation – that is, the creation, diffusion and economic application of knowledge – resulting in a substantial body of literature that has produced a number of studies that have essentially broadened the understanding of the spatial and sectoral patterns of knowledge flows and spillovers. Surprisingly, despite the steadily increasing number of inquiries that utilize patent data in their analysis, relatively few publications exist that discuss actual patent databases (Hall et al., 2001; Dernis and Khan, 2004; Goto and Motohashi, 2007), methodological issues as they relate to the analysis of patents (Pavit, 1985; Griliches, 1990; Archibugi, 1992; Schmoch, 1999) or the various applications of patent analysis (Breitzman and Mogee, 2002). In contrast, Jaffe and Trajtenberg (2002) and Moed et al. (2004) offer excellent insight into patent-related analysis issues, through a collection of papers that are highly relevant to this stream of research.

Throughout the literature, the method – in which patent-related measures are applied in research concerning economic gain as a result of technological change – is broken into two broad categories. First, there are studies that utilize patent data to analyze the inventive behavior of individuals, organizations and/or spatial entities. This approach largely relies on absolute or relative patent counts to measure knowledge creation and inventive output, which are considered proxies for innovation. The second category of studies employs co-inventor data and/or patent citation-related indicators in order to gain a better understanding of knowledge creation and diffusion processes. In this approach knowledge flows, that is, the deliberate exchange of knowledge between individuals, are studied by means of co-inventorship analysis. On the other hand, knowledge spillovers (that is, the unintended exchange of knowledge) are investigated by looking into the citations patents make to previous patents, or receive from future ones. Patent counts definitely provide a good measure of the intensity of knowledge creation itself, and if disaggregated into spatial and sectoral sub-units they have the potential to delineate varying degrees of inventive capacity and levels of output among the entities under investigation. Patents are typically geographically concentrated, which subsequently reflects a concentration of R&D activity. Feldman (1994: 31) finds correlation coefficients of 0.93 between the location of new products introduced to the market and broad patent categories, and 0.88 between patent counts and corporate R&D expenditures, at the state level, across the United States.

An analysis of a sample of publicly traded firms in the pharmaceuticals, electronics and chemicals sectors by Schoenecker and Swanson (2002: 41) derives similar results in terms of patent output and R&D spending, with correlation coefficients of 0.8 or above for all three industries. However, the same study also finds that patent counts and new product introductions into the market only correlate at 0.3 or less in the pharmaceuticals and electronics industry, while the value in the chemical industry is 0.63. It has been stressed in the past (Griliches, 1990) that patent counts should be considered a very noisy indicator in the assessment of the actual value of individual inventions because of the highly skewed distribution of patents that are actually successfully commercialized.

Van Zeebroeck (2009) offers insight into the puzzle of patent value indicators, and a detailed discussion of the literature in this stream of patent analysis. Overall, statistics based on the number of patents granted over a specified period of time, and in a particular industry sector, provide a good indicator of the intensity of knowledge creation and subsequent innovation within a specific territory.

MEASURING KNOWLEDGE FLOWS: PATENT CITATIONS

The first use of patent citation analysis as a method for studying knowledge spillovers appears to date back to the late 1970s (Oppenheim, 1997, 2003). However, it is the findings and methodology developed by Jaffe et al. in their seminal paper published in 1993 that have especially inspired both geographers and economists to follow this line of research. This type of patent analysis is the approach that has attracted more attention in the literature, as opposed to the alternative method that uses patent counts. In a proposed new theory of economic geography, Paul Krugman (1991: 5) suggests that "the most striking feature of the geography of economic activity" is "concentration," and further that "production in many industries is remarkably concentrated in space." This finding has spurred an increased interest in the economics literature, focusing on implications of the concentration of economic activity for economic growth. In essence, this literature states that externalities derived from increasing returns of scale result in higher growth rates (Romer, 1990). In other words, the larger the extent of economic activity, the more new knowledge will be generated, which in turn will spill over to other economic entities and thus generate economic growth. In addition, these knowledge spillovers are geographically bounded within the region from where the knowledge originated (Feldman, 1994). Following this line of reasoning some economists even propose that the only compelling reason for the existence of cities is grounded in these increasing returns to agglomerations of resources, combined with a higher level of communication that facilitates knowledge spillovers in these localities (Lucas, 1993). Although the geography of innovation (Feldman, 1994; Feldman and Kogler, 2010) and new economic geography (Clark et al., 2000) literatures provide new evidence regarding the agglomeration of economic activities, this is not a new approach to analyzing economic activity in a geographic context, and to some extent only reinforces some of the principal findings already made over a century ago (Marshall, 1920).

There is a broad consensus among scholars in various disciplines that concentration is beneficial for regional development. However, the literature provides two contradictory theories concerning what specific type of economic activity is actually desirable in a particular geographic region in order to spur economic growth. One theory promotes an increased concentration of a particular industry within a specific geographic location, as it is considered that this will facilitate knowledge spillovers across firms (Baptista and Swann, 1998). This line of reasoning follows an earlier debate about the economics of location (Lösch, 1954). In contrast, another theory (Jacobs, 1969) argues that it is specifically the exchange of knowledge across diverse firms and economic agents that generates increased returns of economic knowledge (Feldman and Audretsch, 1999). Simply stated there are two opposing approaches available, one that argues for specialization and, on the opposite end, one that suggests the establishment of a diverse industry base in a specific geographic location. The question surrounding which of the two concepts actually is the applicable one, in facilitating knowledge creation and spillovers, and subsequently should be applied in relevant innovation policies, has created a vibrant discussion in a number of disciplines concerned with regional development and growth (Duranton and Puga, 2000; Barrios et al., 2006; Scott, 2006).[7]

There is no agreement in the literature whether intra-sectoral (Jaffe, 1986) or inter-sectoral (Fung and Chow, 2002) knowledge externalities are dominant in explaining the innovation process, and resulting knowledge flow patterns (Beaudry and Schiffauerova, 2009). It is widely recognized that knowledge exchange takes place at various spatial levels and across multiple technology sectors (Howells, 2002), and therefore a dichotomous framework of these two apparently distinctive categories of knowledge spillover might be deceptive as the derived results are a direct outcome of what actual type of externality is under investigation. For example, Griliches (1979) distinguishes between two types of knowledge spillover: those associated with the exchange of goods, so-called rent spillovers, which is knowledge that is rival and excludable in its character; and those arising purely from the process of R&D, so-called pure or idea-creating spillovers, which is knowledge that is characterized by its non-excludability and non-rivalry as it is freely available and can be utilized by many users at the same time, and therefore has a public good character (Arrow, 1962). Patent citations principally refer to the latter category, and there again to specific technical knowledge, which by any means is only a specific segment of knowledge rather than a comprehensive account of overall creative activity leading to technological change and economic growth.

Endogenous growth theories indicate that both forms of spillover are important in explaining economic growth (Romer, 1990). A general skepticism surrounding the difficulty of measuring idea-creating spillovers is reflected in Paul Krugman's influential book on economic geography: "Knowledge flows . . . are invisible; they leave no paper trail by which they may be measured and tracked, and there is nothing to prevent the theorist from assuming anything about them that she likes" (Krugman, 1991: 53). Contrary to this view, empirical studies presented in Jaffe and Trajtenberg (2002), *Patents, Citations, and Innovations: A Window on the Knowledge Economy*, indicate that patent data, and in particular patent citations, do indeed constitute a "paper trail" of knowledge spillovers, an approach which was originally developed in the seminal paper by Jaffe et al. (1993), demonstrating the research potential of this method for the study of complex webs of knowledge spillovers across locations, technologies and time. The goal set out in the original experiment was to test whether knowledge spillovers are localized, by comparing the geographic location of patent citations with that of the cited patents. Further, the aim was also to measure the extent that localization was present, if it were present at all. In order to adjust for uneven patent output growth, and varying levels of technological focus between spatial entities, a control sample was constructed. The Jaffe et al. "control technique," which in its own class has become the standard methodology used in the exploration of knowledge spillovers through patent data, has been employed in numerous similar studies since its inception (Almeida, 1996; Almeida and Kogut, 1997; Hicks et al., 2001; Sonn and Storper, 2008), but has also been subject to substantial critique, in particular by Thomson and Fox-Kean (2005), who make a strong call for a reassessment of this approach.

PATENT DATA: SOURCES

The creation of a database of U.S. patents, which can be utilized for research purposes, dates back to the early 1990s, and over the next decade this long-term research and data creation effort involved a range of researchers, institutions and financial resources. By 2001, this extensive task had been completed and subsequently the National Bureau of Economic Research (NBER) made the database freely available for researchers.[8] The main NBER U.S. Patent Data File includes a number of variables describing individual characteristics of patents granted by the USPTO.[9] In addition, the NBER also provides a second data file that includes a complete list of references made by the patents in the first data set, indicating which prior knowledge, in the form of patents and scientific literature, has been considered significant in the development of an invention. Both data files have been updated to include patents granted up to the end of December 2002. These data, which to some extent have been used in the development of other patent databases that include additional variables, are available for download at Bronwyn Hall's webpage at the University of California, Berkeley.[10] More recently, the NBER has been working on a major National Science Foundation-funded update and extension of its original patent database, which will bring the data up to date through December 2004. Parallel to this initiative there is also a new NBER patent data project (PDP) underway, and although there is limited information available now, the project seems to be mainly concerned with the update of USPTO technology classes over time, and the matching of PDP patent data to actual corporate or other entities to which these patents are assigned by linking them to Standard & Poor's Compustat[11] data (Bessen, 2009).

The vast majority of past research efforts regarding knowledge externalities utilize the NBER U.S. Patent Citations Data File (Hall et al., 2001),[12] and the original method presented in the paper by Jaffe et al. (1993). However, the conflicting results, the limited scale and scope, and the largely static investigation that neglects the dynamic nature of knowledge creation and spillover processes found in most contemporary studies that utilize patent statistics and patent citation analysis, combined with the desire to describe and analyze spatial (geography) and sectoral (industry) patterns of invention at the subnational level, have lately prompted the creation of several patent databases. Examples include, PATENTSCOPE hosted by the World Intellectual Property Organization,[13] and the esp@cenet patent database run by the EPO.[14] Many of these data can be accessed on fee-based websites, but some of them are also publicly available. One of these databases that also features a number of generated variables, such as a disambiguation of inventor names and addresses, can be downloaded for research purposes from the "Patent Network Dataverse" website at the Institute for Quantitative Social Science at Harvard University (Lai et al., 2011, 2013).

More recently, investigations into the sectoral and spatial patterns of invention have also applied data that originate from national patent offices other than the USPTO, such as the EPO and the Japan Patent Office (Goto and Motohashi, 2007), or data that cover triadic patent families, that is, patents that share one or more priorities found in all three major patent offices (Khan and Dernis, 2006).[15] While studies that use triadic patent families employ patents of high value, and their use eliminates the "home advantage" bias in a cross-country analysis that is based on data from only one national office, they also prohibit a comprehensive analysis of the full extent of knowledge creation and spillovers as inventions of apparently less value are not considered, despite their potential impact on the overall rate and direction of knowledge creation.

The Canadian database on patents that will be applied in the case study presented in the following section was developed utilizing the "Patent Network Dataverse" data (Lai et al., 2011, 2013). For a detailed discussion of how this patent database, also referred to as PATDAT, was compiled, along with a description of the spatial and sectoral definitions employed, please refer to Kogler (2010). Thus far, at least to the best of our knowledge, no publicly available database exists that contains detailed information on patents that specifically pertain to Canadian inventors, such as their place of residence, technological focus or timing (i.e. application and grant dates of inventions).[16] One of the main deficiencies of current patent databases is that they do not provide a consistent geography or technology classification over time. Thus, in order to aggregate inventors into meaningful spatial units of socio-economic relevance, Canadian inventors were allocated to the respective Census Metropolitan Area (CMA) or Census Agglomeration (CA) in which they resided at the time the invention was created. Kogler (2010) provides a detailed description of the Canadian patent database that will be applied in the subsequent section.

The use of U.S. data to investigate the spatial and sectoral patterns of invention in Canada may initially appear counter-intuitive. However, considering that Canadian inventions are frequently first, and only, filed as patents in the United States, makes this the preferred approach in terms of an accurate representation of the quantity and timing of invention in Canada (Trajtenberg, 2002). Rafiquzzaman and Whewell (1998: 5) provide evidence that "Canada has one of the lowest propensities to file patents at home of any of the major industrialized countries, with only 6.6 percent of national patent applications originating from residents in 1992", and further, that the "United States . . . receives the largest share of attention from Canadian inventors" (1998: 11). In a more recent context, a comparison of patents granted at the Canadian Intellectual Property Organization (CIPO) and at the USPTO to inventors residing in Canada reveals that more patents are granted to these individuals in the U.S. than in Canada. In absolute terms, in the time period 2005 to 2007, the USPTO granted on average twice as many patents to residents of Canada than the CIPO.[17] Overall, there are several strong arguments that suggest that USPTO patent data are a better fit than CIPO data to investigate invention in Canada, an approach that has also been applied in other recent studies (Aharonson et al., 2008).

While the use of CIPO data would constitute the paradox of a "negative home-advantage bias" in the context of Canadian inventive output, there is the concern that a foreign data source to investigate a domestic socio-economic phenomenon will constitute only a partial representation of the full extent of output, and, in particular, the linkages between the various spatial and sectoral dimensions of invention in Canada. However, the high propensity of Canadian inventors to apply for patents in the U.S. subsequently transpires into the tendency to mainly cite patents within this system, which includes, in addition to U.S. and Canadian inventors, a significant share of patents granted to inventors residing in other countries, and thus creates a somewhat closed system of an international inventory of available technical knowledge that may be built upon in the creation of novel products and processes.[18]

INVENTIVE OUTPUT OF AN ADVANCED MANUFACTURING ECONOMY: A CANADIAN CASE STUDY

The use of patent data for the purpose of generating innovation rankings of companies or jurisdictions is common, not only in the academic literature but in popular media outlets in particular. However, one has to exercise caution when relying on such measures, for reasons mentioned previously. One of the most frequent pitfalls in this regard is that such rankings are based solely on simple patent counts, not taking into consideration that there are variations in the intensity of patenting or the number of co-inventors among industry sectors. The following case study aims to clarify this. Table 11.1 shows the top 35 CMAs and CAs in terms of their total patent output over the entire time period 1983–2007, and in the first and last five years under investigation, respectively. In addition, Table 11.1 displays the corresponding growth rates that occurred between these two time periods. The six largest Canadian cities, based on the number of residents as indicated by the 2006 census of population, are also those ranked the highest, and in the same order, in absolute number of patents produced. All geographic units experienced positive growth rates, although to a varying degree, with the exceptions of the Kingston CMA and the Sarnia CA. Growth rates in top ranked jurisdictions, that is, CMAs and CAs ranked at the top, certainly carry more weight as these geographic units produce patents at a much higher magnitude than the cities that are ranked at the bottom. Considering this, the most notable locations, which exhibited growth rates above 300 percent over the 25-year time period, are the Vancouver, Ottawa-Gatineau, Calgary, Chatham-Kent and Barrie CMAs.

The national and cumulative share of each of these jurisdictions provide further insight into the overall distribution of invention within the national economy. The national share of patents produced in the two largest Canadian cities, Toronto and Montreal, diminished, by the same magnitude, between the two time periods 1983–1987 and 2003–2007. The CMAs ranked third through to seventh all increased their national share, and in the case of the Vancouver and Ottawa-Gatineau CMAs, this increase was quite significant. If we consider the next column in Table 11.1, that is, the cumulative share of national patent output as measured by inventors' place of residency, it is apparent that the top four ranked spatial units account for almost 60 percent of all patents in the Canadian national economy, again a share that increased by about 4 percent over the observed time period. Overall, the top 35 patent producing CMAs and CAs account for about 87 percent of all inventors, if patents granted between the years 1983 and 1987 are considered. This share had increased by 2.5 percent in 2003–2007, only two decades later. If population ranking, based on the 2006 census of population, is compared with patent output ranking as indicated in Table 11.1, about half of the CMAs and CAs are on par, that is, both their rankings match. The geographic units that underperform – that is, produce fewer patents than their population ranking would suggest – are mostly located in the eastern part of Canada; these include the cities of St. John's, Halifax and Saint John. Some of the larger cities that significantly overperform when their patent output ranking is compared with their population ranking are Kingston, Peterborough, Guelph and Chatham-Kent, all of which are located in the province of Ontario.

The final two columns in Table 11.1 refer to patent output per population in the final observed year, that is, patents granted in 2007. Absolute patent output, to a certain

Table 11.1 Patent growth, national share and patents per population in 2007 in 35 metropolitan areas

	Metropolitan area and agglomeration in rank order[a]	Patents (fractional count of inventors)				National share		Cumulative share		Patents per 100k residents	
		1983–07	1983–87	2003–07	growth (%)	1983–87	2003–07	1983–87	2003–07	2003–07 (annual avg.)	Ranking
1	Toronto, ON CMA	14180	1570	3585	128	24.1	20.7	24.1	20.7	14.7	10
2	Montréal, QC CMA	7460	969	1977	104	14.9	11.4	38.9	32.1	9.8	16
3	Vancouver, BC CMA	7079	527	2319	340	8.1	13.4	47.0	45.5	22.3	4
4	Ottawa-Gatineau, ON CMA	6316	583	2438	318	8.9	14.1	56.0	59.6	41.5	1
5	Calgary, AB CMA	2403	183	736	301	2.8	4.3	58.8	63.9	13.6	11
6	Edmonton, AB CMA	2169	210	598	185	3.2	3.5	62.0	67.3	11.1	14
7	Kitchener, ON CMA	1770	191	610	220	2.9	3.5	64.9	70.8	41.2	2
8	Hamilton, ON CMA	1538	181	361	99	2.8	2.1	67.7	72.9	7.9	24
9	Québec, QC CMA	1107	101	345	240	1.6	2.0	69.2	74.9	9.1	20
10	Winnipeg, MB CMA	1075	130	225	73	2.0	1.3	71.2	76.2	7.5	26
11	London, ON CMA	959	103	275	166	1.6	1.6	72.8	77.8	8.2	23
12	Windsor, ON CMA	894	108	273	153	1.7	1.6	74.5	79.4	16.5	7
13	Kingston, ON CMA	866	144	140	−3	2.2	0.8	76.7	80.2	15.8	8
14	St. Catharines-Niagara, ON CMA	603	103	124	21	1.6	0.7	78.3	80.9	6.4	27
15	Saskatoon, SK CMA	590	47	145	209	0.7	0.8	79.0	81.7	11.4	13
16	Victoria, BC CMA	540	59	160	173	0.9	0.9	79.9	82.7	9.4	18
17	Chatham-Kent, ON CA	445	34	173	404	0.5	1.0	80.4	83.7	17.8	6

Table 11.1 (continued)

	Metropolitan area and agglomeration in rank order[a]	Patents (fractional count of inventors)				National share		Cumulative share		Patents per 100k residents	
		1983–07	1983–87	2003–07	growth (%)	1983–87	2003–07	1983–87	2003–07	2003–07 (annual avg.)	Ranking
18	Sarnia, ON CA	444	107	36	−66	1.6	0.2	82.0	83.9	8.7	22
19	Guelph, ON CMA	385	36	91	155	0.5	0.5	82.6	84.4	19.3	5
20	Oshawa, ON CMA	357	29	76	166	0.4	0.4	83.0	84.8	2.8	32
21	Sherbrooke, QC CMA	325	35	106	201	0.5	0.6	83.6	85.5	7.6	25
22	Halifax, NS CMA	325	33	86	164	0.5	0.5	84.1	86.0	4.9	30
23	Barrie, ON CMA	225	12	77	537	0.2	0.4	84.3	86.4	8.8	21
24	Peterborough, ON CMA	224	36	55	52	0.6	0.3	84.8	86.7	5.0	29
25	Brantford, ON CMA	220	23	53	128	0.4	0.3	85.2	87.0	9.9	15
26	Kelowna, BC CMA	214	9	98	1051	0.1	0.6	85.3	87.6	14.8	9
27	Regina, SK CMA	195	15	39	152	0.2	0.2	85.5	87.8	1.7	35
28	Saguenay, QC CA	160	16	29	85	0.2	0.2	85.8	88.0	3.4	31
29	Greater Sudbury, ON CMA	155	14	27	91	0.2	0.2	86.0	88.1	2.5	33
30	Midland, ON CA	154	8	58	622	0.1	0.3	86.1	88.5	24.0	3
31	Granby, QC CA	152	8	61	638	0.1	0.4	86.2	88.8	9.2	19
32	Abbotsford, BC CMA	139	11	39	262	0.2	0.2	86.4	89.0	5.4	28
33	St. John's, NL CMA	121	16	23	39	0.3	0.1	86.6	89.2	1.9	34
34	Drummondville, QC CA	120	8	45	443	0.1	0.3	86.8	89.4	9.5	17
35	Fredericton, NB CA	105	6	29	371	0.1	0.2	86.9	89.6	12.4	12

Note: [a]CMA – Census Metropolitan Area; CA – Census Agglomeration

extent, follows a rank size rule – that is, Zipf's law (Zipf, 1949), determined by the total population in a given geographic unit. However, this measurement is actually not meaningful if the goal is to analyze the actual productivity of individual urban areas. An indicator that is frequently used in this context is to calculate annual patents per 100k population, and based on these results to rank places accordingly. This illustrates the efficiency of each geographic unit, as outlined in the final two columns in Table 11.1. Based on the numbers of patents granted in 2007, and 2006 census of population figures for each of the 35 CMAs and CAs, the annual output of patents per 100,000 residents was first calculated and the results were ranked. This productivity ranking differs significantly from the initial ranking of absolute patent output, which determined the original order of geographic units in the table. The Ottawa-Gatineau and Kitchener CMAs are the top-performing geographic units, with almost twice as many patents per population as the next two places in the ranking, the Midland CA and the Vancouver CMA. However, similar to the comparison of growth rates among all spatial units investigated, this ranking should also be interpreted with caution. The calculations for some highly productive locations, such as the Midland and the Chatham-Kent CAs, are based on much smaller population figures than some of the larger cities, and therefore the results obtained are less significant overall than those derived for cities with a much higher share of the total population.

In order to obtain a more accurate picture of the most productive patent producing geographic units in the Canadian economy over the past two decades, a more detailed calculation of patents per 100k population over the time period 1988 to 2007 was performed. Based on the annual number of patents granted, and the corresponding annual population figures two years prior to the specific grant year,[19] for each of the top 25 patent producing CMAs and CAs as well as the national total, the annual number of patents per 100k was calculated. Figure 11.1 shows how the top ranked CMAs, as indicated by their productivity in 2007, performed in terms of patents per 100k population over the time period 1988 to 2007 compared with the national average.

In addition to the varying levels of productivity between the six places outlined, it is also apparent that there have been considerable changes over time. For the first half of the observed time period the Kingston CMA had by far the highest level of inventive productivity, whereas in the second half the Ottawa-Gatineau and Kitchener CMAs, with about 40 patents per 100,000 population, were almost twice as productive as the cities that follow next in the ranking. The national average increased from about five to more than 10 patents per 100k population over the 20 years.

Patents per population – that is, patents per capita – measures are frequently applied in the literature as an indicator of economic growth and prosperity in either a national (Lee and Kim, 2009), or sub-regional context (Carlino et al., 2007). However, rarely are the limitations of these indicators discussed in more detail, leading, in certain cases, to a misinterpretation due to the assumption of causality that there is a linear relationship between inventive capacity and the innovativeness of a particular place that subsequently translates into economic well-being. Henderson (2007: 505) further invigorates the uncertainty attached to this particular indicator by stating that "patents per population peak at medium densities in medium size cities. Despite the idea that the spatial aspects of R&D spillovers are critical, in fact R&D activities are not willing to pay the high wages and rents to be in the densest locations." Therefore, rankings based on patents

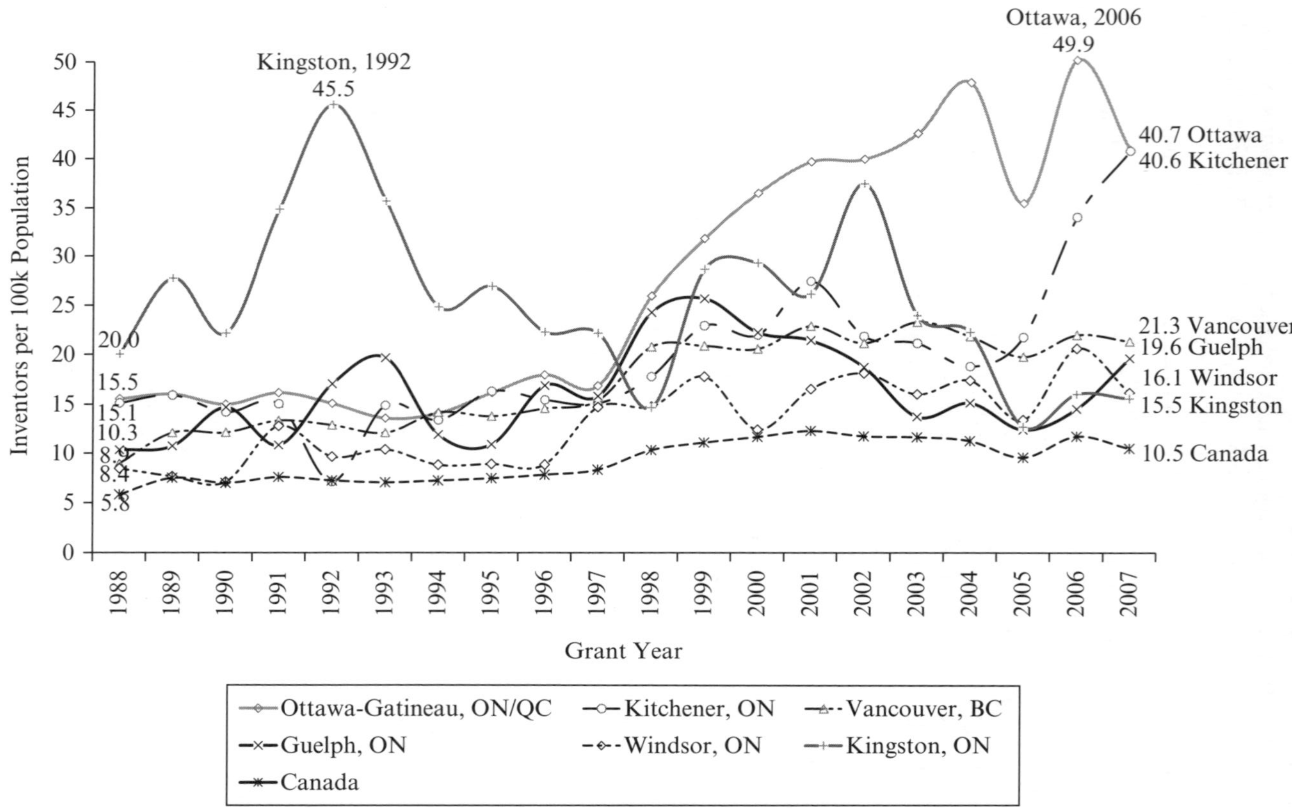

Figure 11.1 *Top performing six Census Metropolitan Areas and national average of patents per 100k population as per grant year 2007 from 1988–2007*

per population, as presented in Table 11.1 and Figure 11.1, have to be interpreted with caution, and primarily have to be acknowledged, as described earlier, as a pure measurement of total inventive output expressed by patent counts. In order to obtain a more accurate, and specifically a more comparable, measurement of inventive productivity among metropolitan areas, an alternative method is suggested in the present study.

There are two central ideas, which may significantly influence the patent output capacity at a particular place, at the center of this approach. The first focuses on the varying propensity to patent among industries. In essence, the importance assigned to production of patents is not the same in each industry, which results in different average employee to patent ratios. In addition, these ratios are not fixed values but change over time. The second important feature concerns the cross-sectoral composition of the employed workforce among industries and geographic units. The sectoral distribution and absolute number of employees at a particular urban setting, due to the varying propensities to patent across industries, will certainly influence the actual patent output observed. Table 11.2 illustrates the propensity to patent among industry sectors in 2007.

The values presented are calculated based on the national annual average of patents granted from 2003 through 2007, and the national sectoral composition of employees as indicated in the 2006 Canadian census of population.[20] The initial task was to transform the employment data, which are classified according to the North American Industry Classification System (NAICS), into the Classification of Economic Activities in the European Community (NACE) industry categorization utilizing a concordance table available from Statistics Canada.[21] Next, the obtained employment figures, now in NACE format, were allocated to the corresponding 44 sectoral fields employed in PATDAT, based on the approach suggested by Schmoch et al. (2003, Annex 1, p. 63). Finally, the actual sectoral patent per employee rates (ppe_s) were calculated by applying Formula 11.1, below:

$$ppe_s = \frac{p_s}{e_s(1/100{,}000)} \tag{11.1}$$

where p_s represents the average annual number of patents granted in sector s, as determined from the 2003–2007 patent cohort, and e_s represents the number of employees in sector s, as originally indicated in the 2006 census of population.

Utilizing the national sectoral patent per employee rates outlined in Table 11.2, which represent a measurement of the propensity to patent among industries, the projected annual patent output for each of the top 25 patent producing Canadian CMA/CAs, as indicated by the total number of patents granted from 2003 through 2007, was calculated based on the sectoral distribution of the employed workforce in each geographic unit. The total projected annual patent output, illustrated in Table 11.3, was then compared with the actual number of patents granted to inventors residing in the CMA/CA of interest. In contrast to Table 11.1, the ranking is based on the annual average of patents granted from 2003 through 2007, rather than a single year – that is, patents granted in 2007 as illustrated in Table 11.1 – in order to adjust for annual fluctuations in patent output. The disparity between the projected and actual value indicates whether a particular place has produced patents above or below the level suggested by the patent per employee rates, and the sectoral composition of the labor force, within a particular CMA/CA. Based on

Table 11.2 Annual sectoral patents per employee rates

	Industry sector	Patents per 100k employees		Industry sector	Patents per 100k employees
1	Food, Beverages	10.4	23	Agricultural and Forestry Machinery	588.9
2	Tobacco Products	58.3	24	Machine-Tools	357.4
3	Textiles	4.8	25	Special Purpose Machinery	520.4
4	Wearing Apparel	17.8	26	Weapons and Ammunition	n/a
5	Leather Articles	87.6	27	Domestic Appliances	409.7
6	Wood Products	6.4	28	Office Machinery and Computers	6565.0
7	Paper	10.6	29	Electric Motors, Generators, Transformers	331.3
8	Publishing, Printing	n/a	30	Electric Distribution, Control, Wire, Cable	290.2
9	Petroleum Products, Nuclear Fuel	66.3	31	Accumulators, Battery	988.5
10	Basic Chemicals	2404.2	32	Lightening Equipment	310.2
11	Pesticides, Agro-Chemicals Products	812.9	33	Other Electrical Equipment	1333.8
12	Paints Varnishes	1.8	34	Electronic Components	550.8
13	Pharmaceuticals	1092.4	35	Signal Transmission,Telecommunications	2060.0
14	Soaps, Detergents, Toilet Preparations	40.1	36	Television and Radio Receivers, AV Electr.	3945.8
15	Other Chemicals	116.7	37	Medical Equipment	594.4
16	Man-Made Fibers	5.7	38	Measuring Instruments	3469.9
17	Rubber and Plastic Products	106.3	39	Industrial Process Control Equipment	375.4
18	Non-Metallic Mineral Products	144.8	40	Optical Instruments	4861.9
19	Basic Metals	52.4	41	Watches, Clicks	39.3
20	Fabricated Metal Products	58.5	42	Motor Vehicles	140.0
21	Energy Machinery	168.8	43	Other Transport Equipment	105.4
22	Non-Specific Purpose Machinery	318.2	44	Furniture, Consumer Goods	80.3

Table 11.3 Patents per population and propensity to patent adjusted ranking of top 25 patent producing CMA/CAs

	Metropolitan area and agglomeration in rank order[a]	Projected annual patent output (2007)	Acutal annual patent output (2007)	Disparity between projected and actual values	Actual patent output above (+) or below (−) projections	Patents per 100k population ranking	Propensity to patent adjusted ranking	Disparity between population and propensity based rankings
1	Ottawa-Gatineau, ON CMA	191.4	487.7	296.3	154.8%	1	1	0
2	Kitchener, ON CMA	206.0	121.9	−84.1	−40.8%	2	20	−18
3	Vancouver, BC CMA	217.8	463.7	245.9	112.9%	3	3	0
4	Guelph, ON CMA	35.1	18.2	−16.9	−48.1%	4	21	−17
5	Chatham-Kent, ON CA	20.0	34.7	14.7	73.5%	5	5	0
6	Windsor, ON CMA	74.4	54.7	−19.7	−26.5%	6	14	−8
7	Kingston, ON CMA	13.4	27.9	14.5	108.2%	7	4	3
8	Toronto, ON CMA	1,019.1	717.1	−302.0	−29.6%	8	16	−8
9	Calgary, AB CMA	131.3	147.1	15.8	12.0%	9	7	2
10	Saskatoon, SK CMA	26.4	28.9	2.5	9.4%	10	8	2
11	Edmonton, AB CMA	91.8	119.5	27.8	30.2%	11	6	5
12	Brantford, ON CMA	26.2	10.5	−15.7	−59.9%	12	23	−11
13	Montréal, QC CMA	588.0	395.4	−192.5	−32.7%	13	17	−4
14	Victoria, BC CMA	14.3	32.1	17.7	123.9%	14	2	12
15	Québec, QC CMA	72.9	68.9	−4.0	−5.5%	15	9	6
16	Barrie, ON CMA	30.0	15.4	−14.6	−48.7%	16	22	−6
17	Sarnia, ON CA	23.3	7.2	−16.1	−69.2%	17	24	−7
18	London, ON CMA	58.8	55.0	−3.8	−6.4%	18	10	8
19	Hamilton, ON CMA	90.9	72.1	−18.7	−20.6%	19	11	8
20	Sherbrooke, QC CMA	27.9	21.3	−6.7	−23.9%	20	13	7
21	Winnipeg, MB CMA	71.6	45.0	−26.6	−37.1%	21	18	3
22	St. Catharines-Niagara, ON CMA	40.8	24.9	−16.0	−39.1%	22	19	3
23	Peterborough, ON CMA	14.3	11.0	−3.3	−22.8%	23	12	11
24	Oshawa, ON CMA	24.5	17.3	−7.2	−29.4%	24	15	9
25	Halifax, NS CMA	56.9	15.2	−41.7	−73.2%	25	25	0

Notes:
[a] CMA – Census Metropolitan Area; CA – Census Agglomeration;
Top 25 patent producing geographic units ranked according to patents per 100k population in 2007.

this an adjusted propensity to patent ranking for the 25 metropolitan areas of interest was generated and evaluated against the original patents per population ranking.

The Ottawa-Gatineau CMA ranks first in both cases. However, the Kitchener CMA, which was originally in second place according to the patents per population ranking, drops to 20th place among the 25 geographic units examined, once the propensity to patent adjusted ranking method is applied. Similarly, the Vancouver CMA maintains its position, whereas the Guelph and Windsor CMAs, which were originally situated at the top of the list, fall well behind other metropolitan areas in the revised ranking. The list of geographic units which have increased their ranking by at least six places based on the new ranking method includes the Kingston, Victoria, Québec, London, Hamilton, Sherbrooke and Peterborough CMAs. From the perspective of the larger cities investigated, the Toronto and Montreal CMAs have lost, while the Calgary and Edmonton CMAs have gained, positions in propensity to patent based on the adjusted ranking.

Although a ranking based on patents per employee rates that considers sub-regional sectoral employment composition is indeed a more accurate approach to measure inventive output than the common patents per population indicators used in the literature, it should be noted that such an indicator is only representative for manufacturing sectors, rather than the entire spectrum of industries that might exist at a particular place. The concordance used in the present study as well as other available ones, such as the concordance between U.S. patent classes and NAICS codes available from the USPTO, only allocate patents to manufacturing sectors.[22] Therefore, any type of ranking based on patent data is not representative of the total innovativeness of a specific jurisdiction, but rather provides insight into one particular, although significant, segment of the economy.[23]

CONCLUDING REMARKS

With the advent of IPR protection, knowledge production has become increasingly commodified (Lamoreaux and Sokoloff, 1996), and a critical dimension of competition (Lichtenberg and Philipson, 2002). In the contemporary "knowledge based" economy, the role of intellectual property and knowledge transfer mechanisms is of particular interest. This is also confirmed by the heightened importance that is attributed to this by science and technology policy instruments (Audretsch et al., 2002). Ideas and knowledge are very complex goods and their appreciation and comprehension largely depend on the absorptive capacity of agents (Cohen and Levinthal, 1990), as well as the places. Thus, the broader ecology of places potentially plays a significant role in this context (Cooke et al., 1997; Morgan, 1997; Storper, 1997; Gertler, 2003; Asheim and Gertler, 2005). The notion of industry towns built around a product-based spatial division of labor, which is well documented through historic data, certainly validates this (Marshall, 1920; Piore and Sabel, 1984; Scott, 1986).

IPRs are essential when it comes to the translation of ideas and knowledge into inventions and subsequent innovations, which is usually accompanied by risk and high costs, and also depends on the resources and knowledge capabilities available to individual economic agents and their partners (Wenerfelt, 1984; Barney, 1991; Kogut and Zander, 1992). Without IPRs an inventor has no incentive to invest time and money into something that can be easily copied by its competitors without incurring significant R&D

costs. IPRs therefore correct this market failure by providing a temporary monopoly to the inventor. For manufacturing industries, patents are of particular importance. A study by Cockburn and Henderson (2003) found that IP managers from a range of industries estimated that patents, trade secrets and know-how account for about three-quarters of the value of IP in their respective entities. Similarly, Arora et al. (2001) and Robbins (2006) found that the same kinds of IPR account for over 70 percent of all licensing royalties. This implies that patents and related knowledge sources are essential IP assets to a wide range of industries.

While the importance of IPRs, and in particular patents, for the economy and its various entities have been outlined in detail in this chapter, it should be noted that their applicability varies among industry sectors. The Canadian case study presented above clearly demonstrates this. In this context, the composition of industry sectors also strongly determines what kinds of IPR are relevant in a locality, which in turn will strongly determine what kind of data should be used to evaluate the inventive potential and innovative output of a particular venture or jurisdiction.

NOTES

1. Of course, this question is only relevant if there is a profit-motive behind the development of an invention. Public entities are not necessarily driven by the same profit-motives as private entities and thus there might be strong variations in the intensity of these questions among the two kinds of entity.
2. More recently, copyright licenses known as the Creative Commons (CC) licenses have been released free of charge to the public by the CC non-profit organization. These licenses intend to provide an easy, but structured and legal way, for creators to express what rights they want to reserve while waiving other rights pertaining to their own work for the benefit of consumers, or other individuals or entities. These standardized licenses aim to lower the overhead cost of copyright management, and have become popular with creators that seek IP protection without commercial compensation. CC licenses do not replace the traditional copyright, but are rather based upon it. For a discussion of the legal framework and the ongoing debate of CC versus copyright, see Loren (2007).
3. Ideas are not patentable; nor are algorithms or computer programs, though a chip with a particular program coded into it might be. The definition of a device includes genetically engineered organisms.
4. These and other specific definitions pertaining to patents and used in USPTO publications are summarized in the online glossary available at http://www.uspto.gov/main/glossary/index.html (accessed November 1, 2013).
5. According to the USPTO (2013), "A utility or plant patent in force on June 8, 1995, is subject to either the 17 year term from grant or the 20 year term from earliest effective U.S. filing date, whichever is longer. A design patent term is 14 years from patent grant. The right conferred by the patent grant extends throughout the United States."
6. Both of these are examples where firms are able to continue extracting value from their respective innovations well beyond the 20-year mark, which is the time span of intellectual property protection they would have received had they applied for a patent at the time these inventions were initially made.
7. What constitutes specialization or diversity to a certain extent depends on the level of industry aggregation that is applied in the investigation. For example, measures that use an industry aggregation based on either 2-, 4- or 6-digit North American Industry Classification System (NAICS) codes, would analyze either 21, 325 or 929 different industry sectors; thus results of what constitutes specialization or diversity would vary accordingly. Nevertheless, considering that NAICS codes mimic an industry structure along the value chain – that is, there are strong communalities in terms of the technological and scientific base among adjacent industry sectors in the system (Griliches, 1979: 104) – a somewhat consistent measure of what constitutes specialization or diversity should arise regardless of the level of aggregation that is applied in the analysis.
8. NBER webpage: http://www.nber.org/patents/ (accessed November 1, 2013).
9. The USPTO is a federal agency in the Department of Commerce. Since its foundation 200 years ago the basic role of the USPTO has remained the same: to promote the progress of science and the useful arts by

securing for limited times to inventors, the exclusive right to their respective discoveries (Article 1, Section 8 of the United States Constitution). For further information about the agency, its mission, and other related details please refer to the USPTO webpage at http://www.uspto.gov (accessed November 1, 2013).
10. Bronwyn H. Hall, Department of Economics at the University of California, Berkeley webpage: http://elsa.berkeley.edu/~bhhall/patents.html (accessed November 1, 2013).
11. Standard & Poor's Compustat database includes financial, statistical and market time-series data on active and inactive companies throughout the world; see http://www.compustat.com (accessed November 1, 2013).
12. The NBER is working on a major update and extension of these data funded by the National Science Foundation; see: http://nber.org/patents/ (accessed November 1, 2013).
13. See http://www.wipo.int/patentscope/en/ (accessed November 1, 2013).
14. Available at http://worldwide.espacenet.com/ (accessed November 1, 2013).
15. Triadic patents are a series of corresponding patents filed at the EPO, the USPTO and the Japan Patent Office, for the same invention, by the same applicant or inventor (OECD, 2008).
16. The Observatoire des sciences et des technologies, which is a research support unit of the Centre interuniversitaire de recherche sur la science et la technologie at the Université du Québec à Montréal, has compiled data about Canadian inventors, some of which is publicly accessible at the Institut de la statistique Québec webpages at http://www.stat.gouv.qc.ca/statistiques/science-technologie-innovation/brevets/index_an.html (accessed November 1, 2013).
17. CIPO data were obtained from the 2005–2006, 2006–2007 and 2007–2008 annual reports: table 10 – patent applications filed and granted to residents of Canada and foreign countries. USPTO data include only patents where the primary inventor is a resident of Canada in order to generate a conservative comparison.
18. The annual share of USPTO patents granted that are of foreign origin increased from 45 to 51 percent between 1995 and 2008 (USPTO, 2009).
19. The two-year offset between annual population and patents granted figures is applied in order to adjust for the average time lag of around two years that persists between the time a patent is applied for and its actual grant date. The revised methodology produces slightly different patents per 100k population values than those presented for 2007 in Table 11.1.
20. Statistics Canada – 2006 Census Statistics Canada 97–559-XCB2006009.
21. Concordance between the NAICS 2002 and NACE, Revision 1.1 available at: http://www.statcan.gc.ca/subjects-sujets/standard-norme/concordances/revision11-eng.htm (accessed November 1 2013).
22. The national share of the labor force employed in the manufacturing sector was 12 percent in 2006; however, strong regional variations exist. For example, in the Toronto CMA about 13.5 percent of people were employed in the manufacturing sector, whereas in the Kitchener CMA this share was 22.7 percent. Because patents mainly correspond to the manufacturing segment of total regional employment, the innovative performance of this sector has of course an amplified impact on total regional growth in places where the share of the manufacturing labor force is high.
23. Firm size should also be considered, along with the presence of certain large-scale operations situated in smaller population centers. For example, Research in Motion (RIM) accounted for well over 50 percent of all patents annually in the Kitchener CMA in the years 2006 and 2007; a trend that most likely will continue due to RIM's patent strategy.

REFERENCES

Afuah, A.N. and J.M. Utterback (1997), "Responding to structural industry changes: a technological evolution perspective," *Industrial and Corporate Change*, **6** (1), 183–202.

Aharonson, B.S., J.A.C. Baum and A. Plunket (2008), "Inventive and uninventive clusters: the case of Canadian biotechnology," *Research Policy*, **37**, 1108–1131.

Almeida, P. (1996), "Knowledge sourcing by foreign multinationals: patent citation analysis in the U.S. semiconductor industry," *Strategic Management Journal*, **17**, 421–431.

Almeida, P. and B. Kogut (1997), "The exploration of technological diversity and the geographic localization of innovation," *Small Business Economics*, **9**, 21–31.

Archibugi, D. (1992), "Patenting as an indicator of technological innovation: a review," *Science and Public Policy*, **10**, 357–368.

Arora, A., A. Fosfuri and A. Gambardella (2001), *Markets for Technology: The Economics of Innovation and Corporate Strategy*, Cambridge, MA: MIT Press.

Arrow, K. (1962), "Economic welfare and the allocation of resources for inventions," in R.R. Nelson (ed.), *The Rate and Direction of Innovative Activity*, Princeton, NJ: Princeton University Press, pp. 609–625.

Asheim, B. and M.S. Gertler (2005), "The geography of innovation: regional innovation systems," in J. Fagerberg, D. Mowery and R.R. Nelson (eds) *The Oxford Handbook of Innovation*, Oxford: Oxford University Press, pp. 291–317.
Audretsch, D.B., B. Bozeman, K.L. Combs, M.P. Feldman, A.N. Link, D.S. Siegel, P. Stephan, G. Tassey and C. Wessner (2002), "The economics of science and technology," *Journal of Technology Transfer*, **27** (2), 155–203.
Baptista, R. and P. Swann (1998), "Do firms in clusters innovate more?" *Research Policy*, **27**, 525–540.
Barney, J.B. (1991), "Firm resources and sustained competitive advantage," *Journal of Management*, **17** (1), 99–120.
Barrios, S., H. Gorg and E. Strobl (2006), "Multinationals' location choice, agglomeration economies, and public incentives," *International Regional Science Review*, **29**, 81–107.
Baumol, W.J. (2002), *The Free-Market Innovation Machine: Analyzing the Growth Miracle of Capitalism*, Princeton, NJ: Princeton University Press.
Beaudry, C. and A. Schiffauerova (2009), "Who's right, Marshall or Jacobs? The localization versus urbanization debate," *Research Policy*, **38**, 318–337.
Bessen, J. (2009), "NBER PDP Project – user documentation: matching patent data to Compustat firms, version: Beta, May 2009," available at https://sites.google.com/site/patentdataproject/ (accessed November 1, 2013).
Blind, K., J. Edler, R. Frietsch and U. Schmoch (2006), "Motives to patent: empirical evidence from Germany," *Research Policy*, **35** (3), 655–672.
Breitzman, A.F. and M.E. Mogee (2002), "The many applications of patent analysis," *Journal of Information Science*, **22**, 187–205.
Carlino, G.A., S. Chatterjee and R.M. Hunt (2007), "Urban density and the rate of invention," *Journal of Urban Economics*, **61**, 389–419.
Clark, G.L., M.P. Feldman and M.S. Gertler (eds) (2000), *Handbook of Economic Geography*, Oxford: Oxford University Press.
Cockburn, I.M. and R. Henderson (2003), "The IPO Survey on Strategic Management of Intellectual Property, conference report," Intellectual Property Owners' Association, Washington, D.C., November.
Cohen, W.M. and D.A. Levinthal (1990), "Absorptive capacity: a new perspective on learning and innovation," *Administrative Science Quarterly*, **35**, 128–152.
Cohen, W.M., R.R. Nelson and J. Walsh (2000), "Protecting their intellectual assets: appropriability conditions and why U.S. manufacturing firms patent (or not)," NBER working paper no. 7552, Cambridge, MA: National Bureau of Economic Research.
Cooke, P., M. Uranga and G. Etxebarria (1997), "Regional innovation systems: institutional and organizational dimensions," *Research Policy*, **26** (4–5), 475–491.
Corrigan, R. and M. Rogers (2005), "The economics of copyright," *World Economics: The Journal of Current Economic Analysis and Policy*, **6** (3), 53–174.
Dernis, H. and M. Khan (2004), "Triadic patent families methodology," Science, Technology and Industry Working Paper 2004/2, Paris: OECD.
Dess, G.G. and R.B. Robinson Jr. (1984), "Measuring organizational performance in the absence of objective measures: the case of the privately-held firm and conglomerate business unit," *Strategic Management Journal*, **5** (3), 265–273.
Dosi G., C. Freeman, R.R. Nelson, G. Silverberg and L. Soete (eds) (1988), *Technical Change and Economic Theory*, London: Pinter.
Duranton, G. and D. Puga (2000), "Diversity and specialisation in cities: why, where and when does it matter?" *Urban Studies*, **37**, 533–555.
Feldman, M.P. (1994), *The Geography of Innovation*, Dordrecht: Kluwer Academic.
Feldman, M.P. and D.B. Audretsch (1999), "Innovation in cities: science-based diversity, specialization, and localized competition," *European Economic Review*, **43**, 409–429.
Feldman, M.P. and D.F. Kogler (2008), "The contribution of public entities to innovation and technological change," in S. Shane (ed), *Handbook of Technology and Innovation Management*, Chichester: John Wiley & Sons, pp. 431–459.
Feldman, M.P. and D.F. Kogler (2010), "Stylized facts in the geography of innovation," in B. Hall and N. Rosenberg (eds), *Handbook of the Economics of Innovation*, Oxford: Elsevier, pp. 381–410.
Fung, M.K. and W.W. Chow (2002), "Measuring the intensity of knowledge flows with patent statistics," *Economics Letters*, **74**, 353–358.
Geels, F.W. (2004), "From sectoral systems of innovation to socio-technical systems: insights about dynamics and change from sociology and institutional theory," *Research Policy*, **33** (6–7), 897–920.
Gertler, M.S. (2003), "Tacit knowledge and the economic geography of context, or the undefinable tacitness of being (there) ," *Journal of Economic Geography*, **3** (1), 75–99.
Goldstein, P. (2001), *International Copyright: Principles, Law, and Practice*, Oxford: Oxford University Press.

Goto, A. and K. Motohashi (2007), "Construction of a Japanese patent database and a first look at Japanese patenting activities," *Research Policy*, **36**, 1431–1442.
Gould, D.M. and W.C. Gruben (1996), "The role of intellectual property rights in economic growth," *Journal of Development Economics*, **48** (2), 323–350.
Greenhalgh, C. and M. Rogers (2010), *Innovation, Intellectual Property, and Economic Growth*, Princeton, NJ: Princeton University Press.
Griliches, Z. (1979), "Issues in assessing the contribution of R&D to productivity growth," *Bell Journal of Economics*, **10**, 92–116.
Griliches, Z. (ed.) (1984), *R&D, Patents, and Productivity*, Chicago: University of Chicago Press.
Griliches, Z. (1990), "Patent statistics as economic indicators: a survey," *Journal of Economic Literature*, **28**, 1661–1707.
Hall, B.H., C. Helmers, M. Rogers and V. Sena (2012), "The choice between formal and informal intellectual property: a literature review," NBER working paper no. 17983, Cambridge, MA: National Bureau of Economic Research.
Hall, B.H., A.B. Jaffe and M. Trajtenberg (2001), "The NBER patent citations data file: lessons, insights and methodological tools," NBER working paper no. 8498, Cambridge, MA: National Bureau of Economic Research.
Hall, P.A. and D. Soskice (eds) (2001) *Varieties of Capitalism: The Institutional Foundations of Comparative Advantage*, Oxford: Oxford University Press.
Henderson, J.V. (2007), "Understanding knowledge spillovers," *Regional Science and Urban Economics*, **37**, 497–508.
Hicks, D., T. Breitzman, D. Olivastro and K. Hamilton (2001), "The changing composition of innovative activity in the US – a portrait based on patent analysis," *Research Policy*, **30**, 681–703.
Hollis, M. and E. Nell (1975), *Rational Economic Man: A Philosophical Critique of Neo-classical Economics*, Cambridge: Cambridge University Press.
Howells, J.R.L. (2002), "Tacit knowledge, innovation and economic geography," *Urban Studies*, **39**, 871–884.
Hurst, N. (2013), "How the America Invents Act will change patenting forever," available from Wired Magazine Online at http://www.wired.com/design/2013/03/america-invents-act/ (accessed November 1, 2013).
Jacobs, J.M. (1969), *The Economy of Cities*, New York: Random House.
Jaffe, A.B. (1986), "Technological opportunity and spillovers of R&D: evidence from firms' patents, profits, and market value," *American Economic Review*, **76**, 984–1001.
Jaffe, A.B. and M. Trajtenberg (2002), *Patents, Citations, and Innovations: A Window on the Knowledge Economy*, Cambridge, MA: MIT Press.
Jaffe, A.B., M. Trajtenberg and R. Henderson (1993), "Geographic localization of knowledge spillovers as evidenced by patent citations," *Quarterly Journal of Economics*, **108**, 577–598.
Khan, M. and H. Dernis (2006), "Global overview of innovative activities from the patent indicators perspective," Science, Technology and Industry Working Paper 2006/3, Paris: OECD.
Klevorick, A.K., R. Levin, R.R. Nelson and S. Winter (1995), "On the sources and significance of interindustry differences in technological opportunities," *Research Policy*, **24**, 195–205.
Kogler, D.F. (2010), "The geography of knowledge formation: spatial and sectoral aspects of technological change in the Canadian economy as indicated by patent citation analysis, 1983–2007," Department of Geography and Planning. Unpublished PhD, University of Toronto, available from the University of Toronto at https://tspace.library.utoronto.ca/bitstream/1807/24787/1/Kogler_Dieter_Franz_201006_PhD_thesis.pdf (accessed October 1, 2014).
Kogut, B. and U. Zander (1992), "Knowledge of the firm, combinative capabilities, and the replication of technologies," *Organization Science*, **3**, 383–397.
Kortum, S. and J. Lerner (1998), "Stronger protection or technological revolution: what is behind the recent surge in patenting?" *Carnegie–Rochester Conference Series on Public Policy*, **48**, 247–304.
Kortum, S. and J. Lerner (1999), "What is behind the recent surge in patenting?" *Research Policy*, **28** (1), 1–22.
Krugman, P. (1991), *Geography and Trade*, Cambridge, MA: MIT Press.
Lai, R., A. D'Amour, A. Yu, Y. Sun and L. Fleming (2011), "Disambiguation and co-authorship networks of the U.S. Patent Inventor Database (1975–2010)," Working Paper, Harvard University.
Lai, R., A. D'Amour, A. Yu, Y. Sun and L. Fleming (2013), "Disambiguation and Co-authorship networks of the U.S. Patent Inventor Database (1975–2010)," http://hdl.handle.net/1902.1/15705 UNF:5:RqsI3LsQEYLHkkg5jG/jRg== the Harvard Dataverse Network [Distributor] V5 [Version].
Lamoreaux, N. and K. Sokoloff (1996), "Long-term change in the organization of inventive activity," *Proceedings of the National Academy of Sciences*, **93** (23), 12686–12692.
Lee, K. and B.-Y. Kim (2009), "Both institutions and policies matter but differently for different income groups of countries: determinants of long-run economic growth revisited," *World Development*, **37**, 533–549.

Levin, R.C., A.K. Klevorick, R.R. Nelson, S.G. Winter (1987), "Appropriating the return from industrial R&D," *Brookings Papers on Economic Activity*, **3**, 783–820.
Lichtenberg, F.R. and T.J. Philipson (2002), "The dual effects of intellectual property regulations: within- and between-patent competition in the US pharmaceuticals industry", NBER working paper no. 9303, Cambridge, MA: National Bureau of Economic Research.
Loren, L.P. (2007), "Building a reliable semicommons of creative works: enforcement of creative commons licenses and limited abandonment of copyright," *George Mason Law Review*, **14** (2), 271–328.
Lösch, A. (1954), *The Economics of Location*, New Haven, CT: Yale University Press.
Lucas, R.E. Jr. (1993), "Making a miracle," *Econometrica*, **61**, 251–272.
Marshall, A. (1920), *The Principles of Economics*, 8th edition, London: Macmillan.
Marx, M., D. Strumsky and L. Fleming (2009), "Mobility, skills, and the Michigan non-compete experiment," *Management Science*, **55** (6), 875–889.
Maskus, K.E. (2000), *Intellectual Property Rights in the Global Economy*, Washington, D.C.: Institute of International Economics.
Michel, J. and B. Bettels (2001), "Patent citation analysis –a closer look at the basic input data from patent search reports," *Scientometrics*, **51** (1), 185–201.
Moed, H.F., W. Glänzel and U. Schmoch (eds) (2004), *Handbook of Quantitative Science and Technology Research: The Use of Publication and Patent Statistics in Studies of S&T Systems*, Dordrecht: Kluwer Academic Publishers.
Mokyr, J. (2009), "Intellectual property rights, the Industrial Revolution, and the beginnings of modern economic growth," *American Economic Review*, **99** (2), 349–355.
Morgan, K. (1997), "The learning region: institutions, innovation and regional renewal," *Regional Studies*, **31** (5), 491–503.
Nelson, R.R. (2001), "Observations on the post-Bayh–Dole rise of patenting at American universities," *Journal of Technology Transfer*, **26** (1–2), 13–19.
Nelson, R.R. and E.S. Phelps (1966), "Investment in humans, technological diffusion, and economic growth," *American Economic Review*, **56**, 69–75.
North, D.C. (1991), *Institutions, Institutional Change and Economic Performance*, Cambridge: Cambridge University Press.
North, D.C. (2005), *Understanding the Process of Economic Change*, Princeton, NJ: Princeton University Press.
OECD (2008), Compendium of Patent Statistics, Paris: OECD.
Oppenheim, C. (1997), "Patent citation analysis – letter to the editor," *Scientometrics*, **39**, 141.
Oppenheim, C. (2003), "Patents, innovation and citations – book review," *Journal of Documentation*, **59**, 369–371.
Pavitt, K. (1985), "Patent statistics as indicators of innovative activities: possibilities and problems," *Scientometrics*, **7**, 77–99.
Piore, M.J. and C.F. Sabel (1984), *The Second Industrial Divide: Prospects for Prosperity*, New York: Basic Books.
Rafiquzzaman, M. and L. Whewell (1998), "Recent jumps in patenting activities: comparative innovative performance of major industrial countries, patterns and explanations," Industry Canada Research Publications Program, working paper no. 27, Ottawa: Industry Canada.
Rigby, D.L. (2000), "Geography and technological change," in E. Sheppard and T.J. Barnes (eds), *A Companion to Economic Geography*, Oxford: Blackwell, pp. 202–223.
Robbins, C. (2006), *Measuring Payments for the Supply and Use of Intellectual Property*, Bureau of Economic Analysis, Washington, D.C.: U.S. Department of Commerce.
Rockett, K. (2010), "Property rights and invention," in B. Hall and N. Rosenberg (eds), *Handbook of the Economics of Innovation*, Oxford: Elsevier, pp. 315–380.
Rockman, H.B. (2004), *Intellectual Property Law for Engineers and Scientists*, Hoboken, NJ: Wiley-IEEE Press.
Romer, P. M. (1990), "Endogenous technological change," *Journal of Political Economy*, **98**, S71–S102.
Rosenberg, N. and L.E. Birdzell (1986), *How the West Grew Rich: The Economic Transformation of the Industrial World*, New York: Basic Books.
Scherer, F.M. (1965), "Firm size, market structure, opportunity, and the output of patented inventions," *American Economic Review*, **55**, 1097–1125.
Scherer, F.M. (1982), "Inter-industry technology flows and productivity growth," *Review of Economics and Statistics*, **64**, 627–634.
Scherer, F.M. (1983), "The propensity to patent," *International Journal of Industrial Organization*, **1**, 107–128.
Schmoch, U. (1999), "Eignen sich Patente als Innovationsindikatoren?" in R. Boch (ed.), *Patentschutz und Innovation in Geschichte und Gegenwart*, Frankfurt a. M.: Verlag Peter Lang, pp. 113–126.
Schmoch, U., F. Laville, P. Patel and R. Frietsch (2003), "Linking technology areas to industrial sectorsm" final report to the European Commission, DG Research, Karlsruhe: Fraunhofer ISI, Paris, Brighton,

available at ftp://ftp.cordis.europa.eu/pub/indicators/docs/ind_report_isi_ost_spru.pdf (accessed November 1, 2013).
Schmookler, J. (1962), "Economic sources of inventive activity," *Journal of Economic History*, **22**, 1–20.
Schmookler, J. (1966), *Invention and Economic Growth*, Cambridge, MA: Harvard University Press.
Schoenecker, T. and L. Swanson (2002), "Indicators of firm technological capability: validity and performance implications," *IEEE Transactions on Engineering Management*, **49**, 36–44.
Schumpeter, J. (1942), *Capitalism, Socialism and Democracy*, New York: Harper.
Scotchmer, S. (2004), *Innovation and Incentives*, Cambridge, MA: MIT Press.
Scott, A.J. (1986), "Industrial organization and location: division of labor, the firm and spatial process," *Economic Geography*, **62** (3), 215–231.
Scott, A.J. (2006), "Entrepreneurship, innovation and industrial development: geography and the creative field revisited," *Small Business Economics*, **26**, 1–24.
Sonn, J.W. and M. Storper (2008), "The increasing importance of geographical proximity in knowledge production: an analysis of US patent citations, 1975–1997," *Environment and Planning A*, **40**, 1020–1039.
Storper, M. (1997), *The Regional World: Territorial Development in a Global Economy*, New York: Guilford.
Tadelis, S. (1999), "What's in a name? Reputation as a tradeable asset," *American Economic Review*, **89** (3), 548–563.
Teece, D.J. (1986), "Profiting from technological innovation: implications for integration, collaboration, licensing and public policy," *Research Policy*, **15**, 286–305.
Teece, D.J. (2010), "Technological innovation and the theory of the firm: the role of enterprise-level knowledge, complementarities, and (dynamic) capabilities," in B. Hall and N. Rosenberg (eds), *Handbook of the Economics of Innovation*, Oxford: Elsevier, pp. 679–730.
Ter Wal, A.L.J. and R.A. Boschma (2009), "Applying social network analysis in economic geography: framing some key analytical issues," *Annals of Regional Science*, **43** (3), 739–756.
Thompson, P. and M. Fox-Kean (2005), "Patent citations and the geography of knowledge spillovers: a reassessment," *American Economic Review*, **95**, 450–460.
Trajtenberg, M. (2002), "Is Canada missing the 'technology boat'? Evidence from patent data," in R. Someshwar and A. Sharpe (eds), *Productivity Issues in Canada*, Calgary: University of Calgary Press, pp. 245–280.
USPTO (2000), "Technology assessment and forecast database, patent description," Alexandria, VA: USPTO, available at http://www.uspto.gov/web/offices/ac/ido/oeip/taf/patdesc.htm (accessed November 1, 2013).
USPTO (2009), "Patent counts by country/state and year: utility patents, January 1, 1963–December 31, 2008," Alexandria, VA: USPTO, available at http://www.uspto.gov/patents/stats/index.jsp (accessed November 1, 2013).
USPTO (2013), "United States Patent and Trademark Office – general questions," Alexandria, VA: USPTO, available at www.uspto.gov/main/faq/ (accessed November 1, 2013).
Van Zeebroeck, N. (2009), "The puzzle of patent value indicators," working paper no. 07/023, Solvay Brussels School of Economics and Management, Centre Emile Bernheim, Université Libre de Bruxelles, Brussels.
Wernerfelt, B. (1984), "A resource based view of the firm," *Strategic Management*, **5**, 171–180.
Zipf, G.K. (1949), *Human Behavior and the Principle of Least Effort*, Cambridge, MA: Addison-Wesley.

FURTHER READING

Feldman, Maryann P. and Dieter Franz Kogler (2010), "Stylized facts in the geography of innovation," in B. Hall and N. Rosenberg (eds), *Handbook of the Economics of Innovation*, Oxford: Elsevier, pp. 381–410. (The geography of innovation describes the importance of proximity and location to innovative activity; this is a comprehensive overview of recent work in this line of inquiry.)
Greenhalgh, C. and M. Rogers (2010), *Innovation, Intellectual Property, and Economic Growth*, Princeton, NJ: Princeton University Press. (A comprehensive overview of micro- and macro-economic issues related to the causes and effects of innovation.)
Hall, B. and N. Rosenberg (eds) (2010), *Handbook of the Economics of Innovation*, Oxford: Elsevier. (A detailed outline of the factors and conditions that lead to innovation and their importance in an economic context.)
Scotchmer, S. (2004), *Innovation and Incentives*, Cambridge, MA: MIT Press. (An extended outline of the historical, legal and institutional contexts in which innovation takes place.)

PART III

INDUSTRY AND FIRM CASE STUDIES

12. Manufacturing textile futures: innovation, adaptation and the UK textiles industry

Megan Ronayne

> The significance of manufacturing to its region of location is a question of constant concern to geographers. (Miller, 1962:2)

INTRODUCTION

Economic geography used to be dominated by studies of manufacturing activities, but over the past 30 years geographers began to turn their attention to understanding the rise and role of business and professional services, financial services and retailing. The study of manufacturing has been largely neglected in the UK as it was considered that manufacturing was an activity in decline and was something undertaken in low-cost economies (Bell, 1974). Globalization resulted in a fundamental shift in the geographies of manufacturing; some manufacturing activity was decentralized to low-cost locations. Low-cost locations have a comparative advantage in producing labour-intensive goods more cheaply and in higher volumes (Dicken, 2011), due to lower wage costs, longer working hours and reduced environmental regulations (Daniels, 2005). This resulted in an unprecedented growth in price-based competition for developed market economies. But this did not mean the decline of manufacturing in high-cost locations (Bryson and Rusten, 2011).

UK manufacturing activities have remained resilient to this price-based competition as firms pursue competitive manufacturing strategies and an increase in the technical composition of production. UK manufacturing activity has therefore declined in terms of employment since its peak in 1966 (Select Committee on Trade and Industry, 2002), but primarily as a result of greater productivity and outputs (Bryson and Daniels, 2007; Hudson, 2010; Dicken, 2011). Manufacturing firms have evolved to meet the competitive demands of the global economy in the 21st century, but this remains to be woven into the social science and political debates. Bryson et al. (2008:184) state, 'the alterations have been so radical that it is possible to argue that economic geographers no longer understand manufacturing'. Further, the UK is 'waking up to the realization that manufacturing still matters' (Christopherson, 2011:2); as manufacturing contributes 12 per cent of the UK's gross domestic product (BDO, 2010), 50 per cent of exports and 75 per cent of research and development (R&D) investment (Bryson et al., 2008).

The global shift of manufacturing has had its biggest impact on the textile industry. This industry was one of 'the first manufacturing industries to take on a global dimension' and it 'exemplifies many of the intractable issues facing today's world economy' (Dicken, 1998:283). The industry has experienced a continuous decline since 1913 from price-based competition (Miller, 1962). This demise and the changes in the fashion industry have been widely documented in the geographical literature (see Miller, 1962;

Dicken, 1986; Gibbs, 1988; Glasmeier et al., 1992; Sunley, 1992; Jones and Haynes, 2004; Scott, 2006; Hassink, 2007; Tokatli, 2012). However, the evolution of the industry has been missed under its own invisible cloak. Research into the economic geography of the textile industry in developed market economies has been replaced by a focus on research into the manufacturing of clothing in low wage economies. The manufacture of textiles is one of the oldest industries (Miller, 1962) but continues to play a strategic role in the UK economy in providing highly innovative and novel solutions, such as the cooling system developed by the StayCool project. This is a research initiative to develop and prototype a novel system for cooling the body that is lightweight, has low energy consumption and so can be worn by firefighters for prolonged periods of time. It is anticipated this will lengthen the time firefighters can stay at a fire and therefore improve the efficiency of firefighting (Kay, 2011).

The analysis of this chapter is part of a PhD project that began in 2011 and will conclude in 2015. This project focuses on the advanced and technical segment of textiles: technical textiles. The analysis is based on ten manufacturers located in the North West of England, the East Midlands, Northern Ireland and Scotland. These companies were identified using 20 public directories and LinkedIn.[1] A telephone survey was also undertaken to ensure companies involved in the project manufactured textile fabric rather than assembled products. The data were acquired through semi-structured face-to-face interviews with managing directors and/or production managers, and supplemented by data obtained from a placement with NWtexnet between April and June 2012.[2] The interviews used for the purpose of this chapter took place between September 2012 and March 2013. All the interviews have been transcribed and, to maintain anonymity, each company has been assigned a randomized number.

This chapter will explore the survival and evolution of UK textile manufacturers in the context of changing geographies of manufacturing in a global marketplace. It is divided into five main sections. The first section provides an historical overview of the demise and evolution of the UK textile industry, highlighting the development of technical textiles. The second section provides an overview of the UK technical textile industry, highlighting the importance of non-priced-based competition (Bryson et al., 2008). The third section will examine the importance of innovation as a competitive strategy and how firms have implemented this. This section also explores how technical textile manufacturers are an example of Christopherson's (2009) 'Phoenix Industries' that play an integral role in developed market economies. The fourth section presents a definition of technical textiles that conceptualizes the firms creating value as advanced manufacturers. The fifth and final section will conclude the analysis.

THE TRANSITION FROM COTTON TO CARBON

The manufacture of textiles is one of the oldest forms of manufacturing, with cotton manufacturing originating before 3000 BC as a home handicraft industry in the Western Hemisphere (Miller, 1962). The importance and growth of the industry spun from its role of fulfilling everyday human needs, such as clothing for body protection and agriculture (Ellis, 2008). In the UK, textiles remained a cottage industry until the late 18th century (Chapman, 1987), when mechanization, especially the invention of the spinning jenny,

Arkwright's water frame and Crompton's mule loom, dramatically altered the organization of the industry. It is debatable that this technological progress spearheaded the industrial revolution (Miller, 1962; Fowler, 2003), as Peter Dicken (1986:221) describes:

> The textile industry was the archetypical industry of the Industrial Revolution of the eighteenth and nineteenth centuries in Britain. In some senses that industrial revolution was a textile revolution: the cotton textile industry was the primary engine of growth and its major geographical centre of production – Lancashire – became the exemplar of the nineteenth century industrial landscape with its oft-described 'dark satanic mills'. Its marketing capital, Manchester, became perhaps the first world industrial city – the 'Cottonopolis' of an industrial system whose tentacles spread across the globe.

The industry was particularly geographical; factories located where there was available waterpower, mild temperatures and high humidity; as 'it was believed that the damp atmosphere made spinning and weaving easier' (Fowler, 2003:2). These areas included Nottingham, Lancashire, the West Riding of Yorkshire, Wales, the Scottish Lowlands and Northern Ireland. However, the industry rapidly centralized in the Lancashire region of the North West of England due to its natural advantages in climate and proximity to transportation links of canals and the port of Liverpool for importing raw cotton and exporting cotton fabric to many parts of the world. The textile industry began to gain great momentum in the 19th century from a strong export market, which equated to half of Britain's export output (Chapman, 1987). This trend continued into the early 20th century as growth in the developing market economies of Latin America, Asia and India drove a huge world market demand for cheap cloth (Miller, 1962; Chapman, 1987). The textile industry reached its peak in 1913, employing over half a million people and exporting approximately seven billion metres of cloth; dominating the world's cotton exports (Miller, 1962; Leunig, 1996).

The demise began during World War One as the industry was viewed as unessential to the war effort and consequently production capacity was reduced by 40 per cent (Fowler, 2003). Between 1918 and 1940 industrializing and developing nations began to develop their own textile industries (Miller, 1962; Chapman, 1987) and the impact on the UK was profound. During this time period, 800 cotton mills were closed, 21 million spindles destroyed and 360,000 looms abandoned as the UK experienced a dramatic decline in exports; exports to India decreased from three billion square metres in 1913 to less than 200 million square metres in 1938 (Miller, 1962).

The global shift had begun. The textile industry was the first to take on a global dimension due to its low barriers to entry as a relatively low-skilled production process, and ideally suited to the early stages of industrialization (Rostow, 1960; Dicken, 1986). As the textile industry is the most labour-intensive industry in modern economies, with labour costs in clothing for example equating to around 60 per cent of the total, it became inevitable that international competition from developing as well as developed market economies would grow substantially (Dicken, 1986). This epitomized the 'new international division of labour' (Dicken, 2011) as companies organized their production activities over different geographical scales; 'in which each part of the process is a specialized activity connected to the next' (Bryson et al., 2004). By the 1960s, imports of cotton into the UK had outpaced exports for the first time and mills were closing at the rate of almost one a week (Fowler, 2003).

The UK attempted to protect the industry from this increasing competition through the implementation of the Multi-Fibre Agreement (MFA). The textile industry was the only industry in the world economy to which special international trade restrictions applied. Significantly, over 85 per cent of all trade was covered by this agreement (Dicken, 1986). The MFA was enforced in 1974 by the US and UK to 'create an "orderly" development of trade in textiles and clothing which would benefit both the developed and developing industries' (Dicken, 1986:249). Separate 'voluntary' agreements with Asian producers were outlined to restrict low-cost imports for a limited time period of four years (Dicken, 1992). The MFA continued until 1 January 2005 and during its existence had become more restrictive, with tighter import quotas (Dicken, 2011). The impact on global trade was immense, as this had only slowed down the demise of the UK textile industry. Inevitably after this date, the influx of price-based competition reduced the UK's manufacturing capacity in textiles and Asia's dominance grew. Asia now generates over 50 per cent of the world total of textile exports, with China as the leading producer, employing 2.7 million workers (Dicken, 2011).

However, during the protectionist era of the industry, a group of companies evolved and developed new fibres, fabrics and innovative process technologies; from the 1950s automobile tyre fabrics and new methods of finishing fabrics were beginning to be developed (Miller, 1962). Significantly, the blurring of textiles and the petrochemical industries derived new man-made fibres that altered the industry; nylon was the first truly synthetic fibre, developed by Wallace Carothers from research on polymers (very large molecules with repeating chemical structures) in Dupont's experimental station (DuPont, 2013). Nylon was commercialized in 1939 for ladies' hosiery. However, its properties of high strength, elasticity and resistance to moisture meant it was quickly channeled into alternative applications, for example military parachutes (Byrne, 2000). Since the 1970s a range of high-performance fibres have been developed which have 'provided some of the most significant and dramatic impulses to the evolution of the textiles industry' (Byrne, 2000:8). For example, aramids were commercially introduced in the 1970s as high-temperature-resistant fibres for protective clothing (Byrne, 2000).

Today, textiles accounts for 2.6 per cent of the UK manufacturing industry (Institution of Engineering and Technology, 2011) and has an infinite variety of end products (Dinsdale, 2012). The composition of the industry is extremely diverse; from lightweight and ultra strong carbon fibre reinforcements for wind-generating turbine blades to complex three-dimensional embroidered structures used as stents to repair damaged blood vessels (Byrne, 2000; O'Mahony, 2007). This demonstrates the industry's economic strength and strategic importance to society, and highlights its evolution rather than extinction (Jones and Haynes, 2004).

TECHNICAL TEXTILES: THE COMPETITIVENESS OF AN INNOVATIVE INDUSTRY

The interconnectedness of the global economy has enforced two major changes to the nature of firm-based competition (Bryson and Taylor, 2010), as manufacturing and export capabilities, especially in textiles, have been 'dispersed to an unprecedented number of developing as well as industrialized countries' (Gereffi, 2005:163). First,

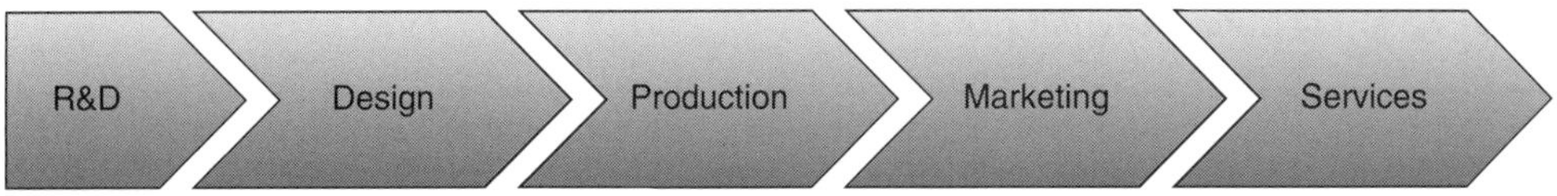

Source: Livesey 2006: Figure 2.

Figure 12.1 The advanced manufacturing process

there has been a substantial growth in firms located in developed market economies experiencing price-based competition from new competitors based in low-cost production locations. Second, firms are experiencing price-based competition from 'virtual' manufacturers that outsource the production component of the manufacturing process to low-cost producers (Figure 12.1) (Bryson and Taylor, 2010:583).

The expansion of manufacturing activity to low-cost producers has therefore challenged the survival of manufacturing firms in developed market economies (Bryson and Taylor, 2010). For their long-term survival in a global economic environment, firms have withdrawn from price-based competition and have undertaken a complex process of restructuring (Hudson, 2010). Manufacturers have moved up the value chain, becoming advanced manufacturers by developing highly innovative products and processes to reap the benefits of skilled, knowledge-intensive and higher-value manufacturing options (Livesey, 2006). Firms now undertake a 'full cycle of activities from research and development, through design, production, logistics and services' (Livesey, 2006:6) to increase the economic value of their outputs (Daniels and Bryson, 2002) and compete on the basis of expertise and 'product differentiation'; where firms seek to produce different products with different characteristics, so that each product is an imperfect substitute for competing products (Bryson et al., 2008; Bryson and Rusten, 2011).

The survival of the UK textile industry is based on the evolution of firms towards a high-value-added segment of the industry: technical textiles (Hassink, 2007). Technical textiles are not a new phenomenon, as fibres, yarns and fabrics have always been used for applications other than clothing (Byrne, 2000). However, the term 'technical textiles' was coined in the 1980s by Techtextil[3] to describe 'the growing diversity of textiles for their technical properties rather than their aesthetic characteristics' (Macbeth and Miln, 2012). This term superseded the earlier term 'industrial textiles', previously used to describe all textile products other than those intended for apparel, household and furnishing end uses. 'Industrial textiles' became inappropriate, with wider technological developments in textiles for medical, sports and transportation applications (Byrne, 2000). Industrial textiles are now a subgroup of the 12 application categories of technical textiles, developed by Techtextil (Table 12.1).

Technical textiles is an extremely dynamic industry consisting of products and processes that are either traditional, replacing well-established materials or techniques, or unique and purely created from the innovative processes and capabilities of material science today (McCarthy, 2012). For example, changes in the production process include the development of multi-axial fabrics, comprising 'one or more layers of fibres with different orientations stitched together' (O'Mahony, 2007:68). The strength of the fabric is in the direction of the fibre, and multi-axial fabrics are commonly weaved or

Table 12.1 Technical textiles taxonomy

1	Agrotech	Agriculture, Horticulture, Forestry and Aquaculture Textiles
2	Buildtech	Building and Construction Textiles
3	Clothtech	Technical Components of Shoes and Clothing, e.g. Linings
4	Geotech	Geotextiles and Civil Engineering Materials
5	Hometech	Technical Components of Furniture, Household Textiles and Floor Coverings
6	Indutech	Textiles for Industrial Applications. Filtration, Conveying, Cleaning etc.
7	Medtech	Hygiene and Medical Products
8	Mobiltech	Automobiles, Shipping, Railways and Aerospace
9	Oekotech	Environmental Protection
10	Packtech	Packaging Materials
11	Protech	Personal and Property Protection
12	Sporttech	Sport and Leisure

Source: McCarthy 2012.

stitch bonded with glass or carbon. This pre-preg material is then coated with 'resin for further reinforcement, filler to improve dimensional stability and additives for additional functionality such as flame retardency' to produce a composite (O'Mahony, 2007:72). Additionally, UK firms producing woven material for advanced carbon fibre structural reinforcements are developing 3D weaving processes; the textile is first woven incorporating multiple uni-directional layers as a flat fabric, before being cut open at one side to produce a 'T' section profile fabric (O'Mahony, 2007).

These technological developments demonstrate how textiles have become an industry of high-value niche applications. This evolution of textile firms has been driven by their competitive strategies and fundamentally by their innovation treadmill maintenance; 'to survive, a firm must continually challenge its existing processes and products and engage in constant innovation' (Bryson and Taylor, 2010:583; Chistopherson and Clarke, 2007).

The resource-based view (RBV) of the firm has become one of the standard theories on the origins of competitive advantage in the field of organizational strategy (Hoopes et al., 2003). The RBV of the firm addresses one primary research question – 'Why does the performance of firms in the same sector vary over time?' (Hoopes et al., 2003:880) –and assumes a 'firm's strategy is intended to achieve some form of sustainable competitive advantage' (SCA) that will deliver enhanced profits and resilience to imitation (Bryson and Taylor, 2010:588). Imitation is a 'form of knowledge spill-over that arises as knowledge has some form of the characteristics of a public good' (Bryson and Taylor, 2010:587). All firms are exposed to various forms of copying behaviour; for example, a product designed and developed by Nike in one year may be copied by rival companies in the next (Bryson and Taylor, 2010). An emphasis on inimitability is therefore central to understanding non-priced-based competition that enables firms to continue to manufacture in a high-cost location (Bryson and Rusten, 2011). Inimitability is achieved through firms mobilizing their own unique set of resources, capabilities and non-price-based competitiveness. Bryson and Taylor (2010) have identified ten related manufacturing strategies that a firm may deploy and combine in different and unique ways to achieve SCA (Table 12.2).

Table 12.2 Ten related manufacturing strategies

	Non-Priced-Based Competitive Strategies	
1	Bundled Products	Products that hybridize products and services together. The service element is delivered locally, and becomes difficult or too expensive to imitate by a competitors based in a low-cost location
2	Customization	Product is co-produced between the producer and consumer, requiring close contact between a producer and a consumer
3	Customer Flexibility	Producer develops flexible systems to respond to customer requirements
4	Manufacturing Experiences	The transformation of the manufacturing process into an experience
5	Skill Development	The development of skills and expertise that are difficult to replicate or transfer
6	Innovation	Continuous ongoing process, product, service, and design innovation
7	Logistics	Competing on delivery times and proximity to market for speed and ability to co-produce products
8	Corporate Social Responsibility	Green products, inclusive products, pink products
9	Place-Based Strategies	'Made in the United States' rather than 'Made in China' or 'Designed in the United States and Made in China'
10	Unrepeatable Processes	Creation of production processes that are difficult to replicate and often protected by patents and registered designs

Source: Bryson and Taylor 2010: 588.

Innovation is one of the most important manufacturing strategies to ensure a 'firm's competitive position is resistant to imitation or inhibits the speed by which imitation can take place' (Maskell and Malmberg, 1999:176). For example, by the time rival companies have copied Nike products, Nike will have introduced a new product (Bryson and Taylor, 2010). Innovation therefore enables firms to be agile, introducing a continuous stream of new, improved or redesigned products and services and developing efficient production processes for resilience in a competitive global environment (Kakkad, 2009; Dicken, 2011). The new currency in the textile industry is 'ideas' (Day, 2012). UK textile manufacturers' innovativeness has been derived from their 'initial advantage' in pre-existing personal networks, technical skills and market knowledge developed over a long time period. Susan Christopherson (2009) has coined these innovative industries as 'Phoenix Industries'; the product of R&D and workforce investments made in the heyday of manufacturing that have evolved into advanced technology firms producing highly sophisticated components.

TECHNICAL TEXTILES: AN EXAMPLE OF PHOENIX MANUFACTURERS

Phoenix industries will never reverse the demise of UK manufacturing but they have an integral role in renewing the economy (Christopherson, 2009). These industries contribute to the diversification of the economy in supporting a number of different industries with innovative components and employing a diversity of people. The technical textile industry is growing four times faster than conventional and apparel textiles, accounting for £3.5 billion (Technology Strategy Board, 2008). Significantly, the value of technical textiles has been notable in the current economic recession (Christopherson, 2009), as firms have stated they have experienced positive growth: 'It's been a very difficult two or three years in manufacturing and we've had our best three years in the last fifteen' (Company 39). A significant contribution to this has been firms' growing export markets; at least 20 per cent of their output is for expanding world markets (Christopherson, 2009; McGovern, 2013):

> We believe we've got an excess of 60 per cent market share across the whole of Europe . . . we now sell to North America, Japan, China . . . and because we're solution providers and we give very good technical responses to commercial vehicle companies, they don't only come to us for the niche products that they start developing, they say 'well we need to have this, this and this', so our portfolio of products into that client base has grown dramatically, those as clients are global clients. (Company 83)

Technical textile manufacturers are primarily small to medium sized enterprises (SMEs) with an average of 68 employees. This is a fundamental transition from when over half of all textile employment was concentrated and dominated by four employers in 1983 (Dicken, 1986). These technical textiles firms were established in the late 1980s and the 1990s, with a considerable amount of firms started by employees of these oligopolistic firms:

> My father founded the company about 30 years ago, he used to work for a big mill for a large textile consortium in the UK; a lot of the textiles after the second world war began to decline and consolidated, so you had companies like Coats Viyella and Courtaulds that swallowed up a whole heap of small businesses. He worked in a fairly large mill which during the consolidation process was closed down and . . . he decided to set up his own business, he saw the opportunity, so he took his skilled guys, half a dozen of them and they set this place up. (Company 89)

Technical textiles firms are located in the traditional textile locations of the North West of England, the East Midlands, East Coast of Scotland and Northern Ireland due to the employees – labor forces are becoming one of the most important elements in firms' location choices to achieve competitive advantage (Christopherson and Clarke, 2007). The skills required by these firms include those of highly skilled textile technologists with advanced degrees as well as traditional weaving and knitting skills with specialist knowledge and experience: 'There isn't any economic reason for being here apart from the expertise and the skill base that we have' (Company 89).

These SMEs serve diversified intermediate markets, producing technical fabrics for an array of applications; for example, M. Wright & Sons produce narrow fabrics and elastic webbings for safety, aerospace, military, leisure and medical applications. Phoenix industries can therefore be described as 'enabling industries'; firms now research, develop and produce technologies that are used in many different industries instead of just one (Christopherson, 2009:26). This has been a key innovation strategy for firms, as it subsequently increases their market share:

> A great way to innovate is through cross-fertilization, from one industry to another; solving problems from one particular fabric from one industry might be applicable in something completely unrelated, in a different industry. [For example] a fabric that was developed for divers' legs started life as a bra wing fabric for a lingerie firm. (Company 89)

As part of serving intermediate markets, firms have become more research oriented and 'constantly engaged in incremental process as well as product innovation' (Christopherson, 2009:26). Innovation in these advanced manufacturing firms is no longer about technological breakthroughs but about various forms of innovation in all areas of their manufacturing operations: 'from how they work with customers and manage globally complex supply chains to how they market themselves in tough, competitive environments' (Kakkad, 2009:1). For example, all the firms interviewed have changed or are in the process of changing their working practices 'to look and act like a modern manufacturing outfit' (Company 97). As Company 24 further described:

> We were a traditional manufacturer in all senses of the word . . . when we decided to focus on a particular brand, we had to focus on everything else, to support what we were doing, to support our quality and improve efficiencies, we reviewed the entire structure of how we did things, and we changed everything . . . everybody is now focused on something which is actually adding value rather than monitoring things, much more responsibility. . . they're consulted a lot more, they get more involved in looking at new processes. (Company 24)

The most important driving force identified for firms' innovation strategies is their customers; 'there is now much more emphasis on engineering products for specific applications rather than being technology-led by fibre and machine developments' (David Rigby Associates, 2001). Increasingly, products are produced through the process of 'user innovation', where firms' ideas and their consumers' needs are blended to produce new products (Bryson and Rusten, 2011). This has been a major contributor to firm survival, as they become solution providers of service-enhanced products (Bryson and Taylor, 2010):

> I would say the majority of our production is specific materials made for specific customers and processes . . . typically let's say somebody approaches us and they want a material for orthopaedic bandages . . . they'll give us some criteria, we would then present products from our established range to see if there's a match between what we currently have and their requirements . . . if none of our standard references are suitable we can make them a development. (Company 31)

Firms are also engaging in the opportunity to co-produce products with their customers as partners. This process of 'open innovation' is increasingly occurring between producers and customers' engineers. This develops strategic long-term relationships with lead customers, which can encourage ideas to be shared regarding product modifications (Rusten and Bryson, 2010; von Hippel, 1988 in Bryson and Rusten, 2011). This process reveals that innovation in manufacturing firms is no longer a linear process, but involves many different feedback loops (Bryson and Rusten, 2011:195):

> In the early 1970s, we had a bit of fortune that Rolls Royce were developing at the time the new current range of modern large aerospace engines, and they had a weight problem on the engine, everything had to get lighter and more efficient . . . one of our directors at the time spent a lot of time playing bowls and ran into a Rolls Royce engineer on a bowling green in Derby, and the guy turned round and said 'look we've got a strength problem, we've got a weight problem, we've heard about this Kevlar stuff, do you reckon there's anything in it for us?' and as a result of that conversation we got involved in working with Rolls Royce to develop, technically to trial test, make examples of the Kevlar containment system which is now obviously current on various of their engines. (Company 97)

In the past five years, firms have introduced high-technology research units to run alongside their production facilities. This formal innovation strategy of R&D laboratories enables firms to be proactive in technological developments as well as being able to provide their customers with a technical service in testing and developing products. Manufacturing companies are therefore now implementing competitive strategies of producing products that hybridize products and services (bundled products) and difficult-to-replicate products and processes by developing and testing 'in-house' to maintain their inimitability (Bryson and Taylor, 2010, table 2):

> We have non-manufacturing capability which we would never have had in the past because the pre-preggers would do all that work, whereas now we've got a lab, we do some infusion ourselves; you know we can do some mechanical testing of parts ourselves to give customers a technical service; we've got a KTP [Knowledge Transfer Partnership] running with Manchester university where we've got an associate looking at how our fabric behaves under different processes, and how we can change product to emphasize some of the attributes of the fabric which are good for that process, but that's again something we wouldn't have done if we weren't looking broadly across the market and trying to keep pace of the new technology. (Company 75)

DEFINING TECHNICAL TEXTILES

Technical textile manufacturers are a highly innovative and strategic segment of the UK economy. However, it has remained difficult for scholars and industry to construct a precise definition of technical textiles. Technical textiles has never been a single coherent industry (Byrne, 2000), and as textiles are becoming more advanced and diversified, firms are beginning to express their identity crisis:

> Nothing we make stays in just its textile format and you know a dry fabric is not, it might be our product but it's going into solid things . . . I would say we are a textile company, we manufacture solutions, textile solutions, but we're definitely more broadly aligned to the composite industry. (Company 75)

Technical textiles are widely defined as 'textile materials and products manufactured primarily for their technical and performance properties rather than their aesthetic or decorative characteristics' (Textile Institute, 2012). This definition is becoming increasingly ambiguous and has significant boundary issues as 'an increasing number of products are combining both performance and decorative properties and functions in equal measures' (Byrne, 2000:2). A sports shoe, for example, consists of 'technical' shoe components (hidden and functional) and 'non-technical' shoe uppers (visible and largely aesthetic) (David Rigby Associates, 2002).

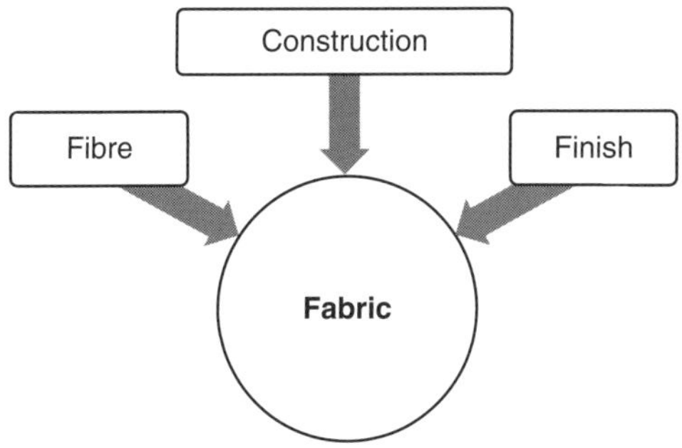

Figure 12.2 *The three main technical attributes of a fabric*

Current definitions are therefore conceptualizing the industry through historical dimensions of the main manufacturing sectors rather than a market-based view of the products involved (Byrne, 2000). This has become a common limitation in current frameworks attempting to classify modern manufacturing industries, as the transformation of manufacturing firms has been radical in the past 20 years (Bryson et al., 2008). The Production Projects and Distributed Tasks approach developed by Bryson and Rusten (2011) is more aligned to develop an understanding of what manufacturers look like today. This approach focuses on the product and the formation of this product; 'conceptualizing the complex inter-relationships that occur between different elements and tasks that come together to create value' (Bryson and Rusten, 2011:165). It is the higher-value fabric manufacturers that are maintaining textile-manufacturing systems in the UK. A focus on the textile fabric produced would therefore identify which firms are producing highly innovative products, creating value. A definition needs to foreground technical fabric producers that have engineered or scientifically advanced the fabric from different combinations of technical attributes to produce a technical textile (Figure 12.2).

Technical attributes of a technical textile fabric therefore consist of high-performance fibres and the different constructions and finishes of the fabric. High-performance fibres are those that 'have exceptional strength, strength-to-weight ratio, chemical or flame resistance, or range of operating temperatures' (Brown, 2005:36). The construction of the fabric includes the different traditional techniques of weaving, knitting and embroidery processes. However, the development of high-performance fibres has resulted in engineers reexamining the structural capabilities of these techniques (Brown, 2005). To illustrate, different fabric structures can now enable a fabric to be breathable or three dimensional for pre-pregs in the composite industry. Finishes and coatings create different effects and can determine what type of fabric is produced; for example, anti-bacterial, stain resistant, waterproof or windproof. An example of combining these technical attributes could include the composition of carbon fibre plus 3D weaving to produce a carbon fibre reinforcement fabric. Technical textiles can therefore be defined as a high-value segment of the textile industry consisting of a group of firms producing textiles:

Textile Fabric + Technical Attribute = Technical Textile

CONCLUSION

Manufacturing still matters to the UK economy. Manufacturing activity has declined in terms of employment since its peak in 1966 but still has a significant role to play in the UK economy; contributing 12 per cent of the UK's gross domestic product (BDO, 2010), 50 per cent of exports and 75 per cent of R&D (Bryson et al., 2008).

Academics no longer understand manufacturing (Livesey, 2006). The focus on manufacturing was displaced by a focus on services as the consequences of global shift witnessed a large amount of manufacturing activities decentralize to low-cost locations (Bryson et al., 2008; Dicken, 2011). This changed the competitive parameters for manufacturers. To survive, firms have evolved into advanced manufacturers, implementing non-priced-based competitive strategies (Bryson and Taylor, 2010). The RBV of the firm provides a theoretical framework to explore the origin of competitive advantages (Hoopes et al., 2003). It assumes that a firm's strategy is intended to develop inimitability for a firm's comparative advantage in the global marketplace (Bryson and Taylor, 2010). Bryson and Taylor (2010) have identified ten non-priced-based strategies deployed by manufacturing firms to enable them to achieve inimitability through mobilizing their own unique set of resources and capabilities.

As a result, firms have become more service oriented and the actual production of a product or component is now only one element within a wider manufacturing process (Bryson et al., 2004; Livesey, 2006). Firms now compete on the basis of expertise, customized products, or in high value-added niche markets (Bryson et al., 2008). The current research has therefore demonstrated that there are still opportunities to manufacture in developed market economies through high-quality, differentiated products and niche markets (Livesey, 2006; Christopherson and Clarke, 2007; Bryson et al., 2008; Christopherson, 2009; Bryson and Taylor, 2010; Hudson, 2010).

The manufacture of textiles is one of the oldest manufacturing industries, and the first to take on a global dimension (Miller, 1962; Dicken, 1998). The UK textile industry has experienced continuous challenges to its competitiveness since its peak in 1913 (Miller, 1962; Dicken, 2011) as industrializing and developing nations between 1918 and 1940 began to develop their own textile industries (Chapman, 1987). Textiles have low barriers to entry as a relatively low-skilled production process and are therefore ideally suited to the early stages of industrialization (Rostow, 1960). The influx of price-based competition from the low-cost developing market economies during the 20th and into the 21st century has dramatically reduced the UK's manufacturing capacity in textiles.

In response to this intensifying price-based competition, the UK has lost its comparative advantage in producing some apparel and textile goods, but has survived and thrived in an advanced segment of textiles (Hassink, 2007). The UK technical textile industry consists of SMEs that develop new fibres, fabrics and innovative processes for high-value niche applications. These SMEs serve intermediate markets, producing technical fabrics used in many different industries. Technical textile manufacturers are extremely research oriented and have survived through a key competitive strategy of constant innovation to ensure their inimitability (Maskell and Malmberg, 1999; Bryson and Taylor, 2010). Their manufacturing focus is to provide innovative solutions for specific customers. Their innovation strategies therefore consist of 'user' and 'open' innovation, where products are produced through blending together firms' ideas and customers' needs (Bryson and Rusten, 2011). This has revealed the importance of developing and managing long-term customer relationships for product development. Formally, firms are beginning to introduce high-technology research units to be proactive in technological developments and expand their technical services for their customers.

The UK technical textile industry is an example of a Phoenix industry – a traditional industry that has evolved to become a key innovator from its initial advantages in pre-existing personal networks, technical skills and market knowledge, developed over a long time period (Christopherson, 2009). Phoenix industries have an integral role to play in developed market economies, as they contribute to supporting the diversification of a number of industries with innovative components and employing a diversity of people (Christopherson, 2009). The value of UK technical textile firms has been notable in the current economic recession as these firms have experienced positive growth and the industry is growing four times faster than conventional and apparel textiles (Technology Strategy Board, 2008; McGovern, 2013).

Technical textiles are not a new sector; however, it is relatively new in terms of its raised profile since the 1980s and new technological developments in fibres and combined processes (Byrne, 2000). Technical textiles have never been a single coherent industry and it has remained difficult for scholars and industry to construct a precise definition of technical textiles (Byrne, 2000). This has been a common limitation in current frameworks attempting to encapsulate and define modern manufacturing industries. The Production Projects and Distributed Tasks approach focuses on products by identifying the individual tasks, and their geographies, that are required to create a product or service (Bryson and Rusten, 2011). A focus on the textile fabric produced will therefore identify firms producing highly innovative products that create value. Technical textiles are defined in this chapter as a high-value segment of the textile industry consisting of a group of firms producing textiles from engineering or scientifically advancing the fabric.

This chapter has explored the survival and evolution of UK textile manufacturers in the context of changing geographies in a global marketplace. This chapter is part of a continuing analysis and therefore elicits preliminary conclusions of the technical textile industry. The evolution of the UK textile industry is based upon individual manufacturing firms' decision making and the competitive strategies of innovation implemented, which have created a highly advanced and niche industry of technical textiles. As part of this current project, further research is needed to explore how firms are utilizing and managing their assets to maintain this competitiveness.

NOTES

1. LinkedIn is a social networking website for people in professional occupations (http://www.linkedin.com; accessed July 2012).
2. NWtexnet is the designated body to support both the advanced flexible materials sector and the wider textiles and clothing sector in the North West of England. This placement allowed information to be obtained from meetings with key people in the industry, including UK and EU textile trade associations and strategic academic staff at UK based and international universities (Ronayne, 2012).
3. Techtextil is the leading international trade exhibition for technical textiles organized annually since the late 1980s by Messe Frankfurt in Germany and in Osaka, Japan (Byrne, 2000:3).

REFERENCES

BDO (2010) The Changing Shape of UK Manufacturing, March. Available at http://static.bdo.uk.com/assets/documents/2010/04/Government_Strategy_Report.pdf (accessed November 2012).

Bell, D. (1974) *The Coming of Post-industrial Society*, Heinemann Educational Books Ltd: London.

Brown, S. (2005) Textiles: Fiber, Structure, and Function. In McQuaid, M. (ed.) *Extreme Textiles: Designing for High Performance*, Princeton Architectural Press: New York.

Bryson, J.R. and Daniels, P.W. (2007) *The Handbook of Service Industries in the Global Economy*, Edward Elgar: Cheltenham and Northampton, MA.

Bryson, J.R. and Rusten, G. (2011) *Design Economies and the Changing World Economy: Innovation, Production and Competitiveness*, Routledge Studies in Human Geography, Routledge: Oxon.

Bryson, J.R. and Taylor, M. (2010) Competitiveness by Design and Inimitability through Service: Understanding the Dynamics of Firm-Based Competition in the West Midlands Jewellery and Lock Industries. *Service Industries Journal*, 30(4), 583–596.

Bryson, J.R., Daniels, P.W. and Warf, B. (2004) *Service Worlds: People, Organizations, Technologies*, Routledge: London.

Bryson, J.R., Taylor, M. and Cooper, R. (2008) Competing by Design, Specialization and Customization: Manufacturing Locks in the West Midlands (UK). *Geografiska Annaler: Series B, Human Geography*, 90(2), 173–186.

Byrne, C. (2000) Technical Textiles Market – An Overview. In Horrocks, A.R. and Anand, S.C. (eds) *Handbook of Technical Textiles*, Woodhead Publishing Limited: Cambridge, 462–489.

Chapman, S.D. (1987) *The Cotton Industry in the Industrial Revolution*, 2nd Edition, Macmillan Education Ltd: London.

Christopherson, S. (2009) Manufacturing: Up from the Ashes. *Democracy Journal*, 25–30.

Christopherson, S. (2011) Riding the Small Wave in Manufacturing to More Good Jobs and a More Diverse Economy. Conference Paper on Big Ideas for Job Creation in a Jobless Recovery. Available at http://aap.cornell.edu/crp/people/faculty-profile.cfm?customel_datapageid_7102=16885 (accessed February 2012).

Christopherson, S. and Clarke, J. (2007) *Remaking Regional Economies: Power, Labor, and Firm Strategies in the Knowledge Economy*, Routledge: Oxon.

Daniels, P.W. (2005) The Geography of the Economy. In Daniels, P., Bradshaw, M., Shaw, D. and Sidaway, J. (eds) *An Introduction to Human Geography: Issues for the 21st Century*, 2nd Edition, Pearson Education Limited: Harlow, 287–312.

Daniels, P.W. and Bryson, J.R. (2002) Manufacturing Services and Servicing Manufacturing: Knowledge-Based Cities and Changing Forms of Production. *Urban Studies*, 39(5–6), 977–991.

David Rigby Associates (2001) North West Textile Cluster Mapping Project. In association with North West Development Agency and NWtexnet. Available at http://www.nwtexnet.co.uk/reports.asp (accessed June 2012).

David Rigby Associates (2002) Technical Textiles and Nonwovens: World Market Forecasts to 2010. Available at http://www.fibre2fashion.com/industry-article/pdffiles/Technical-Textiles-and-Nonwovens.pdf (accessed June 2012).

Day, P. (2012) Lancashire Leads New British Textile Manufacturing Revival. Available at http://www.bbc.co.uk/news/business-17827741 (accessed April 2013).

Dicken, P. (1986) *Global Shift: Industrial Change in a Turbulent World*, 1st Edition, Harper and Row: London.

Dicken, P. (1992) *Global Shift: The Industrialization of Economic Activity*, 2nd Edition, Paul Chapman: London.

Dicken, P. (1998) *Global Shift: Transforming the World Economy*, 3rd Edition, Paul Chapman: London.

Dicken, P. (2007) *Global Shift: Mapping the Changing Contours of the World Economy*, 5th Edition, Sage Publications Ltd: London.

Dicken, P. (2011) *Global Shift: Mapping the Changing Contours of the World Economy*, 6th Edition, Sage Publications Ltd: London.

Dinsdale, P.H. (2012) Can we Forecast the Future using Textiles Studies? *Textiles* 2012/No. 2., Manchester.

DuPont (2013) Innovation Starts Here: 1939 Nylon. Available at http://www2.dupont.com/Phoenix_Heritage/en_US/1939_c_detail.html (accessed January 2013).

Ellis, J. (2008) What Is the Future for Technical Textiles? A paper prepared for NWtexnet, not available for public distribution.

Fowler, A. (2003) *Lancashire Cotton Operatives and Work 1900–1950: A Social History of Lancashire Cotton Operatives in the Twentieth Century*. Aldershot: Ashgate.

Gereffi (2005) The Global Economy: Organization, Governance, and Development. In Swedburg, R. and Smelser, N.J. (eds) *The Handbook of Economic Sociology*, Princeton University Press: Woodstock, 160–182.

Gibbs, D.C. (1988) Restructuring in the Manchester Clothing Industry: Technical Change and Interrelationships between Manufacturers and Retailers. *Environment and Planning A*, 20, 1219–1233.
Glasmeier, A., Thompson, J.W. and Kays, A.J. (1992) The Geography of Trade Policy: Trade Regimes and Location Decisions in the Textile and Apparel Complex. *Transactions of the Institute of British Geographers*, New Series, 18(1), 19–35.
Hassink, R. (2007) The Strength of Weak Lock-Ins: The Renewal of the Westmunsterland Textile Industry. *Environment and Planning A*, 39, 1147–1165.
Hoopes, D.G., Madsen, T.L. and Walker, G. (2003) Guest Editors' Introduction to the Special Issue: Why is there a Resource-Based View? Toward a Theory of Competitive Heterogeneity. *Strategic Management Journal*, 24, 889–902.
Hudson, R. (2010) The Changing Geographies of Manufacturing and Work: Made in the UK? In Coe, N.M. and Jones, A. (eds) *The Economic Geography of the UK*, Sage: London, 139–152.
Institution of Engineering and Technology (2011) The Institution of Engineering and Technology (IET) View: Taking a Team Approach. In Gosney, M. (ed.) *Boardroom Report: Rebuilding UK Manufacturing*. Available at http://fplreflib.findlay.co.uk/jnrlcover/ukms/Boardroom_2011.pdf (accessed September 2011).
Jones, R.M. and Haynes, S.G. (2004) The UK Clothing Industry: Extinction or Evolution? *Journal of Fashion Marketing and Management*, 8(3), 262–278.
Kakkad, J. (2009) *Creating the Playing Field: Innovation Monitor*, EEF (The Manufacturers Organization): London.
Kay, S. (2011) StayCool Project Description. Available at www.staycooleu.com (accessed March 2013).
Leunig, T. (1996) The Myth of the Corporate Economy: Great Britain's Cotton Textile Industry, 1900–1913. Available at http://www.thebhc.org/publications/BEHprint/v025n2/p0053-p0068.pdf (accessed March 2013).
Livesey, F. (2006) Defining High Value Manufacturing. Institute for Manufacturing. Available at http://www.ifm.eng.cam.ac.uk/uploads/Research/CIG/DefiningHVM.pdf (accessed January 2013).
Macbeth, B. and Miln, J. (2012) UK Textiles. Available at www.thegrowthfactory.co.uk/sectors/textiles (accessed April 2013).
Maskell, P. and Malmberg, A. (1999) Localised Learning and Industrial Competitiveness. *Cambridge Journal of Economics*, 23, 167–185.
McCarthy, B. (2012) The Scope of Technical Textiles. Available at http://www.technitex.org/about-us/ (accessed November 2012).
McGovern, S. (2013) GDP Figures: Textile Firm Exporting All Over World. Available at http://www.bbc.co.uk/news/business-21201506 (accessed January 2013).
Miller, E.W. (1962) *A Geography of Manufacturing*, Prentice-Hall Inc.: London.
O'Mahony, M. (2007) Engineered Textiles. In Braddock Clarke, S.E. and O'Mahony, M. (eds) *Techno Textiles 2: Revolutionary Fabrics for Fashion and Design*, Thames & Hudson: London, 52–75.
Ronayne, M. (2012) Trade Organizations, SMEs and the Development of the Technical Textiles Industry. Industrial Placement Report at NWtexnet in Association with EREBUS. Available at http://www.nwtexnet.co.uk/reports.asp (accessed October 2012).
Rostow, W.W. (1960) *The Stages of Economic Growth: A Non-Communist Manifesto*, Cambridge University Press: London.
Rusten, G. and Bryson, J.R. (2010) Industrial Design, Competitiveness, Globalization and Organizational Strategy. In Rusten, G. and Bryson, J.R. (eds) *Industrial Design, Competition and Globalization*, Palgrave Macmillan: Houndmills, 1–22.
Scott, A. (2006) The Changing Global Geography of Low-Technology, Labor-Intensive Industry: Clothing, Footwear, and Furniture. *World Development*, 34(9), 1517–1536.
Select Committee on Trade and Industry (2002) Third Report. Available at http://www.publications.parliament.uk/pa/cm200102/cmselect/cmtrdind/597/59704.htm (accessed April 2009).
Sunley, P. (1992) Marshallian Industrial Districts: The Case of the Lancashire Cotton Industry in the Inter-War Years. *Transactions of the Institute of British Geographers*, New Series, 17(3), 306–320.
Technology Strategy Board (2008) Advanced Materials: Key Technology Areas 2008–2011. Available at http://www.innovateuk.org/_assets/advanced%20materials%20strategy.pdf (accessed January 2012).
Textile Institute (2012) Textile Terms and Definition published by The Textile Institute. Available at http://www.ttandd.org/ (accessed 3 December 2014).
Tokatli, N. (2012) Old Firms, New Tricks and the Quest for Profits: Burberry's Journey from Success to Failure and Back to Success Again. *Journal of Economic Geography*, 12, 55–77.

13. Finding a future for the U.S. furniture industry

Susan Walcott

Don't write us off! We're very passionate about our work. Cut us open and we bleed sawdust.
(North Carolina furniture worker 2010)

INTRODUCTION

Attrition in the ranks of the U.S. furniture industry preceded the national economic downturn by two decades, providing companies in this field a foretaste of the winnowing effects to come on a larger scale. Companies that survived the ensuing creative destruction found competitive niches in targeted custom markets, updated methods and utilized a variety of strategic choices by managers aware of the necessity for change (Schumpeter 1942; Drayse 2008). If the second decade of the 21st century is witnessing a manufacturing moment characterized by re-shoring of formerly outsourced production, it is a long-awaited one for the furniture industry (Helper et al. 2012a,b). This examination of the ongoing industry transition focuses on North Carolina, as a case study of the complex arrangements involved in managing change in response to both exogenous and endogenous shocks. By the end of 2010 the Southeastern region produced 27 percent of U.S.-made furniture, with North Carolina the third largest producer in the U.S. at 7.6 percent, preceded by Michigan (8.2 percent) and California (9.6 percent) (IBIS *World* 2011).

Firm learning and technological change are key parts of the story, but it is individuals who do the learning, often coming across new practices through their own social networks (MacKinnon, et al. 2009). The role of technology as a driving force in adaptation is clear for corporate survival, but choices as to how technology is used are also crucial. Innovations related to lean manufacturing (LM) practices, common in furniture industry survivors, feature cellular configuration of work stations that root workers to a more confined spatial cluster encompassing all production elements, as well as corporate concern with worker welfare demonstrated by on-site medical care, company lunchroom where food is stored and social interaction takes place, rewards for accepted suggestions, worker review committees, and so on. Associated with the "Toyota Way," LM is a process adopted in a variety of industries (Hunter 2008). Lean practices are not a panacea for rescuing floundering firms, but can showcase improved product and process quality.

The South remains central to American furniture manufacturing. The following examination of North Carolina's furniture industry presents an empirical study from which theory and policy implications can be drawn. Furniture formed a major segment of North Carolina's traditional economic base, along with tobacco and textiles. The additional economic pressure from the recession that began in 2008 accelerated the process of winnowing out domestic companies, leaving only those implementing innovations or well into the process of doing so. A central hypothesis is that choices by strong leaders can ameliorate the shift of locations for particular product manufacturing steps along a reconstituting global production network, enabling the retention of traditional strengths in carefully defined niches.

The first section of this examination provides a background of U.S. furniture industry geographic concentrations, followed by a discussion of the industry's flight overseas and ensuing commodity chain arrangements. The next section examines how these two trends intersect in North Carolina and the southern border region of Virginia: the "furniture foothills" (Walcott 2011). The conclusion reviews implications for continuing globalization adjustments, including an expansion in U.S. furniture manufacturing jobs post-2010.

Unless otherwise specified, this research uses the eight North American Industrial Classification System (NAICS) categories associated with the 337 level. Statistical data concerning the furniture industry are supplemented by 40 interviews with major participants in six segments: broadwoven fabrics (8), architectural/cabinetry (2), upholstered furniture (11), wood furniture (7), institutional/office furniture (4), and supporting organizations such as logistics and advocacy organizations (8). Nineteen factories and related businesses were visited to observe innovative production process implementation. Three basic quality categories of high-end, mid-range and mass market furniture prioritize different considerations: 1) reliance on economies of scope, 2) combinations seeking labor cost reduction and 3) extensive mechanization and economies of scale. The location implications of these three categories are discussed in detail.

HISTORY OF THE U.S. FURNITURE INDUSTRY

Spatial clustering long characterized the furniture industry as manufacturers sought agglomerative advantages close to wood inputs, skilled labor and shipping lanes. The center of activity shifted over time, beginning with Boston until the dawn of the 20th century when the Midwest took over, with Cincinnati, Ohio and Grand Rapids, Michigan moving into prominence. New York City and Chicago hosted the first industry-wide showcases. High Point, North Carolina opened its initial furniture market in 1905 to advertise its advantage as an inexpensive site with abundant human and material production elements. By 1921 an expanded, permanent market facility solidified High Point's position on the furniture map as a show-related cluster of businesses (Power and Jansson 2008). The Depression years brought a demand for low-cost furniture that North Carolina fulfilled thanks to Bassett's introduction of mass manufacturing. By the 1980s half of all U.S. furniture came from North Carolina (Lacy 2004; Powell 2006).

Two multi-county clusters capture most of the activity inside North Carolina. The Piedmont cluster specializing in upholstery focuses on High Point, with a second cluster specializing in wood furniture and fabric to the west around Hickory. Cabinetry and architectural woodwork firms feature highly customized pieces largely done by local artisans. The amount of furniture produced overall has greatly diminished under the pressure of global competition.

THE GLOBAL COMMODITY CHAIN

Changes within the U.S. furniture industry since its high point in the 1980s led to major negative consequences throughout the first decade of the 21st century. Several industry insiders portray the demise of domestic furniture manufacturing as largely self-inflicted. The insular, warring clusters in the furniture foothills earned the sobriquet of the "Briar Patch" (Dugan 2009). In some cases inheritance disputes led to irresolvable family fights. In other instances, ownership passed to individuals seen as good managers but unfamiliar with the furniture business and its tight personal networks. The accession of a generation of owners less involved in the business than the founders also added to the sell-off of many formerly family-controlled companies in the late 1980s. One example of the mergers and acquisitions process outcome is the Missouri-based Furniture Brands International (FBN) conglomerate which, during the 1980s, acquired leading furniture companies: Lane, Broyhill, Thomasville, Drexel Heritage, Henredon and Maitland-Smith. Faced with increasingly modernized mechanization by European and Chinese furniture firms, American managers frequently failed to update practices or expand communications with other manufacturers or their own workers, resulting in cascading collapses. The decade from 1990 to 2000 witnessed the demise of 47 North Carolina furniture companies unable to compete with European and Asian companies that turned out goods faster with fewer employees.

As noted in Table 13.1, the Canadian furniture industry suffered a steep decrease in the first decade of the 21st century, dropping as a source of U.S. furniture by almost as great a percentage as it gained in percentage share as a recipient of U.S. furniture exports. From a relatively small base, Vietnam (due to shifts from China) and Poland

Table 13.1 Balances of U.S. furniture import–export destinations, 2008–2012 (YTD Sept by $1 million value)

Country	YTD 2008	YTD 2009	YTD 2010	YTD 2011	YTD 2012
Positive balance					
Canada	−635	−90	15	196	475
Saudi Arabia	67	77	91	103	110
UAE	54	37	52	54	64
Japan	39	44	45	35	57
Australia	53	28	42	45	53
Venezuela	33	29	30	48	52
Negative balance					
China	11,279	9,233	11,497	11,420	12,311
Vietnam	1021	979	1301	1312	1671
Mexico	644	551	721	758	960
Taiwan	641	442	571	589	640
Malaysia	546	460	576	507	560
Italy	607	371	397	432	458
Indonesia	446	362	427	415	452

Source: US Trade Stats Express.

(due to supplying the giant IKEA) recorded large percentage increases in furniture imports to the U.S., while China surged with goods of competitive quality. The Middle East market indicates demand for custom furniture in a newly wealthy region. The rise in furniture exports to Australia and drop in furniture exports to the Mexican market defied distance but paralleled relative prosperity in each country. While U.S. wood furniture exports grew from $2 billion to $5 billion in value from 1991–2011, with more than half of U.S. wood furniture exports destined for Canada, the increase in furniture imports from China dwarfed this trend. Nevertheless, in 2012, 71 percent of furniture purchased in the U.S. was made domestically (Sirkin et al. 2012).

Small firms of fewer than 250 employees characterize this industry, with a few corporations outside the main production regions employing over 10,000 workers. Indiana's giant HillRom, for example, started with caskets, added hospital furniture, and then expanded to the growing institutional and office furniture sectors. Localized agglomerations reflect the usual factor advantages: availability of raw material, appropriately skilled labor (especially former agrarian labor) willing to accept low pay and challenging work conditions, and multi-modal transportation infrastructure (Cater 2005). Danville, Virginia attracted IKEA in mid-2008, re-occupying this traditional location for furniture manufacturers. Reasons cited include existing skilled labor and infrastructure close to the dense east coast market (IKEA 2008). Danville's Cane Creek industrial park contains other suppliers globally linked to IKEA, such as a mattress firm that supplies the company's needs in Poland, the eastern European country to which IKEA likens its Danville facility.

Global production network considerations for furniture include the cost of labor at various stages of manufacturing, components such as metal for structural parts, and shipping, including fuel and timing issues (Rohter 2008). All are influenced by shifting political economic issues such as currency exchange rates and legal structures in the countries involved. A large wage differential, coupled with product quality improvements such as the ability to dry wood components, drove the initial rapid exit of furniture manufacturing from the U.S. to China at the turn of the 21st century (Langfitt 2009; Sirkin et al. 2012). Furniture manufacturing involves assembling basic components, usually from global sources though frequently from the closest distance possible. Time can be important in particular markets, and price balanced with speed of delivery. In the case of Asia, logistical arrangements usually involve from six weeks to two months more than domestically made furniture due to both shipping and warehouse transit involving multimodal transportation switching from ship, train and truck. Schedule disruptions from seasonal demand occur when huge retailers reserve large orders of containers in anticipation of holiday delivery needs. Shortages of labor and electricity capacity in major Asian manufacturing countries hamper order fulfillment. Furniture, like machinery, is a bulky item for containerization and takes up a large amount of interior space. Two key strategies are routinely used in order to increase the amount of furniture shipped in each container. One is the "postponement" method of importing to the U.S. only the most basic wooden components that can be most easily and inexpensively made overseas. More highly customized touches such as detailing, tacks and especially upholstery are usually applied at a U.S. assembly facility (Perry 2010). Ironically, while the amount of trees harvested in major supplier states such as North Carolina remained stable through the early 21st century, the amount shipped as rounds (raw logs), principally to China, increased correspondingly to the increase in wood furniture manufactured in China – and shipped back to the U.S. (Buehlmann and Schuler 2009).

The amount of upholstery in the product is a key consideration for how much and where work is done. Upholstery cut-and-sew kits largely come from China and are assembled in the U.S. A higher degree of customization means more activity in the U.S. for higher-priced design features in fabrics and furniture pieces. North Carolina remains the hub for furniture lobbying, logistics and specialized truck carriers.

RE-SHORING RESURGENCE

> [P]eople create with their hands and their hearts. You can't outsource creativity and craftsmanship. Those hard-working folks are part of a vibrant creative economy . . . vital to North Carolina's future. Secretary, NC Department of Cultural Resources (Evans 2010a)

Shipping costs, time to market and design elements also constitute major production location factors. Managing the offshoring of furniture involved the following related and additional considerations. The rise in oil prices, currently stabilized, was insufficient to offset the wage gap between U.S. and Chinese workers – until recently, when all made-in-China goods faced the combination of worker shortages leading to rising wages, and the cumulative effect of a series of yuan appreciations vs a weaker dollar. While the recently reduced gap in wages between Chinese and American workers to a 5–10 percent level makes manufacturing in the U.S. more attractive, actions by surviving companies in the interim also created new conditions improving overall competitiveness. Distance was overcome by designers flying to China, but intellectual property thefts were a major problem. Time to market remains a competitive advantage for domestically manufactured furniture, particularly during seasonally high demand periods. The additional effect of U.S. trade penalties on companies found guilty of dumping charges related to furniture (e.g., mattresses, wooden bedroom sets) helped to encourage domestic manufacture, a trend accelerated by the ongoing economic recovery. All of these factors led to furniture's inclusion as one of the seven major industries seen by the Boston Consulting Group as most likely to see a re-shoring to the U.S. (Sirkin et al. 2012).

Recent steps indicating a re-shoring of furniture production back to the U.S. as well as domestic expansion of production reflect changes made in response to competitive pressures. Companies focused efforts on a distinct niche for which they were eminently suited, adapted to market conditions and cultivated strong networks with suppliers, retailers and targeted markets. As of first quarter 2014, North Carolina held top rank in furniture employees (33,832), closely followed by California (33,454), Indiana (22,628) and Texas (21,744) (BLS 2014). This represents less than half of the labor force in the year 2000 for North Carolina's furniture and related product manufacturing employment, with similar declines in employment in the other states, but most states also saw a rise in average wage for the remaining workers. Interview evidence indicates that this can be attributed in part to the use of more machinery, and the need to entice a shrinking skilled labor force to remain in the furniture industry. The smallest and most local segment of NAICS 337, the cabinet–vanity–countertop subcategory, declined the most under recessionary pressure on firms with less monetary reserves to withstand the drop in demand. The largest growth came in institutional furniture, characterized by the largest firms and the presence of government and other non-personal orders.

Furniture industry companies in the U.S. fall into three basic categories: companies with headquarters in the U.S. but major manufacturing activities abroad; foreign companies that employ American workers to do work in the U.S.; and American-owned companies that execute the majority of their production in the U.S. The following section considers these three strategies as they are currently practiced in the major furniture industry subsectors. The best strategies integrate global production networks with local regional development, targeting embedded advantages such as appropriate labor skills and infrastructure with market segments both domestic and foreign (Yeung 2009).

Slightly more than one-quarter (26.4 percent) of both household and office furniture companies are located in the southeast, with the majority in North Carolina (IBIS*World* 2010). Furniture companies frequently locate global headquarters in North Carolina for the image and availability of technology-savvy workers, but manufacture overseas for labor cost savings important to their price-sensitive market. Ashley and FBN, the latter created by acquisitions, dominate the U.S. household furniture industry (Evans 2010b; IBIS*World* 2010). Although both remain headquartered in the Midwestern United States, respectively Wisconsin and Missouri, low-cost market giant Ashley recently announced it will open a new, 1000+ employee production operation in North Carolina (Covington 2012). Ashley's prominence relies on leadership in lean process strategies, logistics and IT proficiency sufficient to manage manufacturing in five regional centers throughout the U.S., including a huge upholstery plant in Mississippi, as well as in China and Vietnam (Allegrezza 2010a).

Customized carving work can be done at sufficient quality and more competitive labor rates overseas, but high-end custom-finish detailing work is executed in the U.S. and the Philippines. The most competitive niche for Made-in-the-USA production remains high-end upholstery, prioritizing quality and delivery speed, as well as promoting sensitivity to place of origin and pride in domestic production. Mid-level upholstered furniture firms tend to use choice-limited cut-and-sew kits and simple Chinese wood components. A new, relatively wealthy class in China, able to afford a more spacious dwelling and the means to furnish it, prefers Western-made goods as a sign of prestige. American mid-market brands from Baker to La-Z-Boy and Chinese-owned Ethan Allen are expanding sales in Asia. High-end furniture businesses such as Keeler Brass, a maker of decorative hardware headquartered in Grand Rapids, brought their facilities back from China after being burnt by low-quality copies and unreliable subcontractors (Allegrezza 2010b). Issues with leaks in intellectual property remain a concern for U.S. companies which find their competitive edge undercut by Asian-made goods and push production back to the U.S. Other global firms locate the value-added component of coordinating and design facilities domestically but manufacture in and sell to the developing world.

Middle and low-middle producers are most likely to migrate overseas, using labor cost savings to offset distance factor costs. The relatively commanding size of the U.S. market, and glimmers of a retail re-emergence, pull some foreign headquartered firms to set up branch plants in the U.S. Their intention is to take advantage of the available pool of unemployed but highly skilled and experienced American workers, market proximity and local governments eager to offer economic incentives for new jobs. A major Chinese furniture firm, for example, opened a large upholstery complex in Indiana to serve its

U.S. market (Allegrezza 2010b). The fierce competition reducing firms in the field also indicated a path for new profits in underutilized U.S. facilities (Russell 2010a).

Indications of recovery began to appear by the end of 2009, reinforced by orders in late 2010. Numbers in most categories continued to decline but the rate of descent was less steep, a sign of tentative stabilization. By 2010 almost half of North Carolina's furniture industry employment was in the household and institutional sector, with two-thirds in upholstered furniture (BLS 2011). North Carolina's image as a furniture center draws related companies there, according to many interviewees. "Upholstery is North Carolina," declared one local national brand manufacturer whose company recently acquired a prominent old wood furniture firm that it updated with LM practices as a test case for extending process innovations throughout the main plant headquartered elsewhere. A mixture of large and small upholstery firms engage in partnerships with furniture manufacturers in the statewide clusters using fabric produced by high-tech Valdese Mills, whose story resembles that of wood furniture manufacturers and the success strategies of other industry leaders.

As the largest remaining U.S. broadloom manufacturer, Valdese Weavers in western North Carolina supplies broadloom fabric for upholstered furniture in a very large, highly automated facility with cutting-edge technology. Attention to the market niche of high-end, customized, rapid delivery that customers demand assisted the survival of Valdese. Re-verticalization by acquiring companies that were suppliers ensured continuity by contracting with multiple suppliers for product flow. This in turn promoted an agglomerated cluster in close proximity, feeding business to each other. The presence of Valdese in North Carolina is a major factor for a number of furniture companies involved in upholstery production. The speed and thoroughness with which Valdese acquired cutting-edge machinery – and the business of customers stranded by competitors which collapsed – reflected decisions made by managers who saw the need to transform their industry quickly.

The wood furniture subsector was most impacted by competition from Asian (principally Chinese) manufacturers, but high-quality niches plus world-class process improvements led to retention of even this segment in its traditional North Carolina location. Hickory Chair, named for the western North Carolina town that anchors the furniture cluster in that part of the state, provides the best example of successful adoption involving LM processes. Hickory's path to lean resulted from the sudden demise of its CEO just prior to the critical High Point Market. The head of marketing appointed to carry on rushed to learn business practices from the workers as well as a consultant who introduced him to a major Toyota executive.

These circumstances, and his personal openness to change, led to development of Hickory Chair's "EDGE" lean system transformation by melding company input with Toyota mentoring, relying on an integrated team of suppliers and the manufacturer working together in a process designed for high quality, low waste and rapid response to market demand (Liker 2004; Porto and Smith 2006). Hickory's success as the only FBN affiliate to increase sales and profits early in the recession led to the hiring of LM consultants for the rest of Furniture Brand's affiliated firms to direct and enforce rebuilding during the national economic downturn.

Lean processes are not themselves a panacea. Adoption by firms was slow due to a combination of factors. One set of objections had to do with the correct perception that adoption of such transformative methods required time and costs for retraining in new behavior patterns rather than acquisition of new machinery or personnel. Some managers felt that competitive pressures did not permit flexibility to change. Another group began lean processes, but abandoned some in piecemeal fashion and missed capturing the underlying ethos such as meaningful worker participation.

Institutional and office furniture remains a profitable part of the furniture industry. Key market segments in the healthcare industry and institutional facilities, as well as government contract connections, promote Made-in-USA producers. Focusing on this niche carried successful North Carolina firms through recessionary downturns in segments of the entertainment and hospitality industry. Legacy furniture in Conover, North Carolina stands out due to three familiar factors: modern machinery and lean processes carefully targeted to a competitive niche and incorporation of the talents of a skilled workforce in an advertised team effort. The "laboratory apparatus and furniture" segment is subsumed into this category and provides a profitable market as geriatric care facilities expand in anticipation of increasing demand. Another profitable market segment specializes in supplying customized furniture for office cubicles, the free-standing office sector, by early LM adopters. Major office and institutional makers focus on value-added touches such as design. Design firms tend to be small, with highly mobile talent working overseas doing subcontracting jobs (MacPherson and Vanchan 2010). This can result in a loss of competitive exclusive "look" elements when the designs show up in other furniture.

Lean manufacturing transformation consultants are in high demand, from state university extension agents to individual entrepreneurs such as former corporate employees with a deep understanding of industry practices. Wider knowledge of the time and cost savings that such changes portend is important for the health of the industry. The economic downturn was utilized to position remaining companies for the recovery expansion opportunities. Hickory and High Point cluster companies draw on similar suppliers and communicate through them, as well as using managers with experience in various plants to gauge the difference made by LM-type programs.

Labor skills are concentrated in upholstery and technical skills in the High Point area, with wood working and craft skills in Hickory. Although the labor base is experiencing attrition by aging, workers are still available in all categories needed by companies at both hourly and piecework facilities. Skepticism concerning the future of the furniture industry in North Carolina exists, but a lack of jobs in other areas for which skilled furniture workers have been retrained limits options for this labor base.

Location advantages of North Carolina include transportation and logistics infrastructure as well as supplier proximity that fit just-in-time efficiencies along the high-volume purchasing region of the U.S. east coast, particularly Florida to Boston. Firms tend to use their own trucks to ship for dependability, but appreciate the available furniture logistics capacity. The High Point Market is still seen as viable and vital to maintain a North Carolina furniture industry by concentrating and publicizing the cluster (Power and Jansson 2008). The High Point presence of industry lobbying organization American Home Furnishing Association reflects the region's importance to the industry. Community colleges in the two in-state clusters feature furniture workforce training programs to meet upholstery needs in the Hickory area and logistics training in Greensboro. Company locations indicate a succession of different niches. In one instance an institutional furniture maker took over from a case goods manufacturer. Other firms expanded to larger quarters vacated by less successful companies and added machinery acquired inexpensively from liquidated rivals. Other conglomerates closed plants in outlying locations in order to "realign" operations at facilities in North Carolina. By mid-2012 announcements from several furniture manufacturers portended an imminent re-shoring: Stanley Furniture indicated plans to add 42 jobs, Lexington Home Brands eyed a new market in Brazil, Lincolnton brought business back from China, a Mississippi-based company opened a third plant in North Carolina, and Ashley Furniture (the largest furniture retailer in the U.S., now headquartered in Wisconsin) will build a new facility and hire 550 North Carolina workers. In southern Virginia, a Chinese furniture company announced plans to move into the business park configured for IKEA, hire 300 workers and relocate ten affiliated Chinese furniture companies with a potential gain of 2,000 jobs (Barron 2012). Re-clustering configurations are underway.

CONCLUSION

The major objective of this study involved assessing the impact of globalization's continuing re-ordering of furniture manufacturing strategies and locations, as seen in a case study of North Carolina, a major U.S. production site. Globalization provided the driving force for change in the early 1990s, sending wood furniture production, in particular, overseas. Coping strategies for U.S. domestic survivors included: 1) mergers and acquisitions; 2) a blended strategy of part overseas and part domestic production, with domestic retention of custom work and upholstery in particular; 3) focusing on a specialized market segment (e.g., motion furniture, rare woods); and 4) adapting lean manufacturing processes. Outcomes varied widely, often depending on the skill of company executives, regional development agencies and regulations in foreign locations. Since many companies utilized these strategies during an economic downtown, it is too early to assess the effect by value of goods shipped. Evidence rests on widespread adoption and the repeated belief of managers that therein lays salvation, since early adopters experienced profitable performance prior to the general downturn.

A combination of global and local socio-economic considerations drives corporate decisions prompting location changes. Interviews with more than three dozen industry executives and economic development officials familiar with the furniture cluster all point to a dramatic reassessment of the location of various parts of the manufacturing process in this industry. While survivors rely on new practices such as lean manufacturing and IT-enabled logistics to sharpen their competitive stance, challenges remain. Craftsmen who supply the skills for domestic-made quality furniture are leaving the business through retirement or retraining for fields such as IT, without job guarantees. Growing sectors such as institutional furniture remain in relative demand due to the health of the market with which they are linked (e.g., medical facilities rather than private residences) and the rapid delivery desired by large purchasers.

The global spatial fix mechanism continues to offer alternative global sites in addition to re-populating former U.S. locations with skilled labor and government incentives. Rising wage rates, China's currency adjustments, a new labor retention law and worker unrest, along with an anti-dumping judgment against Chinese bedroom furniture firms in 2002, increasingly sent work to Vietnam, regarded as China's closest competitor (IBIS*World* 2012). Restrictions hampering Vietnam's furniture industry expansion include the small size of the available workforce and challenging infrastructure connectivity that plagues other Southeast Asian and Latin American sites. Problems specifically in nearest U.S. neighbor Mexico, for example, include a shortage of appropriately skilled labor who are more likely to migrate north for jobs within the U.S., inadequate infrastructure and disruptions due to the ongoing drug trade.

Delay in implementing improvements until competition from foreign manufacturers became too strong to resist triggered the decline and reconfiguration of the furniture industry. The exogenous shock set off a stampede to more profitable and market-satisfying centers overseas. The previous cluster locations in the more developed country retain design, infrastructure, market and skilled labor components that more economically supply remaining American market niches. The industry will likely remain highly vertically integrated, drawing on embedded location-specific areas of strength. Global shifts do not mean the end of geographic repositioning domestically, demonstrated by the demise and reconfiguration of the North Carolina Briar Patch.

BIBLIOGRAPHY

Allegrezza, R. (2010a), "Ashley sticks to home run plan," *Furniture Today*, June 7, 2.

Allegrezza, R. (2010b), "Industry finally on comeback trail," *Furniture Today*, June 21, 2.

Barron, R. (2012), "Hard costs drive China to the U.S.," *News and Record*, October, 1.

Buehlmann, U. and A. Schuler (2009), "The U.S. household furniture industry: status and opportunities," *Forest Products Society*, **59**(9), 20–28.

Bureau of Labor Statistics, U.S. Department of Labor. (2008), "Private industry by state and six-digit NAICS industry," available at http://data.bls.gov/PDQ/servlet/SurveyOutputServlet;jsessionid=f0303accd24c$00$0F$3 (accessed June 6, 2009).

Bureau of Labor Statistics, U.S. Department of Labor (2011), "Table 10: private industry by state and six-digit NAICS industry," available at http://www.bls.gov (accessed October 9, 2014).

Bureau of Labor Statistics, U.S. Department of Labor (2014), "Quarterly Census of Employment and Wages (2014)," available at http://www.bls.gov/qcew/apps/table_maker.htm (accessed October 9, 2014).

Cater, J. (2005), "The rise of the furniture manufacturing industry in Western North Carolina and Virginia," *Management Decision*, **43**, 906–924.

Covington, O. (2012), "Ashley Furniture to open facility this year in Davie, creating up to 1,100 jobs," available at www.bizjounals.com/triad/news/2012/04/20/ashley-furniture-to-open-facility-in.html?page=all (accessed April 30, 2012).

Drayse, M. (2008), "Globalization and regional change in the U.S. furniture industry," *Growth and Change*, **39**, 252–282.

Dugan, M. (2009), *The Furniture Wars: How America Lost a Fifty Billion Dollar Industry*, Conover, NC: Goosepen Studio & Press.

Evans, G. (2010a), "Craftique gives desk to N.C. executive mansion," *Furniture Today*, May 10, 30.

Evans, G. (2010b), "Ashley stays on top with its products, services," *Furniture Today*, August 2, 1.

Gereffi, G. (2006), "The global economy might provide some answers to North Carolina's quest to revive traditional industries," *Triad Business Journal*, available at www.dukenews.duke.edu/2006/05/nc_economy (accessed December 1, 2010).

Helper, S., T. Krueger and H. Wial (2012a), "Why does manufacturing matter? Which manufacturing matters? A policy framework," Brookings Metropolitan Policy Program, available at http://www.brookings.edu/research/papers/2012/02/22-manufacturing-helper-krueger-wial (accessed December 13, 2012).

Helper, S., T. Krueger and H. Wial (2012b), "Locating American manufacturing: trends in the geography of production," Brookings Metropolitan Policy Program, available at http://www.brookings.edu/research/reports/2012/05/09-locating-american-manufacturing-wial (accessed December 13, 2012).

Hunter, S. (2008), "The Toyota production system applied to the upholstery furniture manufacturing industry," *Materials and Manufacturing Processes*, **23**, 629–634.

IBIS*World* (2011, 2012), IBIS*World* "Industry report: household furniture manufacturing in US," available at http://clients1.ibisworld.com.libproxy.uncg.edu/reports/us/industry/default.aspx?entid=862 (accessed November 30, 2010, December 15, 2012).

IKEA (2008), "IKEA opens 1st Swedwood factory in United States," available at www.reuters.com/article/pressRelease/idUS161191+21-May-2008+BW20080521 (accessed October 30, 2008).
Koenig, K. (2008), "Inside Swedwood: IKEA's first U.S. plant," available at www.iswonline.com/ArticleLanding/tabid/67.Defaurlt.aspx?tid=1&contentid=11789, (accessed October 30, 2008).
Lacy, R. (2004), "Whither North Carolina furniture manufacturing?" *Federal Reserve Bank of Richmond Working Paper* No. 04–07.
Langfitt, F. (2009), "Furniture work shifts from N.C. to South China," December 18, available at http://www.npr.org/templates/story/story.php?storyId=121576791 (accessed January 2, 2010).
Liker, J. (2004), *The Toyota Way: 14 Management Principles from the World's Greatest Manufacturer*, New York: McGraw-Hill.
MacKinnon, D., A. Cumbers, A. Pike, K. Birch and R. McMaster (2009), "Evolution in economic geography: institutions, political economy, and adaptation," *Economic Geography*, **85**, 129–150.
MacPherson, A. and V. Vanchan (2010), "The outsourcing of industrial design services by large US manufacturing companies," *International Regional Science Review*, **33**, 3–30.
Perry, D. (2010), "Quick thoughts on price hikes, the bedding comeback, China," *Furniture Today*, May 10, 18.
Porto, D. and M. Smith (2006), "Re-making furniture making at Hickory Chair company," *Target*, **22**, 16–34.
Powell, W. (2006), "Furniture," *Encyclopedia of North Carolina*, University of North Carolina Press, available at http://uncpress.unc.edu/nc_encyclopedia/furniture.html (accessed September 10, 2010).
Power, D. and J. Jansson (2008), "Cyclical clusters in global circuits: overlapping spaces in furniture trade fairs," *Economic Geography*, **84**, 423–448.
Rohter, L. (2008), "Shipping costs start to crimp globalization," *The New York Times*, August 3, available at www.nytimes.com/2008/08/03/business/worldbusiness/03global.html?sq=rising (accessed August 1, 2008).
Russell, T. (2010a), "More companies bring production back to U.S.," *Furniture Today*, May 3, 14.
Russell, T. (2010b), "Rolling blackouts hamper some plants in Vietnam," *Furniture Today*, July 5, 3.
Schumpeter, J. (1942), *Capitalism, Socialism and Democracy*, London: Allen and Unwin.Sirkin, H., M. Zinzer, D. Horner and J. Rose (2012), "U.S. manufacturing nears the tipping point: which industries, why, and how much?" The Boston Consulting Group, available at www.bcg.com/documents/file84471.pdf (accessed December 16, 2012).
Sirkin, H.L., M. Zinser, D. Hohner and J. Rose (2012), "U.S. manufacturing nears the tipping point," Boston Consulting Group study, March.
TradeStats Express, International Trade Administration, available at http://tse.export.gov/TSE/MapDisplay.aspx (accessed November 13, 2012).
Walcott, S. (2011), "The furniture foothills and the spatial fix: globalization in the furniture industry," *Southeastern Geographer*, **51**(1), 6–30.
Yeung, H. (2009), "Transnational corporations, global production networks, and urban and regional development: a geographer's perspective on 'Multinational enterprises and the global economy,'" *Growth and Change*, **40**, 197–226.

FURTHER READING

Dugan, M. (2009), *The Furniture Wars: How America Lost a Fifty Billion Dollar Industry*, Conover, NC: Goosepen Studio & Press. (A noted insider's analysis of reasons for the precipitous decline of the U.S. furniture industry.)
Helper, S., T. Krueger and H. Wial (2012), "Locating American manufacturing: trends in the geography of production," Brookings Metropolitan Policy Program, available at http://www.brookings.edu/research/reports/2012/05/09-locating-american-manufacturing-wial (accessed December 13, 2012). (A premier think-tank's take on geographic shifts in the concentration of U.S. manufacturing clusters.)
Liker, J. (2004), *The Toyota Way: 14 Management Principles from the World's Greatest Manufacturer*, New York: McGraw-Hill. (The classic text on "lean manufacturing" as utilized and taught to other companies by Toyota, the premier practitioner.)
Porto, D. and M. Smith (2006), "Re-making furniture making at Hickory Chair company," *Target*, **22**, 16–34. (Key consultants provide a case study of a successful U.S. furniture company turn-around.)
Power, D. and J. Jansson (2008), "Cyclical clusters in global circuits: overlapping spaces in furniture trade fairs," *Economic Geography*, **84**, 423–448. (The major examination of an important furniture industry component.)
Yeung, H. (2009), "Transnational corporations, global production networks, and urban regional development: a geographer's perspective on 'Multinational enterprises and the global economy,' *Growth and Change*, **40**, 197–226. (A prominent theorist's examination of the global industrial context.)

14. New geographies of advanced manufacturing: the case of machine tools

Ronald V. Kalafsky

MANUFACTURING AND THE ROLE OF MACHINE TOOLS

The manufacturing sector endures as an important element of economic development, even as its employment numbers shrink in many countries. Through to the present, manufacturing has served as a means of economic growth, increased international trade, and human capital expansion in a number of emerging economies. In mature market economies, the secondary sector still has considerable impacts, as evidenced by increased value-added per worker and the multiplier effects of this sector. Linked to wider manufacturing activities are machine tools, which are a critical component of most advanced industrial processes, especially within metalworking. A common definition holds that, "A machine tool is a power-driven machine, not hand-held, that is used to cut, form or shape metal" (Association for Manufacturing Technology (AMT) 2002); common examples of machine tools include industrial presses, lathes, machining centers, and grinders. Given this definition, one can find these key capital goods in a wide range of industrial operations, serving numerous critical tasks.

The machine tool industry, however, has often garnered little attention outside advanced industry circles, some business media, and limited policy spheres. Its perceived impacts just do not jump out at the general public, since most of these machines are essentially used "behind the scenes," installed on distant factory floors. The nominal figures for this industry, moreover, are not excessively large or even noteworthy. In any industrialized country, employment in this sector accounts for far less than 1 percent of total employment, and overall the industry's total output pales in comparison with readily visible industries such as aerospace, electronics, motor vehicles, and steel.

At the same time, it must be acknowledged that virtually every manufacturing process, at some level, is impacted by machine tools. Certainly, machine tools are integral to innovation and precision in the key manufacturing industries described above. It would stand to reason that machine tools (and a healthy domestic machine tool industry) could play an important part in a country's economic development, especially in the growth of a robust manufacturing sector. For years, therefore, this industry has been deemed to be a bellwether of a country's industrial competitiveness (Graham 1993), with manufacturers of all types often placed at a competitive disadvantage without a healthy home-grown machine tool industry. By some estimates, manufacturers have waited as long as two years to obtain top-tier, imported metalworking technologies, placing such firms in an unenviable position with regard to their global rivals with far stronger domestic machinery industries (Finegold *et al.* 1994). The fundamental importance of this industrial sector can be illustrated by the fact that in Japan – which has long been among the world's leading machine tool builders (in terms of volume, precision, and innovation) and is home to some of the world's best and largest firms in this field – these vital capital goods are called "mother machines," given their centrality to virtually every manufacturing process (Tsuji *et al.* 1999).

Changes in the geographies of the machine tool industry, then, can act as indicators of international transitions in manufacturing at-large and economic development. Keeping the above in mind, this chapter aims to discuss geographical trends in the global machine tool industry, especially in terms of spatial transitions in purchases, production, and trade. The dynamics of this sector, particularly concerning the geographies of industrial decline or dominance, are critical for understanding manufacturing and in turn raise questions about longer-term development in terms of overall economic trends and also, human capital.

GLOBAL TRENDS IN PURCHASES, PRODUCTION, AND TRADE

Purchase Geographies

As a first step in examining the geographies of the machine tool industry, it is helpful to examine changes in international consumption patterns for this equipment. If a country is to be viewed as a large-scale manufacturer in almost any sense of the word or across virtually any product line, then its producers must purchase and install machine tools on their factory floors. In order to look at relatively recent changes in the industry, Figure 14.1 provides a comparison of machine tool consumption from 2000 to 2011; the list is limited to the largest machine tool-using countries as of 2011. While several trends emerge from the data, what stands out at first glance is the sheer volume of China's machine tool purchases. At over USD 38 billion (one thousand million), this level of consumption is roughly five times greater than the next largest machine tool consumer (Japan) and also greater than the total purchases of the next ten countries combined. From a temporal perspective, the most apparent change is the meteoric increase in Chinese machine tool purchases, expanding by nearly 900 percent. The current volume of China's machine tool consumption and its rapid expansion in little more than a decade provide a gauge of the swift expansion in overall Chinese manufacturing across the past 30 years, particularly in capital goods such as motor vehicles and home appliances.

Beyond the extraordinary volume and increase in China's machine tool purchases, a number of other emerging economies have posted equally impressive growth rates. India's machine tool purchases outpaced China's, 945 to 896 percent. In nominal numbers, admittedly, India's purchases are dwarfed by China's (by over USD 36 billion), yet its rapid growth in machine tool consumption provides some hints that its economic engine is finally emerging, especially in manufacturing. Also noteworthy is Russia, with purchases increasing by over 800 percent. While the growth in both Indian and Russian machinery purchases is impressive, the next section will show that, at this point, it is not being accompanied by matching growth in homegrown machine tool manufacturing. It should also be mentioned that two other newly industrialized economies (NICs), Turkey and Brazil, also post increases in consumption of over 100 percent. Given what was discussed in the first section about the centrality of machine tools to manufacturing

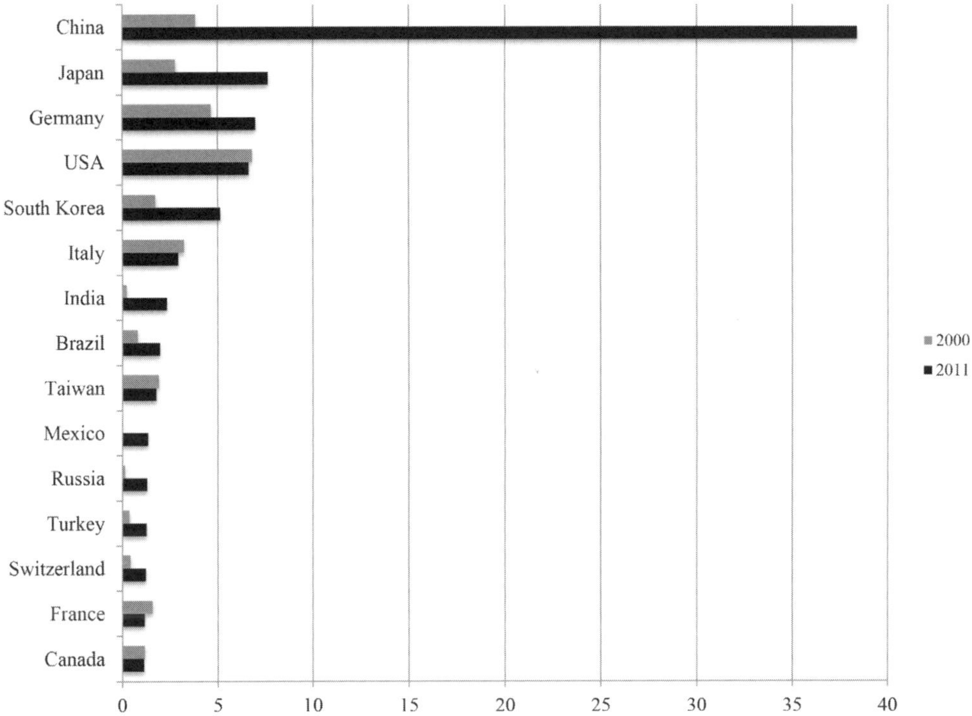

Sources: AMT 2002 and Gardner 2012.

Figure 14.1 Machine tool purchases: 2000 and 2011 (in US$ billions)

at-large, it is not much of a reach to infer that some of the BRIC states (Brazil, Russia, India, China), first referred to in 2001 by Goldman Sachs' Jim O'Neil, are growing in terms of manufacturing, at least as evidenced by the machine tool purchases numbers. And, while a 2000–2011 comparison of growth is not possible,[1] it should be noted that Mexico also remains a large machine tool consumer, placed tenth globally for 2011 (Gardner 2012) and indicating its continued strong position within global manufacturing supply chains.

Japan and Germany place a distant second and third in terms of machine tool consumption for 2011, and yet their continued high rankings and large purchase volumes would suggest that both countries retain considerable manufacturing bases, even in light of higher production costs. These numbers would not necessarily be expected in mature economies, even from industrial dynamos such as these. Note, however, that Germany and Japan (and Switzerland at 190 percent) still also post relatively high growth rates in machine tool consumption. Given that machine tools are often replaced every five to ten years, these numbers at a minimum would indicate that advanced manufacturing continues to be important in these decidedly mature market economies. At this point in time, the above countries continue to excel in higher-end manufacturing beyond machine tools, in industries such as automobiles, medical instruments, and other precision machinery. South Korea, likewise, also must be included in this group of relatively manufacturing-intensive, higher-cost, mature economies. Its growth in machine tool usage continues to be impressive (at 195 percent), illustrating its continued rapid rise among industrialized economies.

The consumption data also suggest that there are a number of countries where machine tool installations are going down, perhaps hinting that manufacturing at-large is playing a smaller role in some economies. The figures for France and Italy, with declining consumption rates of 26 and 9 percent, respectively, point to a hollowing-out of manufacturing activities in general for these countries. The United States, which until just over a decade ago was the world's largest machine tool buyer, has followed a downward trend in purchases, now falling to fourth place and posting an overall decrease in machine tool consumption. Related to the downward trends in manufacturing and machine tool usage of its southern neighbor, Canadian purchases have also declined since 2000.

Note that while Taiwan posted a decline in purchases of this equipment, its case is somewhat unique and related to what is occurring in a nearby market. Taiwan was once a larger machine tool user, but much of its manufacturing has moved to China, a movement not so coincidentally led by many Taiwanese manufacturers establishing operations in nearby Fujian and Guangdong Provinces. At the same time, Taiwanese machine tool production and exports *increased* during the same time period, points that will be discussed further in this chapter.

In terms of machine tool purchases, therefore, several geographies emerge. First and foremost is the emergence of China as the world's largest market for these capital goods, and following China's rise is the advent of several other emerging economies, such as Turkey and India. Several mature economies still maintain high rates of machine tool consumption (e.g. Germany and Japan), highlighting the role of manufacturing at-large in these countries. And finally, declines in machine tool usage are occurring in North America and Western Europe, with the exceptions mentioned previously, suggesting relative decreases in general manufacturing activities in these countries.

Production Geographies

Whereas the machine tool purchase numbers provide some measure of overall manufacturing activity, production numbers can offer insights on leadership in the manufacture of this advanced equipment. Recalling the earlier discussion on the linkages between machine tool leadership and innovation in manufacturing, international trends in production are shown in Figure 14.2. Before going further into temporal and spatial shifts in production, another pattern emerges from the data, even when limited to a select group of countries. First, world machine tool production is concentrated in a relatively select number of countries. The top five countries, in terms of output (China, Japan, Germany, Italy, and South Korea), account for approximately 77 percent of the total global output in machine tools. If this group is extended to encompass the top ten producers, the concentration jumps to 93 percent. In terms of manufacturing at-large, concentration tends to be clustered among the top five countries: China, United States, Japan, Germany, and Italy. In the top five of machine tool production, note that the United States falls from that list, replaced by South Korea, further portending trends in the geographies of advanced manufacturing overall.

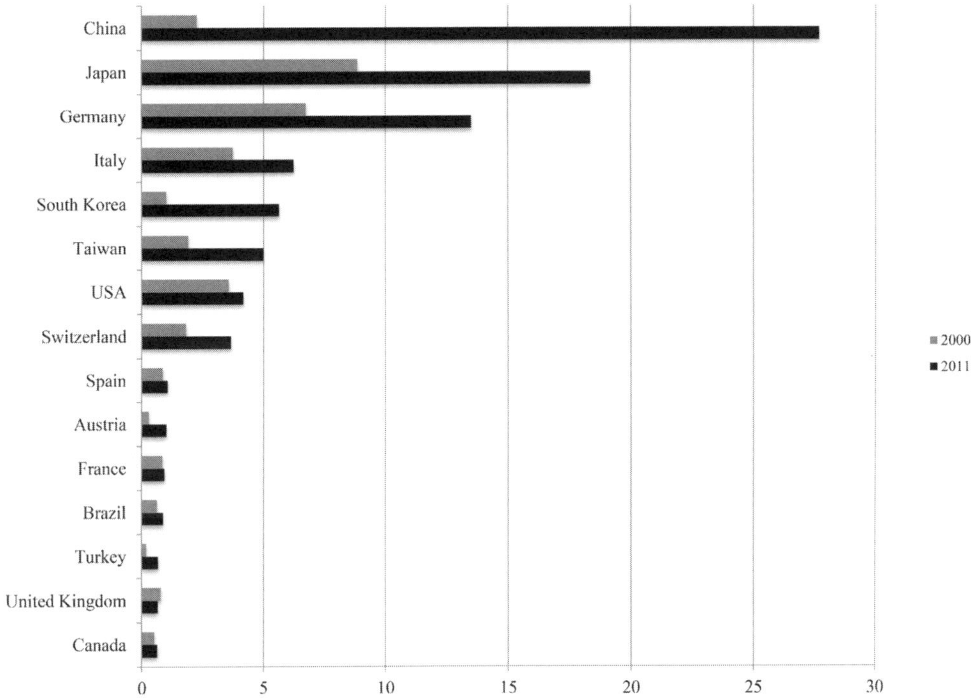

Sources: AMT 2002 and Gardner 2012.

Figure 14.2 Machine tool production: 2000 and 2011 (in US$ billions)

When looking at the machine tool manufacturing data from 2000 to 2011 (as per Figure 14.2), a noteworthy trend emerges: a swift increase in overall production among a handful of states. The most rapid growth comes from China, where machine tool production rose over 1100 percent from 2000 to 2011. In a sense, this signals China's emergence as a global economic power and, more specifically, as a legitimate manufacturing power. Not surprisingly, China recently surpassed the United States as the world's leading *overall* manufacturer as well, now accounting for 18.7 percent of global industrial output (Meckstroth 2012). Its share of international machine tool production, incidentally, accounts for over 29 percent of total global output. Note that China's output of over USD 27 billion is 50 percent greater than second-place Japan, which for years had been locked in a competition with Germany for the top production position. In terms of both production and consumption, then, China is clearly dominant, at least in terms of sheer volume of output.

Another country to ascend the machine tool production ranks is South Korea, which increased output by 465 percent since 2000. This follows South Korea's rise as a leading manufacturing power and conceivably indicates that the country is now building competitive machine tools, an item to be followed up in the next section on trade. Turkey, Austria, and Taiwan also posted growth rates of well over 100 percent since 2000. In the previous section, recall that Taiwan posted decreasing machine tool *consumption*, yet the data indicate an increase in *production* (moving from eighth to sixth globally). Rounding out the list, in terms of growth, are Japan, Germany, and Switzerland, all with growth rates of around 100 percent. While these growth numbers are nowhere near those of many of the aforementioned NICs (or countries that have emerged from NIC status), the fact that many frontrunners in terms of production (from mature industrial states no less) are still growing is impressive and testament to continued leadership in the industry.

When examining these global developments in production, a comparison must be made between two emerging manufacturing economies: Turkey and Brazil. Both national industries produce well under one billion USD in machine tools each, but their growth rates in this industry merit at least brief discussion in terms of manufacturing trends. Since 2000, Turkey has risen from the 20th largest producer in the world to 13th in 2011, posting a production increase of over 260 percent. This hints at its emergence as a regional economic power, especially in manufacturing, and, moreover, this steep upward trend sets it against relatively stagnant or decreasing production across Western Europe, a point that will be expanded upon shortly. Brazil's machine tool industry, instead, has remained at a relative standstill, posting an 11-year gain of just over 41 percent and, moreover, did not increase its world ranking in terms of production, remaining in 12th place. In a similar vein, and while not shown in Figure 14.2, Russia and India must be remarked upon, given that they are often mentioned in the same discussions as other emerging economies discussed throughout this chapter. Recall their rapid rise in machine tool consumption discussed in the previous section. If nothing else, the numbers reveal the rapid rise in sheer production from some economies and a relative lack of production from some NICs. The fact that Russia and India did not even place within the top 15 machine tool manufacturers would indicate, at least at this point, that these economies are not among the leaders in higher-end capital goods production, namely advanced industrial machinery. And while Brazil's machine tool output places 12th globally, its relatively low growth rate in terms of production also sets it apart from some of the prominent (and growing) producers discussed previously.

Other countries, while still exhibiting relatively high production numbers (at least in nominal terms), are slowing. The United States, which four decades ago was the world's largest machine tool producer, now places a distant seventh. This trend has long underscored larger fears over manufacturing at-large in the United States (e.g. DiFilippo 1986; Finegold *et al.* 1994). As far back as the late 1990s, representatives from larger manufacturing firms were testifying before the United States Congress about a diminished domestic machine tool sector (see United States Senate 1999). The trend, however, has continued toward decreased machine tool production. Since 2000, a number of European countries and Canada have posted only small growth rates in machine tool production. Neither France, Spain, nor Canada reported production growth rates exceeding 25 percent since 2000 and, unique among the world's top 15 producers, British machine tool production has actually contracted. Tying these findings to what was discussed earlier about this industry as an indicator of secondary sector health at-large, the indications are that manufacturing is in retreat (or has significantly slowed) in many countries, clustered largely in North America and Western Europe. At the same time, however, one cannot make a blanket statement about manufacturing (whether machine tools or overall) in decline across all mature industrialized economies – witness the continued machine tool dominance of Germany and Japan.

Trade Geographies

The metrics discussed in the previous two sections revealed international trends in machine tool manufacturing and usage. These measures are a reflection of the sheer size of the economies and manufacturing activities of selected countries, as in the case of the consumption numbers for China. The changes from 2000 to 2011 also illustrate emerging or declining manufacturing activities, such as the divergent paths of China and the United States. Regarding international *leadership* in machine tools, however, perhaps more indicative measures are the trade numbers for this industry, particularly exports. When countries (or national industries) begin to move into relatively export-intensive industries, it could indicate several trends. On one hand, perhaps a national industry has certain cost advantages vis-à-vis international competitors. On the other hand, an industry might hold certain competitive advantages (e.g. innovation or precision) over other international rivals. In the case of machine tools, it is likely the latter case, as entry-level machine tool demand is often filled by domestic production and, in any case, the actual value of these types of machines is quite low. In light of what was discussed in the first section, the evidence would suggest, then, that exports are good indicators of a country's relative international standing in this industry.

In a similar vein to the manufacturing and consumption numbers discussed earlier, Figure 14.3 compares global machine tool exports from 2000 and 2011. As alluded to above, the export numbers provide a measure of comparative and competitive advantages in this industry. This is clearly the case in the instance of Japan and Germany. These countries placed first and second, respectively, in both 2000 and 2011, demonstrating clear dominance in the export market for this advanced capital equipment. The total exports of each are clearly much higher than those of any nearby competitors. In the case of Japan, there have been suggestions that its machine tool builders turned to exporting as a means of survival during the country's two-decade economic malaise (Tsuji and Ishikawa 1995). And while this strategy may have indeed been the case, the fact remains that there must be a demand for any exported product and, in this situation, there has been a sustained demand for Japanese-built machine tools. Though certainly embedded within a different economic milieu, Germany has also demonstrated sustained leadership over the past decade. Given that the machine tools produced by both countries are by no means inexpensive, the export numbers suggest that both are in demand from a wider group of customers. And, beyond sheer volume, Germany's machine tool exports grew by almost 90 percent while Japanese exports grew by 65 percent.

Italy, Taiwan, and Switzerland also place within the top five; this group of leading exporters remained the same from 2000 to 2011, with only Switzerland and Taiwan exchanging fourth and fifth places. Italy posted solid gains of 150 percent, staying in third place. Given the current tenuous position of the Italian economy, the data hint that its firms may be employing exports as a means to compensate for lackluster demand, much in the same way as the case of Japan mentioned previously. In a similar light, Spain, with a severe domestic economic crisis, actually moved up one place, to ninth, reporting an increase of over 100 percent.

In the case of exports at least, Taiwan can be considered as having moved beyond its former NIC status, at least in terms of advanced machinery exports, although its situation may be somewhat unique. As mentioned earlier, much of Taiwan's manufacturing

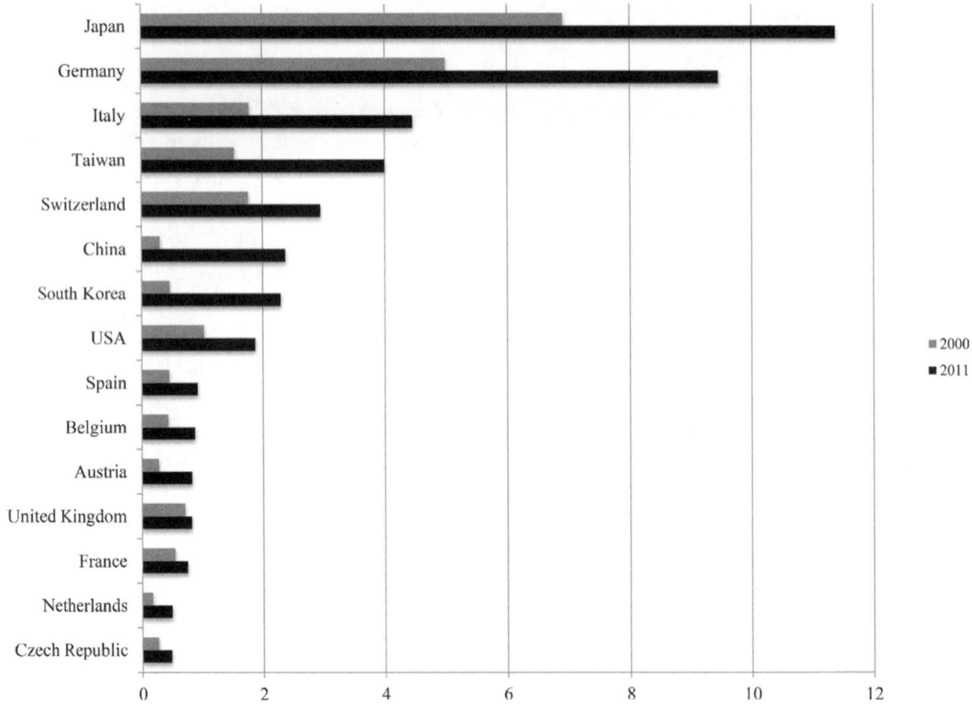

Sources: AMT 2002 and Gardner 2012.

Figure 14.3 Machine tool exports: 2000 and 2011 (in US$ billions)

has moved to China and, accordingly, machine tools are being sent to be installed in these factories: as of 2011, almost 39 percent of Taiwan's machine tool exports are destined for China (Taiwan Association of Machinery Industry 2012). Another question also concerns Taiwan. Its consumption numbers actually declined from 2000, yet production increased, resulting in impressive export numbers, both in nominal and growth terms. If the Chinese economy, however, begins to slow, and recent trends are pointing in that direction, this could have serious knock-on effects on Taiwan's machine tool industry, especially given its apparent lack of geographical market diversification. Put simply, much of its recent growth is tied to increases in manufacturing across the Taiwan Strait.

Two national machine tool industries made noteworthy advances in terms of export growth. Among the groups of the 15 largest exporters as of 2012, China attained the largest gain, moving from 14th place to sixth, posting a gain of almost 700 percent. While at this point, many of these export sales are aimed at the lower-end or entry-level sectors of the machine tool market, the international sales numbers echo large gains in production and consumption for China. South Korea posted the second highest rise in exports, at over 400 percent, moving from ninth to seventh place. In the case of South Korea, at least, its machine tool industry has moved beyond satisfying solely domestic demand and has moved on to the world stage. As mentioned in the discussion on production, one could expect South Korea's machine tool industry to move into the top

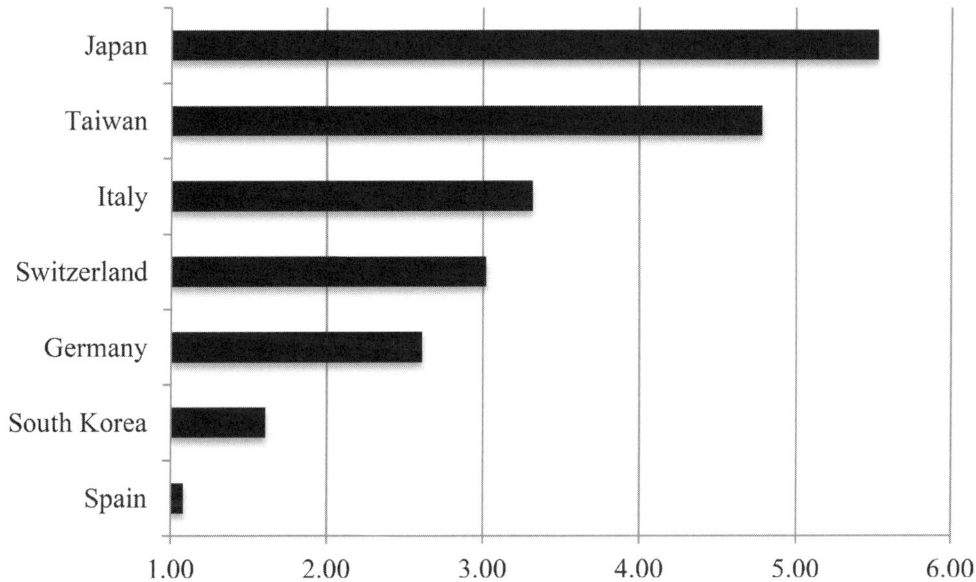

Note: A comparative advantage is "revealed" if RCA>1. If RCA is less than unity, the country is said to have a comparative disadvantage in the commodity or industry.

Sources: Gardner 2012; Taiwan Bureau of Foreign Trade 2013; and World Bank 2013. Calculations by author.

Figure 14.4 Countries with revealed comparative advantage in machine tool exports (2011)

tier of internationally recognized producers, if they have not, in many ways, already achieved this.

Conversely, a number of national machine tool industries moved down the export rankings. The United Kingdom fell from the seventh to the twelfth largest exporter, while France fell from eighth to thirteenth. These countries also posted the lowest export growth over the same time period: 16.4 and 39.7 percent respectively. The United States, as seen with the purchases and production statistics, also showed some signs of waning leadership in this industry. While its manufacturers posted a gain of over 82 percent across the decade, their export ranking fell from sixth to a distant eighth place. Note that while the Netherlands and Belgium both fell within the top 15 machine tool-exporting nations, industry researchers (see Gardner 2012, 4) caution that some numbers can be inflated given that many machines manufactured elsewhere only transit through such countries en route to a final destination.

As a final step in these analyses, global leadership in the machine tool industry will be examined in terms of revealed comparative advantage (RCA), as this measure can be used to compare a country's relative concentration in an industrial activity, especially within international trade. Figure 14.4 shows which countries among the world's top machine tool exporters holds an RCA above one (1.00), which would suggest comparative advantage in machine tool exports.[2] Given these parameters, the list drops to only seven countries. Moreover, clear dominance in machine tool exports is demonstrated by Japan and Taiwan, with RCAs of 5.53 and 4.78 respectively, far beyond third place Italy at 3.32. As discussed when looking solely at the nominal export numbers, Japan's high RCA could be a result of its firms looking beyond stagnant domestic demand, yet this number suggests clear international dominance in this industry. Indeed, six of the world's top ten machine tool makers are headquartered in Japan (Metalworking Insiders' Report 2012). Taiwan's RCA result, as with its export numbers, reflects the large number of shipments destined for China and the continued reliance on this market. Italy, despite (or perhaps as a result of) domestic economic issues, maintains a robust international presence in machine tools, especially industrial presses commonly used in the motor vehicle industry.

Germany, among the world's leaders in production and exports, also posts a high RCA number; its relatively stable economic and manufacturing situations during the recent wider European financial woes may have reduced its RCA somewhat, as domestic manufacturing has remained comparatively robust. Yet, it should be noted that China has been a huge market for German machine tool exports. In many ways but on a smaller scale, what has been said about Germany could be applied to Switzerland. South Korea's appearance on this list again indicates that its machine tool industry has emerged among the world's leaders, moving beyond relying largely on a domestic base of buyers. When this metric is used, China falls far behind the above countries, with an RCA of just 0.50, suggesting that its huge production is still geared largely toward domestic demand and is not yet sought after in global markets. Finally, for comparison purposes, the United States posted an RCA of 0.48, again pointing to the relative decline of the industry. The RCA numbers, overall, provide a clearer picture of global dominance in the machine tool industry.

DISCUSSION

This chapter provided a brief look at the shifting geographies of the global machine tool industry. A number of trends surfaced after looking at the industry in terms of purchases, production, and trade; when taken together, these analyses suggest several developments, divided between emerging and industrialized economies. First and foremost, China has emerged as a dominant consumer *and* producer of machine tools, in terms of both volume and growth, befitting what has become the world's largest overall manufacturer. Whether this trend continues after an expected economic slowdown remains to be seen. For the most part, its somewhat lower export numbers and much smaller RCA result indicate that, at present, its production largely serves a domestic market and exports are aimed at cost-conscious customers. Keep in mind, however, that producers in South Korea and Taiwan were in a similar situation a few decades ago. In the meantime, firms from these two countries have moved up the ranks of the larger and more competitive machine tool builders and exporters.

Another point concerns several economies in terms of their vastly increased machine tool consumption. Countries such as India, Russia, Turkey, and Brazil are purchasing increasingly larger amounts of machine tools, as befits industrializing states with growing manufacturing sectors. In some cases, however, production increases have not matched consumption, underlining that countries such as Brazil, India, and Russia still have long strides to make in terms of advanced manufacturing – production numbers have not responded accordingly, such as has occurred in China. At a minimum, this would suggest that there might be a lag in these countries moving into advanced manufacturing activities. Turkey, however, could be one exception in terms of production but this remains to be seen. Time will tell if several of these industrializing markets will move into more advanced manufacturing, driving a robust, homegrown machine tool industry, or if their demand will continue to be satisfied by imported manufacturing technologies.

While the futures of China and/or other industrializing states may be somewhat open to debate, obvious downward trajectories for several machine tool industries emerged. While it is difficult to make blanket statements, these downward trajectories raise worries about the long-range futures for manufacturing in countries such as France, the United Kingdom, and the United States. The United States machine tool industry, in terms of production leadership, has fallen over the past four decades. Less than a decade ago, moreover, it was the largest buyer of machine tools and has dropped to fourth in just over a decade. And, in fact, actual machine tool consumption decreased from 2000 to 2011. This follows the decline in the relative importance of manufacturing in the United States economy, set against the fact that the United States remains the world's largest economy. In essence, comparatively fewer machine tools are needed in this market; and much of the existing demand is filled by imports. So, concerns over the future of manufacturing, especially innovation in advanced manufacturing, appear to be well founded. The same can be said, albeit on a smaller scale, for the machine tool industries and, long term, the overall manufacturing prowess of France and the United Kingdom. At the same time, other industrialized states such as Germany, Japan, and Switzerland have maintained international dominance in this industry and appear to have retained considerable advanced manufacturing bases. It must be mentioned that South Korea is now securing a position among these world leaders as well. This country and, also, Taiwan, seem to have shorn the NIC label, as both countries have moved higher into the ranks of machine tool production and, maybe more importantly, exports.

The trade figures provide additional insight into the geographies of this industry. While the consumption numbers suggest a broadening in the geography of manufacturing and growth in a number of developing states, the data suggest that export leadership in machine tools remains concentrated in a select number of national industries. And although the dominance of China cannot be denied, at least in terms of sheer volume, it is important to note that Germany and Japan remain firmly ensconced near the top of the metrics discussed in this chapter, especially in terms of exports and RCA. Both countries have encountered economic issues in the past two decades and are decidedly high-wage industrialized states, yet both continue to excel in the production and export of this high-end capital equipment. While a country such as South Korea can ascend to a top position on this list, by and large, international leadership remains ensconced in a select number of countries known for higher-end manufacturing in general, so it would follow that this leadership would extend to machine tool exports.

NOTES

1. Accurate data from Mexico for 2000 were not available, a point confirmed by a Gardner representative. Accordingly, the data for Mexico are not displayed in Figure 14.1.

2. Belgium had an RCA result of 1.02, which is above parity. Given what was explained about the nature of Belgian machinery export statistics in the previous paragraph, however, it does not appear on this list.

REFERENCES

AMT (2002), *Economic Handbook of the Machine Tool Industry (2000–2001)*, McLean, VA: Association for Manufacturing Technology.

DiFilippo, Anthony (1986), *Military Spending and Industrial Decline: A Study of the American Machine Tool Industry*, New York: Greenwood Press.

Finegold, David, Keith W. Brendley, Robert Lempert, Donald Henry, Peter Cannon, Brent Boultinghouse, and Max Nelson (1994), *The Decline of the U.S. Machine-Tool Industry and Prospects for its Sustainable Recovery*, Santa Monica, CA: Rand.

Gardner (2012), *The 2012 World Machine-Tool Output & Consumption Survey*, Cincinnati, OH: Gardner Publications.

Graham, Julie (1993), "Firm and state strategy in a multipolar world: the changing geography of machine tool production and trade," in Heidi Noponen, Julie Graham, and Ann Markusen (eds), *Trading Industries, Trading Regions*, New York: Guilford Press, pp. 140–174.

Meckstroth, Daniel J. (2012), "Is China the largest manufacturer in the world?" Arlington, VA: Manufacturers' Alliance for Productivity and Innovation. http://www.mapi.net/china-largest-manufacturer-world (accessed December 3, 2012).

Metalworking Insiders' Report (2012), *Machine Tool Scoreboard*, ed. Joe Jablonowski, Cincinnati, OH: Gardner Publications. http://www.metalworkinginsider.info/Scoreboard.htm (accessed January 22, 2013).

O'Neill, J. (2001), *Global Economic Paper No. 66. Building Better Global Economic BRICs*, London: Goldman Sachs.

Taiwan Association of Machinery Industry (2012), *Exports of Taiwan Machine Tools in 2010–2011 by Destination*, Taipei: Taiwan Association of Machinery Industry. http://www.tami.org/ (accessed July 15, 2013).

Taiwan Bureau of Foreign Trade (2013), *Trade Statistics*. http://cus93.trade.gov.tw/ENGLISH/FSCE/ (accessed July 15, 2013).

Tsuji, Masatsugu and Mineo Ishikawa (1995), "Technology and industrial transformation: the case of the Japanese machine tool industry," *Oikonomika*, **31**(2): 155–169.

Tsuji, Masatsugu, Shoichi Miyahara, and Mineo Ishikawa (1999), "An empirical analysis of industrial transformation in the Japanese machine tool industry," *Mathematics and Computers in Simulation*, **48**(4): 561–572.

United States Senate (1999), "Challenges confronting the machine tool industry," hearing before the Subcommittee on Manufacturing and Competitiveness. October 28, 1999. Washington, DC: US Government Printing Office.

World Bank (2013), *World DataBank*. http://databank.worldbank.org/ddp/home.do (accessed January 18, 2013).

FURTHER READING

Finegold, David, Keith W. Brendley, Robert Lempert, Donald Henry, Peter Cannon, Brent Boultinghouse, and Max Nelson (1994), *The Decline of the U.S. Machine-Tool Industry and Prospects for its Sustainable Recovery*, Santa Monica, CA: Rand. (A comprehensive analysis of problems confronting the United States machine tool industry.)

Graham, Julie (1993), "Firm and state strategy in a multipolarworld: the changing geography of machine tool production and trade," in Heidi Noponen, Julie Graham, and Ann Markusen (eds), *Trading Industries, Trading Regions*, New York: Guilford Press, pp. 140–174. (An overview of the role of government policy in this industry and the rise of Japan.)

Prestowitz, Clyde V. (1988), *Trading Places: How We Allowed Japan to Take the Lead*, New York: Basic Books. (An examination of the shifts in manufacturing leadership from the United States to Japan.)

15. Farm machinery: a changing path to feed the world

Dawn M. Drake

INTRODUCTION

The farm machinery industry is vitally important to manufacturing economies in many countries as well as the ability of most nations to feed themselves. In many developed countries the farm machinery industry is a key component of the economy despite the fact that it comprises a small portion of gross domestic product (GDP). The industry provides good-paying manufacturing jobs. Farm machinery is also a necessary component of the commercial agro-food industry due to continued declines in the agricultural workforce in the industrialized world. Even as the number of people working in commercial agriculture declines, global population grows. One of the most efficient ways to feed the world, given these conditions, is the use of high-horsepower tractors and combines. In 2009 the average American farmer fed 155 people worldwide, compared with just 27 in the 1960s (Center for Food Integrity 2009). In 2012 US farm machinery manufacturers produced over 35 billion (one thousand million) dollars in products and services to assist in this effort (Boyland 2012), making the industry critical to the success of commercial agriculture throughout the world.

This chapter will examine the farm machinery industry, defining its scope and providing a brief background on several key firms, including both full-line and short-line manufacturers. Further, it will examine key competitive challenges that the industry faces in the future as it helps commercial agriculture feed a growing world population. The introductory section will present a definition of the sector and key changes it has faced in recent years. Section 2 will examine some of the most well-known firms in the industry in the US and internationally. Section 3 will look at two key issues that the industry faces: eroding brand loyalty and government regulations. Finally a concluding section will sum up the farm machinery industry and discuss its future path.

DEFINING THE FARM MACHINERY INDUSTRY

The North American Industrial Classification System (NAICS) defines the farm machinery industry as any firm that is primarily engaged in the production of equipment used in commercial agriculture, including planters, harvesters and grass mowing equipment (not including lawn and garden-type mowers) (NAICS Association 2009). This includes a variety of firms from multi-million dollar companies that market a full line of products to short-line manufacturers that focus on a few core competencies as well as small family operations that produce a few custom pieces per year. This includes more than 1000 companies in the US and abroad (Boyland 2012). In addition to competing among each

Figure 15.1 A high-horsepower tractor in a field in Pennsylvania pulls a disc and cultipacker to prepare the seed bed for a new season. A large tractor like this can pull more implements in tandem than smaller tractors and reduce individual trips across the field (photograph taken by author).

other, US farm machinery manufacturers, which dominate the industry, face increasing competition from foreign companies in both Europe and Asia.

A Changing Sector

The farm machinery industry is important to the production of food to feed a growing world population. As agriculture continues to change, in an effort to feed more people on less land, the farm machinery sector also changes to meet these needs. Increases in horsepower and torque allow operators to pull larger implements, in tandem, across the field, reducing the total number of trips (Wehrspann 2010a) (Figure 15.1). Genetic modifications of crops have led to increased stalk strength and more biomass to be harvested or plowed under, also requiring more horsepower from farm machinery (Wehrspann 2010b). Additionally the growing use of geospatial technologies, such as global positioning systems, has amplified the level of technology integrated into new equipment sold by farm machinery manufacturers (Boyland 2012). All of these trends in commercial agriculture, designed to feed more people with fewer farmers, have led to significant changes in the sector in recent years.

Change for the farm machinery industry also came as a result of recent emissions regulations in both the European Union and North America, which forced competitors in those markets to make a massive reduction in the amount of nitrogen oxides (NOx) and particulate matter (PM) that diesel engines emit (Mowitz 2010a). The technologies each company chose to meet these standards will have implications for future competitiveness in the sector (see discussion later in the chapter).

The 2008 economic downturn had a significant impact on the farm machinery industry. Traditionally, economic events that impacted the automobile industry had similar impacts on farm machinery manufacturers. For instance, Chrysler's near bankruptcy in the 1970s was followed by insolvencies of both International Harvester and the Allis-Chalmers Corporation (Archival and Special Collections, University of Guelph Library 1964). That close association has diminished since 2008. While General Motors and Chrysler suffered bankruptcy and reorganization, farm machinery manufacturers have seen a smaller decline in profits (Boyland 2012) in recent years. One US company even built three new factories in the US and hired 100 new employees. This growth was stimulated by record-high crop prices. Those prices were driven by the use of US corn for ethanol and increased international demand for grain crops to feed meat animals in countries like China and India (D'Amica 2009). Shifts in the economy and technology have led to rapid changes in the farm machinery industry.

FIRMS IN THE FARM MACHINERY INDUSTRY

A simple way to divide the farm machinery industry is into full-line manufacturers and specialty short-line manufacturers. Full-line manufacturers produce a complete line of products, including combines, tractors greater than 100 horsepower (75 kilowatts) and a wide variety of tools for tractors. The president of Massey-Ferguson Limited, A.A. Thornburgh, in his 1958 speech to the Chemical Institute of Canada, summarized best the importance of full-line manufacturers to the farm machinery industry: 'The full-line companies necessarily shape the pattern of the industry. Their operations reflect the basic economic changes that affect the industry as a whole' (Archival and Special Collections, University of Guelph Library 1958).

Short-line manufacturers, on the other hand, focus on a few core competencies in which they can be market leaders. Short-line companies include those, for instance, which exclusively sell harvesting equipment, irrigation equipment or self-propelled sprayers. Table 15.1 lists some of the most well-known US and international companies in the farm machinery industry, classifying them as either full-line or short-line. Both full-line and short-line manufacturers are known companies in the industry and are important to its continued success in helping farmers feed the world.

Major US Full-Line Manufacturers

US full-line manufacturers are behemoths of the industry both in the US and abroad. Three companies, Deere and Company, also known as John Deere (JD), Case New Holland (CNH) and the Allis-Gleaner Corporation (AGCO), dominate more than 62 per cent of world market share (approximately 22 billion dollars of revenue), selling

Table 15.1 Major manufacturers in the farm machinery industry

	US Companies	International Companies
Full-Line Manufacturers	Deere and Company Case New Holland AGCO	Mahindra and Mahindra (India) SAME Deutz-Fahr (Germany) Buhler Industries (Canada) YTO (China)
Short-Line Manufacturers	Equipment technologies – Application tools Kinze – Planting tools Great Plains – Planting tools Reinke Manufacturing – Irrigation Vermeer – Irrigation Pequea – Application tools	CLAAS (Germany) – Harvesting tools Krone (Germany) – Hay tools McCormick International (UK) – Tractors Kuhn (France) – Tillage tools MacDon Industries (Canada) – Harvesting tools Tong Yang Moolsan (South Korea) – Tractors Kubota (Japan) – Tractors Yanmar (Japan) – Tractors

Note: This is only a selected list of the major companies, given there are more than 1000 in the industry.

and manufacturing products across the globe. Each of these Big Three manufacturers varies greatly in experience and current status of competitiveness.

Deere and Company

Deere and Company began its life as a short-line manufacturer, producing steel plows invented by the company's namesake, John Deere, in a small facility in Grand Detour, Illinois. In 1848 the burgeoning company was moved to Moline, Illinois (Mills 1986), where the worldwide headquarters is still located today. Unlike other US full-line manufacturers that have grown and changed via mergers and acquisitions, JD's growth has been more organic, acquiring few companies and entering even fewer partnerships (Gerstner 2000).

Deere is a complex company dedicated to multiple lines of business. In addition to its agriculture line, the company is also the second largest producer of construction equipment in the world (behind Caterpillar), a major producer of lawn and garden equipment for both homeowners and commercial operations, and manages its own financial services and parts distribution businesses (Deere and Company 2006). The largest money maker for JD, outside its core competencies in agricultural and construction machinery manufacturing, is its healthcare division that not only provides for employees of JD but also markets to other companies in the US Midwest (Feder 1994; Barboza 1999). These additional business pursuits are not unusual in an economic sector that is subject to seasonal fluctuations in profits based on farmer demand for equipment (Boyland 2012).

Today JD is the largest company in the agricultural machinery industry, controlling approximately 46 per cent of global market share and a quarter of US sales (over 20 billion dollars annually) (Kanicki 2011; Boyland 2012). It captured the position as global leader in 1958 from International Harvester (today a part of CNH). In 2010 JD faced a new challenge as Indian company, Mahindra and Mahindra, took the lead in worldwide tractor sales from JD (Mahindra 2010).

Despite challenges from new competitors JD continues to grow and expand. The company currently employs around 52,000 people, with manufacturing plants in 13 countries (Boyland 2012; Deere and Company 2012). Around 46 per cent of its manufacturing is located in the US, with other major operations in Mexico, China, Brazil and Europe (Deere and Company 2012). The company maintains distribution channels to 120 countries around the world (Boyland 2012). In 2012 JD was ranked 97th on the Fortune 500 list and 348th on the Global Fortune 500 list, with a revenue exceeding 16 billion dollars for its agricultural division. Furthermore, the company enjoys extreme brand loyalty from farmers. In a recent survey of US commercial agriculture producers, 66 per cent said that they had at least one JD tractor or combine on their operations. Of the 66 per cent, 77 per cent said they would buy another JD (Kanicki 2011). This brand loyalty indicates a bright future for JD, despite mounting challenges from other companies. A loyal following can lead to complacency, however, and one of JD's largest future challenges is how it responds to changing market conditions. This is especially relevant as it relates to emissions standards and whether the company keeps up with the technological curve or depends on its historic momentum to lead it to the future.

Case New Holland

Among the major full-line US manufacturers of farm equipment, CNH perhaps has the most complex and convoluted history. Even its status as a US company can be debated, despite the fact it is traded on the New York Stock Exchange. There are four major companies that make up the core of what today is called CNH. These four companies (J.I. Case Threshing Machine Company, International Harvester, the Tractor Division of Ford Motor Company and New Holland) were brought together through a series of mergers and acquisitions from 1985 to 1999. The agricultural manufacturing portion of CNH has two major nameplates: Case Agriculture and New Holland Agriculture (the company also manufactures construction machinery under Case Construction and New Holland Construction). Case Agriculture was formed by an agribusiness conglomerate, Tenneco, in 1985 by merging its prior acquisition, J.I. Case Threshing Machine Company, with its more recent purchase, International Harvester (Stonehouse and Brumbaugh 1996; Case IH 2010). New Holland Agriculture was formed in a variety of phases. In 1987 the Tractor Division of Ford Motor Company purchased New Holland from the Sperry Corporation in an effort to expand its machinery offerings, which dated back to Henry Ford's assembly line system that made Fordson tractors affordable to the average American farmer (Pripps 1997). By 1990 Ford was seeking a partner to make the production of farm machinery more affordable. It sold a portion of the company to Italian tractor manufacturer FiatAgri and removed its oval logo from the design. The tractors continued to be painted the distinctive Ford blue colour, but are now sold under the New Holland nameplate (*Implement & Tractor* 1999). By 1993 FiatAgri had acquired all of the Tractor Division of Ford Motor Company and in 1999 it also acquired Case Agriculture from Tenneco to form CNH (Gibbard 1999).

How can a company that is a subsidiary of an Italian manufacturer still be considered one of the Big Three US farm machinery manufacturers? I will make an argument for why it should be considered as such. First, FiatAgri formed CNH as a free-standing subsidiary of the company. As of 1 January 2011 CNH was a subsidiary of Fiat and Fiat Industrial owned 88 per cent of the shares of CNH. The remaining 12 per cent of shares are traded on the New York Stock Exchange (Scott 2012). Second, the company is incorporated in the Netherlands but is headquartered in the Chicago suburb of Burr Ridge, Illinois. The president and CEO of the company, as well as the chairman of the board of CNH, are all US residents. All executive decisions for the company are generated in the US. So while Fiat Industrial is a major player in the decisions of CNH, it remains rooted in the US.

Case New Holland is the second largest farm machinery manufacturer in the world, controlling approximately 12 per cent of global market share (Boyland 2012). Furthermore, the company is the leader nationally, selling approximately 37 per cent of all farm equipment purchased in the US and Canada in 2011, which amounted to about 43 per cent of CNH's global sales (Kanicki 2011). The company has factories in 17 countries, employing approximately 30,000 people, and dealer networks in 170 countries (CNH 2012). In 2011, CNH collected almost 19.2 billion dollars of revenue, 73 per cent of which was generated by Case Agriculture and New Holland Agriculture. This translated into 4.45 billion dollars of profits for the agricultural divisions (Boyland 2012). However, the complex ownership and management system, and the continued use of CNH as a bulwark for Fiat's other major US acquisition, Chrysler, raise serious concerns about CNH's future in the farm machinery industry.

Allis-Gleaner Corporation

The Allis-Gleaner Corporation, more commonly identified by its acronym, AGCO, is probably the least known of the Big Three US farm machinery manufacturers. Its products, however, have an international reputation, including brands like Massey-Ferguson, Challenger (that was purchased from Caterpillar) and the German industry leader, Fendt. Despite only controlling 4 per cent of global market share (Boyland 2012), AGCO is the market leader in both South America and Europe (AGCO Corporate 2010) and accounts for approximately 10 per cent of US farm machinery sales (Kanicki 2011).

The history of AGCO is incredibly complicated, starting with the Allis-Chalmers Corporation, based in Milwaukee, Wisconsin, in 1901. From this early time the company was guided by the philosophy that the easiest way to eliminate competitors was to acquire them (Petersen 1978). The current make-up of the corporation includes numerous different companies brought under AGCO's control through a variety of mergers and acquisitions. Many of these brand names are still used as a component of AGCO (Table 15.2). But the history of AGCO has not always been positive. In 1985 the Allis-Chalmers Corporation went bankrupt and was acquired by the German firm Klochner-Humboldt-Deutz Ag (KHD) (Mills 1986). By 1990 North American board members of KHD were able to buy the remnants of Allis-Chalmers and re-establish the company as AGCO with its worldwide headquarters in Duluth, Georgia. Despite being a new company, it continued its philosophy of acquisition, including White Farm Equipment in 1991, Massey-Ferguson in 1994, Hesston in 2001 and Caterpillar's agricultural division in 2002 (AGCO Corporate 2010).

Today AGCO is the third largest farm machinery company in the world, generating revenues of 1.4 billion dollars in 2012 (Boyland 2012). The company employs over 14,000 people in 28 plants around the world. In 2008 AGCO built three new assembly plants in the US and recently completed massive renovations to the Jackson, Minnesota plant to establish a new high-horsepower tractor manufacturing line (AGCO Corporate 2012). In 2012 the company ranked 292nd on the Fortune 500. From 2010 to 2012,

Table 15.2 Brand names marketed by AGCO in 2012

Brand Name	Products
Challenger	Tractors, combines, harvesters, hay and forage tools, seeding and tillage tools, material handling attachments and application equipment
Fendt	Tractors, harvesting machines, and hay and forage tools
Gleaner	Combines
Hesston	Hay and forage tools and manure spreaders
Massey-Ferguson	Tractors, combines, hay and forage tools, seeding and tillage tools, material handling attachments, grounds care equipment and power units
RoGator	Self-propelled application equipment
SpraCoupe	Self-propelled application equipment
Sunflower	Seeding and tillage tools and grain carts
TerraGator	Self-propelled application equipment
Valtra	Tractors, combines, and seeding and tillage tools
White Planters	Seeding tools
Willmar	Dry fertilizer spreaders and tenders

Source: AGCO Corporate 2013.

AGCO experienced a 165 per cent increase in profits, compared with just a 50 per cent increase for JD. Much of the increase came from decisions to streamline the company and to focus innovation expenses on being the industry leader in emissions technology (both of which will be discussed later in the chapter).

US Short-Line Manufacturers

There are a number of short-line manufacturers in the US that focus on a few core competencies, producing a limited range of products. Some of these companies are relatively small, but are a major force in the industry with cutting edge technology in their areas of expertise.

Among those that are leaders in the market is Kinze Manufacturing of Williamsburg, Iowa, one of the largest privately owned agricultural manufacturers in the US today (Kinze Manufacturing 2012), focusing on the production of tillage and seeding tools and grain carts. Kinze got its start in 1965 with one man, Jon Kinzenbaw, and a small welding shop. In 1975 Kinze answered requests from farmers for a planter that would fold for road transport instead of having to be trailered from one site to another because of its width (Kinze Manufacturing 2012) (Figure 15.2). That response to customer needs has driven Kinze's growth to this day and made the company the market leader in efficient planters as well as grain carts (Kinze Manufacturing 2012). Market leadership is maintained with a single 25 acre factory complex and 650 employees producing products for dealers around the world (Kinze Manufacturing 2012).

Kinze is not the only well-known short-line manufacturer in the US. Others include Vermeer of Pella, Iowa that specializes in hay and forage tools; Reinke Manufacturing Company of Deshler, Nebraska that is a major supplier of irrigation equipment to farmers in the US and abroad; and Pequea of New Holland, Pennsylvania that

Figure 15.2 *A Kinze planter folded for transport (as seen here) is approximately 11 feet long compared with more than 40 feet wide when unfolded for work in the field (photograph taken by author)*

manufactures hay tools, manure spreaders and livestock feeding systems. Among other notable short-line manufacturers is Equipment Technologies, parent company of the Apache self-propelled sprayer line, headquartered in Mooresville, Indiana. Apache is the largest independently owned self-propelled sprayer manufacturer in North America. Great Plains, based in Salina, Kansas, is the largest producer of grain drills for seed planting. It specializes not only in planting tools, but also is increasingly making a name for itself in the tillage tool industry.

All of these US-based short-line manufacturers, despite having limited product lines, are crucial to the farm machinery industry. Each is a major contributor to the ability to feed more people with fewer farmers. They are, however, not the only manufacturers in the global industry.

International Full-Line Manufacturers

There are competitors to the Big Three US farm machinery manufacturers that also produce a full line of products, including high-horsepower tractors, implements and combines. Many of these competitors are not major forces in the US marketplace as the general trend is for the US to export more farm machinery than it imports (Boyland 2012). In other regions, especially Asia, however, these companies are well-known and command a loyal following of their own. The industry in general is highly concentrated, with the Big Three controlling over 60 per cent of the sector (almost 22 billion dollars of revenue) and the rest of the top 20 largest firms controlling another 30 per cent (10.6 billion dollars in revenue). This leaves less than 10 per cent of the market (about 2.8 billion dollars of revenue) over which the remaining 1000 firms of various size compete.

Mahindra and Mahindra

The fourth largest farm machinery manufacturer in the world and the leading seller of tractors globally is Mahindra and Mahindra of India. Mahindra got its start in the farm machinery business during World War II by assembling Jeeps for Minneapolis-Moline (today a part of AGCO) to distribute to troops in the Pacific and North African theatres. After the war Minneapolis-Moline rewarded Mahindra for its excellent work by sending tractors for assembly and sale in Asia. When Minneapolis-Moline was bought out by White Farm Equipment in the 1970s, Mahindra took the skills and knowledge gained from this relationship and applied it to manufacturing its own products (Gee 2008). While Mahindra is best known in the US for lower-horsepower tractors used on hobby farms, it is a full-line manufacturer, having broken the all-important 100 horsepower (75 kilowatt) barrier in the last five years to enter the high-horsepower segment of the market that most Asian manufacturers have not cared to broach (ENP Newswire 2009). The company sells tractors in a wide range of sizes as well as implements for those tractors in Europe, Asia and North America. Its combine sales, however, are confined to Asia. Mahindra fills a niche in the market, providing reliable low-cost machines to parts of the world where the price of American innovation is too great for small-scale agricultural operations (Kingsbury 2010). This has allowed the company to continue to grow and prosper, representing the largest threat to the Big Three's position in the global market.

SAME Deutz-Fahr

The fifth largest farm machinery manufacturer in the world is the European full-line producer, SAME Deutz-Fahr (SDF), based in Cologne, Germany. This company is made up of several major components: SAME, an Italian manufacturer of tractors and internal combustion engines, created in the 1940s, and Deutz-Fahr. Deutz-Fahr was best known as a manufacturer of tractors and diesel engines in Germany. While being the fifth largest producer of equipment, SDF manufactures at a much smaller scale than its competitors. In 2008 AGCO, as the smallest of the full-line US farm machinery manufacturers, produced around 1000 combines while SDF produced less than 300 (Tsoneva 2008). While the company sells equipment in all major markets, including North America, Europe and Asia, given its manufacturing capacity, it simply does not have the means to compete with the Big Three or Mahindra.

Versatile

Another major competitor in the farm machinery industry is Versatile, a subsidiary of Buhler Industries, based in Winnipeg, Canada. The Versatile brand was a part of Ford New Holland, but during the 1999 merger with Fiat, US anti-trust laws required the new company to shed components. Versatile was sold to Buhler Industries (*Implement & Tractor* 2000). In addition to the brand name, Buhler also acquired the patents for the Ford Genesis tractors (Wehrspann 2010d). Versatile, at this point, was a short-line manufacturer, focusing on high-horsepower articulated tractors as well as Buhler's line of grain handling equipment. Until 2004 the company produced the highest-horsepower tractors sold in North America (Buhler Versatile 2010).

In 2006 Russian combine manufacturer, Rostselmash, acquired 80 per cent of Buhler Industries. In addition to expanding Versatile's market share around the globe, Rostselmash also diversified the company from a short-line manufacturer to a full-line manufacturer. Today the company not only produces its traditional high-horsepower articulated tractors, but also markets Rostselmash combines and smaller fixed-frame row crop tractors that are affordable to farmers in the US and abroad (Wehrspann 2010d). With the addition of Buhler's grain handling equipment and other implements for fixed-frame tractors, Versatile has made the transition from international short-line competitor to international full-line competitor in just five short years.

International Short-Line Manufacturers

There are also a number of international short-line manufacturers among the competitors in the farm machinery industry. In addition to those that focus on specific implements, there are those that focus on the production of lower-horsepower tractors and implements for the landscaper or hobby farmer. Not surprisingly, many of these firms are located in Asia where market demand is not for the high-horsepower, high-torque monsters built by the Big Three, but for more compact tractors that are manoeuvrable in tight spaces.

Among the most well-known international short-line manufacturers is CLAAS of Harseweinkel, Germany. Its major core competency is harvesting equipment, including combines, forage harvesters, hay balers and mowers. As a result of a buyout of Renault's tractor division, CLAAS also produces a limited line of tractors (CLAAS 2012). The company is not considered a full-line manufacturer because it does not produce any implements other than those for harvesting and its tractors are designed for harvesting operations. The focus on harvesting equipment, however, made CLAAS' products world-renowned for reliability and innovation.

Another major European short-line producer is McCormick International of Britain that produces McCormick and Landini tractors. At one time McCormick International was the British division of Case Agriculture (and takes its name from the inventor of the reaper, Cyrus McCormick). As a part of the merger with New Holland by Fiat, Fiat had to shed the company, which was purchased in 2000 by another Italian firm, ARGO. In 2001 McCormick opened a plant in Pella, Iowa and is currently building a second US plant near Atlanta, Georgia (McCormick International 2012). McCormick produces tractors in a wide range of horsepower ratings, but does not manufacture implements for the tractors or combines, making it a short-line company.

McCormick is not the only international short-line manufacturer to focus on tractors. Several Asian companies limit production to low-horsepower (under 100 horsepower or 75 kilowatts) tractors and implements for those tractors, largely to capture the bulk of demand in Asian markets. Among the most well known in the US is Kubota of Japan, which sells tractors, implements and all-terrain vehicles for travelling around the farm. Another rapidly growing competitor from Asia is Tong Yang Moolsan (TYM) based in South Korea. This company also focuses mostly on small tractors, but has broadened its product range in recent years through a partnership with Mahindra. While not as well known in the US as Kubota, TYM claims to be 'the fastest-growing tractor brand in North America' (TYM 2011: A1). Building on the South Korean automakers' tradition of long warranties, TYM offers customers a five year warranty on all its products, much longer than any of its competitors – full-line or short-line. The focus on quality and reliability in low-cost machines has allowed both Kubota and TYM to become a larger force in the farm machinery industry.

CONDITIONS FACED BY THE FARM MACHINERY INDUSTRY

Despite the fact that farm machinery is one of numerous mature manufacturing sectors in the global economy, its problems are not identical to other mature sectors (such as automobiles or steel). While the issues may sound similar, the farm machinery industry has its own unique methods of dealing with concerns such as brand loyalty and complying with ever more stringent emissions standards. Both issues of brand loyalty and government regulations on emissions continue to have impacts on competitiveness for individual companies in the industry.

Marketing and Brand Loyalty

Brand loyalty among consumers is typical in American agriculture. The average US farmer is more likely to drive a truck manufactured by the Big Three Detroit automakers and to own farm machinery manufactured by JD, CNH and/or AGCO than any international brands. Recent research on the industry found that 'Strong branding runs deep' (Kanicki 2011: 34). Many believe that JD's key advantage is that the brand is considered a part of Americana, just like Chevrolet and apple pie. At the same time additional research related to the farm machinery industry found an increasing amount of 'color-blindness' among farmers (Lessiter 2011). This does not refer to a medical condition, but a consumer decision to focus less on the brand (or colour) of the equipment and more on other aspects. The quality of the equipment, the amount of uptime and the ability of the dealer network to respond to customer needs are increasingly more important than the name on the hood or the colour of the equipment.

This colour-blindness is something on which AGCO is banking. In 2010 AGCO announced it would no longer sell its distinctive orange AGCO tractors. All tractors would be sold as Challengers, Massey-Fergusons, Valtras or Fendts (Fischer 2010). It was an attempt to reduce duplication of efforts by dealers who sold more than one brand. At the same time, however, the orange tractors were a nod to the company's history as Allis-Chalmers and many American farmers were upset with the decision. For AGCO it was a way to refocus a limited R&D budget on brands that had much wider national and international appeal than the AGCO tractors sold primarily in the upper Midwest of the US. The press campaign to promote this strategy has been massive, including a push to market Challenger, also a distinctive colour – Caterpillar yellow – as the 'color of pride' and the 'color of fast-forward farming'. Another new AGCO slogan is 'color isn't the way to buy a tractor'. The company even went a step further in the summer of 2012, bringing together producers from all over the US for a head-to-head test of Challenger against comparable models of JD and CNH tractors in an attempt to prove the innovative superiority of the new colour of farming. All of these demonstrate AGCO's attempts to erode colour and brand loyalty in the US farm machinery industry to encourage farmers to buy not on the name alone, but on superior technology, quality and fuel efficiency.

For JD the strategy is exactly opposite, depending largely on brand loyalty to green, to carry the company through the challenges it faces from other competitors, like CNH, AGCO or even Mahindra. Currently, this 'bet on green' has worked. As mentioned earlier, JD enjoys the greatest brand loyalty among US farmers (Kanicki 2011). A strong dealer network and a reputation for quality products lead many farmers in the US and abroad to depend on JD. If the trend of colour-blindness in American farmers is in fact increasing, however, and potential customers are swayed more by technology than the colour of paint (Kanicki 2011), JD could be facing a major problem in the future. It is not that the company is not innovative. It in fact leads the Big Three in the number of patents obtained. One of the biggest issues for JD is that it does not always market new technology well; leaving farmers to wonder if green is the best option for advanced farming. Given that 'Perception is often more important than reality' (M. Lynch quoted in Lessiter 2011: 51), JD could face serious challenges in the future from other companies that better market technology advances.

For CNH the problem is not brand loyalty or a lack thereof. The bigger concern is how parent company Fiat uses CNH's loyal customer base and its dealer network to support another recent acquisition, Chrysler. Since purchasing Chrysler in 2009, CNH has been used by Fiat to try to bolster the fortunes of the automaker. First, Fiat considered selling CNH to generate funds to support Chrysler (Bertacche and Cinelli 2009). When the company that was interested was Mahindra, this plan was taken off the table. More recently Fiat has used CNH, specifically Case Agriculture, to market the RAM truck line, forcing dealers to replace service trucks with RAMs and offering incentives to farmers who buy a combination of Case combines or tractors and RAM trucks. Given that not all CNH-loyal farmers are RAM-loyal truck drivers, this could have interesting implications for the future.

Increasing Emissions Standards

On 1 January 2011 the farm machinery industry entered what is known to manufacturers as Tier IV Interim (Tier IV(i)) emissions standards, mandated by the US Environmental Protection Agency and the Canadian Environmental Protection Agency. This new set of emissions standards, which applies to all self-propelled farm machinery greater than 75 horsepower (55 kilowatts), brought US and Canadian standards in line with those in the European Union (Deere and Company 2010). Tier IV(i) represents the single largest reduction in NOx and PM in diesel exhaust for farm machinery. Meeting the standards required huge investments in R&D on the part of farm machinery producers. This was the largest step in a process occurring over the past 20 years, not only for farm machinery but also for construction machinery and heavy duty trucks.

Despite the fact that these changes have gradually occurred since the 1980s, Tier IV(i) was the first time that farmers saw appreciable changes in the equipment (Mowitz 2010b), both in terms of cost, with increases anywhere from 3 to 7 per cent being passed on to customers (Mowitz 2010a), and additional care and maintenance for the machines. There are two methods to meet Tier IV(i) standards. Selective catalytic reduction (SCR) is a chemical treatment process used by AGCO and CNH, among others. Cooled exhaust gas recirculation (cEGR) is a thermal treatment process preferred by JD and Cummins (which builds engines for Versatile as well as several other farm machinery producers).

SCR uses a combination of more efficient fuel burn and an urea and de-ionized water mixture known as diesel exhaust fluid (DEF) to treat exhaust as it exits the engine to reduce PM and NOx. Testing by CNH has proven that the SCR system will meet Tier IV Final standards that took effect on 1 January 2014. In order to achieve emissions reductions, the user must add DEF to the tractor or combine approximately once every four tanks of diesel fuel (Wehrspann 2010c). Given the increased fuel efficiency created by burning the fuel at high temperatures, the actual costs of the DEF are negated, but access to the fuel additive can be problematic outside the developed world. In Brazil, for instance, AGCO cannot sell the same technology as it sells in its other major markets because of the lack of access to DEF. In North America and Europe, DEF access is ubiquitous, with most heavy duty and light duty diesel trucks also requiring the additive.

The second option, cEGR, removes PM and NOx by passing a percentage of the initial exhaust from the tractor back through the engine as a cooled air flow that reduces combustion temperatures and the natural oxygen content in the engine. While this leads to less efficient fuel burn-off, when coupled with a diesel oxidation catalyst and diesel particulate filter in the exhaust system, cEGR is able to meet Tier IV(i) standards. It is unclear if cEGR will be able to meet Tier IV Final standards (Xinqun et al. 2010). While cEGR does not require DEF, the cEGR engine requires more frequent oil changes due to high levels of heat deflection in the engine as part of the gas recirculation process. It also consumes approximately 1 per cent more fuel as it recirculates the exhaust (Wehrspann 2012).

Regardless of the method used to achieve Tier IV(i) standards, all low-emissions diesel engines require ultra-low sulphur diesel fuel (ULSF). The majority of PM generated by diesel engines comes from the sulphur particles in the fuel. By burning ULSF, whether in cEGR or SCR engines, an automatic reduction of PM is achieved. The exhaust filtering system on both engine types is designed to handle an already reduced particulate load by burning ULSF. Much like DEF, however, ULSF is largely unavailable in the developing world. This creates complications for companies that usually market a very similar product regardless of geographic location.

The lack of access to DEF and ULSF demonstrates one of the advantages of SCR technology. This technology, in addition to being more fuel-efficient and responsive to demands for torque and horsepower, is also external to the engine. All SCR components bolt onto the engine and exhaust and therefore can be easily removed for sale in countries where access to DEF and/or ULSF is not practical. Manufacturers which use cEGR technology must install completely different engines in equipment sold in the developing world.

CONCLUSIONS

The farm machinery industry includes over 1000 companies, but is dominated by three major US full-line manufacturers: John Deere, Case New Holland and the Allis-Gleaner Corporation. These three firms control over 60 per cent of the 35.4 billion dollars of revenue generated by the industry each year, leaving the rest of demand for the remaining short-line and full-line producers in the US and abroad. The Big Three US manufacturers continue to face growing challenges from international firms that can provide lower-cost alternatives that are accessible to farmers in less developed parts of the world.

In many cases the farm machinery industry is one whose fortunes, challenges and opportunities are very similar to other mature manufacturing sectors, such as automobiles and steel. In recent years, however, the industry has seen its path change, proving to be more solvent than the Detroit automakers during the 2008 global recession. This chapter also identified several challenges facing the industry related to marketing and emissions standards compliance. The farm machinery industry continues to face these challenges head-on.

Given the importance of the farm machinery industry to the world's food supply, there is a continued need for research in the sector. How will manufacturers handle the 2014 final emissions standards? Will certain firms see declines or increases in competitiveness based on how they approached the standards? How will firms continue to compete for a limited customer base in an era when quality matters more than the brand of the equipment? All of these topics, and more, deserve further research.

Despite being a small part of manufacturing GDP for most countries, the farm machinery industry will continue to be crucial to the economic and human well-being of the Earth. Not only does the farm machinery industry create good paying jobs for workers around the world, but mechanization continues to be one of the most promising paths to feed a growing world population with fewer farmers on fewer acres.

REFERENCES

AGCO Corporate (2010), 'History', available at http://www.agcocorp.com/company/history.aspx (accessed 24 October 2010).

AGCO Corporate (2012), 'AGCO Intivity Center and Manufacturing Expansion Officially Open', available at http://www.agcocorp.com/news/media_press_releases.aspx (accessed 25 October 2012).

AGCO Corporate (2013), 'Our Brands', available at http://www.agcocorp.com/products/products_brand.aspx (accessed 22 August 2013).

Archival and Special Collections, University of Guelph Libraries (1958), X41 RHC A0433, Massey–Harris–Ferguson Collection, *Miscellaneous Speeches and Executive Addresses, 1–5*, Box 1, File 4.

Archival and Special Collections, University of Guelph Libraries (1964), X41 RHC A0434, Massey–Harris–Ferguson Collection, *E.P. Neufield Interviews with M-F Executives for 'A Global Corporation'*, Box 1, File 1.

Barboza, David (1999), 'Aiming for Greener Pastures: Farm-Equipment Makers Step Up Efforts to Diversify', *New York Times* 14 April.

Bertacche, Marco and Francesca Cinelli (2009), 'Fiat May Sell CNH to Fund Chrysler, Bernstein Says', available at http://bloomberg.com/apps/news?pid=20670001&sid=af_UR.o.kWAc (accessed 22 April 2009).

Boyland, Kevin (2012), *IBISWorld Industry Report 33311: Tractors & Agricultural Machinery Manufacturing in the US*, Santa Monica, CA, USA: IBISWorld, Inc.

Buhler Versatile (2010), 'History', available at http://www.versatile-ag.ca/about/history.html (accessed 10 October 2010).

Case IH (2010), 'Case IH Web Press Room: History', available at http://pressroom.caseih.com/index.cfm?fuseaction=mediakit.history (accessed 15 October 2010).
Center for Food Integrity (2009), 'Farmers Feed Us Fun Farm Facts', available at http://www.farmersfeedus.org/fun-farm-facts/ (accessed 21 October 2012).
CLAAS (2012), 'CLAAS – Combines, Attachments', available at http://www.claas.com/cl-pw/northamerica/products/start,bpSite=817884.html (accessed 30 October 2012).
CNH (2012), 'CNH Worldwide', available at http://www.cnh.com/Worldwide/Pages/CNHNorthAmerica.aspx (accessed 23 October 2012).
D'Amica, Kenneth (2009), 'The Invisible Hand at Work', available at http://www.aier.org/research/commentaries/265-the-invisible-hand-at-work (accessed 8 March 2009).
Deere and Company (2006), *Growing a Business as Great as Our Products*, Moline, IL, USA: Deere and Company.
Deere and Company (2010), *Emissions Technology*, Moline, IL, USA: Deere and Company.
Deere and Company (2012), 'Worldwide Locations', available at http://www.deere.com/wps/dcom/en_US/corporate/our_company/about_us/worldwide_locations/worldwidelocations.page? (accessed 23 October 2012).
ENP Newswire (2009), 'Mahindra's Farm Equipment Sector Receives Agricultural Leadership Award', *ENP Newswire*, 8 September: n.p.
Feder, Barnaby J. (1994), 'Deere Sees a Future in Health Care', *New York Times* 1 July.
Fischer, Margy (2010), 'AGCO Will Phase Out Orange Tractors', *Farm Journal* 134 (1): 44.
Gee, Marcus (2008), 'Made in India: Meet the Tractor Maker who has John Deere on the Run', *The Globe and Mail* 8 April: B1.
Gerstner, John (ed.) (2000), *Genuine Value: The John Deere Journey*, Moline, IL, USA: Deere and Company.
Gibbard, Stuart (1999), *The Ford Tractor Story Part Two: Basildon to New Holland 1964 to 1999*, E. Yorks and Ipswich, UK: Japonica Press and Old Pond Publishing Ltd.
Implement & Tractor (1999), 'CNH Combines Assets and Products of Case Corporation and New Holland', *Implement & Tractor* 114 (6): 8–9.
Implement & Tractor (2000), 'CNH Divests, Closes and Moves Operations', *Implement & Tractor* 115 (4): 20.
Kanicki, Dave (2011), 'Brand Loyalty: Is It Dead?', *Farm Equipment Showcase* 2011 (January): 34–49.
Kingsbury, Kathleen (2010), 'Deere's Harvest', *Time* 175 (11): 12–14.
Kinze Manufacturing (2012), 'Our Company', available at http://www.kinze.com/our-company/ (accessed 26 October 2012).
Lessiter, Mike (2011), 'Dealers Weigh in on Brand Loyalty Erosion', *Farm Equipment Showcase* 2011 (January): 50–57.
Mahindra (2010), 'Why Mahindra?', available at http://www.mahindrausa.com/Why-mahindra (accessed 22 October 2012).
McCormick International (2012), 'McCormick USA', available at http://www.mccormickusa.com/history.php (accessed 30 October 2012).
Mills, Robert K. (ed.) (1986), *Implement & Tractor: Reflections on 100 Years of Farm Equipment*, Marceline, MO, USA: Walsworth Press.
Mowitz, Dave (2010a), 'Meaner Cleaner Leaner Engines', *Successful Farming* 108 (9): 18–20.
Mowitz, Dave (2010b), 'Tier 4 Trials', *Successful Farming* 108 (9): 18–20.
NAICS Association (2009), 'Farm Machinery and Equipment Manufacturing', available at http://www.naics.com/censusfiles/ND333111.HTM#N333111 (accessed 3 December 2009).
Peterson, Walter (1978), *An Industrial Heritage: Allis-Chalmers Corporation*, Milwaukee, WI: Milwaukee County Historical Society.
Pripps, Robert N. (1997), *Vintage Ford Tractors*, Minneapolis, MN, USA: Voyageur Press, Inc.
Scott, Mark (2012), 'CNH Rejects Merger Deal From Fiat Industrial', *New York Times* 15 October: n.p.
Stonehouse, Tom and Eldon Brumbaugh (1996), *JI Case Agricultural and Construction Equipment 1956–1994, Volume Two*, St. Joseph, MI, USA: American Society of Agricultural Engineers.
Tsoneva, Annie (2008), 'SAME Deutz-Fahr's Croatian Unit Plans to Raise Farm Machine Sales in 2008', *SeeNews* 29 August: n.p.
TYM (2011), 'TYM Tractors: Leading the Agricultural Equipment Industry', *The Eastern Farm and Livestock Directory* 14 (1): A1.
Wehrspann, Jodie (2010a), '2010 Tractor Preview: A Guide to the 150-HP-Plus Row-Crop Models', *Farm Industry News* 42 (11): 21–35.
Wehrspann, Jodie (2010b), '2010 Tractor Preview: A Guide to the 300-HP-Plus 4-WD Tractor Models', *Farm Industry News* 42 (2): 8–19.
Wehrspann, Jodie (2010c), 'Coming Clean: An Insider's Guide to Tier 4 Engines', *Farm Industry News* 43 (9): 10–19.
Wehrspann, Jodie (2010d), 'Versatile Bets Big', *Farm Industry News* 43 (3): 8–11.
Wehrspann, Jodie (2012), 'Fuel Efficiency Redefined', *Farm Industry News* 45 (11): 10–14.

Xinqun, Gui, Danan Dou and Richard Winsor (2010), *Non-Road Diesel Engine Emissions and Technology Options for Meeting Them*, St. Joseph, MI, USA: American Society of Agricultural and Biological Engineers.

FURTHER READING

Boyland, Kevin (2012), *IBISWorld Industry Report 33311: Tractors & Agricultural Machinery Manufacturing in the US*, Santa Monica, CA, USA: IBISWorld, Inc. (A report from a professional consulting firm on the status of the industry.)

Gerstner, John (ed.) (2000), *Genuine Value: The John Deere Journey*, Moline, IL, USA: Deere and Company. (A pictorial trip through the history of John Deere from the company's own perspective.)

Gibbard, Stuart (1999), *The Ford Tractor Story Part Two: Basildon to New Holland 1964 to 1999*, E. Yorks and Ipswich, UK: Japonica Press and Old Pond Publishing Ltd. (A combination of photos and narrative tracing the most recent history of the Tractor Division of the Ford Motor Company.)

Magee, David (2005), *The John Deere Way*, Hoboken, NJ, USA: John Wiley and Sons. (An examination of the corporate structure of Deere and Company, sharing management principles that have kept the company on top.)

Marsh, Barbara (1985), *A Corporate Tragedy: The Agony of International Harvester Company*, Garden City, NJ, USA: Doubleday and Company, Inc. (The definitive story of how International Harvester went from success to bankruptcy under the leadership of the McCormick family.)

Mills, Robert K. (ed.) (1986), *Implement & Tractor: Reflections on 100 Years of Farm Equipment*, Marceline, MO, USA: Walsworth Press. (A collection of articles from the magazine, *Implement & Tractor*, covering an extensive portion of the industry's history.)

Petersen, Walter F. (1978), *An Industrial Heritage: Allis Chalmers Corporation*, Milwaukee, WI, USA: Milwaukee County Historical Society. (A historical look at the Allis-Chalmers Corporation from the perspective of a long-time employee at the West Allis plant.)

Stonehouse, Tom and Eldon Brumbaugh (1996), *JI Case Agricultural and Construction Equipment 1956–1994, Volume Two*, St. Joseph, MI, USA: American Society of Agricultural Engineers. (An examination of the most recent history of JI Case up to its purchase by Tenneco in 1994.)

Swinford, Norm (1999), *The Proud Heritage of AGCO Tractors*, St. Joseph, MI, USA: American Society of Agricultural Engineers. (This book traces the history of the many components of AGCO including Oliver, Minneapolis-Moline and Allis-Chalmers.)

Xinqun, Gui, Danan Dou and Richard Winsor (2010), *Non-Road Diesel Engine Emissions and Technology Options for Meeting Them*, St. Joseph, MI, USA: American Society of Agricultural and Biological Engineers. (A technical evaluation of the different technologies used to meet Tier IV emissions regulations in the US and Canada.)

16. Hidden in plain sight: the North American optics and photonics industry

Jennifer Clark

INTRODUCTION

This chapter describes an established enabling industry: optics and photonics. The photonics industry has existed at the crossroads between science policy and economic development for several decades – falling somewhere along the spectrum from a high-tech emerging industry to an established sector. As a consequence, photonics often fails to garner the state and federal investment directed at seemingly newer and flashier enabling technologies like nanotechnology or biotechnology. It also falls beneath the radar of policy makers focused on new materials like lightweight composites or additive manufacturing. And finally, the definitional issues that make the boundaries of the industry difficult to scope, often leave it off the list of industries targeted for economic development and private investment.

In spite of its advocates and supporters, the photonics industry is largely unknown outside of the scientific research community and a few regional economies. For example, the National Academies of Science has only pursued two studies of the industry, known as *Harnessing Light I* and *II*. The first study was released in 1998 and the second in 2012.[1] This chapter covers both *why* this industry is important and *where* it is important – both in terms of regional and national economies and global supply chains. It also highlights what photonics reveals about working regions and why industries that are so difficult to categorize are also the industries with the most potential for retaining innovation and production (Clark, 2013). Much like interdisciplinary research in academia, it is the innovations at the intersections between established fields that often lead to transformative work.

It is important to note that photonics is not a familiar selection on the popular menu of "regional clusters" assigned to metropolitan areas all over the US and the EU. Instead it is a true *specialization*. For those who follow industry clusters, specializations such as "high-tech," biotech, advanced manufacturing, life sciences, information and communications technology, appear over and over again in lists from chambers of commerce and economic development agencies. Optics and photonics is not unique to the regions where it concentrates, like Rochester, New York or Quebec. However, it is a distinct specialization in a certain, discrete set of regions.

ENABLING TECHNOLOGIES: EVIDENCE FROM THE PHOTONICS INDUSTRY

The photonics industry is an unusual merger of a high-tech enabling technology, *photonics*, and the continuing evolution of a traditional manufacturing industry, *optics and imaging*. Photonics technologies are broadly understood as the science and engineering of generating and transmitting light (photons). The industry involves products, components, and processes associated with this science. The industry is inclusive of a wide range of end products and intermediate components, including optoelectronics. Although the industry is research and development intensive, it also has an established manufacturing capacity (Christopherson and Clark, 2007b; Clark, 2004; Sternberg, 1992). Firms in the photonics industry make a wide variety of products across several markets. This broad impact is part of the industry's strength and resilience. It is also the reason the industry is difficult to identify and assess using traditional metrics.

Photonics research produces materials and components critical to green energy technologies such as solar cells, and photovoltaics, biophotonics, and nanophotonics that have applications in a wide variety of industries, including medical devices and medical research. A wide variety of photonics components contribute to manufacturing processes as well as security and defense applications. In general, the industry can be scoped in two ways: 1) by product and components, or 2) by end product markets. Products and components include: Analysis, Testing, and Measurement Equipment; Cameras and Imaging; Fiber Optics and Communications; Lasers and Light Sources; Optical Components; Optical Design; Optical Materials and Coating; Photonic Manufacturing; and Positioning Equipment.[2] The applicable markets that delineate the industry include: Medicine and Health (examples include imaging technologies like MRIs), Defense and Security (measuring and detection equipment), Telecommunications (fiber optics and data transmission), Energy (photovoltaics) and Consumer Products (digital cameras to Personal Digital Assistants).[3]

Although photonics firms are difficult to capture through traditional firm and occupational categories, there are data on industry trends. In Europe, Canada, and East Asia the industry has held the attention of policymakers since the mid-1990s. Studies have found that the industry has a significant impact in applied research and manufacturing as well as potential for growth and expansion. Photonics21, a European-based industry association, counts over 2000 firms and 700 research organizations among its members.[4] Optics and photonics is or has been a target of science and technology investment in a number of countries, including Canada, Korea, the United Kingdom, China, Germany, France, and several other members of the European Union. In the US, the industry remains the focus of state and local economic strategies almost exclusively. That said, investments in the technologies behind optics and photonics have filtered through the array of national scale funding agencies in the US – the Department of Defense, the National Science Foundation, and the Department of Energy to name but a few. A coordinated policy approach in the US has not reached the levels seen in other countries.

In the US and internationally, the photonics industry tends to be highly agglomerated in city-regions with pre-existing optics and/or imaging production specializations as well as photonics research capacity (Christopherson and Clark, 2007a; Feldman and Lendel, 2010; Hendry and Brown, 2006; Hendry, Brown, and Defillippi, 2000; Miao and Hall, 2012). These regions include Montreal and Quebec City in Canada; Jena in Germany; and Rochester, New York and Tucson, Arizona in the US. This tendency to co-locate in particular regional economies makes photonics an emblematic case of the industrial cluster model as it pertains to technology-intensive industries (Karlsson, 2008; Porter, 2000). Evidence from industry surveys provides employment-level data on the co-location of basic and applied research activities in those urban centers with industry agglomerations (see Appendix, Table 16A.1). Optics and photonics firms depend on embedded regional specializations in the labor market, the research and development infrastructure, and a dynamic small firm network capable of absorbing new technologies, products, and processes.

PHOTONICS REGIONS IN THE WORLD ECONOMY

Photonics has long been categorized as a high-technology industry, overlapping with other information and communication technologies. Analytically, this categorization complicates tracking the industry's overall role in the global economy and the share of production that can be attributed to each national economy. And, because photonics is so intimately tied into established industries like imaging and optoelectronics, it has been difficult for analysts to track its impact over time. Throughout the 1990s, the Advanced Technology Program (ATP), a program under the National Institute for Standards and Technology in the US Department of Commerce, intermittently tracked the production capabilities of US-based photonics firms. ATP essentially defined photonics as synonymous with optoelectronics. In 1999, ATP reported that, "The United States currently controls only 9 percent of the $16 billion photonics component manufacturing market but consumes approximately 40 percent of the products."[5] The ATP's focus was photonics manufacturing of end products. It attributed 75 percent of that market to Japan in the late 1990s. A more recent report on the global photonics market (2006) by Photonics21, one of the EU's formal European Technology Platforms, provides a broader view of the industry and its role in the global economy:

> According to rather conservative estimations, the photonics world market in 2005 amounted to more than EUR 150 billion of which about 40% related to information and communication technologies. The annual growth rates of the optics and photonics industries far exceed those of the economy (e.g. laser sector greater than 14% for the past 10 years; optics and photonics in life sciences exceeding 38%) and also the future prospects are promising (e.g. for OLEDs = organic light-emitting diodes an annual growth rate of about 40% is expected over the next 5 years). The total photonics world market is expected to at least triple within the next 10 years![6]

A 2007 report by Photonics21 estimated the global photonics market at $330 billion (one thousand million). In the US and EU, studies find that the photonics industry is unusual in that employment is concentrated in small and medium sized firms, making the industry a desirable target for economic development purposes. As with other industries, however, the US has taken a "devolution" approach to industry-specific development strategies and a decentralized approach to applied technologies in the last decade, while other countries and other economic communities have taken a coordinated and integrated approach to policies aimed at technology-led economic development. These coordinated strategies outside the US have produced partnerships between government and industry, strong data tracking and analysis, and detailed strategic plans.

Some of these strategic plans are quite sophisticated and they are often grounded in researched, benchmarking reports about the industry.[7] The Optoelectronics Industry and Technology Development Association in Japan releases an annual technical report on the optoelectronics segment of the industry.[8] In the EU, the European Technology Platform Photonics 21 project releases regular reports on the future of the industry and on its applications to new and emerging markets like green production.[9] Also, in Canada, various associations and agencies have regularly released reports on the Canadian photonics industry for more than a decade, including the Canadian Photonics Consortium and the Institute National d'Optique (INO).[10] These reports are coupled with targeted technology strategies aimed at increasing investment in photonics research and, in the case of Quebec, forming and sustaining small firm networks of photonics firms.

In many industrialized countries, there is a policy recognition that innovation often occurs in specialized regional clusters. There is an extensive academic literature on the subject which underpins a nuanced, if problematic, policy agenda (Lovering, 1999; Martin and Sunley, 2003). Rochester, NY is often understood as a critical example of this regional specialization process for photonics technologies, much as Jena, Germany or Quebec City in Canada. However, the influence of these clusters on larger multinational issues, such as trade policy or regional European Union policy, is much more developed and coordinated in the Canadian and EU contexts than in the US. Indeed, the presence of photonics as a prioritized industry in Quebec and a targeted technology in the European Union differs significantly from the policy approach in the US. Notably, the importance of established regional industry clusters is felt at the national, subnational, and supra-national scale in these countries. For example, the president of the Photonics21 strategy group is from Jena and, the president of the Canadian Photonics Industry Consortium is from Quebec City. The point is that regional actors have influence over national and even supra-national scientific and research agendas – thus bringing the policy process closer to ground level and to the networks of firms whose engagement is critical for regional economic development.

OPTICS AND PHOTONICS CLUSTERS IN THE US

As noted above, the tendency of photonics firms to co-locate in particular regional economies makes this industry an important model, and the organizational structure of the firm networks matters a great deal to their viability and credibility over time. Part of that organizational structure is getting the scale of the firm network right. In most cases of photonics clusters in the US, the cluster organization is not organized at the regional level. In fact, most formally organized US photonics clusters are organized at the state level.[11] Examples include the Arizona Optics Industry Association, the Carolinas MicroOptics Triangle and the Carolinas Photonics Consortium, the Florida Photonics Cluster, the New Mexico Optics Industry Association, and the New York Photonics Industry Association, in addition to several others.[12] The exception to this state-level rule is the Rochester Regional Photonics Cluster (RRPC) in Rochester, New York. This cluster aligns with the region that specializes in the technology. Even before the RRPC was formed, economic development agencies from Rochester routinely presented technology-led economic development policies at industry conferences alongside national and state governments (Clarke, 1991).

It is somewhat unusual for industries in the US to take so quickly to the industry clusters model across so many regions. In the case of optics and photonics, these firm networks vary in specialization (ranging from optical equipment to photonics) and in scale (regional and state level). The optics clusters started with the Arizona cluster based in Tucson. Two key factors contributed to this. First, Arizona's relatively early adoption of an industry clusters approach to statewide economic development. By the mid-1990s, Arizona had embraced this approach to both economic and workforce development (Munnich, Love, and Clark, 1999; Munnich, Sheldon, and Clark, 1999; Waits and Vandegrift, 1999). Second, the optics cluster in Tucson saw the opportunity to get on the state's cluster list for targeted economic development. Robert Breault, president and founder of the Tucson-based Breault Research Organization, was particularly instrumental in this effort. Later, Breault brought the clusters model to other regions.[13] In addition, organizations like the International Society for Optics and Photonics (SPIE), one of the major international advocacy groups for photonics research and technologies, have highlighted and encouraged photonics cluster groups.[14]

As noted above, photonics is an enabling industry like biotechnology or nanotechnology. However, photonics is also deeply embedded in an existing set of production industries: imaging, optoelectronics, photographic equipment, and optical equipment. This complicates the use of research performance indicators to analyze the viability of the industry as a whole or the growth and pervasiveness of the technology across markets and production processes.[15]

Even with that lack of consensus, the Rochester cluster is considered highly innovative across several metrics. In terms of the production of patents, Rochester's innovation is concentrated in the optics and photonics industry. The patent activity is unusually high in relative terms (optics, imaging, and photonics-related patent technology accounts for 65 percent of the top 20 patent technology classes granted in Rochester according to the USPTO). Rochester is ranked first for utility patents issued to companies or individuals per 1000 workers.[16] On another per capita measure, Rochester holds 2.33 patents per 1000 workers – the US average is 0.40. Because of its performance and organizational structure, the optics and photonics industry has also garnered an unusual amount of attention from policy analysts both inside and outside the Rochester region.[17]

The Rochester Photonics Industry

The Rochester Regional Photonics Cluster is a not-for-profit organization formed as an "industry cluster" association. It aims to advance the industry, the technology, and the region's capacities in the industry. The RRPC is primarily a small firm membership organization. Its goal is to secure a seat at the table when relevant policy investments in the region are discussed and decided. In particular, the RRPC seeks a role in decisions about research and development, economic development, and workforce development assets. These are regionally embedded factor inputs that affect the competitiveness of small firms' global supply chains and customer markets. The RRPC engages in policy advocacy, networking, and information sharing at the regional level. The stated purpose is to "foster cooperation" between business, academia, and government (Etzkowitz and Leydesdorff, 1997).

Unlike some other firm networks, the RRPC does not count the large anchor firms in the region among its current membership. At points in the organization's history, large firms did participate. And spin-offs and subsidiaries of large firms do participate (e.g. Corning Tropel). It is worth noting that the RRPC retains its small firm character in spite of being in a region long dominated by established transnational firms like Eastman Kodak, Xerox, Bausch and Lomb, and Corning. Support for the cluster comes from several universities, including the Rochester Institute of Technology and the University of Rochester in addition to Cornell University. Notably, Monroe Community College is a long-time participant and partner. Monroe Community College has long offered programs and training in optics manufacturing – an essential program for small manufacturers.[18]

The RRPC itself is made up of 103 member firms and organizations (see Appendix, Table 16A.2). Forty-seven percent of these firms are manufacturers. Another 11 percent are design and engineering firms related to manufacturing. Another 30 percent of member firms provide "industry support services," broadly characterized as business services such as specialized accounting and legal work (including intellectual property and patent lawyers). The remaining members are primarily government and universities and colleges. The membership of the RRPC does not reflect the total number of optics and photonics firms and jobs (estimated around 20,000 jobs) in the region. Clearly, the largest firms are not participants. However, the RRPC membership – like the membership and active participants in the firm network groups for all four regions – reflects the engagement of small and medium sized firms in their regional economy.

The market for the products of both the small and large firms in the region is geographically diverse. A 2002 survey of the small firms in the RRPC indicated that although 16 percent of their customers are in the region, 39 percent of suppliers are within the cluster. The market for small firms was primarily national (57 percent national, 7 percent within greater New York State, and 16 percent in the region), although almost 20 percent of the customers were located outside the US (Christopherson and Clark, 2007a, 2007b). The small firm network does not function as a captive supplier system for the large firms in the region.

As noted above, the RRPC partners with trade associations and national and international advocacy organizations, including SPIE, the Optoelectronics Industry Development Association, and the American Precision Optics Manufacturers' Association. The RRPC works closely with NY Photonics and the photonics cluster groups in other New York regions on statewide policy issues. When, for example, New York State's Empire State Development Agency began its "high tech research and development" agenda in the late 1990s, photonics made a concerted effort to get on the radar of the Centers of Excellence program.

The RRPC also acts as an institutional intermediary, providing information, access, coordination, and credibility to its members and partners. It also lends credibility to other organizations when it collaborates with other regional actors or trade associations. The cluster group's participation checks the box of small firm participation in the design, development, and implementation of policy. Projects tend to be conferences, strategic planning processes, economic and workforce development initiatives (e.g. the US Department of Labor's Workforce Innovation for Regional Economic Development – WIRED – program), and research and development infrastructure projects (e.g. the New York State Infotonics Center of Excellence).

Research Centers as Anchors in a Distributed Network

The question of institutional spaces and distributed networks is particularly interesting in Rochester, for two reasons. First, the contrast between a firm network dominated by small firms and research spaces governed and administered by universities and public–private partnerships in which large firms play central roles. And second, the proximity to Ontario and Montreal, where distributed networks of cooperative research centers are the rule rather than the exception, shining a bright light on the contrasts between the Canadian and US strategies. In fact, when New York State began its Centers of Excellence program, it was largely modeled on an Ontario program (Aaron, 2003).

Originally, when the Centers of Excellence program began in the early 2000s, the Infotonics Center of Excellence was based at Canandaigua, a suburb in the Rochester region.[19] The center was intended as the linchpin of technology-led economic development policy investment in the region, encouraged by state and local government. The facility itself was a former Xerox building. Large firms in the region, including Kodak and Corning, invested in the project. The economic developers suggested that the project would create 5000 jobs (Mullins, 2002). Centers of Excellence also went into other New York cities, including Albany, Buffalo, Syracuse, Binghamton, and Stony Brook. The most successful has been the Center of Excellence in Nanoelectronics and Nanotechnology in Albany, garnering national attention as President Obama focused on manufacturing and innovation policy.[20] And, in 2010, the Infotonics Center in Canandaigua merged with it, essentially transporting the Rochester area center to the growing Albany complex. By early autumn 2011, Governor Cuomo was announcing a $4 billion investment by IBM, Intel, Samsung, Global Foundries, and the Taiwan Semiconductor Manufacturing Company to invest in research in Albany (Kaplan, 2011).

In terms of university-based research centers, Rochester has the advantage of two established research universities with technical expertise in optics and photonics. The Rochester Institute of Technology engages in a wide range of imaging sciences, including electronic imaging, digital image processing, remote sensing, medical imaging, color science, optics, and chemical imaging. The university lists numerous research units related to photonics, including the Chester F. Carlson Center for Imaging, the Laboratory for Advanced Spectral Sensing, the Laboratory for Astrophysics and Photonics, the Laboratory for Imaging Algorithms and Systems, the Liquid Immersion Nanolithography Laboratory, the Munsell Color Science Laboratory, the Photonics Research Laboratory, the Printing Applications Laboratory, the School of Photographic Arts and Sciences, the School of Print Media, and the Ultrasound Imaging Laboratory. Similarly, the University of Rochester has a list of research units that is almost as long, including the Center for Adaptive Optics, the Center for Visual Science, the Institute of Optics, the Laboratory for Laser Energetics, and the Rochester Center for Brain Imaging.[21] This list of research centers demonstrates extensive basic research capacity, but the extent to which the centers directly support applied research or translational work is much less clear.

PHOTONICS CLUSTERS IN CANADA

In the mid-1990s the US and Canada undertook two different strategies aimed at supporting industry clusters and developing a research centers model compatible with innovative, small firm networks. The US took a decentralized state and local approach consistent with the devolution common for the period. Canada took a different approach (Clark, 2010). Instead of simply devolving science and technology policy, Canada attempted to coordinate federal, provincial, and local policies and resources. At the same time, Canadian policymakers crossed policy and scale boundaries between economic development and science, technology, and innovation policy to pursue a broader economic project. The result was a "distributed network approach" to innovation, commercialization, and production, with research and development resources as a centerpiece.

In Canada, of course, the separation between the policy responsibilities of provincial and federal actors is as complex as in any country characterized by a "loose federalism." This tension between state and local responsibilities is not unfamiliar to many non-Canadians who also live with the ever-shifting balance of policy priorities between national and regional/provincial governments (the US and Italy come to mind). However, the case of Quebec policy is unusual in the North American context because the province is differentiated by proactive efforts in social, cultural, political, and economic policy that are distinct to the province. For example, Quebec takes a role in international affairs that is simultaneously in collaboration with and *independent of* Canada's own international engagements.

In the context of Quebec, this unusual level of international engagement stems from a desire to maintain and promote a distinct cultural identity as a Francophone region within a largely English-speaking North American context. However, the Gerin-Lajoie Doctrine has also provided Quebec with a platform for economic engagement in the broader global economy. This proactive attention to Quebec's role in the world has allowed policymakers to focus attention on innovation and economic development policy within the region and outside it – to the benefit of Quebec industries and firm networks. As a consequence, Quebec's targeting of photonics should not be looked at as equivalent state policy in the US. The role of the province, and its ability to invest in and promote the industry significantly, exceeds that of an average "state."

Canada's National Centers of Excellence (NCE) program began as a partnership between Industry Canada and three other federal agencies: the Natural Sciences and Engineering Research Council, the Canadian Institutes of Health Research, and the Social Sciences and Humanities Research Council.[22] The centers are primarily based in universities but also in research hospitals and similar institutions.[23] This network approach pairs a national network of researchers and research centers with a locally embedded network of firms and industry actors. Thus, the Centers of Excellence are both *embedded* in existing regional industrial clusters and *connected* across Canada to a national scientific network (Globerman, 2006). In general, scientific priorities (articulated through funding of targeted technologies) have been set by the federal government and implemented through the university networks and regional institutions (Salazar and Holbrook, 2007). This is the case with photonics. The Canadian Institute for Photonics Innovation (CIPI), located in Quebec City at Laval University, was part of the NCE system from 1999 to March 2012.[24] CIPI coordinated a networked research program across 20 Canadian universities with over 300 graduate students and 90 researchers, with an annual budget in excess of CDN$4 million.[25] In April 2012, CIPI joined with the Canadian Photonics Consortium (CPC) to form PhotonsCanada or Le Consortium Photonique de l'Industrie Canadienne. Also headquartered in Quebec City and led by Robert Corriveau, PhotonsCanada is an industry-led network of research centers, firms, and agencies focused on the Canadian photonics industry as a whole from a base in Quebec.[26]

This model of multi-scalar innovation policy prioritizes horizontal linkages to other institutions as well as vertical linkages to local, regional, provincial, and federal actors. In other words, the distributed network model matches the multi-scalar and multi-locational strategies of firms oriented towards global markets and distributed supply chains. This gives Quebec a particular policy-making advantage. At the same time, the NCE model recognizes the role of proximity, what Meric Gertler called "being there," in the development, commercialization, and ultimately production in technology-intensive industries (Boschma, 2005; Gertler, 2003). Identifying industry clusters, then, is a crucial step towards absorbing enabling technologies and commercializing them in place.

The core of Canada's photonics industry is located in the corridor running from Toronto through Ottawa towards Montreal and ending in Quebec City. All of these cities have photonics cluster groups. However, Quebec City has the greatest policy and organizational capacity. The provinces of Quebec and Ontario together represent 59 percent of the photonics firms in the country. The definitional issues that obscure the "hard numbers" on the industry – which firms count as photonics and which do not – is significantly less of an issue in Canada than the US. In part this is because the existence of these established cluster groups allows for economic development groups like Invest Quebec simply to ask regional and provincial stakeholders to provide counts of everything from employment to investment to primary markets. Current estimates of the number of photonics firms in Canada range between 370 and 450, and current employment estimates are around 2000.[27]

The National Research Center of Canada (NRC) identifies the Ottawa photonics cluster as one of its leading technology clusters, anchored by the Canadian Photonics Fabrication Centre. The Ottawa cluster counts about 70 firms among its membership. In contrast, the Quebec City cluster has several institutions anchoring the firm network, ranging from basic research to commercialization, including: Defense R&D – Valcartier (DRDC Valcartier), INO, CIPI (now PhotonsCanada), and the Optics, Photonics, and Laser Center at Laval. The Ottawa cluster also participates in the Ontario's Photonics Industry Network and the CPC, reinforcing the activity and influence of the firm network at the regional, provincial, and federal scales. In both the Ottawa and the Quebec City firm networks, basic research investment is complemented by organizations with explicit responsibilities for commercialization and scaling-up production.

The Quebec Photonics Industry

The Quebec clusters in Montreal and Quebec City are particularly interesting examples of the combined effects of innovation and development strategies. This is in part because of the active role that Quebec plays in coordinating economic development and science and technology investment at the federal, provincial, and regional scales. Quebec alone claims 105 firms and 10 major research centers, with 5000 employees attributed to the firms and 1800 attributed to the research centers. It is perhaps not surprising that Montreal has about 3300 photonics jobs and 60 companies. The region, at 3.8 million people, is often ranked at the top of world lists of high-tech job centers. What is surprising, however, is that Quebec City, with a smaller regional population of about 750,000, has 2200 photonics jobs and 33 firms.[28] In terms of institutional research and industry networking, Quebec City is the headquarters for Quebec, and arguably for Canada as a whole. For example, Laval University has trained half the optics and photonics researchers in Canada.[29]

The Quebec Photonic Network itself is made up of 59 member firms and organizations (see Appendix, Table 16A.3). This includes Montreal and Quebec City and represents how firm leaders have chosen to organize the network. Forty-four percent of the members are manufacturers while 15 percent are design and engineering firms directly related to manufacturing. Twenty percent of the membership falls into the industry support services category. This profile looks much like the RRPC in terms of the balance between manufacturers, design and engineers, and industry support services.

Like the photonics cluster in Rochester, the university-based and public research community in the region heavily influences the Quebec City cluster. And, like other photonics clusters, the firms in Quebec City identify themselves with a variety of markets: Defense and Security accounts for 18 percent of firms, Life Sciences for 22 percent, Energy for 9 percent, Information Technology for 11 percent, Science for 9 percent, and Industrial for 30 percent. The Montreal cluster has a very similar distribution, with slightly higher numbers for Science and Life Science and lower numbers in Industrial and Defense. The Defense linkages in Quebec City are likely influenced by the presence of DRDC Valcartier.

The Quebec City and Montreal firm networks also benefit from the technology-specific education and training provided by a large network of universities and technical colleges. Like Rochester, the photonics clusters in Quebec benefit from a reliable stream of future employees with specialized skills produced in the region. For example, seven universities in the province offer higher education from undergraduate to doctorate degrees in related fields (Polytechnique Montréal, E.T.S. Engineering School, Concordia University, Laval University, McGill University, Montréal University, and Shebrooke University). Technical schools in Quebec (CEGEPs) also offer specialized programs in fields related to photonics, including John Abbott College and Cégep André-Laurendeau in Montreal, and Cégep de La Pocatière in La Pocatière (known for its Bombardier plant).[30]

Research Centers as Anchors in Distributed Networks

One difference influencing the performance of US and Canadian firm networks is the level of public investment in applied research and commercialization reaching the regions and the capacities that investment builds over time. In 2011, the public investment (federal and provincial) in photonics-related research exceeded CDN$140 million. The bulk of that investment was directed at commercialization. For Quebec City, that investment translated into CDN$15 million for the INO and CDN$4 million for CIPI that coordinated the photonics-focused National Center of Excellence. This, of course, is not a complete accounting of the research and commercialization resources received by institutions in Quebec City, but it does provide an indication of the level of public funding.[31]

Although Quebec City has significant capacity in basic research, public policies do not stop with initial innovations. What the INO refers to as its "flexible business model" actually encompasses a broad approach to product and process technologies as well as an emphasis on small firms. In its annual reports, INO's customer breakdowns are dominated by small and medium sized enterprises (59 percent in 2011 and 53 percent in 2012). Large businesses accounted for 23 percent in 2011 and 16 percent in 2012 of INO's customer base, with government and universities trailing at 13 percent/25 percent and 5 percent/7 percent respectively. And, again turning to the question of the sustainability of such commercialization-focused centers, the breakdown of INO's overall income falls into thirds: 36 percent Canadian funding, 29 percent Quebec, and 35 percent outside revenues in its 2011 annual report.[32] In 2012, the breakdowns were even less dependent on public research sources: 26 percent Canadian funding, 28 percent Quebec, and 46 percent outside revenues. For a translational research center, INO's levels of outside revenue consistently meet or exceed the loose "industry standard" for sustainability set by the Fraunhofer Institutes in Germany of approximately 30 percent outside revenues. Since 2002, annual revenues have ranged between $US 26.5 million and $US 37.6 million with outside revenues (research and development contracts, sales, and transfer considerations) ranging between $US 9.9 million and $US 17.1 million during the same period.[33] In addition, since its inception in the early 1980s, INO has maintained a perfect record of sustainable spin-offs – all are still operating or were acquired. In other words, INO has a spin-off "death rate" of zero.

In part, this success can be attributed to INO's ability to support firms as they shift from innovation to prototype development to commercialization. INO clearly distinguishes technologies by their "technology readiness levels" (TRLs).[34] TRLs are a standard measure ranging from 1 to 9. The first three digits in the range are often the purview of university research centers – innovations where the applications are still unclear. The mid-range, TRLs 4 to 6, have clearer applications but they are not worked out. In the final range, TRLs 7 to 9, are essentially in the workable prototype phase – ready to produce and try out in the market. INO provides small-scale production facilities for the firms in the TRL 7 to 9 stage so they can produce and sell a new product before they invest in the staff and capital equipment. In other words, INO supports firms as they step from innovation into production. In places that retain their innovation and production capacities over time, there are institutions that deliberately target that transition.

In addition to INO and its accessibility to small and medium sized firms, the Quebec City photonics firm network has access to a wide variety of research, development, commercialization, and production tax credits. Provincial economic development programs are tailored to the needs of small, medium, and large firms. This differentiation means that policymakers are more aware of the variation in firm strategies. It also means that large firm priorities cannot completely dominate economic development policies.[35]

One notable program is the Caisse de dépôt et placement du Québec, which manages Quebec's public pension plans as well as some private pensions and insurance funds. As part of the policy initiatives developed during Quebec's "Quiet Revolution," the Caisse has provided an investment resource for Quebec firms since 1965.[36] Although the Caisse invests globally, it maintains specific programs for Quebec-based enterprises to promote small firms and innovation.[37]

CONCLUSIONS

In these two cases, the regions and the industry are highly invested in a *strong branding narrative* that is historically documented and driven by the role of original anchor firms and institutions. These narratives rationalize the sense of community among regional actors and form the basis for cooperative action. Small firms and large firms see a common set of shared interests (usually in research and development infrastructure and a specialized regional labor market) in spite of many underlying adversarial conditions. The historical narrative also allows the firm network and other regional actors to see the industry over the long haul. Thus, static definitions of product lines and technology specializations are harder to maintain. The boundaries of the firm network are more porous, resilient, and adaptable when the story about the industry is seen across a long arc of economic transitions and technological change.

Across regions with strong small firm networks we see the value of a sustained branding narrative tying the industry into the region. In Rochester, New York, the branding of the city as an optics and imaging headquarters supersedes the fortunes and relocations of many of the large firms – Eastman Kodak, Xerox, Corning, Bausch and Lomb – that form the main characters in the story about the city and its optical specializations (Clark, 2004; McKelvey, 1973). As the technology has developed and changed, so have the firms and the products. But this adaption and resilience is a second layer in a story that claims the brand.

Second, *small firms matter*. A dynamic small firm network is a strong subset of the broader branding narrative in all cases. Although large firms may dominate other aspects of the firm network, the value of the small firm network to the viability of the cluster story as both a political and economic story cannot be discounted. As a consequence, policy supports directed to and inclusive of small firm networks are always part of the development strategy. Of course, the magnitude of these investments may pale in comparison to the big advantages leveraged by larger firms. The point remains, however, that small firms have a seat at the table. In order to build a sustainable firm network, the participation of small firms is crucial. And it is important that small firms are not merely captive suppliers of large anchor firms. In many ways, small firms are the glue binding the industry and its operations in the region. If small firms cannot reach markets independently of the large firms in the region, they lose their ability to function as distinct actors and thus lose a degree of stickiness (Markusen, 1996). The case of the optics and photonics industry in Quebec City illustrates how a firm network functions as the regional development counterpart to strong innovation and commercialization institutions.

Finally, *institutions matter*. Policy expertise in shepherding the formation of firm networks is critical. In most of these cases, such expertise is found in partnerships between regional government agencies and university actors. And the process of building the firm network is neither quick nor a one-off. It is deliberative and ongoing. It looks more like a community development process than a business-school project. It also matters which institution is charged with the maintenance and coordination functions of the firm network once the formation process has succeeded. In all these cases, that entity was not a university or economic development agency but instead a new organizational structure over which the firm network could assert and exert ownership. In the US, there is a path dependent cooperative research center model that has continued to follow a different trajectory from those in other industrialized countries. A desire to minimize public investment maximizes the influence of large firms over research assets and minimizes the utility of these resources for smaller firms, which really need them to develop from designers to producers. The Quebec approach offers some significant contrasts even within the same industrial sector and technology arena.

In the US, policymakers have argued that cluster organizations should be industry driven. In the mid-1990s, the implementation of Michael Porter's "industry cluster" framework as a policy strategy for technology-led economic development resulted in a series of state-level targeted development programs loosely coordinated with industry-led regional organizations (Porter, 2000). In the last ten years the industry-driven cluster organizations have become increasingly integrated with university-based research centers (and consequently US science, technology, and innovation policy more broadly). The Rochester Regional Photonics Cluster is unusual in that it is organized and operated by small and medium sized firms located in the region and plays a role in both regional innovation policy and traditional economic development.

The small firm network in Rochester effectively retains a critical level of design and production capacity even in the wake of substantial downsizing and outsourcing by large anchor firms. And the presence of a community college that routinely concerns itself with producing skilled workers creates a specialized labor market capable of the flexible production required. As the embedded research institutions develop new technologies, the region retains capacity to absorb those innovations. In this sense, the Rochester region is working and has worked for some time. The relatively low levels of venture capital in the region have, in part, kept the focus on commercialization rather than the financialization trend (Clark, 2013).

What is striking in Rochester are the ways in which traditional models of economic development and science and technology policy have failed to build on this small firm network. Instead, the state continues to privilege large firm actors. However, it is abundantly clear that many large firms engage in antiquated business models that include consistent outsourcing and are ultimately non-competitive.[38] Measuring the success of research centers in dollar terms that no small firm or small firm network can match creates a hierarchy of firm participants. This becomes an especially crucial economic development issue for science-based industries like photonics that are driven by specialized skills and a combination of process and product innovation strategies. The tendency toward a contract research model rather than an economic development model stands out as a real disadvantage to small firm networks. In the end, it is difficult to build a *cooperative* research center when the rules are pay-to-play.

NOTES

1. See the following reports: "Harnessing Light 2: Capitalizing on Optical Science Trends and Challenges for Future Research," 2012; "Harnessing Light: Optical Science and Engineering for the 21st Century,"1998.
2. For detailed information see "Photonics Buyer's Guide and Directory" (2012).
3. For further details see "About OSA: What are Optics and Photonics?" (2008).
4. See "The Leverage Effect of Photonics Technologies: The European Perspective" (2010).
5. See "ATP Focused Program: Photonics Manufacturing" (1999).
6. See "Photonics21: A European Technology Platform" (2006), "Towards a Bright Future for Europe: Strategic Research Agenda in Photonics" (2006).
7. The website www.photonicspolicy.org (last accessed October 2014) maintains a regularly updated list of the policy reports available online.
8. See the following: "Annual Technical Report: OITDA: Optoelectronics Industry and Technology Development Association" (2009); "Annual Technical Report: Optoelectronics Industry and Technology Development Association" (2010).
9. See the following: "The Leverage Effect of Photonics Technologies: The European Perspective" (2010); "Photonics in Europe: Economic Impact (2007) European Technology Platform Photonics 21 (EU)" (2007); "Towards a Bright Future for Europe: Strategic Research Agenda in Photonics" (2006); "Towards Green Electronics in Europe: Strategic Research Agenda Organic and Large Area Electronics" (2009); "A Vision for Photonics as a Key Enabling Technology in Europe" (2011).
10. See the following reports: "Photonics: Making Light Work for Canada: A Survey of the CPC (Canadian Photonics Consortium)" (2008); "Quebec's Physics, Engineering and Science Programs 2011" (2012). Note that the Canadian Photonics Consortium and the Canadian Institute for Photonic Innovations (CIPI) merged to become the Canadian Photonic Industry Consortium (CPIC or PhotonCanada) in April 2012.
11. See the SPIE Photonics Cluster List: "SPIE's Photonics Cluster List" (2012).
12. See www.photonicspolicy.org (last accessed October 2014) for a list of photonics clusters.
13. Notably in the late 1990s, Breault was one of the leading advocates of cluster models for building small firm networks. Other advocates included Joe Cortright of Impresa and Stu Rosenfeld of Regional Technology Strategies, Inc.
14. See SPIE: "SPIE's Photonics Cluster List" (2012).
15. There is a definitional challenge in assessing the photonics industry using traditional metrics. Patents and scientific publications are blunt indicators of actual performance of this industry and certainly do not point to the production capacity of any regional cluster. For example, the Fresnel lens was invented in the 1820s but it was not until the advent of computer-controlled milling equipment that the production process changed substantially. Both plastic and glass versions of this lens are found in many products, from solar cells to motion picture projectors. The "invention" here is not new but the applications are expanding all the time, enhancing new products as well as new production processes. There is no clear consensus among analysts or researchers as to what patent categories or citation searches are appropriate for photonics. In part this is because each search is defined differently depending on whether the researcher is more interested in matching the basic research or applied research in the field.
16. See "Metropolitan New Economy Index" (2001) (based on 1999 patent data).
17. See "Fanning the Flames of Economic Progress: Igniting Greater Rochester's Entrepreneurial Economy" (2004).
18. See for additional details: "Rochester Regional Photonics Cluster: Membership List" (2012).
19. The New York State Infotonics Center of Excellence started in September 2004. It is a collaborative effort between Kodak, Corning, Xerox, regional universities and colleges, New York State, and the federal government.
20. See media coverage: Seelye (2012) and Aaron (2004, 2005).
21. See the following: "Optics, Imaging, and Photonics: The Greater Rochester New York Region" (2004).
22. See also for details: "Network of Centers of Excellence Program Guide" (2011).
23. See also "Networks of Centres of Excellence Research-driven partnerships: 17 Networks of Centres of Excellence" (2012): "The Networks of Centres of Excellence (NCE) fosters multi-disciplinary, multi-sectoral partnerships between academia, industry, government and not-for-profit organizations. It supports academic research, the commercialization of products and ideas, and the development of significant Canadian business advantages." For reports on program impacts please see "Networks of Centres of Excellence 20th Anniversary Report" (2008); "Networks of Centres of Excellence Secretariat Annual Reports" (1998–2008).
24. See "Canadian Institute for Photonic Innovation" (2012).
25. Annie Breard, Invest Quebec, Quebec Delegation Atlanta.

26. Background on PhotonsCanada, CIPI, and CPC stems from discussions with Robert Corriveau of PhotonsCanada, Michel Tetu of INO, and Pierre Galarneau of INO.
27. Firm counts and employment ranges come from published reports cited here as well as from materials and interviews with principles in PhotonsCanada, INO, and Invest Quebec.
28. See "Directory of Members 2012–2013" (2012).
29. Much of the data on the Quebec Photonics Industry was provided courtesy of Invest Quebec and the Quebec Delegation Atlanta.
30. CEGEP is an acronym for College of General and Professional Education, or a public college, in Quebec.
31. For a complete list of the recipients of photonics funding see "Quebec's Physics, Engineering and Science Programs 2011" (2012).
32. See "INO Annual Report 2011" (2012).
33. Ibid.
34. See Aeronautics and Space Engineering Board (ASEB) (2008).
35. See "Quebec's Physics, Engineering and Science Programs 2011" (2012).
36. The "Quiet Revolution" refers to the transition of Quebec in the mid-1960s to a "modern" state. This includes Loi 101, the legislation that established the primacy of the French language in the province. The term also encompasses the adoption of modern social welfare policies. Many of these related to the preservation of French culture like education and cultural affairs. However, it also included the expansion and development of related social and economic policies like pensions and health care reforms. Although Quebec is often noted for its distinct culture in the arts and languages, there is evidence of this distinct culture across the policy spectrum, including levels of unionization (approximately 40 percent), a pro-Kyoto climate change approach, and the use of public pension funds as an equity investment fund for Quebec-based enterprises.
37. Details on the current state of Quebec social and welfare policy as well as history are derived from lectures by Anotia Maioni, McGill University; Stephane Roussel, Universite du Quebec a Montreal; and Simon Langlois, Universite Laval.
38. This outsourcing by Eastman Kodak and Xerox manifests as both offshored and onshored subcontracting. For a discussion of the onshored subcontracting strategies within Rochester using for-profit labor market intermediaries see Christopherson and Clark (2007b).

REFERENCES

A Vision for Photonics as a Key Enabling Technology in Europe (2011). Düsseldorf, Germany: Photonics21.

Aaron, Kenneth (2003). New York State's Centers of Excellence Win Praise but No Guarantee of New Jobs. *Knight Ridder Tribune Business News*, 1.

Aaron, Kenneth (2004). Commerce Department Official Tours Albany, N.Y., nanotechnology Center. *Knight Ridder Tribune Business News*, 1.

Aaron, Kenneth (2005). Owner of Albany NanoTech Complex Solicits Proposals for New Building. *Knight Ridder Tribune Business News*, 1.

About OSA: What are Optics and Photonics? (2008). Retrieved September 15, 2008, from http://www.osa.org/aboutosa/opticsandphotonics/default.aspx.

Aeronautics and Space Engineering Board, The National Academies of Science (2008). Appendix D: Definitions of Technology Readiness Levels: *A Constrained Space Exploration Technology Program: A Review of NASA's Exploration Technology Development Program*. Washington, DC: National Academies of Science.

Atkinson, R. and Gottlieb, Paul (2001). The Metropolitan New Economy Index: Benchmarking Economic Transformation in the Nation's Metropolitan Areas. Center for Economic Forecasting and Analysis, FSU.

ATP Focused Program: Photonics Manufacturing (1999). Retrieved June 15, 2010, from http://www.atp.nist.gov/focus/photonics.htm.

Boschma, Ron A. (2005). Proximity and Innovation: A Critical Assessment. *Regional Studies*, 39(1), 61.

Butter, M., Leis, M., Sandtke, M., McLean, M., Lincoln, J. and Wilson, A. (2011). *The Leverage Effect of Photonics Technologies: The European Perspective*. Düsseldorf, Germany: Photonics21.

Canadian Institute for Photonic Innovation (2012). Retrieved June 12, 2012, from http://www.cipi.ulaval.ca/index.php?id=1&no_cache=1.

Christopherson, Susan and Clark, Jennifer (2003). *The Optics, Imaging, and Photonics Industry in Rochester, NY: Building a 21st Century Industry*. Ithaca, NY: Cornell University.

Christopherson, Susan and Clark, Jennifer (2007a). Power in Firm Networks: What it Means for Regional Innovation Systems. *Regional Studies*, 41(9), 1223.

Christopherson, Susan and Clark, Jennifer (2007b). *Remaking Regional Economies: Power, Labor, and Firm Strategies in the Knowledge Economy* (1st ed.). New York: Routledge.
Clark, Jennifer (2010). Coordinating a Conscious Geography: The Role of Research Centers in Multi-scalar Innovation Policy and Economic Development in the US and Canada. *Journal of Technology Transfer*, 35(5), 460–474. doi: 10.1007/s10961–009–9137-z.
Clark, Jennifer (2013). *Working Regions: Reconnecting Innovation and Production in the Knowledge Economy*. London: Routledge.
Clark, Jennifer Joy (2004). *Restructuring the Region: The Evolution of the Optics and imaging Industry in Rochester, New York*. (Ph.D.). Cornell University, United States, New York.
Clarke, Lynne (1991). *Centralizing Optics and Imaging: Monroe County Economic Development Strategy*. Paper presented at the SPIE International Competitiveness and Business Techniques, Rochester, NY, 1992.
Council on Competitiveness (2004). *Fanning the Flames of Economic Progress: Igniting Greater Rochester's Entrepreneurial Economy*. Washington, DC: The Council on Competitiveness.
Etzkowitz, Henry and Leydesdorff, L. A. (1997). *Universities and the Global Knowledge Economy: A Triple Helix of University–Industry–Government Relations*. London; New York: Pinter.
Feldman, Maryann and Lendel, Iryna (2010). Under the Lens: The Geography of Optical Science as an Emerging Industry. *Economic Geography*, 86(2), 141–171.
Gertler, Meric S. (2003). Tacit Knowledge and the Economic Geography of Context, or the Undefinable Tacitness of Being (There). *Journal of Economic Geography*, 3(1), 75.
Globerman, S. (2006). Canada's Regional Innovation Systems: The Science-Based Industries. *Canadian Journal of Political Science*, 39(2), 432.
Greater Rochester Enterprise (2004). *Optics, Imaging, and Photonics: The Greater Rochester New York Region*. Rochester, NY: Greater Rochester Enterprise.
Hendry, Chris and Brown, James (2006). Dynamics of Clustering and Performance in the UK Opto-electronics Industry. *Regional Studies*, 40(7), 707.
Hendry, Chris, Brown, James, and Defillippi, Robert (2000). Regional Clustering of High Technology-Based Firms: Opto-electronics in Three Countries. *Regional Studies*, 34(2), 129.
Institut National d'Optique (2012). *INO Annual Report 2011*. Quebec City: Institut National d'Optique.
Kaplan, Thomas (2011, September 27). High-Tech Companies to Invest $4 Billion in New York State, Cuomo Says, *The New York Times*.
Karlsson, Charlie (ed.) (2008). *Handbook of Research in Innovation and Clusters: Cases and Policies*. Cheltenham and Northampton, MA: Edward Elgar.
Lovering, John (1999). Theory Led by Policy: The Inadequacies of the "New Regionalism" (Illustrated from the Case of Wales). *International Journal of Urban and Regional Research*, 23(2), 379.
Markusen, Ann (1996). Sticky Places in Slippery Space: A Typology of Industrial Districts. *Economic Geography*, 72(3), 293.
Martin, Ron and Sunley, Peter (2003). Deconstructing Clusters: Chaotic Concept or Policy Panacea? *Journal of Economic Geography*, 3(1), 5–35. doi: 10.1093/jeg/3.1.5.
McKelvey, Blake (1973). *Rochester on the Genesee: The Growth of a City* (1st ed.). Syracuse, NY: Syracuse University Press.
Miao, Tian and Hall, Peter (2012). Optical Illusion? The Growth and Development of the Optics Valley of China. *Environment and Planning C: Government and Policy*, 32, 863–879.
Mullins, Richard (2002, November 8). Center of Excellence Closes on Canandaigua Building. *Rochester Democrat and Chronicle*.
Munnich, Lee, Love, Patricia, and Clark, Jennifer (1999). *Industry Clusters: An Economic Development Strategy for Minnesota*. Minneapolis, MN: State and Local Policy Program, Humphrey Institute of Public Affairs, University of Minnesota.
Munnich, Lee, Sheldon, Tim, and Clark, Jennifer (1999). *Partnering with Industry to Build Minnesota's 21st Century Workforce: Industry Clusters Assessment Project*. Minnesota State Colleges and Universities System's Targeted Industry Partnership. Minneapolis, MN: State and Local Policy Program, Humphrey Institute of Public Affairs, University of Minnesota.
National Academies of Science (1998). *Harnessing Light: Optical Science and Engineering for the 21st Century*. Washington, DC: The National Academies of Science.
National Academies of Science (2012). *Harnessing Light 2: Capitalizing on Optical Science Trends and Challenges for Future Research*. Washington, DC: The National Academies of Science.
Network of Centres of Excellence Program Guide (2011). Retrieved July 1, 2011, from http://www.nce-rce.gc.ca/ReportsPublications-RapportsPublications/NCE-RCE/ProgramGuide-GuideProgramme_eng.asp.
Networks of Centres of Excellence (2008). *20th Anniversary Report*. Ottawa: Network of Centers of Excellence, Government of Canada.
Networks of Centres of Excellence Research-driven Partnerships (2012). 17 Networks of Centres of Excellence,

Ottawa, Government of Canada. Retrieved July 21, 2012, from http://www.nce-rce.gc.ca/NetworksCentres-CentresReseaux/NCE-RCE_eng.asp.
Networks of Centres of Excellence (1998–2008). *Secretariat Annual Reports*. Ottawa: Network of Centres of Excellence, Government of Canada.
Optoelectronic Industry and Technology Development Association, OITDA (2010). *Annual Technical Report 2009*. Tokyo, Japan: Sekiguchi, Bunkyo.
Optoelectronic Industry and Technology Development Association, OITDA (2011). *Annual Technical Report 2010*. Tokyo, Japan: Sekiguchi, Bunkyo.
Photonics21 (2007). *Photonics in Europe: The Economic Impact*. Düsseldorf, Germany: Photonics21. Retrieved from http://www.photonics21.org/download/Brosch_Photonics_Europe.pdf.
Photonics Buyer's Guide and Directory (2012). Photonics.com. Retrieved June 11, 2012, from http://www.photonics.com/BuyersGuide.aspx.
Photonics: Making Light Work for Canada (2008). Canadian Photonics Consortium.
Porter, Michael E. (2000). Location, Competition, and Economic Development: Local Clusters in a Global Economy. *Economic Development Quarterly*, 14(1), 15.
Quebec Photonics Network (2012). *Directory of Members 2012–2013*. Quebec City: Quebec Photonics Network.
Quebec's Physics, Engineering and Science Programs 2011 (2012). Quebec City: Quebec Photonics Network.
Rochester Regional Photonics Cluster: Membership List (2012). Retrieved from http://www.rrpc-ny.org/membership_list.asp.
Salazar, Monica and Holbrook, Adam (2007). Canadian Science, Technology and Innovation Policy: The Product of Regional Networking? *Regional Studies*, *41*(8), 1129.
Seelye, Katharine Q. (2012, May 8). At Albany Campus, Obama Promotes Manufacturing Success. *The New York Times*. Retrieved from http://thecaucus.blogs.nytimes.com/2012/05/08/at-albany-campus-obama-touts-manufacturing-success/.
SPIE's Photonics Cluster List (2012). Retrieved December 2012 from http://spie.org/x22931.xml.
Sternberg, Ernest (1992). *Photonic Technology and Industrial Policy*. Albany, NY: State University of New York Press.
Towards a Bright Future for Europe: Strategic Research Agenda in Photonics (2006). Düsseldorf Germany: Photonics21.
Towards Green Electronics in Europe: Strategic Research Agenda Organic and Large Area Electronics (2009). Düsseldorf, Germany: Photonics21.
Waits, Mary Jo and Vandegrift, Judith A. (1999). The Achilles Heel of Future Economic Growth: The Workforce Development Challenge: What Is It? How Critical Is It? Who's Doing What To Address It? *Economic Development Review*, 16(3), 63.

APPENDIX

Table 16A.1 Top five photonics industry applications for the top ranked photonics regions and the US, 2011

	San Jose	Los Angeles	Boston	Metro DC	San Francisco	Rochester	New York	Tucson	Orlando	Dallas	USA
1	Semi-conductor Manu-facturing	Defense, Security, Law Enforce-ment	Defense, Security, Law Enforce-ment	Defense, Security, Law Enforce-ment	**Basic Research, Science**	Optics Manuf-acturing	Biomedical, Medical Imaging, Health Care	Astronomy	Defense, Security, Law Enforcement	Defense, Security, Law Enforcement	Defense, Security, Law Enforcement
2	Biomedical, Medical Imaging, Health Care	Biomedical, Medical Imaging, Health Care	Biomedical, Medical Imaging, Health Care	Astronomy	Defense, Security, Law Enforce-ment	Defense, Security, Law Enforce-ment	Semi-conductor Manu-facturing	Defense, Security, Law Enforce-ment	Laser Engineering and Equipment	Semi-conductor Manu-facturing	Biomedical, Medical Imaging, Health Care
3	Consumer Electronics	Astronomy	**Basic Research, Science**	**Basic Research, Science**	Biomedical, Medical Imaging, Health Care	Biomedical, Medical Imaging, Health Care	**Basic Research, Science**	Biomedical, Medical Imaging, Health Care	**Basic Research, Science**	Biomedical, Medical Imaging, Health Care	Semi-conductor Manu-facturing
4	Materials Processing, Lasers in Manu-facturing	**Basic Research, Science**	Semi-conductor Manu-facturing	Earth Sciences, Environ-mental Monitoring, Climate	Other	Other	Defense, Security, Law Enforce-ment	**Basic Research, Science**	Other	Consumer Electronics	**Basic Research, Science**
5	Communi-cations and Networking	Earth Sciences, Environ-mental Monitoring, Climate	Astronomy	Biomedical, Medical Imaging, Health Care	Semi-conductor Manu-facturing	**Basic Research, Science**	Optics Design and Engineering	Optics Design and Engineering	Education and Training	**Basic Research, Science**	Astronomy

Table 16A.2 Members of the Rochester regional photonics cluster, 2012

Name of Member Company or Organization	Type of Organization
Center for Integrated Manufacturing at Rochester Institute of Technology	Academic
Cornell University Center for Materials Research	Academic
Institute of Optics, University of Rochester	Academic
Monroe Community College	Academic
University of Rochester, Laboratory for Laser Energetics	Academic
Arjae Spectral Enterprises, Inc.	Design and engineering
ASE Optics, Inc.	Design and engineering
FluoroLogic	Design and engineering
Imaging Solutions Group	Design and engineering
Lightforce Technology, Inc.	Design and engineering
Moondog Optics, Inc.	Design and engineering
Munro Design & Technologies, LLC	Design and engineering
OES Rochester	Design and engineering
Ontario Technical Design, Inc.	Design and engineering
Oren Sage Technology, LLC	Design and engineering
SMA Optical Technologies, Inc.	Design and engineering
Canadian Consulate General	Government
City of Rochester	Government
Monroe County Dev. Corp.	Government
Air-Craft Environmental Systems, Inc.	Industry Support Services
Bansbach Zoghlin, P.C.	Industry Support Services
Bond, Schoeneck & King, PLLC	Industry Support Services
BrandOrchard	Industry Support Services
Burns Personnel	Industry Support Services
Citizens Bank NA	Industry Support Services
ComTec Solutions	Industry Support Services
Concepts To Revenue, LLC	Industry Support Services
Davie Kaplan	Industry Support Services
EFP Group	Industry Support Services
First Niagara Risk Management, Inc.	Industry Support Services
Harris Beach	Industry Support Services
Jaeckle Fleischmann & Mugel, LLP	Industry Support Services
JK Consulting	Industry Support Services
John S. Herbrand, Attorney	Industry Support Services
M&T Bank	Industry Support Services
Merrill Lynch	Industry Support Services
Mildex, Inc.	Industry Support Services
O'Brien Associates	Industry Support Services
OpticsProfessionals, LLC	Industry Support Services
Optima Benefits Group	Industry Support Services
Patent Innovations LLC	Industry Support Services
PLS/Launch Solutions	Industry Support Services
R. Kraft, Inc.	Industry Support Services
R.A. Smythe, LLC	Industry Support Services
Resource Dynamics Co.	Industry Support Services

Table 16A.2 (continued)

Name of Member Company or Organization	Type of Organization
Rochester Gas & Electric Corp.	Industry Support Services
Seko Logistics	Industry Support Services
Smart System Technology & Commercialization Ctr.	Industry Support Services
Trillium Group, LLC	Industry Support Services
Underberg & Kessler LLP	Industry Support Services
AccuCoat Inc. – Coatings for Optics	Manufacturing
Advanced Glass Industries	Manufacturing
Amplification Technologies, Inc.	Manufacturing
ANGSTROM Precision Optics	Manufacturing
Apollo Optical Systems, Inc.	Manufacturing
Applied Image Group Inc.	Manufacturing
Bridgekey Corporation	Manufacturing
Brinkman Precision, Inc.	Manufacturing
Bristol Instruments, Inc.	Manufacturing
Corning Tropel Corp	Manufacturing
CTM Corporation	Manufacturing
CVI / Melles Griot Optics Group	Manufacturing
EMF Corp.	Manufacturing
G-S PLASTIC OPTICS	Manufacturing
Geospatial Systems, Inc.	Manufacturing
Glass Fab, Inc.	Manufacturing
Heraeus Quartz America LLC	Manufacturing
ITT SSD Precision Optics	Manufacturing
Jackson Welding Supply	Manufacturing
JENOPTIK Optical Systems, Inc.	Manufacturing
JML Optical Industries, Inc.	Manufacturing
Lasermax	Manufacturing
Lumetrics	Manufacturing
Metrology Concepts	Manufacturing
Micropen	Manufacturing
Navitar Inc.	Manufacturing
New Scale Technologies, Inc.	Manufacturing
Optics Technology, Inc.	Manufacturing
Optimax Systems Inc.	Manufacturing
OptiPro Systems	Manufacturing
Photon Gear, Inc.	Manufacturing
Precision Laser Technology	Manufacturing
Precision Optical Imaging	Manufacturing
QED Technologies International	Manufacturing
Qioptiq Imaging Solutions Inc. (Formerly Thales Optem)	Manufacturing
Reflexite Display Optics	Manufacturing
Rochester Precision Optics	Manufacturing
Semrock, Inc.	Manufacturing
Stamper Technology, Inc.	Manufacturing
Swift Glass Company, Inc.	Manufacturing
Sydor Optics, Inc.	Manufacturing

Table 16A.2 (continued)

Name of Member Company or Organization	Type of Organization
Syntec Optics	Manufacturing
TOPTICA Photonics Inc.	Manufacturing
Universal Photonics, Inc.	Manufacturing
Vacuum Innovations LLC	Manufacturing
Vincent Associates	Manufacturing
Vuzix	Manufacturing
GG&C Imaging	N/C
Infinite Optics	N/C
Laird Industries, Inc.	N/C
MiniOptics	N/C
Quantum Technology Associates	N/C

Table 16A.3 Member list of the Quebec photonics cluster

Name of Organization or Firm Member of the Photonics Cluster	Type of Organization/Firm
Cégep André Laurendeau (M)	Academic
Centre for Optics, Photonics and Lasers (QC)	Academic
Centre de robotique et de vision industrielles (QC)	Academic
École de technologie supérieure de Montréal (M)	Academic
Institut national de la recherche scientifique, Énergie Matériaux Télécommunications (Energy Materials Telecommunications) Research Centre (M)	Academic
GiGa Concept Inc. (M)	Design and Engineering
Lacoptic (QC)	Design and Engineering
Leddartech (QC)	Design and Engineering
OMP Inc. (QC)	Design and Engineering
Phytronix Technologies (QC)	Design and Engineering
QPS Photronics (M)	Design and Engineering
Side by Side Optics (QC)	Design and Engineering
Solart Group (M)	Design and Engineering
TeraXion (QC)	Design and Engineering
Defence R&D Canada – Valcartier (QC)	Government
Développement économique LaSalle (M)	Government
National Research Council Industrial Research Assistance Program (M)	Government
Canadian Photonic Industry Consortium (QC)	Industry support organization
Centre de recherche industrielle du Québec (QC)	Industry support organization
Centre de recherche informatique de Montréal (M)	Industry support organization
INO (National Institute of Optics) (QC)	Industry support organization
DigiPlan Tec Inc. (M)	Industry support services
Gestion Sovar (QC)	Industry support services
iFact Consulting (M)	Industry support services
Institut de recherche ¥•g .. ¥·–Québec (M)	Industry support services

Table 16A.3 (continued)

Name of Organization or Firm Member of the Photonics Cluster	Type of Organization/Firm
Norton Rose (QC)	Industry support services
Optech Montréal (M)	Industry support services
R & D Capital (M)	Industry support services
ROBIC L.L.P. (QC)	Industry support services
Allied Scientific Pro	Industry support services (logistics)
Averna Technologies (M)	Industry support services (testing)
EXFO (QC)	Industry support services (testing)
Polyrix Inc. (QC)	Industry support services (testing)
3SPGroup	Manufacturer
ABB Inc. (QC)	Manufacturer
CMC Electronics (Avionics) (M)	Manufacturer
CorActive Hightech Inc. (QC)	Manufacturer
Creaform (QC)	Manufacturer
Doric Lenses Inc. (QC)	Manufacturer
ESI Pyrophotonics (M)	Manufacturer
Fibertech Optica (M)	Manufacturer
Fiso Technologies (QC)	Manufacturer
Genia Photonics (QC)	Manufacturer
Gentec-EO (QC)	Manufacturer
IRphotonics (M)	Manufacturer
Laser Components Canada Inc. (M)	Manufacturer
MPB Communications (M)	Manufacturer
Novacam Technologies Inc. (M)	Manufacturer
Nüvü Caméras Inc. (M)	Manufacturer
Obzerv Technologies Inc. (QC)	Manufacturer
O/E Land Inc. (M)	Manufacturer
O-M6 Technologies (M)	Manufacturer
Opsens (QC)	Manufacturer
Opsun Technologies (QC)	Manufacturer
Photon etc. Inc. (M)	Manufacturer
RDG2 Technologies (M)	Manufacturer
TechMed 3D (QC)	Manufacturer
Telops (QC)	Manufacturer
EH2 Solar	N/C

17. Traditional and emerging markets in the global steel supply chain

Carey Durkin Treado

INTRODUCTION

Over the last 15 years, global steel consumption has more than doubled from 765 million metric tons in 1998 to 1.584 billion metric tons in 2013.[1] Not surprisingly, most of this increase was due to rapid industrialization in China and, to a lesser extent, India. These nations have been rapidly expanding their infrastructure and industrial output, driving up both their demand for steel and their productive capacity.

In combination with technological change, the forces of globalization have significantly transformed the industrial organization of manufacturing, as illustrated in the other case studies in this handbook. The steel industry is no exception. Rapid changes in the technology of production and the size of global markets have restructured both the supply chain that supports the production of steel and the location and ownership of the major players within the steel industry.

A common concern about this transformation is that it reduces the competitiveness of traditional steel-making regions. Over the past 30 years, the U.S. steel industry has experienced changes to productive capacity that have been devastating for particular cities and regions. Despite these losses, the United States continues to participate in and benefit from the current formation of the steel global value chain. Even those regions that typified the most devastating losses, such as Pittsburgh, Pennsylvania, continue to have a role in the global value chain. In this chapter, we will discuss how the twin forces of technological change and global market expansion have contributed to the de-verticalization of the steel industry, the rise of emerging market players, and the transformation of the role of traditional steel-making regions.

DEFINING DE-VERTICALIZATION

The image of the large, vertically integrated firm became standard in the later half of the 20th century with the rise of the "Fordist" production model. However, economic theory does not support this model of the firm as either inevitable or unchanging. Coase (1937) was one of the first scholars to identify the forces that would define whether a firm would be vertically integrated or not. According to Coase, the key determining factor in firm size was the trade-off between the organizational costs of internally producing inputs and the transaction costs of buying inputs (the "price mechanism"). Coase did not view the choice between "making" and "buying" as a static one, but rather one that would respond to changing conditions. In particular, he identified new technology as a source of changing cost structures and thus changing degrees of vertical integration.

Technology that reduced the cost of management, such as the telephone in Coase's time, would tend to increase vertical integration. Technological developments that reduced transaction costs would tend to favor de-verticalization and outsourcing of production. Certainly, advances in electronic communication, including the Internet, have led to significant reductions in the cost of transacting business between firms.

Several more recent studies of the make–buy decision have highlighted reductions in transaction costs as a motivating factor in the rise of outsourcing (McLaren 2000). The reduction in transaction costs occurs through the "thickening" of markets (Grossman and Helpman 2002), which increases the supplier opportunities for a purchasing firm and thus reduces the search and contracting costs associated with buying. As search costs decrease relative to the cost of governance, buying (or "outsourcing") becomes more cost effective than making. Grossman and Helpman (2002) point out that increases in both the number of suppliers and purchasers improve the potential for a successful match and lead to increasing returns to the search process. They conclude (p. 111) that "an industry equilibrium with pervasive outsourcing is more likely to exist, the larger is the industry and the larger is the aggregate economy." The expansion of geographic horizons outside domestic markets as well as the emergence of rapidly industrializing nations have contributed to significant market thickening over the past 20 years.

Industrial organization observers have also found that the de-verticalization process is driven by the pace of technological change (Herrigel and Zeitlin 2010). Large firms find it more cost effective to outsource technology to specialist suppliers rather than attempting to stay current on all necessary technology simultaneously. As a result, collaboration between customers and suppliers becomes part of both product and process developments. According to these theories of firm structure and de-verticalization, the development of an intermediate supplier cluster is enabled by technological change and downstream market expansion. These economic forces, which we will see were impacting the steel industry throughout the close of the 20th century, encourage outsourcing and thus strengthen the supplier base.

TECHNOLOGICAL CHANGE IN U.S. STEEL PRODUCTION

In addition to the ubiquitous changes initiated by the digital age, the steel industry has experienced industry-specific technological developments. Technological advances in the steel-making process have led to a decline in the man-hours required to make a ton of steel from over 12 hours in 1970 to just over 2 hours today, representing a decline in required labor of 83 percent. This radical improvement in the productivity of labor has led to a corresponding decline in industry employment. The U.S. steel industry employed more than 500,000 workers to ship 91 million tons of steel in 1970, but was able to ship almost 92 million tons of steel in 2011 with only 95,000 workers (American Iron and Steel Institute (AISI) 2013). Thus, although the size of the industry has been fairly stable when measured by shipments, the steel industry has experienced one of the steepest declines of any industry in the United States (Collard-Wexler and De Loecker 2013). The combination of declining overall employment and local mill closures has often been misconstrued as an indicator of the waning competitiveness of U.S. steel. However, these trends are largely driven by technological changes that radically improved labor productivity rates and transformed the material input requirements of the steel-making process (Treado 2004; Collard-Wexler and De Loecker 2013).

Part of these productivity gains is due to technological advances within traditional steel mills that integrate the process of making pig iron from iron ore with the process of making steel from pig iron (Giarratani, Gruver and Jackson 2007). Mills that use this process are referred to as "integrated" mills and require large scale capacity for efficient production. The average capacity of an integrated mill is over 3 million tons, with the largest mill (U.S. Steel in Gary, Indiana) producing nearly 8 million tons annually. Additional productivity gains were also made by transitioning significant amounts of capacity from integrated mills to smaller, less labor intensive, "minimills" that rely on recycled scrap steel and continuous casting technology. True to their name, minimills can operate efficiently with a much smaller productive capacity. The average capacity of a minimill is 875,000 tons, although minimill capacity ranges from 250,000 to 2.5 million tons.[2] Their smaller scale translates into lower start-up costs, a smaller installed base, and fewer labor requirements than the large integrated mills. Minimills use less than one-half of a labor-hour to produce a ton of steel, in comparison with nearly three labor-hours per ton of steel at an integrated mill.

Most U.S. integrated mills were built before 1950 and none have been built in the United States after 1979. All of the steel mills built in the United States during the 1980s and 1990s were minimills and many were opened in regions without any tradition of steel production. In the early 1990s, technological advances in thin-slab casting, ladle metallurgy, and rolling mills allowed minimills to produce flat-rolled steel as well as the long products that had been their previous focus. Flat-rolled steel is typically used in applications with more stringent quality requirements, such as autos and appliances, while long products (such as rails and bars) are often supplied to construction industries. By 2001, minimills represented over 90 percent of U.S. capacity in long products and nearly one-third of the U.S. capacity in flat-rolled steel.[3]

As these new mills were opening, the steel industry was also shaken by a series of bankruptcy filings and mill closures (Treado 2004). By 2001, over 30 million tons of steel-making capacity was at bankrupt or closed mills – all but one of which were integrated mills. Pointing to dramatic rises in imported steel quantities that were coupled with excess capacity in European and Japanese mills, the U.S. steel industry successfully argued for Section 201 trade protection and relief in October 2001 (U.S. International Trade Commission 2001). However, empirical analysis by both Treado (2004) and Collard-Wexler and De Loecker (2013) provides evidence that minimill competition played a more critical role than trade in the closure of integrated mills.

Through a rigorous analysis of plant-level data in the steel industry, Collard-Wexler and De Loecker (2013) find that 87 percent of total productivity growth in steel production can be attributed to minimill technological advances. Minimill production raised productivity rates both by replacing older integrated production and by causing a shift within the integrated sector from the least efficient plants, which closed, to more efficient plants. Collard-Wexler and De Loecker note that the surviving integrated plants were able to match the productivity rates of the minimill plants by 2002.

The economic impact of the changes in the structure of the U.S. steel industry was exacerbated by significant changes in the location of production within the United States. In addition to lowering the labor requirements of steel-making, minimill production technology changed the location requirements of steel production by freeing the mills from the necessity of locating near iron ore. Prior to 1980, the majority of steel production in the United States was located in a band that stretched from Chicago to Pittsburgh. This location afforded both water access to the iron ore concentration of the upper Great Lake region and close proximity to the bituminous coal deposits in the Appalachian Mountains (Treado 2004).

With more flexible location requirements, the newly installed mills were able to take advantage of southern locations that offered proximity to new auto plants and regional construction booms as well as low unionization rates. In 1972, the Chicago–Pittsburgh states of Illinois, Indiana, Michigan, Ohio, and Pennsylvania produced over 70 percent of U.S. steel. By 2003, this share had fallen to just over half of U.S. steel production. Meanwhile, the share of steel production in southern states rose from less than 10 percent in 1972 to over one-third in 2003.[4]

DE-VERTICALIZATION OF THE STEEL INDUSTRY

The restructuring of steel production also led to profound changes in the role of steel suppliers. The structure of a supply chain is dependent on the flow of technological knowledge along that chain. In the case of component supply chains, the technological complexity of the product increases along the chain, so that intermediate goods (pistons and engines) are less complex than the final product (automobiles). In the case of process supply chains, such as the steel industry, the flow of technological complexity is often reversed, so that intermediate goods (electronic controllers and rolling machines) can be more complex than the final product (raw steel).

The organizational charts in Figures 17.1 and 17.2 present a simplified illustration of the de-verticalization process in the steel industry. Figure 17.1 illustrates the structure of the vertically integrated steel industry of the 1960s and 1970s, in which suppliers offered steel-making equipment, parts, or raw materials – generally as stand-alone inputs to the steel producers, which were most often located within the same region as the suppliers. Most of the advanced coordination, technical planning, repair, and services associated with these inputs were handled internally by the steel producers, as shown by their placement within the box containing steel mill production. As aspects of mill production moved from "make" to "buy," many of the activities previously performed by mill employees were transferred to outside vendors, which coordinated their activities and output with each other as well as with the steel producers, as illustrated by Figure 17.2. At the same time, the installation of technologically advanced mills at greenfield sites increased demand for a range of goods and services from steel suppliers (Giarratani, Gruver, and Jackson 2007).

By outsourcing stages of production to supplier firms, the de-verticalization process transformed the suppliers as well as their role in the supply chain. Arm's-length transactions of finished products and a few services were replaced by collaborative efforts to provide customized products, new technology, and an expanded array of services. Tasks that had been performed by employees of the steel firms, such as repair, quality control, and design, became services provided by suppliers. Between 1977 and 1997, U.S. steel mills more than doubled their expenditures on contract work, increasing its share of

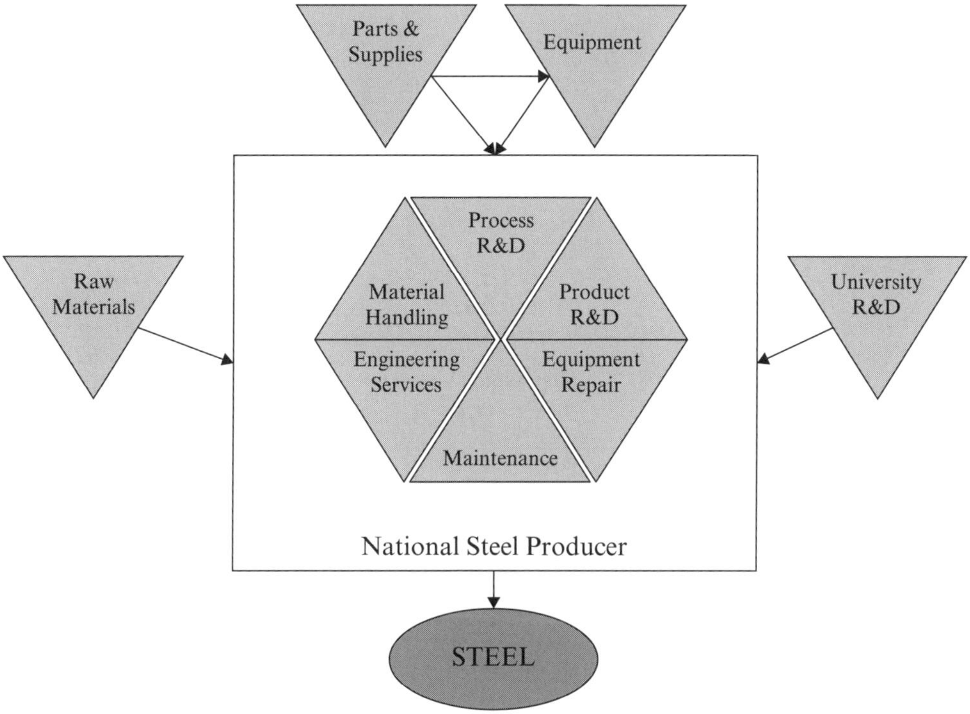

Source: Author.

Figure 17.1 Vertically integrated, domestically based steel production

value-added from 3.6 to 7.7 percent. At the fully integrated mills, the rise in expenditure on contract work was even more pronounced, rising from 2 to 12 percent of value-added. Contract work as a share of wages rose from 8 percent in 1972 to 30 percent in 2002.[5]

In many cases, these services were provided by the former employees of the steel firms themselves. This industrial reorganization released a flow of resources from the steel firm into the intermediate supplier base. The influx of skilled labor enabled many intermediate suppliers to transition from being a commodity supplier to being a more technologically sophisticated supplier of a variety of goods and services (Treado 2009). The result was a significant shift of knowledge from inside the steel firm to outside the steel firm, enabling the inter-firm fragmentation of steel technology and steel production. Once these aspects of the supply chain were separated by firm, it became easier to separate them geographically and it was no longer necessary for the upstream, technical know-how of steel-making to be co-located with the steel-making itself.

The forces of de-verticalization have thus had a profound impact on the role of traditional steel regions and emerging markets in the global steel value chain. As the following sections will detail, de-verticalization of steel technology from steel production has led to a global steel industry in which technology suppliers remain centered in traditional steel production regions, even as the center of steel production has relocated to emerging markets.

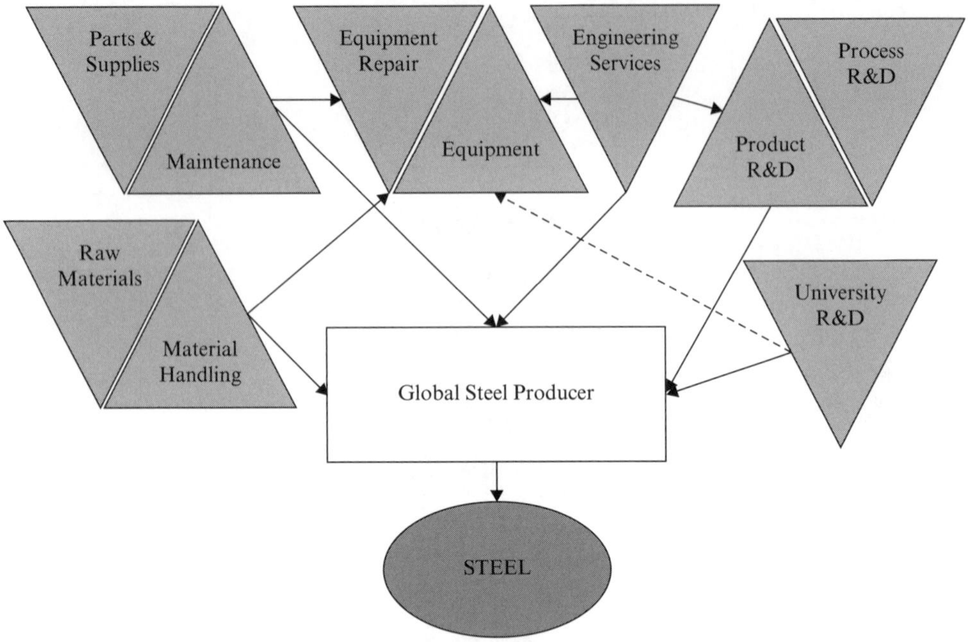

Source: Author.

Figure 17.2 De-verticalized global steel supply chain

EMERGING STEEL MARKETS AND STEEL PRODUCTION

The shifting center of steel production is a direct result of rapid industrialization rates in emerging markets, which have created a growing need for steel and a corresponding increase in per capita consumption rates (Kumar and Chadha 2009). China's per capita consumption grew nearly 150 percent between 2001 and 2007, approaching North American and European consumption rates.[6] Not surprisingly, most of the growth in new steel mills has occurred in emerging markets. The Chinese steel industry tripled output and doubled the number of firms between 2000 and 2006 (Brandt and Rawski 2008). In 2011, the Ministry of Steel of the Government of India announced plans to build or expand 222 mills in India, increasing capacity by 276 million tons over a five year period. In comparison, with the exception of the minimills built during the 1990s, new steel mills are no longer being regularly constructed in Europe, Japan, or the United States. In a stark example of the global relocation of steel production, a steel mill in Germany was recently disassembled and reassembled in China to support rapidly expanding demand (Hudson and Swanton 2012).

At the close of the 20th century, the steel industry exhibited relatively low measures of concentration among manufacturing industries, particularly those that relied on economies of scale in production (Giarratani, Madhavan, and Gruver 2013). A contributing factor to the dispersed ownership structure was that steel production was often considered to be strategically important to domestic security and economic stability. In many nations, the steel industry was subject to government controls or even ownership, contributing to the proliferation of nation-specific steel producers. In 1990, the largest ten steel producers accounted for only 20 percent of global steel production, and the share of the largest single producer (Nippon Steel of Japan) was only 4 percent.[7]

Several economic events led to an unprecedented opportunity for consolidation. During the 1990s, countries with state-owned steel firms began to privatize their industries, creating firms that were facing the challenges of the free market for the first time. Usinor-Salicor of France was privatized in 1995; ENSIDESA of Spain was reorganized and privatized between 1994 and 1997; and POSCO of South Korea was privatized in 2000. In emerging markets, like South Korea, state-owned firms were often privatized during a period of growing steel demand. In traditional steel markets, like Europe, the newly privatized firms faced stagnant or declining demand and often struggled to compete successfully. In the United States, industrial restructuring and rising import levels led to numerous bankruptcy filings by integrated steel producers between 1998 and 2001.

The combination of rapid growth of steel demand in emerging markets, the privatization of government-owned steel industries globally, and the failing financial health of steel mills in traditional steel areas created an opportunity for consolidation . Giarratani, Madhavan, and Gruver (2013) have tracked the resulting merger and acquisition activity in the period between 2003 and 2011. They identified over 1000 cross-border deals during this period and 500 domestic deals, with a total peak value in 2006 of $78.6 billion. In one of the starker examples of globalization, all of the Canadian steel producers were acquired by non-Canadian firms between 2006 and 2008 (Warrian 2010).

Although much of the restructuring during this period involved consolidation of steel firms within the traditional steel regions of North America, Europe, and Japan, one of the most intriguing developments has been the rise of the multinational steel firms based in the emerging market regions. Seven of the ten firms that acquired Canadian steel producers were from emerging markets in South America, South Africa, Russia, and India (Warrian 2010). One of the more significant deals was the 2007 acquisition of Corus Steel, which was formed by the privatization of British Steel in the 1980s, by Tata Steel of India in the largest Indian acquisition of a foreign firm to date.[8]

By 2010, the ownership of the largest steel producers had shifted decisively away from the traditional steel-making regions of the United States, Europe, and Japan and toward producers from emerging markets, as illustrated in Figure 17.3. In 1982 South Korea became the first steel producer from an emerging market to crack the top ten, with Pohang Iron and Steel Company (later POSCO). Over the next 20 years, there was a significant decline in the role of companies from traditional steel-making regions of the United States, Europe, and Japan, and a corresponding increase in the relative contribution and total market size of producers from the emerging markets of China, India, Brazil, and South Korea.

Currently, the largest single contributor to the top ten producer list is Arcelor Mittal, a firm that cannot easily be claimed by any one nation. It was formed in 2006 from the merger of Arcelor and Mittal Steel. Arcelor resulted from the consolidation of three European companies in 2002 and was the second largest steel company in the world at the time of the merger. Mittal Steel was created from a consolidation of Lakshmi Mittal's steel ventures, which began with an Indonesian greenfield plant called PT Ispat Indo in 1976. Ispat established itself in Trinidad and Tobago in 1989 and began to acquire steel

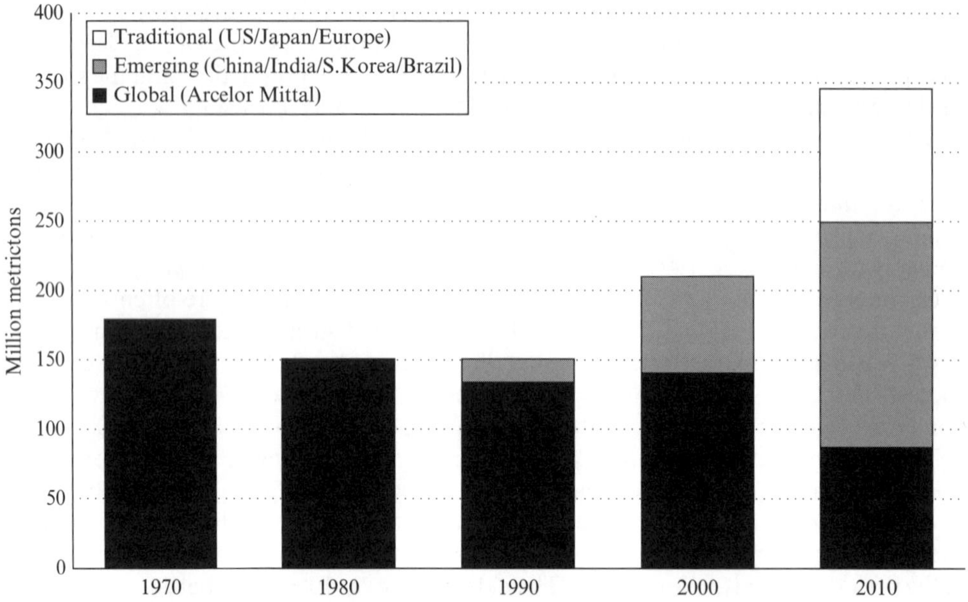

Source: Author's presentation of data from the World Steel Association.

Figure 17.3 Top ten steel-producing companies by region

mills around the world. Ispat International was established in the United Kingdom in 1995 and merged with the LMN Group (another steel venture from Lakshmi Mittal) and the International Steel Group of North America in 2004.[9] In 2010, Arcelor Mittal facilities produced over 98 million tons of steel, representing almost triple the amount of steel produced by the former leader, Nippon Steel.[10]

More so than any other steel company, Arcelor Mittal embodies the globalization of the steel industry. Not only did Arcelor Mittal move the steel industry toward a more efficient scale of operations, but the impetus for the restructuring arose from the developing world. While Mittal Steel may have been the most ambitious of emerging market firms to acquire steel industry assets in the developed world, it was certainly not alone in this business strategy. Gerdau of Brazil acquired Ameristeel in 1999.[11] The iconic Rouge Steel of Michigan was acquired by Severstal Steel of Russia in 2004. Sahaviriya Steel Industries Plc of Thailand purchased Teesside Cast Products from the center of England's steelmaking region in 2011 in order to secure a source of steel slabs for its plants in Thailand.[12]

According to the United Nations' World Investment Report of 2006, the frequency of international investments made by emerging-market multinationals has increased significantly since 2000 and the largest mergers and acquisitions are of firms in developed countries. This investment strategy is also referred to as "outward foreign direct investment" (OFDI) to distinguish it from the long standing trend of investment by multinationals from developed countries into developing markets. The focus of much of this OFDI has been in primary industries, such as oil and gas, and in resource-based manufacturing, such as steel.

In comparison, surprisingly little foreign direct investment in steel production has flowed from traditional steel regions into emerging markets. In 2010, the list of top ten steel producers included three companies from traditional steel regions: JFE Steel and Nippon Steel of Japan and U.S. Steel. JFE Steel, which was formed from the merger of NKK Steel and Kawasaki in 2002, only has production facilities within Japan. Nippon Steel, which merged with Sumitomo Metals of Japan in October 2012, has invested in downstream processing and service locations in China, Thailand, Indonesia, and Vietnam and in a steel mill in Brazil, but does not operate steel mills outside Japan.[13] However, the company has stated its intention of using the combined market power of Nippon Steel and Sumitomo Metals to begin investing in emerging markets.[14] U.S. Steel was the only major U.S. steel-maker to acquire steel mills in emerging markets, purchasing a Slovak steel mill in 2000 and a Serbian steel mill in 2003. Its investments have not been as successful as expected.[15] The Serbian mill was sold back to the Serbian government in January 2012 for one dollar after the mill lost $200 million in 2011.[16]

TRADITIONAL STEEL REGIONS AND STEEL TECHNOLOGY SUPPLIERS

While emerging markets have been playing a larger role in the location of and ownership of steel production facilities, traditional steel regions have remained a significant source of steel technology and know-how. Successful clusters of steel suppliers have been identified in several former centers of steel production, including the Teesside region in the United Kingdom (Sadler 2004) and Pittsburgh in the United States (Treado 2009). In addition, the dominant participants in the supply of steel technology, the "mill builder" firms, continue to be based in regions that were former steel powerhouses, such as SMS-Siemag of Germany, Danieli of Italy, Siemens VAI of Austria, and Mitsubishi of Japan. Rather than dispersing steel technology suppliers globally, the fragmentation of the steel process supply chain and thickening markets created by globalization have actually supported the survival of steel technology suppliers in traditional steel locations. The three classic agglomerating forces of knowledge spillovers, labor pooling, and transaction costs have been strengthened by de-verticalization and globalization.

First, although advances in electronic communication have drastically lowered the cost of sharing information across distances, some forms of knowledge-intensive activities continue to depend on geographically based knowledge spillovers. Location-specific technical knowledge can become a basis for international comparative advantage in what Dohse and his co-authors refer to as the "micro geography of investment" (Dohse, Hassink and Klaerdin 2012). They find that in a production environment characterized by a disaggregated supply chain, tacit or "sticky" knowledge is a more crucial factor in sourcing decisions than commonly available knowledge. In the global division of resources, there is some "stickiness" to technological steel know-how that keeps its supply anchored in traditional steel production locations. Investments by emerging-market multinationals in the industrial sectors of the developed world are often found to be part of a strategy that is knowledge seeking, rather than the type of resource-seeking that characterized investments that flowed in the opposite direction (Dohse, Hassink and Klaerdin 2012; Guillen and Garcia-Canal 2009). For example, although Arcelor and Mittal were producing roughly the same amount of steel at the time of their merger,

Arcelor's R&D budget was approximately ten times that of Mittal Steel (Giarratani, Madhavan, and Gruver 2013).

The stickiness of technical knowledge is closely linked to the advantage of access to a local pool of labor with specialized industrial knowledge. A regional labor force with steel-making expertise is thus the second agglomerating force in traditional steel regions. Not only do suppliers in these locations benefit from long experience with steel technology, but they have also benefited from the release of technical knowledge to the regional firms as mills downsized and shuttered. In some cases, new firms were created by former employees of the steel producers. For example, a scientific instrumentation company (RJ Lee Group) was formed in the early 1980s by the departure of several scientists from U.S. Steel's research labs, which were downsizing at the time (Treado 2009). That company, which is still headquartered in the Pittsburgh region, now employs over 300 people and serves as a supplier to a dozen different industries.[17] The role of industrial spin-offs has been found to be critical to the creation of supplier clusters as disparate as automotive design firms in Detroit and semiconductor firms in Silicon Valley (Klepper 2009).

Of course, the creation of an entirely new firm is not a pre-requisite for the transfer of knowledge from the steel producers to the steel suppliers. In many cases, as described in the section on de-verticalization, key employees were absorbed by existing supplier firms and became the basis for an expansion of the traditional role of intermediate suppliers. In either case, the availability of industry-specific expertise within a cluster of firms tends to attract new investment. In a survey of suppliers in the Pittsburgh region, Giarratani and Treado (2008) found that nearly half of the respondents had entered the Pittsburgh region after 1990, by which point most of the major Pittsburgh mills had closed. The new entrants to the region included both start-ups and mergers and acquisitions of current firms.

The third advantage offered by agglomeration is the reduction of transaction costs. Although globalization increases the number of potential sources of service and product supply, those options are not without transaction costs, particularly search costs. Several scholars have found that agglomerations of suppliers offer increasing returns to search costs (Grossman and Helpman 2005; Helsley and Strange 2007). In part, this is due to the reputation that a region develops for a particular industrial specialization. The location of a given supplier within a well-known regional cluster offers a signal to the purchasing firm that the supplier represents a valid sourcing option, narrowing the purchasing decision.

These three factors together have supported the ability of steel suppliers to continue to thrive in regions that are no longer the center of steel production. Clusters of steel suppliers have formed in traditional steel regions where proximity to the final customer is no longer as important as proximity to other suppliers. These clusters typically include a regional presence for one or more of the large mill builder firms, which often serve as a middle man between steel producers and smaller firms further up the supply chain (Treado 2009). Surrounding these mill builders are the numerous firms that provide raw materials, equipment, supplies, and technical services either directly to distant steel mills or indirectly through the mill builders.

In addition to improving their access to the global steel supply chain, collaborations among technology suppliers have enabled many firms to expand their reach into new markets (Sadler 2004). Giarratani and Treado (2008) similarly found market diversification to be more prevalent among the steel suppliers in the Pittsburgh region with significant local linkages. Contrary to common concerns about the destructive capacity of outsourcing and globalization, in the steel supply chain these forces were producing a more robust and secure demand for the output of the developed world, although paradoxically with lower labor requirements (Herrigel, Volker, and Voskamp 2013). In fact, the reorganization of the steel supply chain appears to be closer to Schumpeter's (1975) process of "creative destruction" in which the weakening of one industry is intrinsically linked to the strengthening of another.

Although agglomerating forces are currently supporting clusters of suppliers in traditional steel producing regions, those same forces have the potential to support one or more supplier clusters in emerging markets through the "upgrading" of steel technology suppliers in those markets. While most mill projects in emerging markets rely on global sourcing to access the latest technology and most experienced builders, other projects seek out low-cost local suppliers. This type of regional sourcing project facilitates technology transfer to developing regions as well as the transfer of high wage jobs from traditional to emerging steel markets. India, for example, has the largest participation of firms in the Association of Iron and Steel Technology (AIST) – a key trade association for steel technology suppliers – outside North America. Considerable progress has been made in the development of a steel supplier base in India through the AIST India chapter, which was formed in 2008 and has been recognized as the fastest-growing chapter in three of the last six years.[18] However, its membership is only a fraction of the over 1400 members in the Pittsburgh chapter of the AIST, a city that often serves as an example of the rapid decline of steel-making in the United States.[i]

CONCLUSION

The forces of globalization and outsourcing are often assumed to lead to an economic gain for emerging industrial regions at the expense of traditional industrial regions. In particular, regions dependent on traditional industries, like steel, are further assumed to be trapped in a downward economic cycle as the rest of the world innovates and expands. In contrast to this assumption, a closer look at regional shifts in the global steel supply chain illustrates how traditional industrial regions can be transformed into a global provider of advanced technology and thus share in the growth of emerging markets. The roots of this transformation are found in technology-led de-verticalization and globalization-led thickening of market opportunities in the steel industry.

The relationship between the traditional steel producing regions of Japan, Europe, and North America and global steel markets can no longer be accurately represented by focusing solely on the market for raw or finished steel. Powerful structural changes in the organization of steel production have enabled those traditional regions to focus on upstream, knowledge-intensive supply of technology and services, while emerging markets focus on the production of raw steel. In addition, producers in emerging markets are able to access the expertise and advanced technology of the traditional steel producers through international mergers, acquisitions, and investments. As classic trade theory predicts, this exchange has created benefits for both traditional and emerging steel regions. However, it is important to remember that the current structure of the global steel supply chain is no more permanent than the one it replaced.

NOTES

1. Data compiled from publications of the World Steel Association, www.worldsteel.org (last accessed October 27, 2014).
2. Capacity data are from the Steel Plant Database of the Center for Industry Studies at the University of Pittsburgh.
3. Steel Plant Database, Center for Industry Studies.
4. AISI, *Annual Statistical Report*, "Raw Steel Production by States," various years.
5. U.S. Bureau of the Census, Economic Census of Manufacturing, Industry Series, Iron and Steel Mills, 2002.
6. World Steel Association, *World Steel in Figures 2008*, as reported in (Kumar and Chadha 2009).
7. World Steel Association, *Top Steel Producing Companies*, various years, Brussels.
8. See "Indian Firms Move to World Stage," http://news.bbc.co.uk/2/hi/business/6944461.stm (last accessed October 27, 2014).
9. See the Arcelor Mittal website for additional details: http://corporate.arcelormittal.com/who-we-are/our-history (last accessed October 27, 2014).
10. World Steel Association, *Top Steel Producing Companies*, various years, Brussels.
11. http://www.gerdau.com/sobre-gerdau/memoria-gerdau-linha-do-tempo-home.aspx (last accessed October 27, 2014).
12. See "SSI begins to reap benefits of strategic UK acquisition," http://www.bangkokpost.com/business/economics/322669/ssi-begins-to-reap-benefits-of-strategic-uk-acquisition (last accessed December 5, 2012).
13. Company information can be found at www.nssmc.com (last accessed 10 February, 2013).
14. See "Nippon Steel & Sumitomo to Push Cost Cuts Amid Competition," http://www.businessweek.com/news/2012-09-30/nippon-steel-and-sumitomo-to-push-cost-savings-amid-competition (last accessed October 1, 2012).
15. http://www.ussteel.com/uss/portal/home/aboutus/history (last accessed October 27, 2014).
16. See "'Nominal' Sale Small Price for U.S. Steel's Serbia Pullout," http://www.post-gazette.com/business/heard-off-the-street/2012/02/05/Nominal-sale-small-price-for-U-S-Steel-s-Serbia-pullout/stories/201202050360 (last accessed 6 December 2012).
17. http://www.rjlg.com/about-us.aspx (last accessed October 27, 2014).
18. https://www.aist.org/local-member-chapters/find-a-chapter/india/chapter-history (last accessed 28 October, 2014).

i. According to the India Member Chapter Bulletin, there were 151 members in 2011: http://www.aist.org/chapters/IndiaMemberChapterBulletin_2011.pdf (last accessed 21 March, 2013). Information on the membership of the Pittsburgh chapter can be found at https://www.aist.org/local-member-chapters/find-a-chapter/pittsburgh/chapter-history (last accessed 28 October, 2014).

REFERENCES

AISI. *Annual Statistical Report*. Washington, DC: American Iron and Steel Institute, 2011.

Brandt, Loren and Thomas G. Rawski. *China's Great Economic Transformation.* New York: Cambridge University Press, 2008.

Coase, R.H. "The Nature of the Firm." *Economica* 4, no. 16 (November 1937): 386–405.

Collard-Wexler, Allan and Jan De Loecker. "Reallocation and Technology: Evidence from the U.S. Steel Industry." *NBER Working Paper* no. 18739 (January 2013).

Dohse, Dirk, Robert Hassink, and Claudia Klaerdin. "Emerging Multinations, International Knowledge Flows and economic Geography: A Research Agenda." *Kiel Working Papers* no. 1776 (2012).

Giarratani, Frank and Carey Durkin Treado. "Intermediate Steel Industry Suppliers in the Pittsburgh Region: A Cluster-Based Analysis of Regional Economic Resilience." *Economic Development Quarterly* 22, no. 1 (February 2008): 63–75.

Giarratani, Frank, Gene Gruver, and Randall Jackson. "Clusters, Agglomeration, and Economic Development Potential: Empirical Evidence Based on the Advent of Slab Casting by US Steel Minimills." *Economic Development Quarterly* 21, no. 2 (May 2007): 148–164.

Giarratani, Frank, Ravi Madhavan, and Gene Gruver. "Steel Industry Restructuring and Location." In *Handbook of Economic Geography and Industry Studies*, eds Frank Giarratani, Geoffrey Hewings, and Philip McCann, 11–37. Cheltenham and Northampton, MA: Edward Elgar Publishing, 2013.

Grossman, Gene M. and Elhanan Helpman. "Integration versus Outsourcing in Industry Equilibrium." *Quarterly Journal of Economics* 117 (February 2002): 85–120.
Grossman, Gene M. and Elhanan Helpman. "Outsourcing in a Global Economy." *Review of Economic Studies* 72, no. 1 (2005) 135–159.
Guillen, Mauro F. and Esteban Garcia-Canal. "The American Model of the Multinational Firm and the 'New' Multinationals From Emerging Economies." *Academy of Management Perspectives* 23, no. 2 (May 2009): 23–35.
Helsley, Robert W. and William C. Strange. "Agglomeration, Opportunism, and the Organization of Production." *Journal of Urban Economics* 62 (2007): 55–75.
Herrigel, Gary and Jonathan Zeitlin. "Inter-firm Relations in Global Manufacturing: Disintegrated Production and its Globalization." In *The Oxford Handbook of Comparative Institutional Analysis*, by G. Morgan, J. Campbell, C. Crouch, O.K. Pederson, and R. Whitely, 527–561. Oxford: Oxford University Press, 2010.
Herrigel, Gary, Wittke Volker, and Ulrich Voskamp. "The Process of Chinese Manufacturing Upgrading: Transitioning from Learning from to Recursive Mutual Learning Relations." *Global Strategy Journal* 3, no. 1 (February 2013): 109–125.
Hudson, Ray and Dan Swanton. "Global Shifts in Contemporary Times: The Changing Trajectories of Steel Towns in China, Germany and the United Kingdom." *European Urban and Regional Studies* 19, no. 1 (2012): 6–19.
Klepper, Steven. "Silicon Valley – A Chip Off the Old Detroit Bloc." In *Entrepreneurship, Growth, and Public Policy*, eds Zoltan Acs, David B. Audretsch and Robert J. Strom, 79–115. New York: Cambridge University Press, 2009.
Kumar, Nagesh and Alka Chadha. "India's Outward Foreign Direct Investments in Steel Industry in a Chinese Comparative Perspective." *Industrial and Corporate Change* 18, no. 2 (2009): 249–267.
McLaren, John. "'Globalization' and Vertical Structure." *American Economic Review* 90, no. 5 (December 2000): 1239–1254.
Sadler, David. "Cluster Evolution, the Transformation of Old Industrial Regions and the Steel Industry Supply Chain in North East England." *Regional Studies* 38, no. 1 (2004): 55–66.
Shumpeter, Joseph A. *Capitalism, Socialism, and Democracy*. New York: Harper, 1975.
Treado, Carey Durkin. "Imports, Technology, and the Success of the American Steel Industry." *Sloan Industry Studies Working Papers* no. WP-2004-02 (2004).
Treado, Carey Durkin. "Pittsburgh's Evolving Steel Legacy and the Steel Technology Cluster." *Cambridge Journal of Regions, Economy and Society* (2009): 1–16.
United Nations Conference on Trade and Development. *FDI from Developing and Transition Economies: Implications for Development.* World Investment Report, New York and Geneva: United Nations, 2006.
US International Trade Commission. *ITC Details Determinations Concerning Impact of Imports of Steel on U.S. Industry*. Press Release, October 23, 2001.
Warrian, Peter. "The Importance of Steel Manufacturing in Canada." Munk Centre for Global Affairs, University of Toronto, 2010.

FURTHER READING

Coase, R.H. "The Nature of the Firm." *Economica* 4, no. 16 (November 1937): 386–405. (Classic exposition of the firm in economics.)
Collard-Wexler, Allan and Jan De Loecker. "Reallocation and Technology: Evidence from the U.S. Steel Industry." *NBER Working Paper* no. 18739 (January 2013). (Theoretical model of the impact of minimill technology on steel industry productivity.)
Giarratani, Frank, Ravi Madhavan, and Gene Gruver. "Steel Industry Restructuring and Location." In *Handbook of Economic Geography and Industry Studies*, eds Frank Giarratani, Geoffrey Hewings, and Philip McCann. Cheltenham and Northampton, MA: Edward Elgar Publishing, 2013. (Detailed history of technological and economic changes in the global steel industry and overview of current players.)
Grossman, Gene M. and Elhanan Helpman. "Integration Versus Outsourcing in Industry Equilibrium." *Quarterly Journal of Economics* (February 2002). (Considers equilibrium conditions for make–buy decisions, including industry size, cost of governance, and search costs.)
Grossman, Gene M. and Elhanan Helpman. "Outsourcing in a Global Economy." *Review of Economic Studies* 72, no. 1 (January 2005): 135–159. (Develops a framework to study firms' decisions about *where* to outsource.)
Helsley, Robert W. and William C. Strange. "Agglomeration, Opportunism, and the Organization of

Production." *Journal of Urban Economics* 62 (2007): 55–75. (Builds on McLaren (2000) and Grossman and Helpman (2002) by focusing on local outsourcing rather than outsourcing at the national or global level.)

McLaren, John. "'Globalization' and Vertical Structure." *American Economic Review* 90, no. 5 (December 2000): 1239–1254. (Game theoretic model of factors that influence the make–buy decision. Finds key role for thickness of markets.)

Sadler, David. "Cluster Evolution, the Transformation of Old Industrial Regions and the Steel Industry Supply Chain in North East England." *Regional Studies* 38, no. 1 (2004): 55–66. (Assessment of the role of technology-based and market-based clusters in regional economic development using detailed survey of steel suppliers in Teesside region of England.)

Treado, Carey Durkin. "Imports, Technology, and the Success of the American Steel Industry." *Sloan Industry Studies Working Papers* no. WP-2004-02 (2004). (Comparison of relative impact of international trade and minimill technology on competitiveness of U.S. steel industry.)

18. Intermediate manufacturing: profit, dependency and value attainment in supply chains

Rachel Mulhall

INTRODUCTION

The intermediate metal processing (IMP) industry manufactures metal components that are incorporated into a range of end-user markets. The industry comprises two principal activities: forging (Standard Industrial Classification (SIC)[1] 2003 code 28.4) and casting (SIC 2003 code 27.5) of metal. Both industries produce a range of metal products, from basic shapes to complex fabrications of multiple components, and for a diverse set of markets and production volumes, from one-off custom products to mass production. The industry plays a critical role in supporting other manufacturing activities, with 51.1 per cent of outputs from the basic and fabricated metals industries (SIC 2003 group codes 27 and 28) used as inputs in further manufacturing (Department for Business Innovation & Skills (BIS), 2010). In addition, IMP firms provide key capabilities and services to United Kingdom (UK)-based manufacturers through the provision of development work, low volume production and customization of components for the automotive, aerospace, marine, construction and engineering markets. These industries are significant contributors to UK exports and value-added products, together accounting for 5.85 per cent of the UK economy's gross value added (GVA) (Office for National Statistics (ONS), 2011).

The IMP industry has three distinct characteristics. First, as an intermediate manufacturer, the firms have limited direct access to end-user markets. Products are sold to further manufacturers, which undertake additional production activities and assembly of larger parts before sale to the end manufacturer. Second, the products of the IMP industry are specifically produced to customer designs, with the industry essentially selling its productive capacity. The customized nature of the products means that IMP firms have little formal ownership right to product designs – and, therefore, are highly dependent on the fortunes of their customer base – and have a wide portfolio of products, production tasks and markets. Lastly, the industry has experienced a prolonged erosion of profitability. Historically, there have been periods of rationalization (particularly during the 1970s (Massey and Meegan, 1982) and the recession of the 1980s (Baden-Fuller, 1989)) and there continues to be an ongoing decline in enterprise and employment numbers in the industry. Overall profitability in the industry is low. For this reason, volume of production is an important element to survival. Firms need a sustained level of business across their customer portfolio in order to remain viable.

This mix of characteristics creates distinct challenges for the IMP industry. Profit erosion over the last few decades has reduced the financial stability of firms, leaving the industry vulnerable to cash flow difficulties in the short term and limited investment capacity for productivity improvements and growth in the long term. This chapter seeks to examine the elements behind the profit crisis and the methods of adjustment the industry is undertaking. Complex forms of dependency between the IMP firm and its customers shape the timing, duration and scale of value attainment for IMP firms. By adjusting these dependencies IMP firms are attempting to manage profit generation across their customer portfolio in order to survive.

The following discussion is based on fieldwork data collected from September 2009 to October 2010 in the West Midlands region of the UK. Qualitative interviews were undertaken in 55 firms (both in the IMP industry and its direct customers), transcribed and coded. The region has a long history of industrial activity, for which it became known as the 'workshop of the world' during the 19th century, and even today 11 per cent of the region's workforce is employed in the manufacturing sector (Medland, 2011). Metal industries have traditionally been a specialization of the area (Allen, 1929; Florence, 1948) and the region retains one of the highest concentrations of IMP firms in the UK, despite also suffering the largest decline in employment and enterprise numbers since 1998 (Eurostat, 2011). In response to this, the industry has undergone extensive restructuring and a niche metal industry has developed in the region producing higher value products and diversifying away from a previous dependency on the automotive industry (Bryson and Taylor, 2006).

PROFIT EROSION IN THE IMP INDUSTRY

The IMP sub-industries, casting and forging, are distinctly different production processes. Overall, the casting process lends itself to the production of lower volumes of more complex shapes and materials due to the ability to specify the metal composition and create more intricate moulds for complex shapes. In contrast, the forging process is capable of producing more standardized and consistent components but of a simpler design and metal composition. These capabilities, although different, each serve a wide variety of markets and an IMP firm will typically have a portfolio of product and process competencies. Although the range of techniques varies, both industries are capable of producing (and increasingly orientated towards) high value-added products, albeit for different markets or product types, based on their ability to produce and customize complex products.

From the mid-1990s onwards the IMP industry has continued to decline to a greater degree than the manufacturing industry average in the UK (ONS, 2009). The manufacturing industry overall in the UK has seen growth in both turnover and GVA (ONS, 2009). In contrast, the IMP industry has experienced a larger reduction in GVA than turnover (by around 7 per cent), indicating a reduction in operating surplus (an indicator for profitability). Firms in the West Midlands have an average profit margin of only 2.5 per cent (interview data, 2009–10), with over 90 per cent of interviewed firms (of those with known profit levels) having a single digit profit margin or lower.

The reduction in operating surplus can also be seen in an examination of the cost–price squeeze on the industry (Lawrence, 1987). The index of output prices to input costs in Figure 18.1 shows that from the mid-1990s to early 2000s the cost–price ratio was above 1, indicating that the industry's outputs sold at a higher price than the inputs cost to make them. However, since this point there has been a drop in ratio to approximately

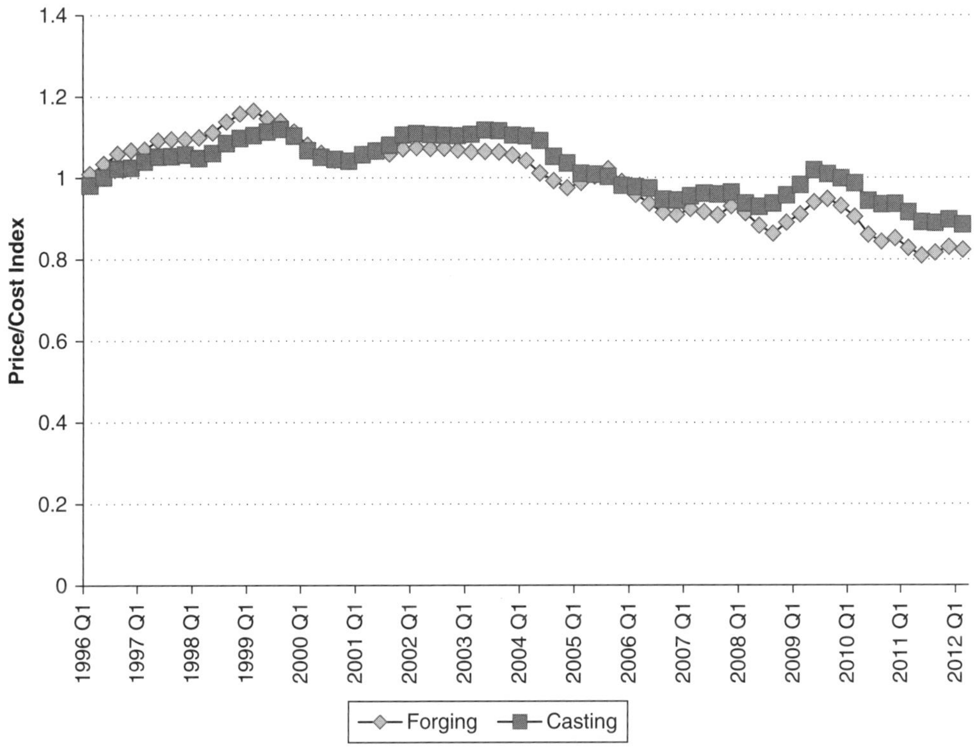

Source: Producer Prices Index (ONS, 2012).

Figure 18.1 Ratio of index of prices received for outputs to the index of prices paid for inputs by IMP sub-industry, UK, 1996–2012

0.8 (2010), indicating a relative loss of operating surplus. From Figure 18.2 it can be seen that there are two inter-related factors contributing to this loss. First, input costs have risen significantly since the mid-2000s. Second, this rise has not been associated with an equal increase in output values. Although the value of outputs in both industries has steadily increased over recent years, input costs have increased at a faster rate. In addition, the values of casting outputs, which were higher than forging output values in the early 2000s, have not increased to the same extent as forging output values since the mid-2000s. It should be noted that this is only a broad indication of trends in the industry because the range and proportion of inputs vary considerably between firms and their product portfolio. It does however, illustrate a wider trend for rising input costs and fluctuating output values.

Input Cost Changes

The cost base of IMP firms has changed. Although labour costs continue to be the main element (on average 43 per cent of selling cost), non-labour costs are now a

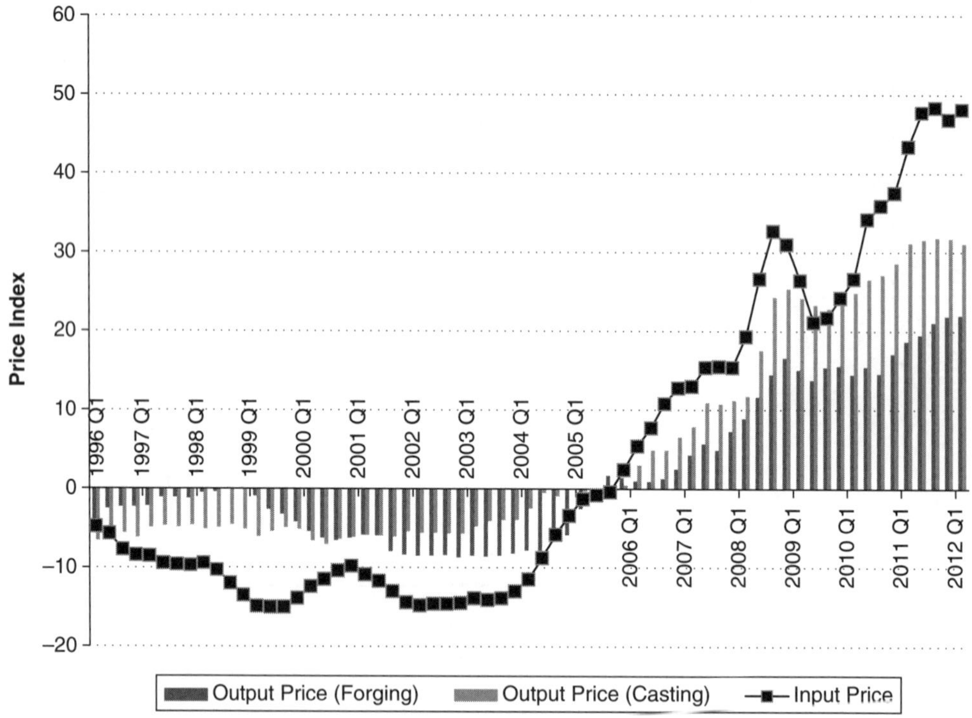

Source: Producer Prices Index (ONS, 2012).

Figure 18.2 Price–cost ratio: input and output price index, UK, 1996–2012

more significant part of the cost base than a decade ago (interview data, 2009–10). This is a reflection of both increased automation in the production process and the rising costs of other inputs, particularly metal and energy. This has important ramifications for the industry and individual firm. Labour costs are largely structured at the regional scale (Christopherson and Clark, 2007) and provide relatively stable costs. Influences on labour costs, such as the level of mechanization, skill shortages and unionization (Bluestone and Harrison, 1982), tend to occur over long temporal periods. In contrast, the non-labour costs considered here – metal and energy – are commodities: they are structured through more complex procurement practices and have a far greater rate of change in factor prices. This price volatility can alter the structure of the cost base in very short periods of time. Recovery of these costs is more difficult as it cannot be accurately incorporated into the cost base of the firm and additional measures are required to recoup price changes. Industry conventions have been developed to manage price volatility in metal markets through surcharge[2] systems; however, energy is a relatively new challenge (with volatility in prices since 2005) and IMP firms are experiencing significant difficulties in managing the cost (Mulhall and Bryson, 2013).

Output Price Changes

The IMP industry has undergone a continued transition to higher value products and services to meet the increasing demands of its customer base and differentiate it from international competitors, and in particular those from lower labour cost regions. Innovations in the industry are largely based on process improvements rather than product innovations (Hansen, 2010; Heidenreich, 2009) due to the limited product ownership and the critical volumes required, which encourage process improvements to achieve efficiency and a higher rate of return. This transition to value-added products has also involved developing additional services that allow firms to orientate themselves to customers and engage in co-innovations (Bryson and Taylor, 2010; Bryson *et al.*, 2008). IMP firms have developed new services but are also repackaging existing activities that were previously undertaken informally and for no charge. These activities, such as design modifications and stock management, are increasingly being incorporated as an explicit aspect of the producer's offering to its client base. Working closely with the customer to develop products and produce specialized, customized and low volume products is a key aspect of the competitiveness of the IMP industry in the UK. Links between firms based on co-development practices and the need to be relationally and geographically close to customers remain a key competitive advantage for the metal industry (Littunen and Tohmo, 2003; Tully and Berkeley, 2004; Watts *et al.*, 2006).

The competitive position of the IMP industry is underpinned by the strengthening of customer relationships through the absorption of additional tasks to increase the portfolio of capabilities and services provided by suppliers. Despite this, profitability in the IMP industry remains low and in some cases has declined over recent years (separately from the recessionary impact of 2008). This indicates that the relationship between IMP firms and their customers is a key aspect of long term stability in the industry and involves a complex negotiation of value within the relationship. In addition, the increasing significance of commodity inputs has generated distinct challenges of cost management within the supply chain. The inability to transfer input price changes through the supply chain can severely impact the profitability of a product by absorbing operating surplus. As such, the industry faces complex and technical issues related to its long term survival and competitiveness that are highly influenced by the nature of the relationships IMP firms have with their customers. The following section will focus on the types of relationship IMP firms have developed and the implications for value attainment in output prices rather than explicitly exploring how input cost changes are transferred through the supply chain (for a full examination of this aspect of profit erosion see Mulhall and Bryson, 2013).

INTERDEPENDENCY IN CUSTOMER RELATIONSHIPS

The customer–supplier relationship is characterized by a series of decision points in the production process that determine the type of product to be manufactured, standards, the order type and agreement, method of production and associated investments. At each of these points various forms of investment are made, whether financial (Williamson, 1979) or relational (Murphy, 2011). These levels of investment generate distinct and differentiated forms of sunk costs (Clark and Wrigley, 1995). Sunk costs imply a preference to continue a production relation while prior investments in equipment or knowledge retain value by maintaining the current production context (Cox, 1996). Multiple areas of sunk costs generate a complex system of investment structures between transaction partners and may entail conflicts between different sets of investment. These layers can generate forms of interdependency between firms. Although sunk costs are not the only dependency element in production relations (others include access to resources (Pfeffer and Salancik, 1978)), they form distinct inter-firm asymmetries.

The configuration of sunk costs has generated four distinct types of customer–supplier relationship in the IMP industry:

1. *Vulnerable.* In this relationship type there is little asset-specific investment by either party as production volumes are low and infrequent. Although these firms are highly skilled, they provide generic production capacity for a wide range of markets and products.
2. *Interdependent.* Both the IMP and customer firm are highly invested in this form of relationship due to ongoing specialization of skills, equipment and production methods to meet the demands of a small group of customers in a particular market.
3. *Lock-In.* The IMP firms in this relationship type provide discrete capabilities orientated to the development and pre-production stages. Investment is in developing a strong collaborative relationship with customers and enhancing design-based skills.
4. *Boundary Spanning.* This relationship type specifically seeks to generate formal ownership rights from investment in collaborative development projects with customers.

Interdependency between customers and suppliers is a central characteristic of these forms of relationship. Three of the above relationship types illustrate instances of interdependency between buyer and supplier: Interdependent, Lock-In and Boundary Spanning. These relationship types have different forms of interdependency and security. The Interdependent approach uses investment in equipment and knowledge and often sole manufacturing capability of strategic parts to retain production. This approach is used with all major customers as these IMP firms have a small set of customers that together represent around 85–90 per cent of turnover. The dependency is formalized through continued production agreements, although not necessarily formal contracts. The use of agreements in this approach is important because profit is generated towards the end of the product life-cycle; therefore the supplier needs to retain the whole life of production in order to generate profit. The customer is also keen to maintain the relationship to prevent additional tooling and set-up costs with other suppliers during the life of the product. Here, the interdependency is generated through the *contract* life as both parties can generate cost efficiencies over the life of the production contract by maintaining the supply relationship.

The Lock-In approach specifically attempts to create dependency through investment in knowledge but without any formal security to tie customers into continued production. Instead IMP firms 'package' the development work as a service that will hopefully encourage customers to keep the next stage of production with them. Critically, this approach attempts to generate interdependency with a specific group of customers within their client portfolio. Although the agreement covers only one product, the sunk costs encourage ongoing collaboration between the transaction partners on multiple projects (Grabher, 2002; Grabher and Ibert, 2006). Here, value is attached to retaining the *customer* for future projects.

In the Boundary Spanning approach the ownership of design rights acts as a formalized and permanent securitization of the supplier's involvement in design and production. The customer is tied to the supplier because of these rights and, as such, the interdependency is created for a specific *product*. Here the switching costs are related to a particular product and improvement in the supplier's skill set, although investment generates generic skills and not transactional-specific knowledge bases.

The types of relationship are not mutually exclusive and firms may engage in different relationships across their customer portfolio and during different temporal periods of a specific customer relationship. The relationship characteristics are shaped around the key decision points in the production process and the individual IMP firm's strategic orientation and resources (particularly financial). How the customer and supplier firms choose, or are able, to orientate the relationship and develop forms of interdependency has direct implications for the nature and duration of additional value in production activities.

ATTAINING AND MAINTAINING VALUE IN PRODUCTION CONTRACTS

The areas of profit generation differ between the relationship types according to how the capabilities and services of IMP firms are orientated towards their customer base. The nature of adjustment to high value manufacturing determines the periods in the production process when firms can undertake profitable production. In addition, the associated forms of interdependency generated from this adjustment have direct implications for how IMP firms can retain this profitability. The periods of profitability for each relationship type are outlined in Figure 18.3.

In Vulnerable relationships the IMP firm generates its profits from price premiums on low production volumes and outsourcing of finishing activities. As jobbers[3] these firms often provide additional finishing services to provide a complete product to the customer. Due to the small volumes and high capital expenditure required to invest in such facilities the IMP firm outsources these activities. A premium is charged to cover the logistical expense and low numbers, again generating a significant profit stream for some firms. Overall the profit levels are variable due to fluctuating demand.

The profitability of the Interdependent group varies over time. IMP firms usually manufacture the component for the entire life of the product (up to seven years) because tooling equipment is particularly expensive. Initially the IMP firm will make a loss due to production difficulties; this is funded by the supplier under fixed sales prices from the customer. The drive by the customer for continual cost reductions encourages investment by the IMP firm in process efficiency. Once these difficulties have been managed the IMP firm can start to generate profit from the contract, which supplements the losses made early on in the production contract.

The development work in Lock-In relationships generates the highest levels of profit, albeit for a short period of time as projects are finite and once production reaches a critical level it will be transferred to a high volume supplier which can provide cost

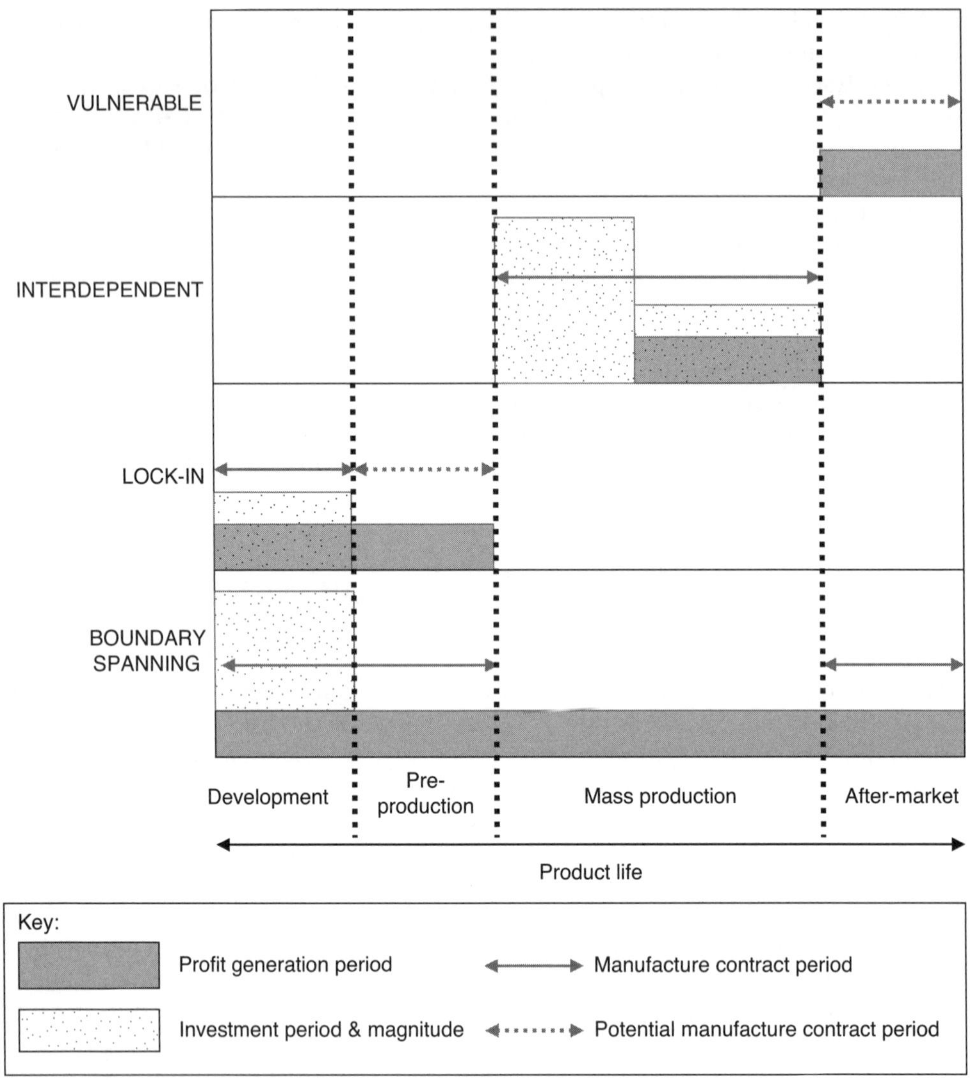

Source: Author 2012.

Figure 18.3 Value in contract structures

efficiencies to the customer. Due to the limited temporal period to capture profit, these IMP firms attempt to extract all the production work prior to this move.

The Boundary Spanning firms generate profitability from property rights for design involvement. Where this is through a joint venture the IMP firm is better able to stabilize their profit levels and generate profit for the entire product life as production must be undertaken by the IMP firm. In instances where the IMP firms have their own products, although the firm is entitled to all of the profits, routes to market are difficult and therefore this generates a relatively small revenue stream (around 5 per cent of turnover).

The variability in profitability over time within the contract is a key aspect of firm adjustment to higher value-added manufacture. Different aspects of 'higher value added' production have associated periods of profitability, both contractual and temporal. This concept builds upon the industry life-cycle approach to profitability in production cycles. Markusen's (1985) work on the profitability of industries characterizes profit cycles primarily on investment, output and employment at distinct points in the industry's life-cycle. Here, however, the profitability is related to the *product* life-cycle and is also influenced by the distribution of profit between buyer and supplier.

CAPTURING VALUE

The ability to capture value differs between relationship types from two primary influences. The first is the formalization of IMP firms' involvement in product design. By formalizing production activities the IMP firms are better able to justify the additional value they bring to the manufacturing process and are more likely to capture this value. Both the Lock-In and Boundary Spanning group have formalized their capacity and involvement in product development but in different ways. In the Lock-In approach the involvement does not generate any product design rights and, as a result, profit has to be negotiated between buyer and supplier. This formalization has in some cases generated an additional revenue stream as a design service. However, the profit is not secured and is generated only during their involvement with manufacturing. In these situations, relative power in negotiating the distribution of value is reduced as the IMP firm has no ownership rights to the product to secure its manufacture. In contrast, the Boundary Spanning group has enhanced the formalization of design involvement by insisting on intellectual property. This has ensured production is not undertaken without their involvement, thereby securing a revenue stream from part ownership. In instances where the IMP firm does not hold any direct product-based intellectual property right, which may be the case in investment consortiums, the IMP firm is again unable to secure involvement in production, and the firm is in a position similar to that of Lock-In firms which have to negotiate profit based on sunk cost dependencies.

The second factor is IMP firms' access to potential markets. As an intermediate supplier they are characterized by supplying further manufacturers and, as such, they have limited, if any, direct routes to market. Consequently, the demand and price for their production process are resultant on the marketing and distribution practices of their customers, or even customers' customers. The creation, or enhancement, of added value by the suppliers only generates profit if prices are supported by the customer. Increases in production costs and price pressures from customers can slowly erode the value added achieved by the IMP firm. This is particularly an issue for ongoing buyer–supplier relationships where production price changes are not necessarily reflected in sales prices. The Interdependent group is particularly susceptible to this because of their long term production agreements. The enhanced value created during the contract can be absorbed through production cost increases and cost pressures from customers. The lack of product ownership or direct end-user sales leaves these IMP firms unable to avoid this erosion of value over time. Although profitability increases during the contract, this is a result of increased efficiencies by the supplier. Their ability to capture a secure level of added value dissipates during the production agreement. Product ownership in the Boundary Spanning group, however, where access to market is secured through joint ownership, can generate a steady profit level as customers are tied to that particular manufacturer. There has been ongoing consideration of the movement of value through firm ownership structures (Lawrence, 1987) but the increased fragmentation of the production process has made product ownership a key determinant in capturing value (Angel and Engstrom, 1995; Dedrick *et al.*, 2010).

Despite IMP firms increasing their value-added activities they remain a low profitability industry because additional value added is not always captured by the firm, and profits in one area of production are often used to support less profitable, and sometimes loss making, activities. This supports Ruigrok and van Tulder's (1995) suggestion that involvement in design tasks does not necessarily capture value for the supplier. Although all firms are engaged with value-added activities (both design and finishing based), only the firms with some intellectual property rights are actually able to capture a continued and secure level of profit on their production activities. In the other relationships the distribution of value varies over the course of the production agreement (Figure 18.3) and the supplier's capacity to negotiate profit margins (Dedrick *et al.*, 2010; Murphy and Schindler, 2011; Rutherford and Holmes, 2008; Starosta, 2010). The Boundary Spanning group is characterized by its strategic direction towards product ownership to secure revenue streams and negotiation power within the buyer–supplier relationship. The Lock-In firms instead use complementary assets (Teece, 1986), such as specialized design and production skills, to capture value because they are unable to generate formal ownership structures or enforce them. The majority of IMP firms are engaged in high value-added activities; however, the distribution of value from this is complex and varies between firms and between relationships within firms according to strategic intent, investment capacity and inter-firm power dynamics.

CONCLUSIONS: HIGH VALUE MANUFACTURING AS A ROUTE TO SURVIVAL?

The IMP industry has undergone a continued transition to higher value-added products and services to remain competitive. Despite this transition, the industry continues to achieve only a very low profit margin. This is a result of both rising input prices and failure to capture additional value or cover input cost increases from the customer. The introduction of additional design, procurement and stock management activities, as well as the advancement of material and process capabilities, have helped strengthen the significance and nature of the relationship with their customers. However, these adjustments have generated complex dependencies in the client–supplier relationship, for which IMP firms are then forced to manage and adapt in order to attain the additional value that they provide and enhance their profitability.

The methods to do this are diverse and success is varied. The many levels of sunk costs within the buyer–supplier relationships and production agreement have generated intricate forms of interdependency. These areas of sunk costs have different temporalities of influence and value properties, which is illustrated through the different forms of relationship evident in the IMP industry. This has an important ramification when thinking about the transition of firms in high cost locations towards value-added manufacture. Value upgrading is not solely about the type of manufacture (product or process) but is also linked to the nature of the exchange relationship. Agreements, and specifically contracts, form distinct dependencies based on sunk costs and product ownerships. These have direct implications for *how much* and *for what periods of time* suppliers can capture value. The conceptualization of high cost locations as only undertaking high value-added manufacture is also questionable (Herrigel, 2010). Low value manufacture continues to form a considerable part of the activities in the industry precisely because 'higher value' activities only form distinct aspects of production and for discrete periods of time in the production agreement. Additional services, which have traditionally been undertaken informally, are a means of supplementing profitability across the customer portfolio but attaining this value remains difficult. The 'higher value' activities rely on investments in production contracts, as well as ongoing relationships, in order to generate enhanced profitability in the IMP industry.

The IMP industry has illustrated the continued significance of traditional manufacturing industries to global supply chains and the maintenance of advanced manufacturing in the UK. Although there is a large level of imports of casting and forging products into the UK, the UK IMP industry continues to be a significant supplier to UK engineering markets (with 20 per cent of castings and 33 per cent of forgings traded directly) and is the fifth largest exporter within the European Union (CBI, 2010). Exports during 2004–08 by the industry increased by 5.1 per cent annually and generated trade surpluses in specific product types, notably machinery components, railway equipment and vehicles (CBI, 2010). The industry has begun to absorb core tasks from its customer base and now provides significant design and pre-production capabilities, which are vital to the maintenance of manufacturing in the UK. Firms have undergone innovative and complex adjustments in order to remain competitive and survive in an increasingly international economy. A critical element of their success has been the ability to tailor the additional services and capabilities to capture additional value in the production process and manage the nature of their relationship and commitment structures with individual customers to ensure their profitability, and hence viability, over their entire customer portfolio.

ACKNOWLEDGEMENTS

The arguments in this chapter are based on research undertaken during an Economic and Social Research Council funded PhD scholarship and are included in the unpublished thesis held at the University of Birmingham.

NOTES

1. The Standard Industrial Classification is a classification system used for business activity in the UK. The system classifies business entities based on the activities undertaken. Each type of activity covered by the

system is assigned a numerical code in a framework system. The current system is the 2007 framework, replacing the 2003 framework in 2008.
2. A surcharge is an additional payment for the difference in actual price paid for an input to the base price (a price pre-agreed between transaction partners). These are usually undertaken at designated temporal periods (e.g. quarterly) to recover the changes in the commodity price.
3. A 'jobbing' firm is orientated to manufacturing very low volumes (typically 'one offs') for an extensive range of markets. As such, these firms have a wide portfolio of skills and capabilities to meet the demands of multiple markets.

REFERENCES

Allen, G.C. (1929), *The Industrial Development of Birmingham and the Black Country, 1860–1927*, London, UK: George Allen & Unwin.
Angel, D.P. and Engstrom, J. (1995), 'Manufacturing Systems and Technological Change: The U.S. Personal Computer Industry', *Economic Geography*, **71**, 79–102.
Baden-Fuller, C.W.F. (1989), 'Exit From Declining Industries and the Case of Steel Castings', *Economic Journal*, **99**, 949–961.
BIS (2010), *Manufacturing in the UK: An Economic Analysis of the Sector*, London, UK: Department for Business, Innovation and Skills.
Bluestone, B. and Harrison, B. (1982), *The Deindustrialization of America: Plant Closings, Community Abandonment, and the Dismantling of Basic Industry*, New York, US: Basic Books.
Bryson, J.R. and Taylor, M. (2006), *The Functioning Economic Geography of the West Midlands*, Birmingham, UK: West Midlands Regional Observatory.
Bryson, J.R. and Taylor, M. (2010), 'Competitiveness by Design and Inimitability through Service: Understanding the Dynamics of Firm-Based Competition in the West Midlands Jewellery and Lock Industries', *Service Industries Journal*, **30**, 583–596.
Bryson, J.R., Taylor, M. and Cooper, R. (2008), 'Competing by Design, Specialization and Customization: Manufacturing Locks in the West Midlands (UK)', *Geografiska Annaler: Series B, Human Geography*, **90**, 173–186.
CBI (2010), 'CBI Market Survey: The Castings and Forgings Market in the UK', CBI Ministry of Foreign Affairs of the Netherlands. Available at: http://www.cbi.eu/system/files/marketintel/201020–20castings20and20forgings20–20the20Netherlands1.pdf (accessed: 2 October 2013).
Christopherson, S. and Clark, J. (2007), *Remaking Regional Economies: Power, Labor, and Firm Strategies in the Knowledge Economy*, London, UK and New York, US: Routledge.
Clark, G.L. and Wrigley, N. (1995), 'Sunk Costs: A Framework for Economic Geography', *Transactions of the Institute of British Geographers*, **20**, 204–223.
Cox, A. (1996), 'Relational Competence and Strategic Procurement Management: Towards an Entrepreneurial and Contractual Theory of the Firm', *European Journal of Purchasing & Supply Management*, **2**, 57–70.
Dedrick, J., Kraemer, K.L. and Linden, G. (2010), 'Who Profits from Innovation in Global Value Chains? A Study of the iPod and Notebook PCs', *Industrial and Corporate Change*, **19**, 81–116.
Eurostat (2011), *Structural Business Statistics: Multi Yearly Regional Statistics (NUTS 06)*, European Commission. Available at: http://epp.eurostat.ec.europa.eu/portal/page/portal/european_business/data/database (accessed 2 October 2012).
Florence, P.S. (1948), *Investment, Location, and Size of Plant: A Realistic Inquiry into the Structure of British and American Industries*, Cambridge, UK: Cambridge University Press.
Grabher, G. (2002), 'The Project Ecology of Advertising: Tasks, Talents and Teams', *Regional Studies*, **36**, 245–262.
Grabher, G. and Ibert, O. (2006), 'Bad Company? The Ambiguity of Personal Knowledge Networks', *Journal of Economic Geography*, **6**, 251–271.
Hansen, T. (2010), 'The Danish Fabricated Metal Industry: A Competitive Medium-Low-Tech Industry in a High-Wage Country', *Geografisk Tidsskrift – Danish Journal of Geography*, **110**, 64–80.
Heidenreich, M. (2009), 'Innovation Patterns and Location of European Low- and Medium-Technology Industries', *Research Policy*, **38**, 483–494.
Herrigel, G. (2010), *Manufacturing Possibilities: Creative Action and Industrial Recomposition in the United States, Germany, and Japan*, New York, USA: Oxford University Press.
Lawrence, G. (1987), *Capitalism and the Countryside: The Rural Crisis in Australia*, Sydney, Australia and London, UK: Pluto Press.

Littunen, H. and Tohmo, T. (2003), 'The High Growth in New Metal-Based Manufacturing and Business Service Firms in Finland', *Small Business Economics*, **21**, 187–200.
Markusen, A. (1985), *Profit Cycles, Oligopoly, and Regional Development*, Cambridge, MA, US and London, UK: The MIT Press.
Massey, D. and Meegan, R. (1982), *The Anatomy of Job Loss: The How, Why and Where of Employment Decline*, New York, USA: Methuen & Co.
Medland, A. (2011), *Regional Trends: Portrait of the West Midlands*, London, UK: Office for National Statistics.
Mulhall, R.A. and Bryson, J.R. (2013), 'The Energy Hot Potato and Governance of Value Chains: Power, Risk and Organizational Adjustment in Intermediate Manufacturing Firms', *Economic Geography*, **89**, 395–419.
Murphy, J.T. (2011), 'Global Production Networks, Relational Proximity, and the Sociospatial Dynamics of Market Internationalization in Bolivia's Wood Products Sector', *Annals of the Association of American Geographers*, **102**, 208–233.
Murphy, J.T. and Schindler, S. (2011), 'Globalizing Development in Bolivia? Alternative Networks and Value-Capture Challenges in the Wood Products Industry', *Journal of Economic Geography*, **11**, 61–85.
ONS (2009), *Annual Business Inquiry*, London, UK: Office for National Statistics.
ONS (2011), *Annual Business Survey 2009 Revised Results*, London, UK: Office for National Statistics.
ONS (2012), *MM22 Producer Prices Indices (2005=100, SIC07)*, London, UK: Office for National Statistics.
Pfeffer, J. and Salancik, G.R. (1978), *The External Control of Organizations: A Resource Dependence Perspective*, New York, US and London, UK: Harper & Row.
Ruigrok, W. and van Tulder, R. (1995), *The Logic of International Restructuring*, London, UK: Routledge.
Rutherford, T. and Holmes, J. (2008), '"The Flea on the Tail of the Dog": Power in Global Production Networks and the Restructuring of Canadian Automotive Clusters', *Journal of Economic Geography*, **8**, 519–544.
Starosta, G. (2010), 'Global Commodity Chains and the Marxian Law of Value', *Antipode*, **42**, 433–465.
Teece, D.J. (1986), 'Profiting from Technological Innovation: Implications for Integration, Collaboration, Licensing and Public Policy', *Research Policy*, **15**, 285–305.
Tully, J. and Berkeley, N. (2004), 'Visualising the Operating Behaviour of SMEs in Sector and Cluster: Evidence from the West Midlands', *Local Economy*, **19**, 38–54.
Watts, H.D., Wood, A.M. and Wardle, P. (2006), 'Owner-Managers, Clusters and Local Embeddedness: Small Firms in the Sheffield (UK) Metal-Working Cluster', *Entrepreneurship & Regional Development*, **18**, 185–205.
Williamson, O.E. (1979), 'Transaction-Cost Economics: The Governance of Contractual Relations', *Journal of Law and Economics*, **22**, 233–261.

FURTHER READING

Buckley, P.J. and Ghauri, P.N. (2004), 'Globalisation, Economic Geography and the Strategy of Multinational Enterprises', *Journal of International Business Studies*, **35**, 81–98. (Identifies a deepening of the division of labour through the increasing fragmentation of production activities by multinational enterprises and the resultant ownership structures for production activities.)
Gereffi, G., Humphrey, J. and Sturgeon, T. (2005), 'The Governance of Global Value Chains', *Review of International Political Economy*, **12** (1), 78–104. (Decomposition of interdependency in global value chains.)
Herrigel, G. (2010), *Manufacturing Possibilities: Creative Action and Industrial Recomposition in the United States, Germany, and Japan*, New York, USA: Oxford University Press. (Additional discussion of the role of dependency in inter-firm relationships in the manufacturing sector.)
Mulhall, R.A. and Bryson, J. (2013), 'The Energy Hot Potato and Governance of Value Chains: Power, Risk and Organizational Adjustment in Intermediate Manufacturing Firms', *Economic Geography*, **89**, 395–419. (A detailed examination of the maintenance of value in production contracts through a case study of volatile energy input costs in the IMP industry and the attempted transfer of price rises to customers.)
Özatağan, G. (2011), 'Dynamics of Value Chain Governance: Increasing Supplier Competence and Changing Power Relations in the Periphery of Automotive Production – Evidence from Bursa, Turkey', *European Planning Studies*, **19**, 77–95. (An examination of the changing nature of supplier relationships.)
Rutherford, T. and Holmes, J. (2008), '"The Flea on the Tail of the Dog": Power in Global Production Networks and the Restructuring of Canadian Automotive Clusters', *Journal of Economic Geography*, **8**, 519–544. (The changing nature of supplier relationships and the development of supplier influence in production systems.)

19. Aerospace manufacturing: past, present and future

Colin G. Drury

1. UNIQUE FACTORS OF AEROSPACE MANUFACTURING

This chapter provides an overview of aerospace manufacturing. Aviation hardware, like most products, has aspects of product and history that make it uniquely challenging, so these are presented first. The service requirements for aerospace hardware determine to a large extent the manufacturing requirements. Thus in order to understand how modern manufacturing issues, such as supply chains, design and outsourcing, affect manufacturing, a current overview is presented before these effects are considered in more detail.

(a) Flying is Hard

The building of heavier-than-air flying machines has about a century of history, during which time the product has been transformed from a single-seat wood and fabric biplane capable of about 50 kph to 300+-seat carbon fiber composite airliners capable of cruising at 900 kph over 15,000 km stage lengths. Flying is inherently difficult, and indeed there were hundreds of years between the desire to fly and its achievement (Gibbs-Smith 1966). Any aircraft design requires knowledge in the areas of aerodynamics, materials, structures, power production and increasingly electronics, plus a manufacturing organization capable of design integration and efficient production. Unlike other products, particularly transportation products, the design requirements are very stringent as the finished product must actually fly safely in all flight regimes. Aircraft structures must be light and strong enough to fly with the desired payload, as the total weight of the aircraft must be supported by the aerodynamic lift generated by the airframe and by the thrust developed by the engines. The final external form of the product is determined by a deep knowledge of physics and the desired mission characteristics, rather than by considerations of market aesthetics. Early aircraft manufacturing history abounds with examples of insufficient strength, for example the shedding of wings or wing covering in World War I (WWI) fighters (Weyl 1965), or the twisting and flutter in monoplane wings of the 1930s due to insufficient rigidity (Jarrett 1997a). Even current airliners occasionally require strengthening after full-scale testing to meet structural loads thought to have been covered in the design phase (e.g. Airbus A-380, Boeing 787). Finally, aircraft currently cost billions of dollars and there are many years between conception and arrival in service, e.g. F-35, B-787. In this respect, a new aircraft is more comparable in scope to a large building than a more traditional product, such as a computer or an automobile.

(b) Flying is Highly Visible and Highly Regulated

Aircraft (apart from unmanned aerial vehicles or UAVs) carry people and thus any mishap is a matter of great public concern. Aircraft accidents involve a large amount of accumulated energy, from kinetic energy of many metric tonnes of machine moving at over 150 kph to the large mass of chemical energy carried in the tonnes of fuel. Thus accidents are so spectacular in public consciousness, due to the energy release and the presence of the traveling public, that safety is the prime consideration in design, manufacturing and operation. Despite the fact that accidents have decreased by orders of magnitude over the century of aviation history (Boeing 2012), accidents still cause detailed investigation and lead to design and operational changes. A recent example was the crash of the Colgan Air Q-400 at Buffalo NY in 2009 (National Transportation Safety Board (NTSB 2010a)). The NTSB cited "flight crew monitoring failures, pilot professionalism, fatigue, remedial training, pilot training records, airspeed selection procedures, stall training" in the crash, of which stall training changes and increased co-pilot minimum experience have been finalized by the Federal Aviation Administration (FAA) to date (2014). Because of this public visibility of aviation, design, manufacturing and operations are highly regulated industries in every country. In the United States, the FAA maintains thousands of Federal Air Regulations, covering all aspects of aviation from design and manufacture to training and maintenance. Companies manufacturing airliners, for example, must comply with airworthiness standards for all components of the aircraft, to show that no single component failure can lead to a hull-loss accident.

(c) Aircraft are Now Long-Lived and Expensive

In contrast to the early years of aviation, current aircraft are expected to last for many years. Until WWII, most aircraft were lost to accidents, or replaced as obsolete, within a few years. This was partly due to the inherent difficulty of flight, especially in poor weather conditions, but also due to the rapid development cycle of aircraft. In WWI, aircraft were designed in a matter of weeks and even in WWII there is the famous example of the P-51 that went from order to first flight in 117 days to meet a customer requirement (Jarrett 1997b). Now both civil and military aircraft are expected to last for decades in service. Aircraft must often be used for that long to repay their high purchase price, typically around $US80 million for a B-737 or around $US 300 million for a B-747. General aviation aircraft, such as trainers or sport aircraft, can cost around $US100,000, while military aircraft have ranged up to $US500M for the B-2. Civil airliners and general aviation aircraft can be 25 to 50 years old if maintained correctly, with such compendia as *Jane's All the World Aircraft* still listing airframes designed in the 1960s, such as early versions of the B-737 and B-747 families. Military aircraft can be even longer-lived, with the B-52 and Mig-21 being 1950s designs still in service, with of course many upgrades to engines and avionics (National Research Council (NRC) 1997). Note also that in almost all wars, more aircraft are lost to accidents than to enemy action. This long fleet life is achieved at the cost of continuous maintenance, again making aircraft a product closer to a large building than to a consumer artifact.

(d) Aviation is International

The design and manufacture of aircraft have always occurred across countries and continents, with the USA, France, UK, Germany and Russia having led the industry since the earliest times. Within each country, and now across continents, mergers and consolidations have consistently reduced the number of aircraft manufacturers. Currently there are only three major airframe manufacturers in the USA: Boeing, Lockheed-Martin and Northrop-Grumman. Obviously these comprise five individual names of aircraft companies, showing at least the later merger structure. All of these have large military customers, plus Boeing has a large commercial division: Boeing Commercial Aircraft. In Europe, the obvious equivalent is the Airbus Group, formed from companies in France, Germany, the UK, Italy and Spain. Airbus has military and helicopter product lines in addition to its more famous Airbus airliners. Canada (with Bombardier Aerospace), China (with Commercial Aircraft Corporation of China, Comac) and Russia (with United Aircraft Corporation) have had similar consolidation, so as to be able to meet the large development costs of new large aircraft. The supply chains are even more international, with major components of aircraft being manufactured, and indeed partially designed, around the globe. As just one example, fuselage sections for Boeing airliners are made by Mitsubishi in Japan. Smaller manufacturers, and manufacturers of small aircraft, still abound in most developed countries, ranging from light plane kits to business jets.

(e) Aviation is a Pioneer of Design Innovation

While not universally true, many new techniques for design and manufacturing started, or received initial large-scale use, in aviation. An obvious example is the use of an integrated design tool such as CATIA, developed by and for Airbus but used successfully in what was claimed to be the first paperless aircraft, the Boeing B-777 (Sabbagh 1996). This technique, also used in other industries such as automotive, allows many business partners to provide their own design expertise to the overall aircraft design. It has not been without its problems, but has in general ensured that parts are manufactured using the same code as their design, with consequently better fit at assembly time. More recent examples of design innovation are the use of friction-stir welding instead of riveting to join fuselage sections of the Eclipse very light jet and the extensive use of composite structures in fuselage and wing of the Boeing 787 and Airbus A350.

(f) Space Multiplies the Challenges

Some of the above factors are increased in importance when the product is a spacecraft rather than an aircraft. Given that even to low earth orbit 1 kg of payload may need 10 kg of rocket and fuel, then payload mass is a huge driver of the feasibility and cost of space launches. Thus lightness and strength (c.f. Space Shuttle Orbiter failures) are even more important in spacecraft than in aircraft. We are moving towards privatized, and potentially cheaper, orbital capability but the cost of payload to orbit was still about $US4000 to $US15,000 per kg in 2000 (Futron 2002), depending on rocket size and whether or not the cost was subsidized by government. Launch costs to the much higher geo-stationary orbit are proportionally much greater. When a spacecraft is in orbit, whether autonomous or manned, the environment is one of the harshest known and the potential interventions to mitigate unforeseen problems are necessarily limited. This leads to highly controlled manufacturing, with clean rooms, test cells and comprehensive procedures all designed to prevent the launch of defective spacecraft.

2. THE STRUCTURE OF AEROSPACE MANUFACTURING

While a typical aerospace manufacturer probably does not exist, the concept can still be useful academically as a case study. Here we divide the industry, in a way that it generally labels itself, into those who assemble the complete vehicle and those who design and build specialist components. As we shall discuss later, the component manufacturers are playing an increasing role in the complete aircraft design and manufacture process.

(a) Airframes

Manufacturers range from small companies producing general aviation aircraft (e.g. Opdyke 1999), often providing kits that owners can build for themselves, through medium-sized companies specializing in a niche market (e.g. Beechcraft Corporation for corporate jets, or Bombardier, making smaller regional airliners) to very large "airframers" such as Boeing or Airbus which concentrate on large civil and military aircraft. When companies become very large, they can become prestige symbols for governments – for example, Boeing for the USA, Airbus for the European Union and Comac for China – leading to considerable political as well as marketing and manufacturing pressures. Thus many sales of military aircraft across national borders must offer participation of local manufacturing to gain government backing for the contract.

Flying has traditionally attracted gifted scientists, designers, engineers and pilots, and, at least in the pioneer days (Opdyke 1999), turned them into de facto manufacturers. Obvious examples are Glen Curtiss in the USA, A. V. Roe in the UK, Louis Bleriot in France and Hugo Junkers in Germany, who all founded aircraft design and manufacturing companies bearing their name. Some pioneer designers crossed borders with their companies, such as Anthony Fokker relocating from Germany to the Netherlands after WWI (Weyl 1965), or Igor Sikorsky having his company in Russia until the revolution of 1917 and later in the USA (Cochrane, Hardesty and Lee 1989).

Aerospace companies also have histories that affect their current operations, often a history of mergers and acquisitions. Excellent histories on many individual manufacturers in publications by Putnam (34 volumes) and Docavia (5 volumes, in French) cover companies in the UK, USA, Germany, Sweden, Russia (Putnam) and in France and the USA (Docavia). They range in date from De Havilland aircraft (Jackson 1962), through the 1980s (Liron 1984) to Junkers aircraft (Kay 2004). These histories cover a chronological record of the company and its aircraft, but also include photographs of the manufacturing facilities in different eras. Each of these includes earlier manufacturers absorbed into the later company; for example, Wegg (1990) details General Dynamics aircraft including eight absorbed companies from Thomas Morse to Convair. Of course, subsequent to that publication, the aviation interests of General Dynamics became part of Lockheed-Martin in 1993. The history of aviation manufacturing in the former Soviet Union (subsequently mainly Russia and Ukraine) and China is quite different, with government taking a much stronger hand in all aspects of design, manufacturing and sales.

Manufacturing by smaller companies often has had to diversify in order for them to stay in business through lean times. A typical example is the Schweizer company, a maker of high-performance sailplanes in Elmira, New York, which has at various times manufactured aircraft parts under sub-contract and non-aviation items such as truck bodies and even lathe guards (Schweizer 1991). They also built much of the structure of the Bell 47J2 helicopter, control surfaces for the Grumman Gulfstream, and even complete Grumman Ag-Cat agricultural aircraft and Hughes 300 helicopters, the latter evolving into their own 330 turbine powered helicopter in 1988. (Note that Schweizer (1991) is an excellent business as well as aeronautical company history, giving for example, detailed tables of annual sales and employment.) Even major manufacturers have weathered hard times by using their aviation skills to produce non-aviation products. After WWII, Grumman's orders for naval aircraft dropped rapidly but they used their aluminum monocoque construction expertise to build light-weight canoes for the growing consumer market.

(b) Engines

Typically, airframe manufacturers do not produce their own engines, relying on specialist engine manufacturers to produce competitive engines. This has not always been the case: for example, Curtiss, Bristol and Junkers were all major engine manufacturers as well as airframers up to WWII. Currently the major engine manufacturers of turbine engines (turbofans, turbojets and turboprops) are Pratt & Whitney, General Electric and Rolls Royce, with various subsidiaries producing smaller engines, for example Pratt & Whitney Canada produces small turbofans and turboprops. Engine supply to both civil and military aircraft is fiercely competitive, as evidenced by the fights in the US Congress in the past ten years over an alternative engine to power the F-35. In civil aviation, the demand is for ever-higher levels of fuel efficiency because fuel costs are either the largest single airline operating expense or the second after payroll. Reducing fuel consumption by a few percent is highly significant, but reduced fuel consumption must be accompanied by reduced emissions and reduced noise levels. Noise levels have decreased from about 100 decibels (dBA) for an early B-747 to 78 dBA for a late B-777 (FAA 2002), a reduction of over 100 times the sound intensity because a reduction of 3 dBA halves the sound intensity. Emissions reduction has become increasingly important, with some countries and regions (e.g. the European Union) introducing a carbon tax on all aircraft, domestic and foreign, arriving and departing. These beneficial changes have taken place through extremely careful design and very tight manufacturing tolerances. Often, aircraft engine manufacturers will form consortia to share the very high development costs of new and progressively developed engines. Examples are CFM International (comprising General Electric and Snecma), MTR (a joint venture between MTU Aero Engines of Germany, Turbomeca of France and Rolls Royce of the UK) and Engine Alliance (a joint venture between General Electric and Pratt & Whitney). As engine manufacturers combine in different ways there is the added challenge of putting firewalls between the stand-alone business and these joint ventures.

(c) Components and Avionics

Most airframers are primarily designers and assemblers of aircraft, or spacecraft, from components produced by others as much as by themselves. This applies to small manufacturers as it has for most of the 20th century, where standard components (e.g. wheels, propellers and cockpit instruments, as well as engines) came from more specialized manufacturers. This certainly continues today where there are specialist companies producing components for many different airframers. An example would be Dowty, which started as a manufacturer of internally sprung wheels (e.g. for the Gloster Gladiator fighter of the late 1930s) and continued as undercarriage (landing gear in the USA) specialists. A related example is Dowty-Rotol, which produces propellers for many turboprop airliners such as those from Bombardier as well as for the Lockheed-Martin C-130J military transport. Dowty bought the Rotol group of propeller manufacturers, itself formed from propeller divisions of Rolls Royce and Bristol Aircraft.

Aviation electronics, known as avionics, are of increasing importance in aircraft design. They grew from the simple autopilots developed before WWII into a range of control, communications and signal detection/processing equipment in civil and military aircraft as well as spacecraft. Over the past 50 years the primary flight control of the aircraft has changed from rods and cables between cockpit controls and control surfaces, through power-assisted controls that lower the pilot forces required, to digital controls that move control surfaces by electric motor or hydraulics in response to electrical input from the pilot controls. Doing this has reduced both aircraft weight and stick/rudder forces, but mainly has enabled new control laws to be inserted between pilot controls and control surfaces. These can be modified by flight regime – for example, speed, altitude, angle of attack – to improve control and safety, although there are differences of opinion over the relative authority of pilot and computer. Engine controls have similarly evolved from direct mechanical links to Full-Authority Digital Engine Controls, with similar increases in ease of engine handling and improved safety, for example reduction of flame-out conditions.

Avionics has transformed the cockpit from individual dial indicators ("steam gauges" in aviation jargon) into flat-panel digital displays integrating many functions ("glass cockpits"). As with primary flight and engine controls, the insertion of a powerful computer between pilot and display allows for both increased control capability and decreased pilot workload. Thus civil aircraft have primary flight displays showing at least attitude, altitude and heading. More typical displays show navigation (usually via GPS), terrain maps and airfield information. These displays have become the normal displays for general aviation, although they may not have increased flight safety (NTSB 2010b) as well as for airliners and the military. Manufacturers of avionics are specialists, for example Honeywell, Rockwell-Collins or Thales for larger aircraft and Garmin for general aviation. Airframers specify either off-the-shelf avionics or, for more complex aircraft, they work with the avionics manufacturer to develop a specific set of displays integrated into a Flight Management System. In the military, avionics has advanced to encompass targeting, weapons systems management and digital communications between inter-connected vehicles, including UAVs.

(d) Maintenance and Repair

As noted above, aircraft last long enough to require considerable maintenance and repair to ensure what the FAA terms "continuing airworthiness." The scale of maintenance is large enough to warrant its inclusion in manufacturing, as inspection and repair costs can exceed the initial purchase price of an aircraft. According to the International Air Transport Association (IATA) the costs for maintenance ranged from $US461 per flight hour for regional jets to $US1430 per flight hour for wide-body jets with more than two engines (IATA 2011), amounting to several millions of dollars per year for each airliner. The worldwide total of maintenance and repair spending in 2010 was $US42.3 billion (IATA 2011). All commercial aircraft are designed for maintenance using a design process termed MSG-3 from the Maintenance Safety Group of the FAA. This considers each component (airframe parts, engine parts, avionics, etc.) and determines its failure modes. It then estimates failure rates over time and the probability of detecting each failure, combining these to provide a set of inspection and repair dates to ensure that no single failure can result in a hull-loss accident. For example, with cracks in airframes (a problem with aging aircraft fleets), MSG-3 determines the crack growth rate for each failure mode and the inspection reliability for cracks as they become larger. It then specifies a set of inspection intervals such that there are multiple opportunities for detection of each crack between it growing to detectable size and it threatening component failure (Drury and Spencer 1997).

Maintenance and repair are carried out by Maintenance and Repair Organizations (MROs), which may or may not be part of the airline operating the aircraft. MROs comprise line maintenance performed on the flight line (about 17 percent of MRO costs), heavy base maintenance conducted at special facilities (about 20 percent), engine maintenance (about 43 percent) and component maintenance (about 20 percent) (all estimates from IATA (2011)). For all except line maintenance, the work involves specialized hangars (except in the tropics, where work is often performed outside), specialized tools for inspection and repair, specialized fixtures for access to large aircraft, and specialized and licensed mechanics, known as Aviation Maintenance Technicians (AMTs) in the USA. The work is carried out using strict and highly regulated guidelines. For example, maintenance operations have AMTs following documents called "task cards," signing off on each step that the work was completed, to provide a comprehensive paper trail paralleling the actual maintenance and repair work (e.g. Patel, Drury and Lofgren, 1994). Heavy engine repair is usually performed in a dedicated engine facility, often operated by the engine manufacturer which sells "power by the hour" to the airline. Similarly, component repair is performed in "back shops" at airlines or other MROs, or often outsourced to a lower tier of specialists, for example for avionics. In these cases, AMTs typically replace a component that appears to be defective and send the old component out to the back shop for diagnosis and repair. Increasingly, maintenance is software supported with portable or tablet computers being used for presenting the task card to the AMT and recording the sign-offs (Drury, Patel and Prabhu 2000). In addition, the software itself can be in need of repair or upgrading, and software changes can be used to provide functional repairs. This is particularly the case for space vehicles, where physical repair is exceedingly expensive. The Hubble Space Telescope cost about $US2.5 billion to build and launch in 1990, but has had five upgrade missions in the 23 years of operation since then, each costing $US1.7 to $2.4 billion (General Accounting Office (GAO) 2004) for a total cost of over $US10 billion to date. Satellite servicing and repair by software and reconfiguration is more common than physical repair. For example, the award-winning QuickSCAT ocean observation satellite launched in 1999 has required many software changes to reconfigure operational status as batteries fail and antennae cease to move correctly (NASA 2009). It had a design life of two years but operated as planned for over ten years and even after 13 years can still be configured from the ground to provide useful data on ocean winds.

3. CHALLENGES FACING AEROSPACE MANUFACTURING

Aerospace manufacturing is a complex industry; even if nothing changed, it would still face challenges. But the ongoing and coming challenges leave it anything but static. Although the ultimate speeds of most aircraft classes have not changed much in 50 years, the changes to operating capability and cost/environmental impact have been very large. For example, each generation of jet airliner improves the safety record and lowers fuel and maintenance costs (impacting seat km costs) while reducing noise and environmental pollution. However, the continuing expansion of passenger and cargo traffic has kept overall environmental impact relatively steady. Aircraft design is not the only factor in the environmental impact of aviation: improved operational procedures also aid in impact reduction. For example, the move away from radar-based air traffic control to GPS-based operations allows for what the FAA terms "Trajectory-Based Operations" that reduce fuel consumption and environmental costs by providing more direct flight paths (Joint Planning and Development Office (JPDO) 2011). A specific example is tests of Required Navigation Performance for landings at Phoenix airport in 2006, which showed a $US2 million cost saving and a 2500 tonne carbon dioxide reduction (FAA 2009). These savings are not possible without aircraft design and manufacturing changes; in this case the addition of appropriate avionics to enable the new procedures.

In this section we will survey some of the challenges to aerospace manufacturing beyond those relating to sustainability.

(a) Design

Although airframers have traditionally performed their own design work, relying on outside contractors for engines and components only (see Section 2c above), the trend is towards more design autonomy within the overall airframer's specifications. As noted above (Section 1e), the advent of comprehensive design software has allowed a number of design innovations of interest in manufacturing. First, it allows second and third tiers of the supply chain to contribute to the design process in a way that captures their unique design and manufacturing expertise. Thus outsourcing design of the auxiliary power unit (APU), which is a small gas turbine engine providing electrical power when the main engines are not operating, enables APU specialists (e.g. Hamilton Sundstrand or Honeywell) to design and manufacture their optimal unit within the airframer's specifications. These will include power output, fuel consumption, weight, space constraints and operating temperatures, so that the design fits into the overall aircraft package. Thus not only is the production of components outsourced (see 3b below), but so is the design expertise. This is a challenge (as well as an opportunity) because it requires another level of communication within the design team so that integration produces no surprises in terms of space, weight or function. This is particularly true for multiple partners in the design who must coordinate with each other as well as the more natural coordination with the airframer. Initial troubles with Airbus's A-380 and Boeing's B-787 have hinted at lack of design integration, but there have been no definitive answers so far. Recent papers have noted problems with distributed design, for example at Boeing, and Dell (Amaral, Anderson and Parker 2011) and attempted to model optimal partial-outsourcing design strategies (Anderson, Feng and Parker 2011).

Another opportunity comes with the use of integrated software for design: the freedom from mock-ups as the design progresses. With detailed three-dimensional design in software, fit and tolerances can be predetermined without the need to build a physical mock-up for testing. Such mock-ups were an integral part of aircraft design stretching back before WWII, for example the 1946 mock-up shown on page 273 of Brown (1970) of the M.52 prototype supersonic test aircraft by Miles Aircraft that was never built due to government policy changes. They were used to reassure the customer that the aircraft would be physically accessible to the flight and ground crews, and that such items as cockpit display layouts fitted customer needs. As Sabbagh (1996) records (pp. 57 *et seq.*), the Boeing 777 used CATIA for design and EPIC (Electronic Preassembly In the Computer) in place of mock-ups. This technique allowed, for example, the early detection of a torque tube that would fit the wing correctly but which had insufficient clearance to allow assembly. Similarly, the wiring harness could be designed to fit through the structure by software, instead of hand-assembly in the mock-up and subsequent recording of wire lengths and paths. On later aircraft such as the Boeing 787, computer representations of human manikins are used to ensure that AMTs can reach points requiring maintenance and have sufficient clearance to use the correct tools (e.g. McMullin et al. 2008).

New initiatives in design are constantly being tried. The whole Design for X movement began with Design for Manufacturability (see Boothroyd 1996), and has now embraced Design for Quality, Reliability, Serviceability and even Design for Inspectability (Drury 1996). More recently Washington (2012) outlines the Defense Advanced Research Projects Agency's (DARPA's) efforts to reduce the design-to-service cycle from decades (e.g. F-35) to potentially months using a more integrated approach to design. This is made possible by IT tools that can help handle the complexity inherent in any modern large-scale system. Instead of the design–build–test–redesign cycle the aim is to use valid subsystem models integrated into a system model to consider design trade-offs in parallel and perform virtual testing of hundreds of potential configurations, again eliminating the need for mock-ups and prototypes.

(b) Outsourcing and Supply Chain Logistics

We have already seen that design can be outsourced (3a above), but outsourcing is far more extensive in aerospace manufacturing. An obvious starting point is engine and avionics, which are almost always outsourced, but we are now seeing even major assemblies produced elsewhere and transported to the final airframe assembly plant. From the outset, Airbus Industrie was designed to involve plants in constituent countries of the consortium. Thus wings are manufactured in the UK at the former plants of Bristol Aircraft (Filton) and De Havilland Aircraft (Broughton), while fuselage sections are sourced in Hamburg, Germany and Saint-Nazaire, France; tail sections in Japan and Spain. All are assembled in Toulouse (France), Hamburg (Germany) or Tianjin (China). Engines and avionics come from the suppliers listed earlier. Similarly, Boeing extensively outsources assemblies with composite fuselage sections from the USA, Japan and Italy; wings from Japan and tail assemblies either from Italy or Boeing's own plant. Assembly is performed in parallel at the main plant (Everett, Washington) and the new plant (North Charleston, South Carolina) in the United States.

This activity is not new. In each world war, all combatant countries had components manufactured by second-tier companies, as well as what the UK called "shadow factories" built to provide parallel sources for each aircraft type. In WWI the wooden-framed Sopwith Camel could be extensively sub-contracted component by component to local woodworking shops, while metal fittings were manufactured in quantity by Britain's reservoir of machine shops. Full-scale production was undertaken not only by Sopwith, but by Boulton & Paul (Norwich) and Fairey Aviation (Hayes), each airframers in their own right with their own designers (Pudney 1964). Similar outsourcing to meet wartime needs was just as evident in France and Germany. In WWII, Supermarine Spitfires were built by the thousands at a shadow factory at Castle Bromwich near Birmingham (Andrews and Morgan, 1981).

Manufacturing outsourcing presents challenges of supply chain management, particularly regarding quality (e.g. Foster 2007), which can have literally life-or-death consequences in aviation and space vehicles. It also exposes the airframer to the classic problems of supply chains: bottlenecks and disruptions. These can be manifested in acute form by the sheer size of some of the components outsourced. Fuselage barrel sections can be over 6 m in diameter and 12 m long: wing components can be up to 20 m long, and all need inter-country or even intercontinental transportation. Boeing and Airbus have used ships and outsize aircraft (Airbus A300–600ST Beluga and Boeing 747 Dreamlifter) to transport such huge components, with the aircraft reducing transport time from weeks to days, for example between Japan and Seattle. There is clear potential for supply disruption by weather or maintenance delays. There is also the potential for outsourced manufacturing delays, widely publicized for the B-787, when second-tier suppliers do not have the production expertise or capacity to manage ramp-up of production. As with design outsourcing, manufacturing outsourcing has been blamed (without specific evidence) for initial problems with the A-380 and B-787 (Plumer 2013).

Finally, note that aircraft maintenance is largely outsourced by aircraft operators. Whereas 25 years ago each airline had extensive ramp and heavy maintenance facilities, cost pressures have driven them to look for lower overheads and reduced labor costs through outsourcing. Over 64 percent of aircraft maintenance was outsourced in 2007 (McFadden and Worrells 2012), and this includes line, heavy and component maintenance. As Drury, Guy and Wenner (2010) noted, this introduces added complexity of interaction between the aircraft operator, the MRO and the regulatory authorities. However, after three separate studies the authors concluded that the errors tended to remain potential and that outsourced maintenance did not appear to be any more error prone than in-house maintenance. In a follow-up study on the issue of outsourcing to non-English-speaking countries (Drury, Ma and Marin 2009), an extensive experimental data collection on four continents gave quite encouraging results. MROs in Spanish- and Chinese-speaking countries were able to comprehend work documents, such as task cards, written in English, even if they did it more slowly than native English speakers. Training AMTs in technical English and use of well-designed task cards both proved effective in reducing comprehension errors.

(c) Regulation

As noted above (Section 1b), aviation and space flight are highly regulated activities in almost all countries of the world. In the USA, the FAA is the regulatory body, while in the EU the regulatory authority is the European Aviation Safety Agency, which replaced the earlier Joint Aviation Authorities. Each of these agencies, and their equivalents in other jurisdictions, provides detailed regulations that all in aviation must follow in that country. These range from certification of aircrew, AMTs and air traffic controllers, through specifics of design and construction of aircraft and components, to the rules under which aircraft users (e.g. airlines, businesses) must operate. The stated aim of each regulatory authority is safety for those flying and those on the ground, but at least in the USA a part of the FAA's mission is to ensure the "most efficient aerospace system in the world." This means having a dual mission of effectiveness and efficiency, which could potentially be in conflict; for example, in separation standards for commercial aircraft where safety demands larger separations while efficiency demands a smaller value. This is a challenge for regulators, but can also impact design by specifying the equipment that needs to be installed in each aircraft. For aircraft designers and manufacturers, these national regulations provide a consistent design and verification framework to ensure that all manufacturers meet the same minimum standards. Because many regulatory authorities work together, these standards are typically worldwide, although each reserves the right to its own rules. Thus an airliner may be built in the USA (although sourced globally) and certified there, but the manufacturers and many regulatory authorities combine to ensure simultaneous certification in the EU with a minimum of repeated testing. Some jurisdictions, such as the Civil Aviation Administration of China, are newer to the task of certifying major indigenous aircraft (e.g. C-919 or ARJ21) so delays have been experienced. There is always the possibility of government interference, making global certification subject to political pressures.

(d) Materials

Because aerospace continues to move towards higher aircraft performance, it has a vested interest in new materials. The example given here is of aircraft structures, but similar examples come from heat-resistant materials for turbine engines, fireproof interior materials and new battery technology. Metal structures replaced wooden airframes throughout WWI, finding almost universal acceptance from the late 1930s onwards. Light alloy structures demanded careful design where the skin became a load-bearing element, in contrast to an underlying wooden or steel structure with fabric as an aerodynamic covering. This monocoque construction is now standard, but the materials themselves are now changing. Plastics have been used in aircraft structures since WWII, when thermoset plastics (e.g. Bakelite) were used for covering the radar antennae in an aircraft nose. In the 1960s more advanced plastics became available, which have seen use in airframes and other components since the 1980s (GAO 2011). Starting with General Aviation aircraft, composite structures from glass- or carbon-reinforced plastics initially provided tail sections for Airbus (A-320 in 1988) and Boeing (B-777 in 1995) airliners, and currently have been used for whole fuselages and wings (B-787 and A-350). Such composites reduce structure weight for the same strength as aircraft-aluminum alloys (Kalanchiam and Chinnasamy 2012), plus they are corrosion-free and have better fatigue resistance (Ilcewicz 2010). However, such structures bring their own challenges, primarily because they fail in different ways from more traditional alloys – for example, de-lamination between layers, disbanding, water infiltration (GAO 2011) – and new techniques for in-service inspection are required (Drury 2005). These techniques (e.g. ultra-sonic non-destructive inspection) are not new to the aviation maintenance industry, but need to move to a new level of reliability and productivity to be applicable to large-scale structures.

Future materials changes in metals production (e.g. additive manufacturing and explosive forming for complex curved parts (Warwick and Norris 2013)) and in composites (e.g. large stitched carbon fiber without an autoclave (Warwick 2013b)) are coming to airframe manufacturing.

(e) Finance

Design and manufacturing of aircraft are very expensive, and thus a primary challenge for aviation manufacturing. This is obvious from the asking prices of major military and commercial aircraft quoted above (Section 1c), but designing, building, certifying, selling and providing customer service for the smallest general aviation aircraft are certainly not cheap. The start of Airspeed Aircraft in 1930 has been well documented by N. S. Norway, one of its founders, better known as the novelist Neville Shute (Shute 1954). A small number of aircraft designers, out of work when British airship manufacturing was abruptly cancelled, decided to form their own aircraft design and manufacturing company, Airspeed. Norway spent most of one winter soliciting funds for subscription to a share offering in the new company, but only £5000 was raised against the £10,000 minimum thought necessary. The board went ahead, first producing a glider to break records (to publicize the new company), then receiving a powered aircraft order from a close friend. The company went on to manufacture some notable light aircraft, over 8000 Oxford trainers for the Royal Air Force and a number of airliners post-WWII before being taken over by a larger aircraft manufacturer. A more current and unusually financed startup corporation, Quest Aviation, was formed in the 1990s to design and build a rugged, single-turboprop bush plane transport for use in the developing world (Garvey 2013). From the idea to the start of design work in 2002, donations were received from a large number of charities which saw a need for such an aircraft. By 2004 a prototype was flying, with commercial production starting in 2007 and over 80 of these Kodiak aircraft have now been delivered. Uniquely, one in ten of these aircraft have been donated to charities.

The large airframers rely on both commercial and military contracts; thus finance comes both from governments and the commercial world. Both sources are subject to large and rapid swings in financial fortunes with changes of government and economic cycles. Commercial aircraft are particularly vulnerable as they require gestation and payback periods of many years while their customers are driven by immediate financial pressures. The disputes before the World Trade Organization between Boeing and Airbus (Disputes DS316 and DS347), each claiming that the other receives illegal government subsidies, are perhaps the most famous examples of the complexity of the financing of large commercial aircraft. For spacecraft, the recent financing by NASA in the USA of commercial vehicles to supply low earth orbit in place of government-owned vehicles is seen as a breakthrough worldwide, as governments have dominated space travel since its inception. US space vehicles have always been built by commercial enterprises (Boeing, Lockheed-Martin, etc.) but now the emphasis is on vehicles such as the Falcon/Dragon system manufactured by Space Exploration Technologies and flown by them on commercial contract for International Space Station re-supply missions (Warwick and Ostrove 2013).

4. CONCLUSIONS

While aircraft have been in service for a century and spacecraft for half a century, their design, manufacture and continued safe operation represent significant challenges to the enterprises involved. If customers for these vehicles, their operators and passengers, and the world's governments were content with stasis, aviation would be relatively simple. However, despite the fact that aircraft continue in service for many decades, constant improvement is demanded. Although the nature of that improvement might change, embracing at different times production rate, passenger safety, operating economics and environmental concerns, the imperative is still innovation. With such a technically difficult endeavor as flying, the fact that an impressive safety record is now commonplace is in itself a major achievement. New materials, design and manufacturing methods, ever-changing regulations and global production have to be integrated into financially viable enterprises. With five-year forecasts for 9287 commercial transports (Warwick 2013a, p. 100) and 4474 business jets (Warwick, 2013a p. 137), the commercial side of aviation appears set to experience continued growth. The challenges are not diminishing in the near future.

REFERENCES

Amaral, J., Anderson, G.G. and Parker, G.G. (2011), "Putting it Together: How to Succeed in Distributed Product Development," *MIT Sloan Management Review*, **52**(2), 51–58.

Anderson, G.G., Feng, Q. and Parker, G.G. (2011), "Partial Outsourcing and Linked Learning Processes." Available at SSRN: http://ssrn.com/abstract=298624 or http://dx.doi.org/10.2139/ssrn.298624, access date October 13, 2014.

Andrews, C.F. and Morgan, E.B. (1981), *Supermarine Aircraft Since 1914*, London, Putnam.

Boeing (2012), *Statistical Summary of Commercial Jet Airplane Accidents: Worldwide Operations 1959–2013*. Available at http://www.boeing.com/news/techissues/pdf/statsum.pdf, access date October 13, 2014.

Boothroyd, G. (1996), "Design for Manufacture and Assembly: The B-D Experience," in G. Q. Huang (ed.), *Design for X: Concurrent Engineering Imperatives*, London, Chapman & Hall, pp. 19–40.

Brown, D.L. (1970), *Miles Aircraft Since 1925*, London, Putnam & Co.

Cochrane, D., Hardesty, V. and Lee, R. (1989), *The Aviation Careers of Igor Sikorsky*, Seattle, WA, University of Washington Press.

Drury, C.G. (1996), "Design for Inspectability," in G.Q. Huang (ed.), *Design for X: Concurrent Engineering Imperatives*, London, Chapman & Hall, pp. 216–229.

Drury, C.G. (2005), *Handbook of System Reliability in Airframe and Engine Inspection*. Available at http://www.tc.faa.gov/LOGISTICS/grants/pdf/2004/04-G-037.pdf/Table%20of%20Contents.doc, access date October 13, 2014.

Drury, C.G. and Spencer, F.W. (1997), "Human Factors and the Reliability of Airframe Visual Inspection," *Proceedings of the 1997 SAE Airframe/Engine Maintenance & Repair Conference (AEMR'97)*, also SAE Technical Paper 972592, August 1997.

Drury, C.G., Guy, K. and Wenner, C.A. (2010), "Outsourcing Aviation Maintenance: Human Factors Implications, Specifically for Communications," *International Journal of Aviation Psychology*, **20**(2), 124–143.

Drury, C.G., Ma, J. and Marin, C.V. (2009), "Language Error in Aviation Maintenance: Quantifying the Issues and Interventions in Four World Regions," *International Journal of Aviation Psychology*, **20**(1), 25–47.

Drury, C.G., Patel, S.C. and Prabhu, P.V. (2000), "Relative Advantage of Portable Computer-Based Workcards for Aircraft Inspection," *International Journal of Industrial Ergonomics*, **26**(2), 163–176.
FAA (2002), *Estimated Airplane Noise Levels in A-Weighted Decibels*, Advisory Circular 36–3H, Washington, DC, US Department of Transportation.
FAA (2009), *Fact Sheet – NextGen Goal: Performance-Based Navigation*, Washington DC. Available at http://www.faa.gov/news/fact_sheets/news_story.cfm?newsid=8768, access date October 13, 2014.
Foster, S.T. (2007), *Managing Quality: Integrating the Supply Chain* 3rd Edition, Upper Saddle River, NJ, Pearson Education Inc.
Futron (2002), *Space Transportation Costs: Trends in Price Per Pound to Orbit 1990–2000*. Available at www.Futron.com, access date October 13, 2014.
GAO (2004), *Space Shuttle Costs for Hubble Servicing Mission and Implementation of Safety Recommendations Not Yet Definitive*, Report GAO-05–34, Washington, DC, Government Accountability Office (d0534.pdf).
GAO (2011), *Status of FAA's Actions to Oversee the Safety of Composite Airplanes*, Report GAO-11–849, Washington, DC, Government Accountability Office (d11849.pdf).
Garvey, W. (2013), "A Higher Mission," *Aviation Week and Space Technology*, **174**(47), 20.
Gibbs-Smith, C.H. (1966), *The Invention of the Aeroplane (1799–1909)*, London, Faber and Faber.
IATA (2011), *Airline Maintenance Cost, Executive Commentary*, AMC_ExecComment_FY09.pdf, Montreal, Canada.
Ilcewicz, L. (2010), *FAA Composite Safety Administration and Certification Initiatives*, FAA Composite Safety & Certification Initiatives, JAMS Meeting (Seattle, Washington), May 19, 2010.
Jackson, A.J. (1962), *De Havilland Aircraft Since 1909*, London, Putnam.
Jarrett, P. (1997a), *Biplane to Monoplane: Aircraft Development 1919–39*, London, Putnam Aeronautical Books.
Jarrett, P. (1997b), *Aircraft of the Second World War: The Development of the Warplane 1939–45*, London, Putnam Aeronautical Books.
JPDO (2011), *Trajectory-Based Operations (TBO) Study Team Report*, Washington, DC, Joint Planning and Development Office (TBO Study Team Report.pdf).
Kalanchiam, M. and Chinnasamy, M. (2012), "Advantages of Composite Materials in Aircraft Structures," *World Academy of Science, Engineering and Technology*, **71**, 597–601.
Kay, A.L. (2004), *Junkers Aircraft and Engines, 1913–1945*, London, Putnam Aeronautical Books.
Liron, J. (1984), *Les Avions Farman*, Paris, Collection Docavia, Editions Lariviere.
McFadden, M. and Worrells, D.S. (2012), "Global Outsourcing of Aircraft Maintenance," *Journal of Aviation Technology and Engineering*, **1**(2), 63–73.
McMullin, D.L., Jacobsen, A.R., Carvan, D.C., Gardner, R.J. Goegan, J.A. and Koehn, M.S. (2008), "The Boeing 787 Dreamliner – a Case Study in Large-Scale Design Integration," Santa Monica CA, *Proceedings of the Human Factors and Ergonomics Society 52nd Annual Meeting – 2008*, 1670–1671.
NASA (2009), "NASA Assessing New Roles for Ailing QuikScat Satellite." Available at http://winds.jpl.nasa.gov/news/index.cfm?FuseAction=ShowNews&NewsID=1), access date October 13, 2014.
NRC (1997), *Aging of U.S. Air Force Aircraft: Final Report*, National Research Council publications, NMAB-488–2, Washington DC, National Academy Press.
NTSB (2010a), *Loss of Control on Approach, Colgan Air, Inc. Operating as Continental Connection Flight 3407, Bombardier DHC-8–400, N200WQ, Clarence Center, New York, February 12, 2009*, Washington, DC, National Transportation Safety Board (AAR1001.pdf).
NTSB (2010b), *Introduction of Glass Cockpit Avionics into Light Aircraft*, NTSB/SS-10/01, Washington, DC, National Transportation Safety Board (SS1001.pdf).
Opdyke, L.E. (1999), *French Aeroplanes Before The Great War*, Atgen PA, Schiffer Publishing.
Patel, S., Drury, C.G. and Lofgren, J. (1994), "Design of Workcards for Aircraft Inspection," *Applied Ergonomics*, **25**(5), 283–293.
Plumer, B. (2013), "Is Outsourcing to Blame for Boeing's 787 Dreamliner Woes?" *Washington Post*, January 18.
Pudney, J. (1964), *The Camel Fighter*, London, Hamish Hamilton.
Sabbagh, K. (1996), *21st Century Jet*, New York, Scribner.
Schweizer, W. (1991), *Soaring with the Schweizers*, Falls Church, VA, Rivilo Books.
Shute, N. (1954), *Slide Rule: The Autobiography of an Engineer*, New York, Ballentine Books.
Warwick, G. (2013a), "Aerospace 2013," *Aviation Week & Space Technology*, **174**(47), 42–137.
Warwick, G. (2013b), "Future Carbon," *Aviation Week & Space Technology*, **175**(15), 53–54.
Warwick, G. and Norris, G. (2013), "Better Metal," *Aviation Week & Space Technology*, **175**(15), 42–45.
Warwick, G. and Ostrove, W.N. (2013), "Launch Space: Market Shifts, New Competitors Bringing Changes to the Launch-Vehicle Business," *Aviation Week & Space Technology*, **174**(47), 94–95.
Washington, W.G. (2012), "Make or Break: Arguing Complexity is Killing Aerospace, DARPA Seeks Simpler and Faster Design Process," *Aviation Week & Space Technology*, **174**(40), 76–77.

Wegg, J. (1990), *General Dynamics Aircraft and their Predecessors*, London, Putnam.
Weyl, A.R. (1965), *Fokker: The Creative Years*, London, Putnam.

FURTHER READING

Aviation Week & Space Technology. (This weekly magazine is the standard for news of the aviation and space industries. It has regular articles on design innovation, manufacturing, operation and maintenance of aircraft and spacecraft.)

Collection Docavia – Editions Lariviere (in French). (The many volumes in this series include at least five on French aircraft manufacturers which are the best source for French manufacturing history.)

Foster, S.T. (2007), *Managing Quality: Integrating the Supply Chain* 3rd Edition, NJ, Pearson Education Inc. (The third edition of a standard text on supply chain quality.)

Huang, G.Q. (ed.) (1996), *Design for X: Concurrent Engineering Imperatives*, London, Chapman & Hall. (This older compilation covers the relationship between design and manufacturing, inspection, quality and maintenance.)

Khanna, R. (2012), *Entrepreneurial Nation: Why Manufacturing is Still Key to America's Future*, New York, McGraw-Hill. (A useful recent book with good data on US manufacturing.)

Putnam Aeronautical Books, London, Putnam. (The 34 volumes by Putnam constitute the definitive historical record of aviation manufacturers in the UK, USA, Germany and the Soviet Union/Russia.)

Sirkin, H.L., Rose, J. and Zinser, M. (2012), *The US Manufacturing Renaissance: How Shifting Global Economics Are Creating an American Comeback*, Philadelphia, PA, Knowledge@wharton. (A readable (and free) download from amazon.com giving the results of Boston Consulting Group showing the outlines of a potential revival of American manufacturing.)

20. Manufacturing stoke: emergence, transformation and consolidation in the surfboard industry

Andrew Warren and Chris Gibson

INTRODUCTION

This chapter traces the emergence, transformation and consolidation of the surfboard manufacturing industry. With historical antecedents in ancient Polynesian cultures, surfboard making has evolved from a small-scale do-it-yourself (DIY) activity into a formalized commercial industry. Since the late 1950s surfboards have been manufactured in industrial-style workshops located adjacent to popular surfing places. In these settings surfboard making combines a unique range of human skills: an eye for design, knowledge and application of hydrodynamics, manual crafting; and close relations between workers, tools, materials and local customers. Drawing from cultural economy theory, we focus on 33 workshops across three global hubs of surfboard production: Southern California, O'ahu, Hawai'i and east coast Australia. In these three regions surfboards represent functional products utterly necessary to pursue surfing.

Our analysis of surfboard making speaks to a rehabilitated interest among economic geographers in analysing the shifts, developments and challenges facing manufacturing regions, industries, firms and workers (Bryson et al. 2008; Doussard et al. 2009; Warren and Gibson 2013). On the one hand, writers of the post-industrial variety emphasize the reorientation of developed market economies around professional services, retail, finance, creative and information-intensive industries (Bell 1973; Florida 2002; Clark 2011). Under the influence of liquid global capital, manufacturing is theorized as inhabiting a state of incessant decline. On the other hand, alternative perspectives offer incisive reconsideration of the vibrant synergies, diversity, complexities and rescaling of manufacturing (Hudson 2002). Rather than all-encompassing decline, manufacturing in advanced economies is more accurately experiencing constant change. Increases in gross value added, patents, research and development (R&D) spending, output and productivity reflect the ongoing importance of manufacturing. Between 2010 and 2012 manufacturing employment in the United States increased 4 per cent, contributing 489,000 new jobs (Langdon and Lehrman 2012). Some commentators have gone so far as to announce a 'renaissance' in American manufacturing (Sirkin et al. 2012). While perhaps an overstatement, clearly changes to manufacturing industries, policy treatment and divisions of labour proceed at different speeds across different economies.

Jettisoning manufacturing from the terrain of advanced economies neglects the functional linkages operating across sectors. Manufacturing connects to a diverse range of economic activities: design services, R&D, branding, retail, finance and resource extraction (Bryson et al. 2008; Gibson et al. 2012). The automotive industry, for example, depends on steel making, engineering services, parts supply, and industrial and interior design. Likewise, the global electronics industry (computers, mobile phones, televisions) depends on a number of firms functionally connected along design, branding, equipment manufacturing and distribution (Yeung 2007). Trafficking rationalist logics to explain industrial decline has come at the expense of economic knowledge more attuned to the realities that manufacturing is a diverse range of activities inhabiting different scales (Gibson et al. 2012). More detailed investigation into the internal dynamics of manufacturing industries and firms, particularly at regional scales, is needed to enlighten ongoing challenges and successes.

Surfboards, as analysed in this chapter, elaborate vibrant connections between customers, creative design, material innovation and physical production. Rather than abrupt division of work tasks, complementary labour relations support the viability of surfboard workshops in regional coastal settings. The story traversing our chapter is consequently about how small and medium-sized firms and their workforces have emerged, transformed and continued to survive in an increasingly globalized, corporatized and competitive industry. Surfboard manufacturing in Southern California, Hawai'i and east coast Australia entails a story of resilience and adaptiveness.

In the next section we conceptually frame our analysis of surfboard manufacturing. Drawing from cultural economy theory we prioritize the way values, relations, practices and skills shape commercial production. After conceptual framing we outline research methodology. Concentrating on the physical sites of production, our ethnographic approach involved close interaction with workers, workshop managers, firm owners and customers. Empirics follow talk of methods and are organized in three sections. The first section charts the emergence of commercial workshops under systems of local, craft-based custom production. We emphasize the close links between workshops (manufacturers), customers (markets) and local surfing spaces. The second empirical section examines transformations to production and customer relations under capital-intensive, automated systems of production. In light of financial hardships over recent years, the third empirical section discusses the consolidation of surfboard manufacturing around customized production, regional markets and close customer relationships. We conclude the chapter by discussing how surfboards (and the industry's internal dynamics) illustrate the diverse skills, values and exchanges that shape viable manufacturing industries.

CULTURAL ECONOMY AND SURFBOARD MANUFACTURING

Surfing subculture is now a $10 billion a year global industry (Global Industry Analysts 2011). Since the 1970s surf brands have internationalized the distribution of a range of consumer products. Companies with local origins, including Rip Curl, Quiksilver, Billabong and Volcom, have turned into corporate entities. Some surf brands are listed publicly on stock exchanges and pursue new ways of producing and selling surfing gear to the masses. The surf industry encapsulates a diversity of image- and design-based products. Tentacles spread into retail industries, tourism and the manufacture of wet suits, apparel, shoes, sunglasses, watches and hats. Global surf brands have helped popularize the sport of surfing, along with its idyllic image. Surfing's international popularity provides board manufacturers with an expanding commercial market. Surfboards also authenticate the efforts of corporate firms by animating claims of 'genuine' surf brands

(Warren and Gibson 2013). However, price squeezes, brand visibility, sophisticated distribution systems and control of retail space mean corporate surf firms are also a threat to the viability of small independent surfboard workshops.

Analysing the surfboard manufacturing industry through classical and orthodox economic geographical theories (of profit and loss, vertical integration, agglomeration dependencies, rational choice theory) cannot enunciate how the industry works, what factors shape production or the key issues that face workshops and workers. Instead we maintain sensitivity to the 'cultural' logics and values of surfboard manufacturing. Our analysis situates the surfboard industry within a cultural economy framework. Cultural economy is used here to examine the human dimensions shaping commercial production; the relationships between (and among) workshops, retailers, suppliers, workers and the 'things' they make.

Over the last two decades the phrase 'cultural economy' has emerged in various guises. Because 'the' cultural economy is most evident in certain sectors it has become an idiom of the cultural industries (Scott 2000), increasingly replaced with 'creative industries' and including a growing list of activities trading on innovation, entrepreneurialism and intellectual property (see O'Connor 2013). These are industries in every sense, consisting of small and large firms with inputs and outputs, workers, workplaces and sometimes unions. They also differ from other forms of capitalist production. Cultural (and creative) industries bank on immaterial and embodied processes such as inventiveness, resourcefulness and imagination. The job of musicians, artists, directors, actors and designers is to create 'symbolic content' for commercial exchange (Reimer 2009).

Surfboard making itself is a cultural industry. Production involves high levels of creative skill; substantial symbolic components (ascribed meanings of boards within surfing subcultures); knowledge of fashions and preferences; constant updates and adaptations of design. Custom board manufacturers offer high value-added products personalized to individual surfers and local breaks. Customized surfboards are infused with artistic and rarity value. Local workshops contribute to and utilize regional reputations within global surfing subcultures to export their products. Surfboards entangle physical geographies, popular local pastimes, tastes, design and material production (see Molotch 2002). But the cultural economy *as* cultural/creative industries approach induces a different epistemology from the original framework. Culture is treated as an economized component of advanced post-industrial societies, as opposed to a constituent feature actively shaping the spaces and relations of economies.

In this chapter we use a cultural economy approach to disrupt dichotomies of 'culture' and 'economy' as discrete 'natural' categories (Gibson-Graham 2008; Gibson 2012). Cultural economy theory stresses the entwined nature of economic, cultural, social and political relations (Barnes 2001). The purpose is to shift the epistemology of economic knowledge away from the premise of underlying, abstract market-driven forces towards theorizing the relations between cultures and economies. Channelling Karl Polanyi (2001 [1944]), it is exigent to place economies in their relevant contexts, accepting history, politics and culture as profound 'shapers' of economies and development trajectories. Cultural economy is, therefore, an approach to unravel how economies are made and remade (Amin and Thrift 2007). In questioning what it is that constitutes the fabric of 'economies' (sectors, industries, firms and workers), a cultural economy framework allocates no ontological privilege. Instead the aim is to consider actors, spaces, activities, knowledge and resources situated outside traditional notions of economy, yet which help bring 'the' economic into being.

The polysemy of cultural economy frameworks is helpful for examining surfboard manufacturing. The surfboard industry has its origins as DIY craftwork performed in coastal suburban tool sheds and organized around waged day jobs. Hobbyist 'backyard' work was motivated by personal desires to go surfing. Human actions were much more than collectivized tactics implemented to gain economic advantage. Cultural values, personal passions, relationships, sporting competition and local geography shape surfboard making as an industry as much as wages, prices and rents.

Nonetheless core 'economic' matters of market share and proximity, competition, exploitation, agglomeration and labour markets still have significance. Cultural economy frameworks cannot simply reverse a binary, making culture primary and economy secondary. Economies are open, porous entities influenced as much by culture as culture is by economies. Cultural economy brings into clarity how economic activity is measured, categorized and constructed (Mitchell 2008). Culture is a constituent feature of economies, used in this chapter for theorizing how small, independent manufacturing workshops create and maintain viability in an era of cheaper, standardized (re)production.

METHODOLOGY: ETHNOGRAPHY AND LOCATING WORKSHOPS

Empirical research presented in this chapter was undertaken inside surfboard workshops. Research centred on the three most renowned surfing regions globally: Southern California, O'ahu, Hawai'i and east coast Australia. Between 2008 and 2013, more than 130 workers from 33 workshops took part in the research. A total of 12 workshops were located along Australia's expansive east coast; 11 in Southern California (between Los Angeles and San Diego) and 10 on O'ahu, Hawai'i. In each region, workshops employed between 2 and 23 workers; ranging from shapers and glassers (the two core labour tasks) to polishers, sanders, sales staff and shopfloor managers. Workshops reflected the different scales of surfboard production globally. Overall, 12 firms exported surfboards internationally, with a further 15 distributing to national retail markets. The six smallest workshops manufactured a few hundred boards a year and serviced surfing communities within their immediate regions.

In each factory ethnography took the form of extended 'workplace tours', designed to help immerse researchers in the activities of workshops and workers (see Burawoy 1998). In workshops time was spent actively participating in jobs alongside board makers – unloading trucks, cleaning tools, carrying materials and sanding boards. Technical tasks designing, crafting and sealing surfboards were closely observed, as shapers and glassers discussed their work in situ. Semi-structured interviews were conducted with the 33 workshop owners/managers. Extended conversations were also held with materials suppliers and customers. Interviews and discussion during workplace tours were captured with a hand-held audio recorder. Some 600 hours of interview material was collected. A fieldwork diary was used to document observations and discussions not captured on tape.

Ethnography benefited from the residual positioning of surfboard making within wider surfing subculture. Andrew Warren is an active member of that subculture, which enabled a measure of 'insider' status to be established. Positionality became crucial for eliciting and interpreting responses and interactions, which 'outside' researchers might have ignored or misinterpreted. Surfboard making is also highly gendered and hetero-normalized. As a white, heterosexual male, a level of acceptance and access was granted by male board makers; unlikely for non-surfing 'outsiders' or female researchers. A final point: to maintain privacy pseudonyms are used in the chapter when referring to individual participants. Where relevant, workshops are referred to using the initial letter corresponding to their location (C: California; A: Australia and H: Hawai'i). The number following the letter denotes firm size in descending order relative to location. Thus A1 was the largest participating workshop in Australia; A12 the smallest.

EMERGENCE OF THE SURFBOARD MANUFACTURING INDUSTRY

Surfboard making's historical and cultural antecedents are located in ancient Polynesia, especially Hawai'i. Through colonization surfing and board making were maintained as cultural practices by Hawaiians (cf. Walker 2011). In the early 1900s surfing participation spread to California and east coast Australia, influenced by a fledgling tourism industry and demonstrations by visiting Hawaiian surfers.[1] Hawai'i was the surfing centre of the world, yet by the 1920s stand-up surfing was practised along Southern California and in Australia, particularly around Manly and Bondi Beaches in suburban Sydney and further north on the Gold Coast.

Prior to the 1950s surfboard making was a DIY hobbyist activity operating outside the formal economy. The 'hubs' of production were backyard workshops and tool sheds in prosaic coastal towns and suburbs. Work on surfboards was structured around waged day jobs. Australian surfer Barry Bennett started making surfboards from wood in the 1940s. During an interview Bennett outlined the nature of early DIY surfboard making:

> I was an electrician, and it was work your day with the Maritime Service Board, come home, and do a couple of hours on a board before tea. That's how it was. There was someone making them in their garage. No one was dependent on making a surfboard for a living. It was out of a love of surfing that we got into it. . . . We started building them under my mum's house [Waverley]. There would be timber across the yard, half-finished boards leaning against the house, pots of varnish in the driveway. It was a hobby, so we could go surfing.

Surfboards were made purely to enable surfing participation. As with Barry Bennett in Australia, Californian surfer Hobie Alter told how his first surfboard factory was established at Dana Point in 1954. Commercial success arrived gradually. In an interview with surf journalist Drew Kampion, Alter explained how 'people laughed at me for setting up a surfboard shop. . . . They said "once you've sold a surfboard to each of the 250 surfers on the [Californian] coast, you'll be out of business." But the orders kept coming.' Hobie Alter would become one of the surfboard industry's most creative and influential figures.

In the late 1950s two important developments helped catalyse surfboard making into a formal industry: foam and popular culture. In June 1958, after two years of testing, Alter and fellow Californian surfer Gordon Clark discovered the correct chemical mixture and a high-pressure system for casting polyurethane (PU) foam in a variety of lengths and thicknesses suitable for surfboard making. Whereas timber surfboards (the standard material for construction to that point) required three or four days of milling, filing, trimming, gluing, screwing and sanding, foam 'blanks' were much less labour intensive. The varnish used on timber was replaced by fibreglass cloth and liquefied polyester resin to create a waterproof seal. The innovation of PU foam rendered hardwood and hollow timber boards obsolete.

Shortly after the integration of foam, popular culture propagated the beach and surfing to a wide audience. Think *Gidget*, the *Beach Party* films, Elvis's Hawai'i-themed movies and The Beach Boys' music. New markets developed supporting larger-scale commercial production. Timing was everything. At precisely the moment surfing converged with popular film and music, foam was becoming readily available. Growing participation spurred unprecedented demand for surfboards at both ends of the Pacific Ocean. Viable commercial markets congealed and existing surfers making DIY boards recognized emergent business opportunities.

Growing levels of surfing participation, access to foam and ambitions to forge a living from surfboards contributed to a 'surfing boom' that lasted for most of the 1960s (Lawler 2011). One of the first commercial workshops on Australia's famous Gold Coast was started by Joe Larkin, who described how his surfboard making became a sustainable enterprise:

> One thing we did in the 1960s was change the DIY approach. Surfboards were being made in the back shed, slapped together on weekends or after work. But once we could make a better product and at a reasonable price, it became more attractive to just buy them. The big thing that drove this was the foam. See foam reduced the labour involved in making the boards and was a bit of a catalyst. With all these new people surfing, I could see a full-time, profitable thing emerging. . . . I was walking on trawlers [deckhand] in 1962 to supplement my income. But the surfboard making paid its own way in 1964. By 1965, I had nine blokes working for me in the workshop on Kirra Beach. (Interview)

Foam assisted in making boards more efficiently. Popular culture drove surfing participation and consumer demand. Surfboard workshops became small fledgling commercial firms with rents, employees, materials suppliers and customers.

By the 1960s clusters of surfboard workshops had become established in popular surfing towns: Santa Monica, Dana Point and San Clemente in Southern California; O'ahu's North Shore and Waikiki area; and along Australia's east coast from the Gold Coast in the north to Torquay in the south. Long-time workshop owners we interviewed explained how they had borrowed money from friends and family to fund initial commercial ventures. Hobie Alter borrowed $1500 from his father for his Dana Point factory; Joe Larkin also borrowed start-up funds from his family. Only a small number of workshops had credit provided by conventional lending institutions such as banks.

Manual Systems of Custom Production

The commercial strategies for early manufacturers involved offering customized surfboards suited to individual surfing bodies and local breaks. Employed board makers were often skilled local surfers who carried subcultural capital and maintained relationships with customers. The physical process for hand making custom surfboards had changed remarkably little in six decades. Shapers designed and planed surfboards from moulded 'blanks', made from either PU or polystyrene foam. Creating new designs and material production occurred within the labour of the shaper. Hand shapers considered each customer's surfing ability, body size and weight (affecting buoyancy and movement) and the types of wave ridden. The physical geography of a workshop's location influenced the styles of board they manufactured. Large and powerful waves frequent to Hawai'i demanded very different designs and use of materials compared with boards made for the smaller wave point breaks of Australia's Gold Coast. After shaping, workers known as glassers covered boards with fibreglass cloth, coating with liquefied resin and polishing to achieve a smooth and shiny finish. The glasser's labour ensured a waterproof finish.

Hand shaping and glassing were both examples of manual labour involving highly attuned embodied skills: sense of feel and touch over foam and fibreglass, skilled use of hand-held tools (planers, sanders, surface form tools and squeegees) and a sensitive eye for detail. In larger workshops, sanding and polishing were jobs devolved from shaping and glassing. Delegating these jobs could improve efficiency and 'train up' younger apprentice makers for careers in shaping and glassing. Shop managers were employed in larger workshops to ensure business was well organized and customers well serviced. The viability of custom production was supported by the high turnover rates for boards. Regular surfers in each of the three case study locations ordered two or three custom boards a year (costing $600 to $1200 each depending on their design). New boards were replacements for broken models and specific orders for ritualistic surf holidays.

An experienced shaper on the North Shore of O'ahu explained the custom system of surfboard production in H4:

> Customers come into the workshop and me or Blake sit down and go through the style of board they're riding now. We talk about what they are after in a new board. You need to know what sort of surfer they are. How big are they? What waves do they surf? All this is important. I come up with a design applying the different elements that give the board its function: width, length, thickness, tail, fins and rocker. Design features make custom boards unique [Figure 20.1]. I work through the design so when I get my hands on the blank the board is pictured in my mind. Then I go to work shaping it. Custom boards are a personal thing. We've developed a lot of close friendships with customers. (Mark, workshop tour)

Rather than being detached, workers in custom surfboard manufacturing were closely tied to customers. Shapers expertly completed a number of different labour tasks; meeting with customers to discuss their needs; designing functional products and shaping boards into material form. Surfboard makers often shared friendships with customers and shared waves at local breaks. Participation and enmeshing within local surfing subcultures were an enduring legacy of custom surfboard manufacturing. Occupational skills accrued gradually. Workers moved slowly from apprentice to early career and then journeyman maker. Only after 30 years in the industry, personally making and supervising some 30,000 surfboards (boards made by hand were usually numbered, making it possible to 'quantify' expertise), did a shaper gain recognition in the industry as a master craftsman.

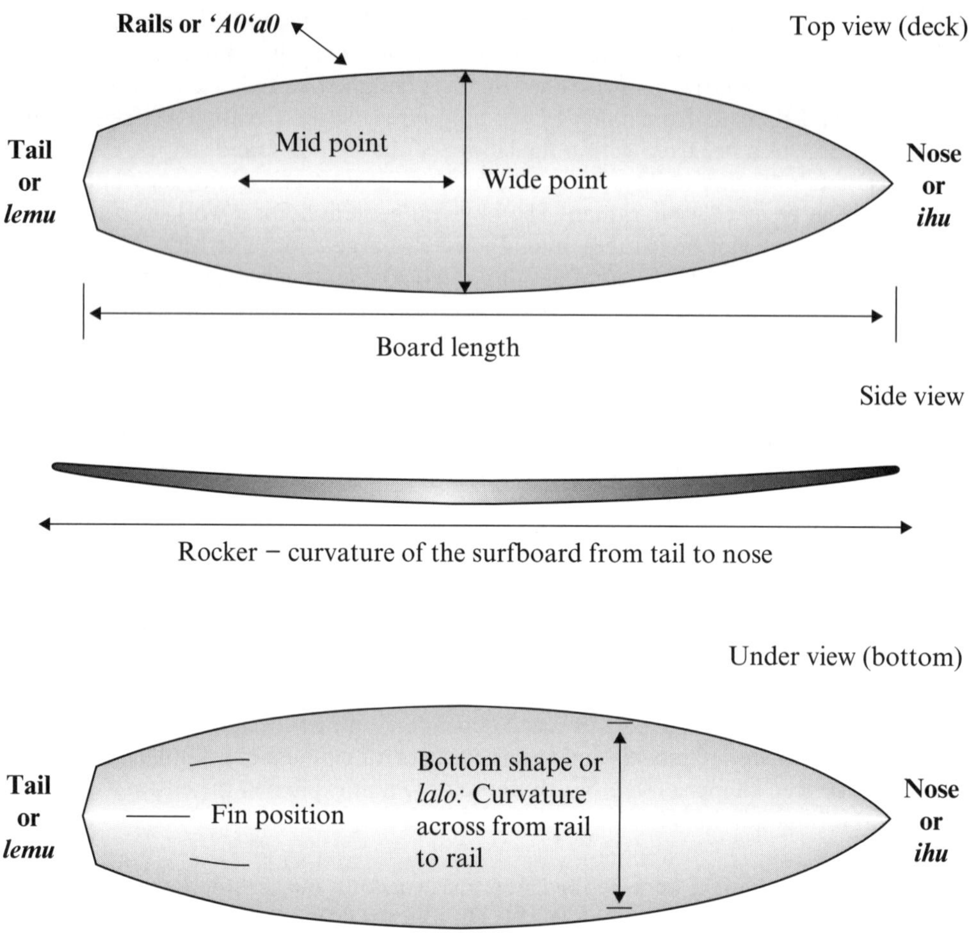

Source: Authors.

Figure 20.1 Profile of a typical surfboard and its design elements

The most popular workshops in California, Hawai'i and Australia produced 1500 handmade custom surfboards a year, turning over around $1 million. Most made more modest incomes producing 500 to 600 custom boards annually. Labour intensiveness and need to focus on regional styles, tastes and waves constrained plans from workshop owners to increase market reach and profitability. Consequently surfboard labels sought out new methods of organizing and structuring operations to fulfil ambitions to up-scale outside immediate regional settings.

TRANSFORMATION: CAPITAL-INTENSIVE AUTOMATED PRODUCTION

In the 1990s surfing received heightened exposure through a well-organized and promoted World Championship Tour. The 'big three' surf brands (Rip Curl, Quiksilver and Billabong) intensified international marketing campaigns and developed corporate profiles. Surfing's economic success mapped onto the intensification of surf retail (Warren and Gibson 2013). In light of heightened marketability, a number of board manufacturers in well-known surfing regions pursued fresh opportunities to increase market share and profitability.

From the late 1990s surfboard manufacturing was transformed by capital-intensive systems of automated mass production. Computer aided design (CAD) programs and computer numerical control (CNC) machines were adapted to automate the shaping of foam blanks. In using automated technologies designs were created with CAD programs, which coordinated the movement of CNC cutting instruments. Surfboard firms widely adapted to CAD/CNC to raise productivity and expand sales. Standardization became crucial for up-scaling. Eddy, the owner of H1, which sold 5000 computer-shaped surfboards a year, summed up the benefits of CAD/CNC:

> Hand shaping limits the number of boards you can make, simply because a hand shaper can only make so many each week. But with the shaping machines, which we started using 15 years ago now, you can copy a standard design and go about producing them over and over for the surfers who don't need a customized product they maybe can't use properly anyway. With the way the technology is now, I can also pump out more refined boards and get better consistency. I can guarantee a level of product because things are standardized. That all adds up to a more profitable business. (Interview)

Unavoidably long production times under manual, customized systems of production were dramatically reduced using CAD/CNC. During busy summer periods waiting times of eight weeks were common for hand-shaped custom boards. Many surfing consumers needed nothing more than a generic design which they could purchase at a surf shop, a user-friendly board that could be tried out at the beach immediately. Automation and standardization seemingly became a solution for small, independent workshops seeking to reach a wider mass market.

In comparison to the 20 or 30 boards that experienced hand shapers could produce in a five-day working week, CAD/CNC systems churned out 200 in that time. Once the CNC machine shaped a blank, it required five minutes of fine hand sanding by a worker known as a 'production' shaper. The surfboard was than ready for glassing, which remained manual, hand-based work. Mechanized production helped workshops increase output to target more lucrative, mass markets outside regional bases. Rip Curl, Quiksilver and Billabong were also at precisely this time offshoring production of surf apparel (to China, Taiwan, India, Vietnam, Guatemala and Mexico), developing extensive retail networks and launching well-funded marketing campaigns.

The owner of A5 outlined the logic underpinning their decision to switch to more capital-intensive automated systems of production:

> I wanted to expand my market. Customization is all very time consuming and with the average surfer we don't need to spend all that time talking about a board, especially when they can't

> surf properly. Or, maybe the customer just wants a standard design that works fine in a variety of conditions. Automation and the computer help with that side of things and open you up to a lot more opportunities. (Jim, interview)

Rather than surfers meeting with and talking to the people making their boards, customers were encouraged to purchase models from supplied retailers. Aspiring workshops utilized CAD/CNC technologies to expand market coverage and profitability. With automated technologies standardizing design the surfboard industry became less like an artisanal craft-based industry and more like other mass producing forms of manufacturing. Hand shaping was downsized and confined mainly to supplying designs and board templates for computerized shaping, functioning something like an R&D division. Close associations between makers and surfers were increasingly disconnected in favour of the surfboard industry's equivalent to mass production.

In total 30 of the 33 participating workshops utilized CAD/CNC systems of production to some level. Two-thirds of the workshops using CAD/CNC technology maintained hand shaping at lower levels to fill specific custom orders. The other third (ten workshops) rarely hand shaped any surfboards – four on O'ahu, three in Southern California and three on the east coast of Australia. A small minority (two workshops in Australia and one in Southern California) continued hand shaping all of their boards, refusing to use computerized shaping technology. Workshops using CAD/CNC technologies increased profit margins by 5 to 6 per cent over manual systems of custom production. Yet the real benefits of automation and standardization were in growing sales volumes through increasing market reach.

New Business Dealings

For surfboard workshops up-scaling production, market access was mediated by retailers, subcontracting and intensified marketing. Instead of shapers and customers meeting in workshops and sharing conversations on the style and design of a new surfboard, CAD/CNC focused on general mass markets. Workshops reproduced easy-to-ride designs or tweaked models that were at one point designed and made by a hand shaper for a local surfer. Wider market access was achieved by utilizing surf retail outlets. The owner of A8 provided his interpretation of the linkages between automated production, standardized design and surf retail:

> In the retail market, the way I look at it, you've got a set of rings, there's an inner ring, there's a middle ring, and then there's an outer ring, and I've always wanted to reach the outer ring. People that just want to get in touch with the culture, how are they going to get a board for little Johnny? They don't necessarily need to take my time asking for measurements and designs. It's all very time consuming, and I've had to try and sort of cut that out too. That's when I struck up a relationship with a surf retail chain. They sell my boards in their shops off the shelf. (Workshop tour)

Surf retailers happily accepted a supply of surfboards for a cut of the final sales price. Despite contributing only 10 per cent of overall sales, surfboards helped reinforce a surf retailer's 'genuine' status and guaranteed a certain level of foot traffic. In each region a small number of surfboard workshops invested in separate retail stores. Alternatively, workshops without the capacity to finance an external retail outlet reconfigured space to incorporate a small retail section, comparable to a car dealership's showroom.

Two methods were followed by manufacturers to supply external retailers. In the first approach, workshops directly freighted finished surfboards. In the second approach, copies of CAD files were emailed to a contracted firm that produced the designs in a final place of sale. The latter approach relied on contractors manufacturing, labelling and delivering an agreed number of boards to licensed retailers. Electronic approaches saved on freight costs, inevitably damaged inventory and was preferred for export. Freight-based approaches were favoured for domestic retail supply.

To increase brand visibility and achieve market reach, workshops invested in marketing campaigns. Some workshops spent 15 per cent of gross monthly income on advertising. Television and print media were used, with the most popular approach involving sponsorship of professional competitive surfers. Branded boards and sponsorship cheques were provided to selected professional surfers who competed in international events and aimed to achieve exposure in print and online surf media. Several workshops managed a 'team' of riders, including world champion surfers. The surfers' job was to promote the label to a broad surfing audience. Workshops also adapted to e-commerce. Customers could visit a central website and purchase an advertised model. After paying with a credit card, surfboards would be shipped within a few business days.

Retail Troubles

In transforming to automated systems of standardized production, surfboard workshops necessarily accrued debt. Initial borrowing of around $150,000 funded the set-up of CAD/CNC systems and reconfiguration of workshops. Borrowing costs added onto increased marketing expenses and bills from material suppliers. Profits had grown in conjunction with market reach, facilitated by strong retail presence and achieving economies of scale. Nevertheless with the global financial crisis workshops were being severely pressured by retailers to lower their wholesale prices. With a number of surfboard firms using CAD/CNC shifting production to China (Boardworks) and Thailand (Global Surf Industries, Firewire), retailers were able to leverage established local labels to achieve lower pricing. In contrast to cash on delivery for supplied boards, retailers stretched out consignment periods, not paying workshops for stocked boards until they sold.

When asked about their relations with surf retailers, workshop owners painted a picture of increasingly exploitative relations:

> We had a pretty good relationship with retailers. We supplied boards and an invoice. The invoice was paid in 30 days. No problem there. But over the last few years the balance of power is all with them [retailers]. With more competition retailers usually go with the lowest price that will give them the best margin. My boards have to sit on consignment and I'm not paid until they've all sold. . . . That is challenging for a small manufacturer. We aren't a big company and pay suppliers and wages every week. (Kurt, H7, interview)

Holding workshops to tight retail deals was the realization that high sales volumes needed to be maintained to meet debt, labour and materials costs.

Exacerbating a difficult retail environment was the informal culture pervading the surfboard manufacturing industry. Signed contracts with retailers were almost unheard of. New surfboard designs were rarely subjected to intellectual property protection or trademarking. Business deals became organized around verbal agreements and handshakes. Transference of design knowledge and hand-crafting skills was also interrupted by CAD/CNC automation. Hand-shaped custom production had a built-in succession process with skills and knowledge passed from one generation of workers to the next.

Laid-back business relations can be traced back to surfing's countercultural days of the 1960s and 1970s. In particular, the lack of intellectual property protection and patenting of new innovations generated a situation where successful designs emerging from one workshop were copied, reproduced and relabelled by competitors. Surfboard makers described this as the 'culture of the industry'. Mike's story was illustrative:

> I stole the design off [shaper's name omitted]. I copied one of his boards [using CAD/CNC] because a couple of mates really wanted one. So I made one, and it went really ******* good. Another guy wanted one, then another guy, and another guy. So I finished up, "Ah ****, all right, I'll make a model," and I called it the [name omitted]. Check out the logo. I put it on Facebook and got a lot of laughs from it. We've always been stealing each other's models anyway [laughs]. (A6, interview)

In creating new board designs and crafting foam blanks, surfboard makers effectively turned their intellectual capital over to workshops for replication under systems of CAD/CNC mass production.

Further complexities encountered in the industry surrounded the oligopolization of surf retail by the three large surf apparel companies, and their expanding use of 'ghost' shaping arrangements. None of the big three remained directly involved in surfboard manufacturing. Instead large surf brands contracted independent workshops to make a set number of surfboards. The boards were ridden by the corporate firm's own professional team riders and sold in their flagship retail stores. Workshops making the boards were rendered 'ghosts' in the production process because their roles remained uncredited. Contracted factories were not permitted to label boards.

The following example was instructive. A2 was a successful label with a strong regional market and exported 2000 boards per year to licensed dealers in Japan. A2 was also the Australian licensee of H2, a larger firm based in Hawai'i and exporting designs to the Australian market. H2 emailed A2 their CAD files. A2 made and labelled H2's boards locally to the set specifications using their own CAD/CNC equipment. But problems arose when the Hawaiian brand, in turn, entered into a production and distribution deal with a large surf company. Suddenly A2 found themselves manufacturing boards as an outsourced third party. This had little benefit for the workshop because the large surf apparel company demanded A2 provide exclusively to their flagship stores in Australia. Those shops subsequently only ordered a small number of boards:

> Because they took them [H1] on as one of the [company name removed] shapers . . . we had a relationship with them as far as supplying boards in Australia. But one of the things they discussed in forming our relationship in the beginning was that if they were prepared to maintain good numbers and keep stock up then we should give them exclusivity and only supply to their stores and no others. Then we said to a whole bunch of shops, sorry we can't supply you because we have a deal with [company name removed]. . . . But they never ordered many boards. We had just flogged off these other shops that would have stocked our boards. (A1 owner, workshop tour)

Not a rare example, the dubious dealings and episodic breakdowns in trust between producers and stores were the source of much discussion and complaint by managers and workers alike. Ghost shaping and uncertain retail dealings had become 'part of the industry' with the integration of standardized production using CAD/CNC automation (Chuck, C3, interview). The situation with surfboard workshops confirms that an important consideration in debates about the future viability of manufacturing in advanced market economies is the shifting role of retailers in specific sectors upstream of production – and how concentration of power at this point in the 'system of provision' for individual goods can complement or jeopardize local manufacturers (see Fine and Leopold 1993).

The problems with retail dealings followed an increasingly insecure employment landscape for surfboard makers. Capital-intensive automated production targeted mass markets and diluted the close, personal relationships between customers and surfboard makers. With CAD/CNC automated production the ability of workers to positively change employment conditions – given a lack of any collectivized union – had also diminished. A job in the surfboard industry was now characterized by casualized, temporary and contract work. Job insecurity was an acute but regular experience. Under casual employment workers made a set amount of money for each board they shaped or glassed. Workers across all three regions explained how the going rates of pay had not changed in a decade.

Despite attempts to improve profitability using CAD/CNC technologies, among the workshops we surveyed market reach had actually been lost to offshore competitors. In 2004, 74 per cent of reported surfboard sales from surf retailers in the United States were for boards made domestically; in 2010, the number had dropped to 63 per cent (Surf Industry Manufacturers' Association 2011). While surfboard sales from retail stores remained about the same size between 2004 and 2010 ($190 million), the market share of imported boards in the United States rose from $49.4 million in 2004 to $70.3 million in 2010 – a 42 per cent increase in six years. Once an intrinsically local form of commercial production, surfboard making was, like many other forms of manufacturing, in danger of becoming more and more a global, low cost, low quality and high turnover affair.

CONSOLIDATING AND MAINTAINING VIABILITY

Since 2008 surfboard workshops in California, Hawai'i and Australia have faced significant economic challenges. Australian workshops using CAD/CNC to export boards have suffered because of the prolonged high value of the national currency. In October 2011 renowned Gold Coast manufacturer BASE Surfboards was declared insolvent, with debts totalling $5.6 million and real assets of less than $1 million (Warren and Gibson 2013). The closure resulted in the direct loss of 50 local jobs. Combined with competition from offshore players in Thailand and China, workshops in California and Hawai'i have experienced low levels of domestic retail spending, flowing into once profitable export markets in Japan and France.

Nonetheless, despite global uncertainty local surfboard manufacturing remains viable, albeit at smaller scales of custom production. Among the workshops toured those doing best had reconsolidated business around customization, maintaining a personal service for regional consumers:

> What we are seeing over the past few years is a real return to the factory. Custom orders have grown 10 per cent this year [2012]. Things are pretty solid actually. Rather than buying a board from a retail store, our long-time customers keep coming back. But we also see a lot of the younger surfers coming in. They've been surfing a few years now and realize they've been riding the same boring thing everyone else is. Custom is great because things at the local level are healthy. (Graham, C5, interview)

Customers continued to place value on the relationships they developed with board makers. Regionally made surfboards also performed better on local waves, a significant incentive to keep buying bespoke boards.

A conversation with a 26-year-old customer in Hawai'i articulated the persistent competitive advantage provided by small workshops trading in local, high quality production:

> Andrew: You were buying your boards from [surf retailer] but now you're here getting a custom board?
> Corey: This is my third. I started surfing and knew nothing about what goes into a surfboard. It was easy to pick one off the shelf in a surf shop. Then I had a bunch of friends who got custom boards made. They were sweet. So I walked in here, spoke with Todd [shaper] and haven't gone back.
> Andrew: Has the price become less important?
> Corey: Custom boards are usually well made. They last longer and so it's a better experience. Compared to buying something cheaper that is the 3000th copy of another model, well I would rather go custom and pay some more.

Labels that had extensively turned to CAD/CNC for standardization were actively realigning business plans. For some there was a perceptible refocusing on 'providing the custom experience' (Charlie, C3, interview). For other workshops a sustainable small business was about continuing to offer quality made customized boards. Dealing directly with local surfers also tempered reliance and participation in external retail networks.

For workshop owners, consolidation of business on customization, personal service and attention to detail revealed the resilience of surfboard manufacturing and regionally based markets:

> I'm thriving on custom orders. That's getting stronger. Surf shops have a bad name now for selling surfboards; you know 18 year old girls and little groovy guys don't put you on a good surfboard. So sales are coming back to the surfboard factory, a little bit more direct like it used to be. . . . Rip Curl, Quiksilver and Billabong making massive profits a few years ago, it gave the surfboard industry money and just a big bubble. (A3, interview)

Despite troubling retail relations and threats from offshore manufacturers, ongoing demand and strengthening of custom markets among local surfers provided a sense of optimism for the surfboard industry's future. Surfboard workshops, makers and markets remained viable in established, regional centres of production.

CONCLUSIONS

This chapter has outlined both the challenges facing the surfboard industry and the features sustaining healthy surfboard firms. Clearly, the informal historical and subcultural origins of surfboard manufacturing have patterned structures of the industry in ways that reverberate into the present. Commercial problems have been aggravated by new global competition and slow retail. Nonetheless, independent workshops in Southern California, O'ahu, Hawai'i and east coast Australia offer high value-added products infused with artistic and rarity value. Surfboard workshops and their workers commit to making expensive, high quality boards personalized to individual riders and local surfing spaces. Board making thus offers important insights into the need to consider the internal dynamics and relations of particular manufacturing industries and firms (Warren and Gibson 2013). Ongoing debates around the future of manufacturing in advanced market economies frequently eschew the linkages and synergies internal to industries, firms and locations. Such neglect is to the detriment of efforts to understand the evolving systems of production, markets and relations supporting, transforming and sustaining particular forms of manufacturing.

It may be that industries such as surfboard making are slighted by regional development policy because of small statistical size, narrow niche markets, marginal profitability and informality. We argue that such conclusions are a mistake. Even as macro-economic conditions conspire to damage exports and retail, surfboard workshops are sustained by local demand, loyal consumers and enduring passions for wave riding. Surfboard making challenges regional policy makers to rethink programmes that condemn manufacturing to the industrial scrap heap. Regional places, their manufacturing firms, industries and workforces can be more viable as centres of production than is often theorized. In surfboard making viability is created through enmeshing and participation within local subculture. Sensitivity to local consumers (and surf breaks) is maintained, along with the succession of creativity and knowledge.

For most of the surfboard industry's century-long history, production has been about bespoke, craft-based labour. Mechanization and low cost imports only came very late (in the 1990s), well after an extensive international commercial sector had been established around local artisanship. Surfboard manufacturing's ongoing focus on quality, craft and creativity has become a valid response to declining Fordism and a means to consolidate regional production and consumption (see also Bryson et al. 2008). Equally significant, surfboard making highlights the limitations and dangers of up-scaling craft-based production from regional subcultural origins in pursuit of global markets and profits. Less a moribund economic activity shifting immovably to cheap labour locations, surfboards demonstrate the cross-fertilization of skills and knowledge essential in achieving productive, enduring and more efficient forms of manufacturing.

NOTE

1. The most famous of the travelling Hawaiian surfers were George Freeth, who toured California in 1908–1909, and Duke Kahanamoku, who first visited California in 1912 and travelled to eastern Australia in 1914–1915. Freeth and Kahanamoku initiated the first wave of surfing popularity outside the Pacific Islands, actively teaching, demonstrating and sharing their surfing knowledge.

REFERENCES

Amin, A. and N. Thrift (2007), 'Cultural economy and cities', *Progress in Human Geography*, **31** (2), 143–161.
Barnes, T.J. (2001), 'Retheorizing economic geography: from the quantitative revolution to the "cultural" turn', *Annals of the Association of American Geographers*, **91** (3), 546–565.
Bell, Daniel (1973), *The Coming of Post-Industrial Society*, New York: Basic Books.
Bryson, J.R., M. Taylor and R. Cooper (2008), 'Competing by design, specialization and customization: manufacturing locks in the West Midlands (UK)', *Geografiska Annaler B*, **90** (2), 173–186.
Burawoy, M. (1998), 'The extended case method', *Sociological Theory*, **16** (1), 4–33.
Clark, David (2011), *Post-Industrial America: A Geographical Perspective*, New York: Routledge.
Doussard, M., J. Peck and N. Theodore (2009), 'After deindustrialization: uneven growth and economic inequality in "postindustrial" Chicago', *Economic Geography*, **85** (2), 183–207.
Fine, Ben and Ellen Leopold (1993), *The World of Consumption*, London: Routledge.
Florida, Richard (2002), *The Rise of the Creative Class. . . and How It's Transforming Work, Leisure, Community and Everyday Life*, New York: Basic Books.
Gibson, C. (2012), 'Cultural economy: achievements, divergences, future prospects', *Geographical Research*, **50** (3), 282–290.
Gibson, C., C. Carr and A. Warren (2012), 'A country that makes things?', *Australian Geographer*, **43** (4), 109–113.
Gibson-Graham, J.K. (2008), 'Diverse economies: performative practices for "other worlds"', *Progress in Human Geography*, **32**, 613–632.
Global Industry Analysts (2011), *Surfing: A Global Strategic Business Report*, available at http://www.strategyr.com/Surfing_Market_Report.asp (accessed 10 June 2013).
Hudson, R. (2002), 'Changing industrial production systems and regional development in the New Europe', *Transactions of the Institute of British Geographers*, **27** (3), 262–281.
Langdon, D. and R. Lehrman (2012), 'The benefits of manufacturing jobs', U.S. Department of Commerce: Economics and Statistics Administration, available at http://www.esa.doc.gov/sites/default/files/news/documents/1the_benefits_of_manufacturing_jobs_final_5_9_12.pdf (accessed 9 June 2013).
Lawler, Kristin (2011), *The American Surfer: Radical Culture and Capitalism*, New York: Routledge.
Mitchell, T. (2008), 'Rethinking economy', *Geoforum*, **39**, 1116–1121.
Molotch, H. (2002), 'Place in product', *International Journal of Urban and Regional Research*, **26**, 665–688.
O'Connor, J. (2013), 'Intermediaries and imaginaries in the cultural and creative industries', *Regional Studies*, Online first, DOI:10.1080/00343404.2012.748982.
Polanyi, Karl (2001 [1944]), *The Great Transformation: The Political and Economic Origins of Our Time*, Boston: Beacon Press.
Reimer, S. (2009), 'Geographies of production III: knowledge, cultural economies and work (revisited)', *Progress in Human Geography*, **33**, 677–684.
Scott, Allen. J (2000), *The Cultural Economy of Cities*, London: Sage.
Sirkin, Harold L., Justin Rose and Michael Zinser (2012), *The US Manufacturing Renaissance: How Shifting Global Economics Are Creating an American Comeback*, Philadelphia: Wharton.
Surf Industry Manufacturers' Association (2011), *2010 Retail Distribution Study*, Custom Data Request, Boulder: Leisure Trends Group.
Walker, Isaiah Helekunihi (2011), *Waves of Resistance: Surfing and History in Twentieth-Century Hawai'i*, Honolulu: University of Hawai'i Press.
Warren, A. and C. Gibson (2013), 'Making things in a high-dollar Australia: the case of the surfboard industry', *Journal of Australian Political Economy*, **71** (2), 26–50.
Yeung, H. (2007), 'From followers to market leaders: Asian electronics firms in the global economy', *Asia-Pacific Viewpoint*, **48** (1), 1–25.

FURTHER READING

Bryson, J.R., M. Taylor and R. Cooper (2008), 'Competing by design, specialization and customization: manufacturing locks in the West Midlands (UK)', *Geografiska Annaler B*, **90** (2), 173–186. (Explores how small lock manufacturers successfully compete on the basis of customization and specialization.)
Gibson, C. (2012), 'Cultural economy: achievements, divergences, future prospects', *Geographical Research*, **50** (3), 282–290. (A critical reflection on the shifting use of cultural economy theory.)
Lawler, Kristin (2011), *The American Surfer: Radical Culture and Capitalism*, New York: Routledge. (Charts the popular rise and commercialization of surf culture in the United States.)

Molotch, H. (2002), 'Place in product', *International Journal of Urban and Regional Research*, **26**, 665–688. (Analysis of the way 'place' becomes embodied in products and commodities.)
Walker, Isaiah Helekunihi (2011), *Waves of Resistance: Surfing and History in Twentieth-Century Hawai'i*, Honolulu: University of Hawai'i Press. (The story of Hawaiian surfing as a platform for resisting colonial intrusion in the surf zone.)

21. Migrant manufacturing: translocal production and the establishment of a Polish bakery in Birmingham, UK

Catherine Harris

INTRODUCTION

Agata owns a bakery in Birmingham, UK that was established in 2003. The business has grown dramatically since its beginnings as a small shop, to include two premises (one of which is a factory where the food products are manufactured). From starting the business with only her husband, Agata now employs 50 members of staff. From selling only through their own shop, the bakery now supplies hundreds of branches of Tesco, the supermarket retailer, throughout the UK. This story of manufacturing and entrepreneurship has another interesting aspect – Agata is Polish, along with the products that her bakery manufactures. She arrived in the UK in December 2002 intending to establish a business in response to the UK government's announcement that it would allow immediate free movement into the UK of the 'Accession 8' (A8) country workers following European Union (EU) accession in 2004 (see Harris *et al.*, 2012; Harris *et al.*, 2014). At this time eight East European countries joined the EU (Czech Republic, Estonia, Hungary, Latvia, Lithuania, Poland, Slovakia and Slovenia). A second phase of accession occurred in 2007, when Bulgaria and Romania joined the EU. The enlargement of the EU led to new migration flows, but initially many EU member states placed restrictions on migrant flows from the new member states (NMS). However, the UK permitted the immediate and unrestricted flow of migrants from the NMS. Consequently, EU accession has brought unprecedented flows of migrants to the UK. Many of these migrants were searching for employment, but some were migrant entrepreneurs, like Agata, who were searching for opportunities to establish their own business. The migrant Polish entrepreneurs established new firms in many sectors.

Those migrants coming from Poland have, arguably, been dominant – the most numerous, and certainly the most visible in the media. Research into Polish migration has reflected its economically driven nature (Drinkwater *et al.*, 2006: 2), the insecure nature of the work undertaken by economic migrants (McDowell *et al.*, 2009) and non-economic motivations for migration and experiences following arrival (Eade *et al.*, 2006; Garapich, 2006; Fomina, 2009; White, 2010; Gill and Bialski, 2011; White, 2011). Much of this research has focused on migrants as employees, and neglected migrants as entrepreneurs. This is to distinguish between Polish migrants who are job takers or job makers. Further research has focused on the distribution of migrants across the UK (Stenning and Dawley, 2009; Harris *et al.*, 2012, 2014). Given the significance of entrepreneurship for other ethnic groups (Ram, 1997) and the sizeable Polish population of the UK, it could be assumed that Polish entrepreneurs are an important feature of ethnic entrepreneurship in the UK. Indeed, the British–Polish Chamber of Commerce estimates that there are currently 40,000 Polish entrepreneurs who have established businesses in the UK (Pollard *et al.*, 2008: 44).

Despite claims of large numbers of self-employed Polish migrants in the UK, there has been very little research into this topic. Instead, the literature largely focuses on the size and distribution of migration flows and the experiences of Polish 'employed' migrants in the UK. There is an important research gap here on self-employed Polish migrants that provides the point of departure for this chapter. As a consequence of the unprecedented flows of Poles to the UK and their impact throughout the labour market, geographers need to consider the role played by Polish entrepreneurs in local economies. Polish migrants brought with them skills as well as contacts with Poland. Many of these entrepreneurs established businesses based on a set of distinctive Polish assets and this has led to the formation of Polish shops but also manufacturing companies.

The literature on ethnic entrepreneurship (Light, 1972; Bonacich, 1973; Borjas, 1986; Aldrich and Waldinger, 1990; Ram, 1997) has focused primarily on domestic conditions; that is, on the immigrant communities themselves and on their relations with the host society. Although reference has been made to connections with the home country for such groups as the Koreans (Light and Bonacich, 1988) and Chinese (Wong and Ng, 2002), the main research focus has remained on the contextual and individual variables that allow enclave entrepreneurs to succeed in their new local environment. Instead, like Flusty (2004), this chapter argues that ties across national borders are the product of relationships between specific persons, in specific locales and it is these ties which are critical to running a migrant manufacturing firm.

Accordingly, the concept of transnationalism opens up a new dimension in the study of immigrant economic adaptation because it focuses explicitly on the significance of resilient cross-border ties. Too much emphasis can be placed on the role of ethnicity and national belonging in the lives of migrants. Migrants have cultural assets or capital but also connections. Migrant entrepreneurs can thus be considered to represent another form of transnational or international business, but perhaps the term that best describes this relation is 'translocal' (White, 2011: 1). The term 'transnational' overlooks the relationships that occur between localities in favour of relationships at a national level. The concept of 'transnational' hides or obscures the importance of place-to-place or local-to-local relations. Migrant entrepreneurs thus may exploit connections with their country of origin, but these connections are place based. The concept of translocalism, however, is neglected in current research, particularly when considering migrant entrepreneurship and manufacturing.

Taking these gaps in the literature into consideration, this chapter considers the significance of translocal exchanges and global value chains in the establishment and evolution of Polish manufacturing enterprises in the UK. The chapter is structured as follows. First, the background to EU entrepreneurship is considered, with attention paid to EU accession and its implications for entrepreneurship. Second, a brief outline of the study on which this chapter is based is provided. A case study of a Polish food manufacturing firm in the UK is then provided as a means to explore the processes involved in migrant manufacturing enterprises. Issues such as translocality and adaptation are of critical importance to this. The conclusion identifies future avenues for research.

TRANSLOCAL RELATIONSHIPS

It is first useful to explore the concept of transnationalism in order to illustrate the utility of the translocal approach over the transnational. Since its 'launch' in 1990 (Glick Schiller *et al.*, 1992), the term 'transnationalism' has been extensively employed in migration scholarship, acquired a myriad of meanings and generated considerable debate (see for example Kivisto, 2001; Portes *et al.*, 2001; Vertovec, 2001). On the broadest level, 'the literature on transnationalism generally underscores the fact that large numbers of people now live in social worlds that are stretched between, or dually located in, physical places and communities in two or more nation-states' (Vertovec, 2001: 578). One of the key components of transnationalism is the 'multiplicity of involvements that transmigrants sustain in both home and host societies' (Basch *et al.*, 1994: 7).

As Vertovec's formulation suggests, it is possible to view migrants' links with their sending communities as being translocal and transnational in equal measure. Some authors (Cartier, 2001; Smart and Lin, 2007) appear not to sense any tension between the terms, and use the term translocal as an occasional alternative to transnational, apparently for stylistic variation. The term translocal is also used more precisely to imply that a distinction is being drawn between translocal and transnational.

Given that many scholars of transnationalism have been particularly interested in the impact of transnational ties on state borders, citizenship and migrant integration, it is hardly surprising that the term is often used with particular focus on the portability of 'national' cultural and political identities. This way of using transnationalism, implying that it is all about 'nationality' and not about 'locality', is so common that it seems helpful to employ the term translocalism to restore the focus on cross-local links. However, different scholars view the relationship between the concepts of transnationalism and translocalism differently (Vertovec, 2001; Oakes and Schein, 2006; White, 2011).

There has been a coalescing of interest surrounding the notion of translocality, which increasingly appears in geographical work on transnationalism (see for example Cartier, 2001; Oakes and Schein, 2006; Smart and Lin, 2007; McFarlane, 2009; Freitag and Von Oppen, 2010). Translocality involves 'a focus on the multiplying forms of mobility. . . without losing sight of the importance of localities in people's lives' (Oakes and Schein, 2006: 1). Translocalism provides a spatial understanding of what situates the migrant experience within/across particular 'locales' without confining it to the territorial boundedness of the nation-state (Brickell and Datta, 2011). Seen as a way of situating earlier deterritorialised notions of transnationalism, translocalism is focused largely on social networks and economic exchanges (Table 21.1).

The translocal can be seen as a sub-set of the transnational, but the translocal emphasises the local scale and that there are differences between different places within a nation (Table 21.1). Collectively, communities of fellow nationals span from nation to nation, but individual migrants experience the sending–receiving country relationship as local to local. For example, Smith (2001: 169) argues, 'recent research on transnationalism illustrates that the specific social space in which transnational actions take place is not merely local but often "translocal" (i.e. local-to-local)' (see Grillo, 2007: 204). Viewed within this relationship, both translocalism and transnationalism are aspects of globalisation.

It is common, however, for transnationalism to be viewed as distinct from translocalism, and for translocalism to be identified with a narrower perspective. For example,

Table 21.1 Transnationalism versus translocalism

Concept	Transnationalism	Translocalism
Social space	Nation to nation	Locality to locality
	Interactions take place between nations	Interactions take place between places
Sense of belonging	Nationality	Locality
	Migrants are defined by their nationality	Migrants are defined by their place of settlement and ties to their localities
Individual	Connection to nation-state, national identity and citizenship	Connection to a specific place and to a group of people
	Migrants feel a connection to the nation-state	Migrants feel a connection to their locality
Dynamism	Change between nations	Change between localities
	Change and movement over time occur between nations	Change and movement over time occur between localities

Source: Harris 2012.

Wessendorf, writing about Italian migrants to Switzerland and their return migrations, prefers translocal to transnational for the following reasons:

> The connections Italians maintain to their homeland are translocal rather than transnational. Italian migrants' relation to place is localised in that they usually travel between the town of settlement in Switzerland and the village of origin in Italy. Most of them feel a strong connection to these places rather than to the nation-states. (Wessendorf, 2007: 110)

For these Italian migrants in Switzerland translocality is a significant concept. These migrants draw attention to multiplying forms of mobility. They have settled in a Swiss town without losing sight of the importance of localities in their lives in their villages of origin in Italy.

Oakes and Schein (2006: 20) argue that translocality 'deliberately confuses the boundaries of the local in an effort to capture the increasingly complicated nature of spatial processes and identities, yet it insists on viewing such processes and identities as place-based rather than exclusively mobile, uprooted or "travelling"'. Taking into account the localised contexts where transnational networks are maintained, negotiated and sustained in everyday urban life, scholars now assert the importance of local–local connections during transnational migration (Smith and Guarnizo, 1998; Smith, 2001; Velayutham and Wise, 2005; McKay, 2006; Gielis, 2009; Lyons and Mandaville, 2010). They suggest that localities need not necessarily be limited to the shared social relations of local histories, experiences and relations, but can connect to wider geographical histories and processes.

These calls to situatedness during mobility, however, still retain national boundaries as the predominant focus of local–local connections. This is perhaps the history of translocality itself, which has emerged from a concern over the disembedded understanding of transnational networks. Research on translocality primarily refers to how social relationships across locales shape transnational migrant networks, economic exchanges and diasporic space. In such an inquiry spatial registers of affiliation that are part of migrants' everyday embodied experiences remain largely unexplored. The effect of this has been to subsume the debates on translocality within a wider notion of transnationality. Translocality is now widely seen to be a form of 'grounded transnationalism' – a space where deterritorialised networks of transnational social relations take shape through migrant agencies. This means that translocality as a form of local–local relations exists primarily within the debates on transnationalism and provides a conceptualised approach for explaining migrant entrepreneurship and other forms of migrant employment.

The use of the translocal approach has clear advantages over the transnational approach, particularly in the case of migrant entrepreneurship, since ties across national borders are the product of relationships between specific persons, in specific locales. The 'transnational' overlooks the relationships occurring between localities in favour of relationships on a national level. Translocalism, on the other hand, pays attention to these local-to-local relationships and it is these local relationships that are critical to the establishment and running of migrant enterprises. In this way the concept is scalar, in that translocal may mean translocal international relationships (i.e. between two or more locations in different countries) or it could mean local translocal (i.e. between two or more locations within the same region).

EU ENTREPRENEURSHIP

To understand the translocal processes involved in EU migrant manufacturing it is first necessary to explore EU accession and ethnic entrepreneurship. This context positions Polish migrant entrepreneurs within wider geopolitical developments that led to the expansion of the EU through accession of NMS.

EU Accession

In the UK, debates surrounding migration have grown considerably following the 2004 enlargement of the EU to the east and again in 2007. The EU is a free trade area, with 'free movement of workers' (FMOW) for its citizens, without visas or work permits (Portes and French, 2005: 3). Previous accessions had caused EU member states (the 'EU-15') considerable concern about the impact of complete liberalisation on their labour markets, and the Accession Treaties gave the EU-15, the existing member states of the EU, the option to delay implementation of full FMOW for up to seven years. Most, including France, Germany and Spain, imposed restrictions on movement in one form or another; the United Kingdom was one of only three countries (along with Ireland and Sweden) to allow migrants from the NMS to enter their labour markets more or less without restriction. The UK government announced in December 2002 that it would allow immediate free movement of A8 country workers following accession.

Amongst the A8 immigrants to the UK, those coming from Poland have dominated the flows. Harris *et al.* (2012) disaggregate National Insurance Number (NINo) allocations[1] to NMS migrants by country of origin (Figure 21.1). Between 2004/05 and 2008/09, Polish nationals were the largest group of migrants to relocate to the UK from

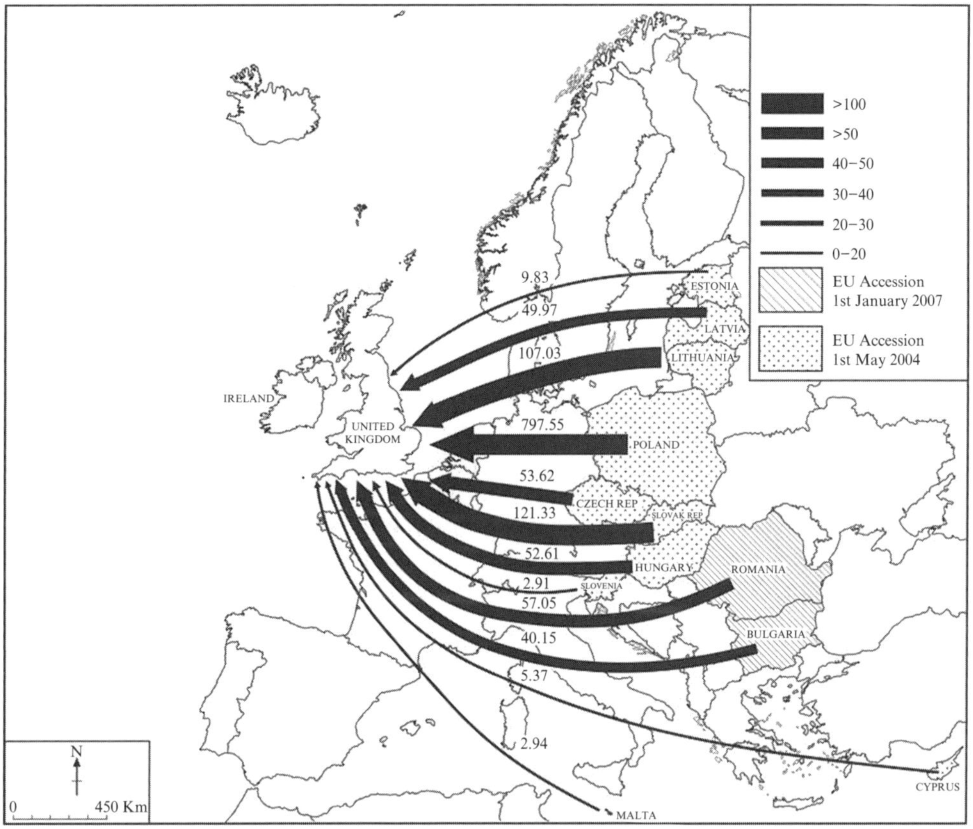

Figure 21.1 *Total NINo registrations to EU accession nationals from April 2004–March 2009 ('000s)*

the accession states. Therefore, Polish migrants in the UK would appear to be the NMS group, which justifies further investigation.

Polish Entrepreneurship

In Poland, persons engaged in self-employment constitute more than 30 per cent of the active workforce (Lommatzsch *et al.*, 2004: 113). Therefore, it is likely that some of this self-employed population will have migrated to the UK as a result of EU enlargement. In contrast, Drinkwater (2008) claims that only 4.4 per cent of the recently arrived (post-accession) Polish population in the UK are self-employed. It is not unexpected that recent immigrants will have very low rates of self-employment, because it is likely that these groups will have difficulty obtaining the capital required to start a business (Borjas, 1986), but capital will have been available from family members remaining in Poland to support migrants wishing to establish a business. In addition, Polish migrants can now freely enter the UK and need not therefore apply for a business visa to work in the UK.

Despite their suggested sizeable presence and the significant literature on Polish migrants in the UK, the political and media debate concerning Polish immigration to the UK has concentrated on employment rather than the role of migrants as entrepreneurs. It has focused on Poles as 'job takers', rather than 'job makers'. This is reflected in academic debates and there is very little research into those Polish migrants who are self-employed.

Nevertheless, there is some limited research into Polish entrepreneurship. Lassalle *et al.* (2011a) investigate Polish entrepreneurial strategies in the UK, and provide case studies of Polish entrepreneurs in Glasgow detailing migration and business strategies (Lassalle *et al.*, 2011b). This builds upon earlier work that focuses on the role of social capital in Polish immigrant businesses in Scotland (Helinska-Hughes *et al.*, 2009). Vershinina *et al.* (2009) investigate ten Polish immigrant entrepreneurs operating in Leicester. Amongst the Polish entrepreneurs they identify three groups – traditional, opportunity and opportunist entrepreneurs – based on when they immigrated and the implications of the differing origins and amounts of capital they were able to access and convert into entrepreneurial activity. They move beyond a discussion of ethnic entrepreneurship in terms of social capital arising from ethnic group membership and show the existence of intra-ethnic variation in the UK Polish entrepreneurial community. The existing research on Polish entrepreneurship (Helinska-Hughes *et al.*, 2009; Vershinina *et al.*, 2009; Lassalle *et al.*, 2011a; Lassalle *et al.*, 2011b) does not consider the role of migrant entrepreneurs who have established businesses in the manufacturing sector.

Translocal Entrepreneurship

The utility of the translocal approach was explored earlier in this chapter. It is particularly relevant in the case of migrant entrepreneurship. A variety of individuals are involved in translocal migrant entrepreneurship. Other than the entrepreneurs themselves, these can be family members, friends or more formal support networks, such as business associations. These individuals can offer their assistance to a business through a variety of resources, including business advice, financial assistance and labour. In the case of businesses owned by migrant entrepreneurs these resources provided by family, friends and support networks are often located in many different cities or locations over two or more countries, including the country of origin and the recipient country.

As transnational ethnic entrepreneurship demonstrates, the relationship between globalisation and locality exists on different levels, many of which are beyond the scope of the chapter. Like Flusty (2004: 8), this chapter argues that ties across national borders are 'the product of specific persons in specific locales' and these ties are critical in establishing and running migrant manufacturing businesses. In common, for example, with researchers who have analysed connections forged by migrants from particular locations in Latin America to cities in the USA (Smith, 2001: 170), the chapter focuses on how Polish manufacturing entrepreneurs create links between places in different countries, and in particular the extent to which they view their migration and business activities as being from locality to locality rather than nation to nation. Many of the linkages that connect migrants with their homeland centre around named individuals, and these relational linkages are place based as well as reflecting a connection between two nations.

Fomina (2009) emphasises that well-educated Polish migrants in Bradford want to integrate into British society, and that – in their eyes – this is a key difference between them and less-educated Poles, as well as British Pakistani Bradfordians. Clearly, less well-educated and confident Polish migrants, who do not speak much English and work in sections of the labour market dominated by migrant workers, with weaker ties to the receiving community, are likely to socialise more with fellow Poles (White and Ryan, 2008; White, 2010). White (2011: 5) suggests that, if 'transnationalism' is used in the sense of 'doing Polish things in the UK' – watching Polish television, eating Polish food, and so on – then working-class labour migrants may possibly be more transnational as well as more translocal, more ethnically oriented as well as more oriented towards their particular dual places of residence, than highly educated Polish migrants.

A more helpful approach to translocalism is to remember that Poles in the UK since 2004, unlike earlier generations of migrants, have retained easy access to Poland, and to Polish goods and services, so that 'they do not particularly need to worry about maintaining their Polishness' (Fomina, 2009: 28). If Polish identity can be taken for granted, then local preoccupations may have space to take precedence over national ones.

A further reason why the national may seem less salient than the local to Polish migrants in the UK is simply to do with travel patterns. In the days of visas and limited air routes, the journey from Poland to the UK required a Polish passport, a UK visa and usually also travel between Warsaw and London, the capital cities and symbols of the nation-state. Today, it is simply necessary to take one's identity card and drive from Pszczyna to Manchester, or fly from Kraków to Birmingham; that is, from regional centre to regional centre, avoiding capital cities. These translocal exchanges can have important implications when such migrants choose to establish a business in the recipient country.

STUDY OUTLINE AND METHODOLOGY

The empirical material used in this chapter comes from a much larger study that investigated entrepreneurship amongst Polish migrants in the West Midlands region of the UK. This research is based on the results of 48 in-depth interviews with Polish entrepreneurs in the West Midlands, who migrated around the time of EU enlargement in May 2004.

This research project identifies the drivers which motivated Polish nationals to migrate to the UK and enter self-employment. The study identified that an important feature, which characterises Polish entrepreneurs, is the utilisation of translocal (local–local) exchanges to support their businesses. The research also considered how Polish enterprises in the West Midlands evolve over time. Polish entrepreneurs adopt adaptation strategies to expand their businesses but also in response to stresses they face from the external environment. The findings are intended to provide an integrated evidence base that will contribute to ongoing debate into migrant entrepreneurship as well as informing European policies and strategies for interacting with and supporting migrant entrepreneurs and the communities within which they are situated.

POLISH FOOD MANUFACTURING

There are many barriers to entry that prevent individuals from establishing firms. These barriers include access to capital, premises, suitable labour and expertise. Some sectors have much higher barriers to entry, including manufacturing. It is much easier to establish a small shop than it is to establish a small manufacturing firm. There are thus differential barriers to entry that are sector dependent. Manufacturing sectors have very different barriers to entry. One of the Polish migrants based in Birmingham had established a bakery. A bakery requires premises, baking skills and an emphasis placed on the production of high-quality goods. The Polish migrant in question focused on the production of Polish bakery products. This provided differentiation in the marketplace, but also enabled them to target the growing population of first and second generation Poles who were interested in purchasing Polish products. Small food firms, such as this one, are considered to play a potentially important role in achieving sustainable economic growth in local economies (McDonagh and Commins, 1999; Murdoch, 2000).

Agata's bakery specialises in manufacturing traditional Polish bread, cakes and pastries. Small food firms such as Agata's produce specialised regional products of a different nature from those produced by large firms that focus on mass produced standard products (Avermaete *et al.*, 2004). Large firms generally have a national or international market approach and consequently focus on products with more of a mass appeal. In this sense, an important component of Europe's highly valued cultural identity is invested in small manufacturing companies that provide culturally distinctive products (Committee of the Regions, 1996; Ilbery and Kneafsey, 1999; Traill, 1995).

In Poland Agata was a recent Business Studies graduate who worked in marketing; her husband worked for a bakery. Establishing a business requires different skill sets or capabilities and the partnership between Agata and her husband went some way to providing this migrant couple with the skills required to establish a firm. It was Agata's husband's passion for baking and the skills he had acquired from his time in the trade that inspired the couple to establish their own bakery. They believed that combining his baking skills with Agata's Business Studies knowledge would lead to the establishment of a successful business. Initially they established a small bakery and this has grown into a thriving manufacturing firm. There are a number of interesting characteristics that have been critical in the development of this bakery, which will now be explored to identify the processes involved in establishing, running and expanding a migrant manufacturing business. These key features are: the timing of the migration, translocal exchanges, business adaptations and localisation.

THE TIMING OF MIGRATION: MANUFACTURING MOTIVATION, ENTREPRENEURIAL LEARNING AND FAMILY FINANCE

Agata and her husband arrived in the UK from Poland in 2001. On doing so Agata worked as a waitress and her husband as a builder. Whilst it is likely that the couple were working illegally as they arrived in the UK before EU accession in 2004, this was not explored in the interview since it was not the focus of the research and may have made the respondent uneasy and reluctant to answer further questions. Like many Polish entrepreneurs who arrived in the UK prior to EU accession, Agata was prepared to leave her job at the time and change her sector of employment to enter into entrepreneurship. Consequently, pre-accession migrant entrepreneurs like Agata, are risk takers. Such Polish entrepreneurs support claims that firm founders objectively have a higher propensity for risk than do members of the general population, but that firm founders do not perceive their actions as risky (Fry, 1993). Agata exemplifies the motivation characteristics of entrepreneurs described by Shane *et al.* (2003), such as need for achievement, risk taking and setting goals.

Agata's change in career, combined with the timing of migration, was motivated by the long-term desire to establish a business. She describes how:

> In Poland we had the idea, the determination, the skill and the passion . . . but we wanted to get here [Birmingham] and see what was going on. We thought that by coming [to the UK] in 2001 that would give us time to see how things worked and what England was like. (Agata, Polish Bakery, 2009)

Agata believes that the business is successful because the couple's migration occurred prior to accession. This allowed them time to save money, become accustomed to life in the UK and to undertake research. This meant that they could gradually establish a solid business and be ready to exploit the opportunities of wealth creation brought about by the flows of Polish migrants following accession. The timing of the migration (in 2001) was a carefully constructed decision of an active nature. It enabled time to be taken to learn and acquire new skills in the region of the business before actually establishing it (in 2003), which is important to a successful business since 'entrepreneurship is a process of learning' (Minniti and Bygrave, 2001: 7). Indeed, Lundvall (1994) has remarked that the term 'learning economy' signifies a society in which the capability to learn is critical to economic success.

Although the couple worked for several years in the UK to save money to fund their business, a significant amount of financial support was provided by Agata's parents in Poland. She spoke not only of her parents' input in terms of capital, but also their financial advice. This capital and advice flowed between the locality of origin in Poland (Gdańsk) and the recipient locality in Birmingham. The bakery was established via a critical translocal relationship that provided access to finance and advice. This also shows how the concept of 'translocality does not only mean people' (Oakes and Schein, 2006: 1). It is 'constituted as well by the circulation of capital, ideas, images, goods and styles, services' (Oakes and Schein, 2006: 1) and identities; that is, a Polish bakery run by Poles.

In an added dimension to these translocal exchanges and the global value chains of the business, Agata sends a percentage of her profits to her parents as a remittance for their financial involvement in the business. She acknowledges that the flow of finance is now from Birmingham in the UK to Gdańsk in Poland, rather than the other way around. Agata pays a small amount of interest on this capital that she was lent by her parents. These translocal flows of capital and financial advice are centred on the local lives of the individuals involved – Agata and her parents. She explains that:

> My parents are very important to my business. They gave me money in the beginning to set up the bakery. Plus they gave me a lot of financial and business advice. They still do give me some [advice] but now the money goes the other way. They have a cut of my profits to repay them for the financial help they gave me at the start. (Agata, Polish Bakery, 2009)

Much of the motivation for migration from Poland to the UK is often attributed to the uneven development between the two countries, with greater prospects perceived to be offered in the UK. This is illustrated by Polish entrepreneurship. For example, Agata spoke of migrating and establishing a business in Birmingham as 'life was a struggle in Poland and the economy was poor', whereas in England she could 'have a better life'.

Agata explained her parents' desire to help finance the venture so that she could establish the business prior to accession in order to capitalise on the flows of migrants arriving in 2004. She also discussed her parents' financial backing in reference to their views on loans and credit cards. As was the case for many entrepreneurs, for her family it is important to enter into a venture without bank loans or credit cards.

The family financial involvement in this case supports De Visscher *et al.* (1995), who describe family firms as having patient capital with the capacity to invest in long-run return opportunities rather than quarterly return requirements or a search for short-term profits (Dreux, 1990). Family firms often experience difficulty in raising capital (Leach, 1991), but this was not the case for the bakery as it was able to draw upon family resources from Poland. For Agata, family and finance and the translocal relationships that they utilise are of critical importance to the success of the manufacturing business.

OPENING THE BAKERY: TRANSLOCAL EXCHANGES

After several months of careful planning and selecting their premises, Agata and her husband opened their bakery in December 2003. The bakery opened in a shop unit where the bread was both baked and sold. Whilst their equipment and ingredients were largely sourced from within the UK, their recipes originated from Poland and some ingredients were imported, illustrating the translocal processes involved in establishing a migrant manufacturing business. When the business opened, in addition to Agata and her husband, it employed three members of staff, who were all Polish. These members of staff migrated from particular localities in Poland to live and work in the English locality of Birmingham. Supporting the motivation to the entrepreneurship approach developed by Shane *et al.* (2003), such as need for achievement and self-efficacy, the couple saw this time before accession as the opportunity to utilise their translocal relationships, to become accustomed to life in Birmingham and to 'get the business right before enlargement' of the EU occurred. The couple were thus anticipating enlargement and the migration that would then occur from Poland to the UK.

At the time of opening, the bakery's customer base was small and again solely Polish. The bakery sold to local Poles who visited the bakery, particularly those from the older generation. This demonstrates how close translocal and intra-local community ties in an owner's network can help get manufacturing businesses started within communities and may help keep businesses solvent through the products and services they provide. The early phase of this business being established supports Light's (1984) research on the Los Angeles liquor store industry, which found that the majority of customers of ethnic businesses are co-ethnics. Agata and her husband also fostered relationships with several Polish delicatessens elsewhere in the UK, to which they sold their products. Agata comments that:

> In the beginning it was important to use our Polishness to develop links which would be good for the business. We had Polish staff who were recent migrants. Since we were making Polish products, using Polish recipes, we thought that this was important. We also made contact with Polish delis in Manchester so we could sell our products outside the local area. Plus we kept in touch with Polish bakeries in Poland to ask for advice and ideas. (Agata, Polish Bakery, 2009)

Agata keeps in contact with food manufacturers in Poland through email interactions and Skype conversations. These businesses are run by contacts her and her husband made before leaving Poland. This highlights translocality through the multiple localities involved in the Polish bakery, both in the UK and Poland. Polish staff, Polish recipes and Polish products are used in the bakery, which are manufactured in Birmingham and then transported to Polish businesses in other sectors (particularly retail) throughout the UK.

Following EU accession in May 2004 the bakery gradually became busier due to the influx of Polish migrants. Not only did the number of customers entering the shop rise substantially, but the customer demographic also changed from an older generation of customers to customers from the new, younger wave of migrants. The couple developed further relationships with new Polish delicatessens that were being established by migrant Poles in the West Midlands and throughout the UK. The bakery recruited more staff, taking their total to ten, but the workforce remained solely Polish. Two of these employees were friends of the couple who had migrated to the region from Poland and were searching for work.

Agata and her husband maintained strong contact with family and friends who remained in Poland. They did this for advice and support but also to promote their business should any of their friends or acquaintances have chosen to migrate to the UK. For early stage Polish manufacturing businesses mutual trust is very important, and such trust is more likely between co-ethnics than in contacts across ethnic boundaries (Light, 1984). Through this dependence on co-ethnics, the couple did not see the localities in the UK and Poland as separate entities. They treated them as interconnected since both localities were critical to the functioning of the business.

THE TIMING OF ACTIVITIES THROUGH ADAPTATION: BREAKING OUT AND SCALING UP

By 2007 the business was flourishing. Agata and her husband chose to adapt the business to achieve growth and therefore acquired new and larger premises, also located in Birmingham. The original premises were retained as a shop with a small production unit, but the new premises were located on an industrial estate on the outskirts of Birmingham. This new facility was going to function as a Polish bakery factory and as the firm's manufacturing base, but it also contained a shop to sell the goods produced.

At this time, Agata demonstrated several characteristics of being a 'searcher' (Eade *et al.*, 2006) for new entrepreneurial opportunities. She sought to expand her business through acquiring new clientele. As well as a solid Polish customer base, the Polish bakery now sold products to customers from many nationalities. This can perhaps be attributed to the sizeable Polish community in the region, leading to an interest in Polish culture and their cuisine, rather than a strategic plan to break out from the confines of targeting the Polish community. It demonstrates localisation as Polish products were increasingly integrated into the local community. Agata commented that:

> Of course a lot of our customers are Polish. When we opened they were all Polish. But now we have more and more English customers, and other nationalities. Often they have Polish friends who they visit and they will offer them some of our bread or cakes. If they like it they will come and buy it. It doesn't matter that we are a Polish bakery because our products taste good to every nationality. (Agata, Polish Bakery, 2009)

Despite the confidence that the bakery's products would 'sell themselves' beyond the Polish community, during 2007 the owners were strategically seeking opportunities to adapt and develop it further by 'scaling up'. To achieve this Agata arranged meetings with various Tesco supermarkets in the region. She attended the first meeting with some of her Polish bread and cakes. She told the store manager about the background of the business and about her vision for the business. She also invited them to try the products she had brought with her. As a result of this successful meeting, the Polish bakery signed a contract with this Tesco store to provide them with a range of products. Agata repeated this process at other branches of Tesco. The bakery now supplies hundreds of Tesco stores throughout the UK, with the products manufactured in their larger factory premises. Although it is likely that these products are largely sold to Polish customers, they are available to an ethnically diverse range of customers within a UK established store.

This English and Polish blend in the production and sale of these products adds an interesting dimension to the translocal relationships involved in break-out and localisation of Polish entrepreneurs. A Polish product is being manufactured in Birmingham by a mixture of Polish and English workers. It is sold in a variety of Polish run and English run shops, to a largely Polish customer base, but it is also being purchased by other nationalities. This shows that to achieve localisation and break-out, the Polish bakery was required to use its translocal connections with localities in Poland whilst fostering new ones in Birmingham and other localities in which their product is sold. The concept of a translocal relationship operates at many different scales – from translocal international relationships to regional and local translocal interactions. There are translocal exchanges between Birmingham and localities in Poland in terms of people, capital and ideas in order to achieve this break-out and localisation. However, as time has progressed from the business being established in 2003 it appears that these translocal international linkages have become less important, with less reliance on family for funding and on Polish migrants for their customer base. The more localised the business has become, the less it uses its translocal linkages. Translocal international exchanges are still present through interactions with family and friends who remain in Poland, and through using Polish ingredients which are imported to Birmingham, supporting Avermaete *et al.* (2004: 475), who suggest that small food firms are increasingly importing raw materials from outside their local region. These exchanges demonstrate that there is a blend between the translocal and the local, and the composition of this blend changes over time.

As with Bagwell's study of Vietnamese nail bars, Agata hinted that the presence of 'innovative members within the family network appeared to be particularly important for encouraging business diversity and thus break-out' (Bagwell, 2008: 390). Agata commented:

> It did take some hard work to get recognised outside the Polish community, but my parents advised and encouraged me. There are only so many Poles in the area so we knew that to expand the business we had to appeal to English people . . . But not only that – we had to move beyond our shop and only selling to Polish delis [delicatessens]. Getting our products stocked in Tesco was the major breakthrough of the business. Now everyone, everywhere in England, can sample our products. (Agata, Polish Bakery, 2009)

Thus, 'rather than simply acquiring [an English] clientele, effective break-out requires attention to market position, appropriate niches, new product development and the management of human resources' (Curran and Blackburn, 1993, quoted in Ram, 1997: 151). The break-out by the Polish bakery supports the work of researchers who suggest that in environments which are hostile or turbulent, there is a tendency for firms to adopt business strategies that are adaptive (Davis *et al.*, 1991; Jennings and Seaman, 1994). The decision to adapt the business and to achieve break-out has been critical in the success and survival of the manufacturing business, which has experienced the hostile and turbulent environments of the 2008 recession and also the return migration of Poles back to Poland (Mostrous and Seib, 2008). By diversifying the customer base, the bakery was less susceptible to these hostile environments and is not now reliant on a small number of local Polish consumers. To Agata, 'our contract [with Tesco] is very important to us. Without it we could have a much smaller pool of customers and be more likely to suffer during difficult times.'

The Polish bakery highlights that ethnic entrepreneurs often seek to break-out from the constraint of servicing a distinctive ethnic cohort to provide services (goods, etc.) to the wider population. This supports Ram and Hillin (1994), who are mainly concerned with inter-sectoral shift, where ethnic minority firms seek to penetrate into higher-order mainstream markets unconstrained by a local or a co-ethnic customer base (Jones *et al.*, 2000). They explain that this may sometimes take the form of existing firms adapting through branching out and diversifying, as is the case with the Polish bakery.

Now that the Polish bakery is a successful business, Agata is keen to offer advice and support to other Polish entrepreneurs in the West Midlands. She is a member of the Polish Business Club and regularly attends meetings to share her experiences, particularly of how to integrate into the local community. She also works closely with the British Polish Chamber of Commerce based in London, which aims to facilitate and develop Polish businesses in the UK and maintains links between Poland and the UK. Through this she has had discussions with the advisor to the Finance Minister of Poland about how to encourage Polish manufacturing and entrepreneurship in the UK. Agata and many of her fellow Polish entrepreneurs rely on this diverse range of social structures and institutions to support their entrepreneurial activities whilst also aiding those of others.

THE POLISH LOCAL BAKERY

At the time of the interview (November 2009) the business employed 50 members of staff, across its two premises, from a range of nationalities, including both English and Polish. This success and localisation are the reasons why Agata and her husband plan to remain in the Birmingham area running their bakery. They also have Polish and English friends in the area that they have 'grown close to' and this demonstrates a further aspect to their localisation. Their 'life now belongs in the UK, and particularly in Birmingham'. However, they can easily return to see family in Poland due to the availability of cheap and frequent flights brought about by the 'myriad technologies' (Oakes and Schein, 2006: 1) which facilitate translocal relationships. The significance of establishing a life beyond the business through the transgression of spatial boundaries and movement between scales of translocality demonstrates that only part of their reason for remaining in Birmingham is market driven (Oakes and Schein, 2006: 2).

For the time being Agata and her husband are happy to continue the business as they are, but with 'an eye on new opportunities'. They are planning to maintain their original shop premises in Birmingham but relocate their main manufacturing operations to larger premises in Tamworth, Staffordshire, to increase their output and workforce. Tamworth offers cheap premises with excellent transportation links, and is close to Birmingham, making it an ideal choice to establish a larger factory. Critically, as Agata explains, they do not rule out a change of business direction in the future if the market requires it. Thus,

> Our future is here [Birmingham] with the business. We love it. We have friends and are thought of as good business people . . . We are happy as we are right now . . . with life and the business, and we have our new premises on the way. But, we do not take this for granted. We have ideas in our minds and we are aware that times could become tough again. We need to be ready to respond if they do. (Agata, Polish Bakery, 2009)

These comments support strategic choice theory, which assumes that the key to entrepreneurial success lies in the decisions of individual entrepreneurs who identify opportunities, develop strategies, assemble resources and demonstrate initiative. Not only do the 'more fit' survive, but the 'more fit' are those who are best able to read and interpret patterns in the environment and adapt over time. Agata's bakery illustrates that the ability to adapt would seem less a function of luck, being in the right place at the right time, or having access to resources or information that others do not have, and more a function of the individual entrepreneur, his/her team and the ability to learn (Levinthal, 1991; Venkataraman and Van de Ven, 1998). Through a carefully timed migration decision, passion, hard work, forward thinking, adaptation, and using translocal relationships and resources, Agata and her husband have created a successful Polish bakery in Birmingham, which now serves an ethnically diverse range of customers throughout the UK.

As the business has developed it is now less reliant on the translocal linkages which it used when it was being established. Family finance and advice have been replaced with support from local institutions. The customer base has diversified, particularly through gaining a contract with Tesco. Furthermore, Agata and her husband see their future and the future of the business in the West Midlands. All of this suggests that the business has become a Polish local manufacturing business. Nevertheless, the concept of translocalism remains an important moment in the evolution of the business. Translocal exchanges are still present through importing Polish ingredients, and interactions with family and friends in Poland. As the bakery has evolved Agata and her manufacturing business have become increasingly localised within the surrounding local community, but these local features blend with the translocal.

CONCLUSIONS

This chapter contributes to research into migrant entrepreneurship and also to studies of manufacturing, since an element of the chapter is concerned with both individual and firm adaptation to circumstances by drawing upon the capabilities of the individual and also the place of business. This is achieved through the local client base and the possibility of developing a wider and more distributed client base. The chapter also provides an introduction to migrant entrepreneurs who have established businesses in the manufacturing sector, in this case Polish migrants in the UK. The need to consider migrant manufacturing, particularly amongst East European migrants, is highlighted by the unprecedented influx of migrants to the UK as a result of EU accession. This has had a significant impact on the UK labour market but the entrepreneurial ventures of these migrants, including those with businesses in the manufacturing sector, are an under-researched topic suggesting the need for further inquiry.

The example of Agata's bakery demonstrates the importance of translocal exchanges in establishing a manufacturing business as a migrant. These translocal exchanges are particularly evident through the provision of family finance and also advice. This is critical to this business since it relies on its Polish identity and some Polish supplies. Nevertheless, there is evidence of a weakening of translocal international relationships as the business has evolved and an increasing localisation has emerged with the development of the bakery business. The local Polish business developed through a set of critical local translocal relationships. These link the original bakery with the new manufacturing facility and the company with a set of strategic but increasingly distributed customer relationships. This comes from the clients and the bakery. The bakery was initially focused on one locality and then there were a number of attempts through distributed inputs to develop a distributed client base. Through this notion of distributed consumers, the firm broke with local dependency on Polish clients and developed relationships with specialist shops and larger stores such as Tesco. Critical to the success and growth of this business has been a 'scaling up' and 'break-out' from serving only Polish customers to manufacturing products for the UK supermarket giant Tesco. Therefore, it can be suggested that for migrant manufacturers to succeed, they need to look beyond their co-ethnic customer base, even if they are manufacturing a product with a distinct national identity.

Whilst this research investigates the UK aspect of migrant manufacturing and suggests its translocal nature, it does not explore the lives of other individuals who remain in Poland but are still involved in such businesses, such as family or suppliers. This points to an avenue for future research into the variety and network of individuals involved in migrant manufacturing. It provides the basis to explore the other businesses which are critical to the manufacturers' production and supply chains, and whether these other businesses are located in Poland, if they are Polish run but UK based, or if businesses from the host country are part of the manufacturers' networks. Such an investigation would highlight the extent of the translocal international networks involved in migrant manufacturing and the global value chains utilised by these businesses.

NOTE

1. The NINo allocations data are collected by the UK Department for Work and Pensions. NINo registrations give an indication of the number of *working* migrants in the UK, since having a NINo indicates that an individual is highly likely to be employed, or seeking employment.

REFERENCES

Aldrich, H.E. and R. Waldinger (1990), 'Ethnicity and entrepreneurship', *Annual Review of Sociology*, 16, 111–135.

Avermaete, T., J. Viaene, E.J. Morgan, E. Pitts, N. Crawford and D. Mahon (2004), 'Determinants of product and process innovation in small food manufacturing firms', *Trends in Food Science & Technology*, 15 (10), 474–483.

Bagwell, S. (2008), 'Transnational family networks and ethnic minority business development: the case of Vietnamese nail-shops in the UK', *International Journal of Entrepreneurial Behaviour & Research*, 14 (6), 377–394.

Basch, L.G., N. Glick Schiller and C. Cristina Blanc-Szanton (1994), *Nations Unbound: Transnational Projects, Post-Colonial Predicaments, and De-territorialized Nation States*, Langhorne, PA: Gordon and Breach.

Bonacich, E. (1973), 'The theory of middleman minorities', *American Sociological Review*, 37, 547–559.

Borjas, G. (1986), 'The self-employment experience of immigrants', *Journal of Human Resources*, 21, 485–506.

Brickell, K. and A. Datta (eds) (2011), *Translocal Geographies: Spaces, Places and Connections*, Farnham: Ashgate.

Cartier, C. (2001), Globalizing South China, Oxford: Blackwell.

Committee of the Regions (1996), *Promoting and Protecting Local Products: A Trumpcard for the Regions*, Brussels: Committee of the Regions.

Curran, J. and R. Blackburn (1993), *Ethnic Enterprise and the High Street Bank*, London: Kingston Business School, Kingston University.

Davis, D., M. Morris and J. Allen (1991), 'Perceived environmental turbulence and its effect on selected entrepreneurship, marketing, and organizational characteristics in industrial firms', *Journal of the Academy of Marketing Science*, 19, Spring, 43–51.

De Visscher, F.M., C.F. Aronoff and J.L. Ward (1995), *Financing Transitions: Managing Capital and Liquidity in the Family Business. Family Business Leadership Series*, Marietta, GA: Business Owner Resources.

Dreux, D.R. (1990), 'Financing family business: alternatives to selling out or going public', *Family Business Review*, 3 (3), 225–243.

Drinkwater, S. (2008), 'Recent immigration to the UK', available at: www.parliament.uk/documents/upload/EA206 per cent20Drinkwater.doc (accessed 14 December 2007).

Drinkwater, S., J. Eade and M. Garapich (2006), 'Poles apart? EU enlargement and the labour market outcomes of immigrants in the UK', available at http://ftp.iza.org/dp2410.pdf (accessed 14 December 2007).

Eade, J., S. Drinkwater and M. Garapich (2006), *Class and Ethnicity – Polish Migrants in London*, Guildford: Universities of Surrey and Roehampton.

Flusty, S. (2004), *De-Coca-Colonization: Making the Globe from the Inside Out*, New York: Routledge.

Fomina, J. (2009), '*Światy równoległe – wizerunek własny Polaków w Wielkiej Brytanii*', available at: http://www.isporg.pl/files/1975607530144791001254742838.pdf (accessed 19 August 2010).

Freitag, U. and A. Von Oppen (eds) (2010), *Translocality: The Study of Globalising Processes from a Southern Perspective*, the Netherlands: Brill Academic Publishers.

Fry, F.L. (1993), *Entrepreneurship: A Planning Approach*, Minneapolis–St. Paul, MN: West Publishing.

Garapich, M. (2006), 'London's Polish borders: class and ethnicity among global city migrants', http://www.surrey.ac.uk/Arts/CRONEM/LONDON-Polish-Borders-interim-report.pdf (accessed 2 March 2010).

Gielis, R. (2009), 'A global sense of migrant places: towards a place perspective in the study of migrant transnationalism', *Global Networks*, 9, 2, 271–287.
Gill, N. and P. Bialski (2011), 'New friends in new places: network formation during the migration process among Poles in the UK', *Geoforum*, 42 (2), 241–249.
Glick Schiller, N., L. Basch and C. Szanton Blanc (1992), *Toward a Transnational Perspective on Migration: Race, Class, Ethnicity and Nationalism Reconsidered*, New York: Annals of the New York Academy of Sciences.
Grillo, R. (2007), 'Betwixt and between: trajectories and projects of transmigration', *Journal of Ethnic and Migration Studies*, 33 (2), 199–217.
Harris, C. (2012), *Entrepreneurship amongst Polish Migrants in the West Midlands, UK*, unpublished PhD thesis, Birmingham: University of Birmingham
Harris, C., D. Moran and J.R. Bryson (2012), 'EU accession migration: national insurance number allocations and the geographies of Polish labour immigration to the UK', *Tijdschrift voor Economische en Sociale Geografie*, 103 (2), 209–221.
Harris, C., D. Moran and J.R. Bryson (2014, forthcoming), 'Polish labour migration to the UK: data discrepancies, migrant distributions and indicators of entrepreneurial activity', *Growth and Change*, DOI:10.1111/grow.12087.
Helinska-Hughes, E., M. Hughes, P. Lassalle and I. Skowron (2009), 'The trajectories of Polish immigrant businesses in Scotland and the role of social capital', paper presented at *Entrepreneurship and Growth of Family Firms*, June 2009, Cracow: University of Economics.
Ilbery, B. and M. Kneafsey (1999), 'Niche markets and regional speciality food products in Europe: towards a research agenda', *Environment and Planning A*, 31, 2207–2222.
Jennings, D.F. and S.L. Seaman (1994), 'High and low levels of organizational adaptation: an empirical analysis of strategy, structure, and performance', *Strategic Management Journal*, 15, 459–475.
Jones, T., G. Barrett and D. McEvoy (2000), 'Market potential as a decisive influence on the performance of ethnic minority business', in J. Rath (ed.), *Immigrant Businesses: The Economic, Political and Social Environment*, London: Macmillan.
Kivisto, P. (2001), 'Theorizing transnational immigration: a critical review of current efforts', *Ethnic and Racial Studies*, 24, 549–577.
Lassalle, P., E. Helinska-Hughes and M. Hughes (2011a), *Polish Entrepreneurs in the Glasgow Area: Migratory and Business Strategies*, Warsaw: The Polish Academy of Sciences.
Lassalle, P., E. Helinska-Hughes and M. Hughes (2011b), 'Polish migrant entrepreneurial strategies in the UK', *Challenges of Europe: Growth and Competitiveness – Reversing the Trends*, Split-Bol, Croatia, May.
Leach, P.E. (1991), *The Stoy Hayward Guide to the Family Business*, London: Kogan Page.
Levinthal, D.A. (1991), 'Organizational adaptation and environmental selection – institutional processes of change', *Organization Science*, 2 (1), 140–145.
Light, I.H. (1972), *Ethnic Enterprise in America: Business and Welfare among Chinese, Japanese, and Blacks*, Berkeley, CA: University of California Press.
Light, I.H. (1984), 'Immigrant and ethnic enterprise in North America', *Ethnic and Racial Studies*, 7, 195–216.
Light, I.H. and E. Bonacich (1988), *Immigrant Entrepreneurs: Koreans in Los Angeles*, Los Angeles: University of California Press.
Lommatzsch, K., R. Antczak, M. Markiewicz, K. Pietka and M. Walewski (2004), 'Poland's economy prior to accession', *Economic Bulletin*, 41 (4), 113–118.
Lundvall, B.Å. (1994), 'The learning economy – challenges to economic theory and policy', *EAEPE Conference*, October, Copenhagen.
Lyons, T. and P. Mandaville (2010), 'Think locally, act globally: toward a transnational comparative politics', *International Political Sociology*, 4, 124–141.
McDonagh, P. and P. Commins (1999), *Globalisation and Rural Development: Demographic Revitalisation, Entrepreneurs and Small Business Formation in the West of Ireland*, Athens: Ashgate Publishing Company.
McDowell, L., A. Batnitzky and S. Dyer (2009), 'Precarious work and economic migration: emerging immigrant divisions of labour in Greater London's service sector', *International Journal of Urban and Regional Research*, 33 (1), 3–25.
McFarlane, C. (2009), 'Translocal assemblages: space, power and social movements', *Geoforum*, 40 (4), 561–567.
McKay, D. (2006), 'Translocal circulation: place and subjectivity in an extended Filipin community', *Asia Pacific Journal of Anthropology*, 7 (3), 265–278.
Minniti, M. and W. Bygrave (2001), 'A dynamic model of entrepreneurial learning', *Entrepreneurship: Theory and Practice*, 25 (3), 5–16.
Mostrous, A. and C. Seib (2008), 'Tide turns as Poles end great migration' available at http://www.timesonline.co.uk/tol/news/uk/article3378877.ece (accessed 26 October 2008).
Murdoch, J. (2000), 'Networks: a new paradigm of rural development', *Journal of Rural Studies*, 16 (4), 407–420.

Oakes, T. and L. Schein (2006), *Translocal China: Linkages, Identities and the Reimagining of Space*, London: Routledge.
Pollard, N., M. Latorre and D. Sriskandarajah (2008), 'Floodgates or turnstiles? Post-EU enlargement migrations flows to (and from) the UK', available at www.ippr.org/publicationsandreports/publication.asp?id60=3 (accessed 28 January 2010).
Portes, A., W. Haller and L.E. Guarnizo (2001), 'Transnational entrepreneurs: the emergence and determinants of an alternative form of immigrant economic adaptation', available at http://www.transcomm.ox.ac.uk/working%20papers/ WPTC-01–05%20Portes.pdf (accessed 4 February 2011).
Portes, J. and S. French (2005), *The Impact of Free Movement of Workers from Central and Eastern Europe on the UK Labour Market: Early Evidence*, London: Department for Work and Pensions.
Ram, M. (1997), 'Ethnic minority enterprise: an overview and research agenda', *International Journal of Entrepreneurial Behaviour and Research*, 3 (3), 149–156.
Ram, M. and G. Hillin (1994), 'Achieving "break-out": developing mainstream ethnic minority business', *Small Business Enterprise and Development*, 1 (1), 15–21.
Shane, S., E. Locke and C. Collins (2003), 'Entrepreneurial motivation', *Human Resource Management Review*, 13, 257–279.
Smart, A. and G.C.S. Lin (2007), 'Local capitalisms, local citizenship and translocality: rescaling from below in the Pearl River Delta region, China', *International Journal of Urban and Regional Research*, 31(2), 280–302.
Smith, M.P. (2001), *Transnational Urbanism: Locating Globalization*, Oxford: Blackwell.
Smith, M.P. and L.E. Guarnizo (ed.) (1998), *Transnationalism from Below*, Brunswick, NJ: Transaction Books.
Stenning, A. and S. Dawley (2009), 'Poles to Newcastle: grounding new migrant flows in peripheral regions', *European Urban and Regional Studies*, 16 (3), 273–294.
Traill, B. (1995), *Small and Medium Sized Food Manufacturing Enterprises in the EU*, Reading: University of Reading.
Velayutham, S. and A. Wise (2005), 'Moral economies of a translocal village: obligations and shame among South Indian transnational migrants', *Global Networks*, 5, 1, 27–47.
Venkataraman, S. and A.H. Van de Ven (1998), 'Hostile environmental jolts, transaction set, and new business', *Journal of Business Venturing*, 13, 231–255.
Vershinina, N., R. Barrett and M. Meyer (2009), 'Researching immigrant entrepreneurial development: an ethnographic study of Polish entrepreneurs in Leicester', Research in Entrepreneurship and Small Business Conference, 19–20 November, Budapest, Hungary.
Vertovec, S. (2001), 'Transnationalism and identity', *Journal of Ethnic and Migration Studies*, 27 (4), 573–582.
Wessendorf, S. (2007), 'Roots-migrants: transnationalism and "return" among second-generation Italians in Switzerland', *Journal of Ethnic and Migration Studies*, 33 (7), 1083–1102.
White, A. (2010), *Polish Families and Migration in Poland Since EU Accession*, Bristol: Policy Press.
White, A. (2011), 'The mobility of Polish families in the West of England: translocalism and attitudes to return', *Studia Migracyjne – Przegląd Polonijny*, 1, 11–32.
White, A. and L. Ryan (2008), 'Polish "temporary" migration: the formation and significance of social networks', *Europe-Asia Studies*, 60 (9), 1467–1502.
Wong, L.L. and M. Ng (2002), 'The emergence of small transnational enterprise in Vancouver: the case of Chinese entrepreneur immigrants', *International Journal of Urban and Regional Research*, 26 (3), 508–530.

22. Škoda Auto: the transformation from a domestic to a Tier Two lead firm

Petr Pavlínek

INTRODUCTION

The automotive industry is considered to be a key industrial sector of strategic importance because of its size, links to other industries, employment effects and its potential regional development effects through the development of networks of component suppliers. As such, the governments of both developed and developing countries have supported the development of the automotive industry through various measures and economic policies (Dicken, 2011). Between 1990 and 2012, the production of cars in Central and Eastern Europe (CEE) increased 2.5 times to 5.3 million units and that of East-Central Europe (ECE) quadrupled to 3.3 million units.[1] This rapid development was orchestrated by core-based transnational corporations (TNCs) through large inflows of foreign direct investment (FDI). As a result, the automotive industry has become a leading industrial sector in terms of its share of industrial production, employment and overall exports in countries such as Czechia, Hungary, Poland and Slovakia (e.g. Pavlínek, 2002a, 2002b; Pavlínek *et al.*, 2009).

Škoda Auto (henceforth Škoda) is a Czech-based automobile manufacturer, which is owned by the German Volkswagen (VW) company. Škoda is one of the 12 VW brands.[2] It represents one of the most important and successful manufacturing companies in CEE. According to Ledgard (2005:3), Škoda is 'the most successful and globally oriented manufacturer to emerge in any category from a former iron curtain country'. Since its takeover by VW in 1991, Škoda has been gradually transformed from a small uncompetitive national car maker into a competitive and increasingly internationalized automobile manufacturer. Its output increased more than five times between 1991 and 2012. VW has developed Škoda as a part of its growth strategy based on market segmentation. Škoda plays an important role in the Czech economy and in its regional economic development through the production network of its suppliers. It has profoundly affected the post-1990 transformation of the Czech automotive components industry. Its case is interesting for several reasons: Škoda represents an example of a successful transformation from a state-owned enterprise into a highly competitive capitalist firm; it illustrates the potential impacts of FDI in less developed countries; it represents an example of the incorporation of a less developed country firm into the global production network (GPN) of a core-based global firm; and it is an example of what I term a *Tier Two lead firm.* A Tier Two lead firm is a foreign-owned lead firm whose autonomy and decision-making power are limited by foreign ownership. Despite its success, Škoda's foreign ownership and its dependence on VW also illustrate the potential drawbacks of the acquisition and subsequent foreign control of firms located in less developed economies.

This case study of Škoda builds upon my long-term research interest in Škoda and the Czech automotive industry as a whole (e.g. Pavlínek, 2002a, 2002c, 2003, 2008, 2012; Pavlínek and Janák, 2007; Pavlínek and Ženka, 2011). In addition to various secondary sources, it draws upon interviews undertaken with Škoda managers and union leaders between 1996 and 2011. This chapter is organized as follows. First, it begins with an overview of Škoda's pre-1991 history. Second, it situates Škoda's post-1991 restructuring in a broader context and then briefly discusses changes in the global automotive industry from the 1990s to the 2000s. Third, it examines Škoda's restructuring following its 1991 takeover by VW. Fourth, the chapter identifies and reviews the developments in the 2000s. Fifth is an analysis of Škoda's research and development (R&D) activities and, sixth, a review of Škoda's internationalization of its production network and the expansion of exports. Seventh is a discussion of Škoda's position in VW's GPN as a Tier Two lead firm and the importance of strategic coupling for Škoda's continued success. The main points of the case study are summarized in the conclusion.

THE PRE-1991 DEVELOPMENT OF ŠKODA

Škoda Auto traces its history back to 1895 when its predecessor, Lauring & Klement (L&K), was originally established as a bicycle repair shop in the city of Mladá Boleslav in Central Bohemia. In 1899, L&K began to make motorcycles, and in 1905 they produced their first passenger car (the Voiturette). In 1925, L&K merged with the Pilsen-based Škoda industrial conglomerate, because of L&K's financial difficulties and the declining competitiveness of its products. The merger secured the continuing assembly in Mladá Boleslav under the new Škoda brand. Škoda built a new factory and introduced assembly lines in 1928, which cut production costs and made Škoda automobiles more competitive on the domestic Czechoslovak market. By 1936, it had become the best-selling domestic brand, with 3,013 units sold. During the Second World War, Škoda was incorporated into the German concern Reichswerke-Hermann-Göring A.G. and mainly produced components for military terrain vehicles and planes, cartridge cases and other weapon components. Its passenger car output decreased from 5,672 units in 1939 to 35 in 1944 (Kožíšek and Králík, 1995).

Following the Second World War, the Škoda vehicle factory complex in Mladá Boleslav was nationalized, separated from the Škoda concern and designated as the only producer of small passenger cars in Czechoslovakia, with the task of developing and producing a 'people's passenger car'. The production of trucks was transferred to other firms. Although production was restored after the war, the onset of the Cold War was not favourable for passenger car production, which sharply decreased in the early 1950s. It was not until 1964 that the Škoda factory complex was modernized and expanded to increase its daily production capacity from 120 to 600 passenger cars (Kožíšek and Králík, 1995). Despite this modernization, Škoda's productivity continued to lag behind Western European automakers (see Pavlínek, 2008:69). Škoda continued to struggle under the conditions of a centrally planned economy because, as a fully state-owned company, it was unable to make any autonomous production or investment decisions. The government did not allow Škoda to import components from Western Europe that were unavailable in Czechoslovakia, purchase foreign licences to produce them domestically or finance their domestic development. Despite these difficulties, Škoda managed to export about half of its output, mainly to the other state socialist countries of CEE in the late 1960s (Pavlínek, 2008).

After the 1964 modernization, Škoda did not receive any significant investment for the next 20 years. An ambitious project to modernize Škoda cars was scrapped by the government in 1971 in favour of building a new automobile plant in Bratislava, Slovakia. The 1970s and early 1980s were, therefore, a period of stagnation during which the gap between Škoda and Western European automakers further widened and Škoda's sales in Western Europe declined because its cars were obsolete and increasingly uncompetitive, even though they were amongst the cheapest passenger cars on the market. Eventually, these factors and the need to earn hard foreign currency prompted the government to approve a modernization project in 1982, which mandated Škoda to modernize its cars and build a new assembly factory. The Favorit was introduced in September 1987 and its mass production was launched in August 1988, despite the serious difficulties Škoda had to face (see Pavlínek, 2008:72–76). Škoda's production capacity increased to 180 thousand units annually. This modernization preceding the end of the state socialist period was important since it once again narrowed the gap between Škoda and Western automakers. Škoda continued to produce its own models more or less independently of Western automakers, which put it in a unique position in CEE. At the same time, it made Škoda the most indebted Czech company, which desperately needed additional investment to survive and become truly competitive in Western markets.

In early 1990, only a few months after the collapse of the state socialist system in November 1989, the Czech government announced the public tender to privatize Škoda. Originally, 24 foreign companies registered to participate but only eight were seriously interested. In the final selection between VW and Renault, the German VW company was selected to become Škoda's joint venture partner. VW bought 31 per cent of Škoda's shares in April 1991 and immediately assumed full managerial control. VW gradually increased its stake in Škoda during the 1990s, becoming sole owner in 2000 (see Pavlínek, 2008:79–86 for details).

THE CONTEXT OF ŠKODA'S POST-1990 RESTRUCTURING AND ITS INTEGRATION IN GPNs

The post-1990 development of Škoda needs to be understood in the context of changes that took place in the global automotive industry in the 1990s and 2000s, as well as in the context of post-communist economic transformations in CEE. Global value chains (GVC) and GPN perspectives have been particularly useful in understanding these changes (e.g. Gereffi *et al.*, 2005; Henderson *et al.*, 2002). The contemporary automotive industry is typified by quasi-hierarchical (captive) value chains dominated by large and powerful TNCs (Gereffi *et al.*, 2005; Humphrey and Schmitz, 2004). These lead firms (assemblers) determine the architecture of the final product and of crucial parts and components. They also organize and coordinate the vertical networks of component suppliers by selecting core suppliers, developing and monitoring quality system management throughout their production network, and by setting conditions under which these components are delivered (Humphrey and Schmitz, 2002, 2004; Sturgeon *et al.*, 2008).

The opening of potentially vast markets in CEE, China, India and other less developed world regions in the early 1990s, combined with the liberalization of trade and FDI, opened new profit-making opportunities for core-based automakers. The wave of investment in new assembly plants in 'emerging economies' was based on the market potential of these regions, political pressures on lead firms to assemble vehicles in these large markets, and on the possibilities of low-cost export-oriented production in the peripheral areas adjacent to developed countries' markets and within regional trade agreements (Humphrey *et al.*, 2000; Liu and Dicken, 2006; Liu and Yeung, 2008; Pavlínek, 2002d; Van Biesebroeck and Sturgeon, 2010). To support their operations in less developed economies, lead firms increasingly required their most important suppliers to establish component plants close to vehicle assembly plants to satisfy the demands of lead firms for the quality and timing of delivery of components. Leading core-based suppliers satisfied these demands for global sourcing by rapidly internationalizing their production through a wave of mergers, acquisitions and joint ventures. This led to the emergence of 'global suppliers' that were increasingly specialized in particular modules and/or technologies and capable of supplying lead firms wherever they established their assembly plants (Sturgeon and Lester, 2004). Some 'global suppliers' became larger than the lead firms. In the process, lead firms gradually reduced the number of suppliers (Sadler, 1999) and the suppliers were reorganized into distinct tiers based on their capabilities and their relationship with lead firms. These tiers also differ in the degree of internationalization of suppliers (Humphrey and Memedovic, 2003; Pavlínek and Janák, 2007; Sturgeon and Lester, 2004).

The restructuring of the global supplier base in the automotive industry had important consequences for domestic suppliers in less developed countries (Barnes and Kaplinsky, 2000; Sturgeon and Lester, 2004). The most capable and largest domestic suppliers were taken over by foreign suppliers, which weakened the domestic sector. With the increasing specialization and internationalization of leading suppliers, the majority of the remaining domestic suppliers were relegated to the lowest supplier tier because of their limited capabilities and small size. As such, their opportunities for upgrading in captive value chains were limited.

I have analysed changes in the Czech automotive industry in the context of the post-1990 economic transformation elsewhere (Pavlínek, 2002a, 2002c, 2003, 2008). In the early 1990s, the Czech automotive industry was obsolete after 40 years of organization under a centrally planned economy. Škoda and its 237 Czechoslovak suppliers were investment starved, technologically underdeveloped and inefficient. They were not ready to face foreign competition. Price and trade liberalization associated with the governmental 'shock therapy' of 1991 was a major shock for the entire industry. Hasty privatization in the early 1990s led to the fragmentation of large state-owned enterprises into smaller units. The fragmentation of the Czechoslovak automotive industry was taking place at the time of the consolidation of the global automotive supplier industry through mergers and acquisitions. In this situation, the takeover of Škoda by VW saved not only Škoda but was also crucial for the future of the Czech automotive industry as a whole because it led to its rapid and profound restructuring.

THE POST-1991 RESTRUCTURING OF ŠKODA

The purchase of Škoda in 1991 was part of VW's long-term development strategy. VW was keen to expand into CEE to catch up with Fiat, Renault and Citroën, which had operated in Poland, Romania, the former Yugoslavia and the former Soviet Union through licence agreements since the mid-1950s (Pavlínek, 2002b). VW believed that Škoda could be successfully used as a bridgehead to CEE markets, where it controlled a 35 per cent market share during the state socialist period and, therefore, had been well known and established in the region with the exception of the former Soviet Union. The Czech government required that VW retained the Škoda brand, which fitted with VW's strategy of brand segmentation that it had previously applied with its acquisition of the Spanish car brand SEAT. Additionally, VW believed that a distinct Škoda brand would allow it to sell cheaper cars outside Western Europe without undermining the image of its existing brands (Myant, 1999). Despite Škoda's mounting problems in the late 1980s, VW perceived it to be a relatively successful company, given the conditions in which it had had to operate in the state socialist economy for the past 40 years. Based on the joint venture agreement with the Czech government, VW planned to invest 5.7 billion USD to increase Škoda's production capacity from 180 thousand units to 450 thousand units by 1997 and build a new engine plant.[3] VW hoped to develop Škoda into a low-cost producer that could successfully compete with the Japanese and South Korean firms. In the early 1990s, the wages of Škoda workers were 90 per cent lower than the wages of comparable VW workers (Pavlínek, 2008).

Still, VW was buying a troubled company that desperately needed capital for modernization and had increasing difficulties in financing day-to-day production, and paying for supplies of components and the salaries of its workers, because its domestic sales had plummeted from 123 thousand units in 1990 to 27 thousand units in 1991 (Pavlínek, 2008). Without a major injection of capital Škoda would have most likely ended in bankruptcy and liquidation because it would have been unable to rapidly improve the quality of its cars to keep up with the foreign competitors that flooded the Czechoslovak and CEE markets with new and used passenger cars. Škoda faced major difficulties even after the formation of the joint venture with VW. It accumulated 31 million USD of losses between April and December 1991; the Czech government and VW argued over investment plans, and rapid knowledge transfer from VW to Škoda proved to be more challenging than VW had originally anticipated.

Knowledge Transfer from VW to Škoda

The transfer of knowledge from VW to Škoda was an important part of the initial phase of Škoda's restructuring. After the joint venture with VW had been formed, Škoda's managers were sent to other VW plants to learn about modern methods of production and new management skills; hundreds of Škoda workers were sent to tour VW's factories; and 50 VW managers, mostly German, were sent to work at Škoda. The knowledge transfer from VW to Škoda originally followed VW's established practices without taking specific conditions at Škoda into consideration. This one-way transfer of knowledge initially ignored the accumulated skills and knowledge of Škoda workers and managers because these were developed under the conditions of a centrally planned economy and, as such, were considered to be useless by VW managers. However, the knowledge transfer did not work the way VW had originally planned. The attitude and approach of VW managers caused resentment amongst Škoda workers and managers, and were eventually challenged by Škoda's labour unions. The pressure from Škoda's labour unions, combined with challenges that VW's car production system was facing across the board in the early 1990s, led VW's managers to reconsider their approach to Škoda. A new strategy was to combine VW's and Škoda's strengths, experience and innovation potential through tandem management and coaching in which 400 VW expatriates were teamed up with Czech managers, with a goal of transferring knowledge to Czech managers (Dörr and Kessel, 1996, 1998). Despite problems, tandem management successfully transferred knowledge from VW to Škoda. By 2005 only 40 VW expatriates worked at Škoda while 40 Czech managers worked at other VW plants (Pavlínek, 2008).

Restructuring Strategies

At the same time, VW set out to restructure Škoda's production processes, with the immediate goal of improving the quality of the Favorit, the only model Škoda was assembling in the early 1990s. VW replaced Škoda's hierarchical and centralized structure of production organization with a more horizontal and decentralized system based on teamwork, which was inspired by Japanese lean production. This 'fractal' organization of production was first implemented in 1994 at the new factory for the assembly of the Felicia, a Favorit successor. In addition to team work, it involved so called 'strategic insourcing', in which selected key suppliers of preassembled modules, such as drive units, roofs, cockpit units, doors or seats, set up assembly operations at the Škoda production complex and delivered these modules directly to the assembly line as part of a just-in-time regime. This innovation cut Škoda's inventory costs to a bare minimum. In the next step, Škoda planned the installation of these modules directly into the cars by the suppliers, which would have replaced unionized Škoda workers by lower-paid and non-unionized workers employed by the suppliers on the assembly line. However, this radically new production concept was successfully resisted by the unions. Cost-cutting and rationalization strategies were fully implemented in a new modular plant built to assemble a new mid-size model (the Octavia) in 1997. The Octavia assembly plant was heralded as the most efficient assembly plant in Europe when it began production, and the Octavia became Škoda's most successful model in the late 1990s and 2000s (Pavlínek, 2008). To further rationalize production and increase flexibility, Škoda introduced the 'Škoda production system' throughout the company in 1997 and 1998. The goal was to reinforce the introduction of Toyota-inspired production methods, to maximize simplification and standardization of all production activities, and, by doing so, to further increase production efficiency and product quality (Büsching, 1998). The new system also further increased the pace of work on the shop floor to such an extent that shop floor workers could not take scheduled toilet breaks and had to resort to a warning strike in 2001 to force the management to lower the pace on the assembly line and increase the number of assembly line workers (Pavlínek, 2008). All these measures were designed to develop Škoda's main competitive advantage, which lies in the combination of flexible production organization, a fast pace of work and low-cost skilled labour force.

The quality of the original Favorit and all subsequent Škoda models could not be bettered without significantly improving the quality of the supplied components, and this was achieved through the restructuring of external relations between Škoda and its component suppliers. Before 1990, Škoda sourced the vast majority of supplies from its 237 Czechoslovak (mostly Czech) suppliers. Imports of foreign components were low (worth only 229 million USD in 1988) and of these only 17 per cent came from Western suppliers, such as certain electronic parts the Czechoslovak and CEE industries were unable to produce and specific admixtures for the production of plastic materials unavailable in CEE (Pavlínek, 2008). Overall, the quality of domestic and CEE components was low. VW addressed this problem by encouraging its Western suppliers to establish manufacturing operations in Czechia and by pressuring domestic suppliers to increase the quality of supplied components (Pavlínek, 2003). First, VW asked its important Western suppliers to set up operations in Czechia while pressuring domestic suppliers to form joint ventures with foreign firms. As a result, by the end of 1993, 60 joint ventures and greenfield plants had been established in Czechia to supply Škoda, and this development occurred less than three years after the VW takeover. These Czech-based foreign-controlled firms accounted for more than 50 per cent of Škoda's total supplies by 1993 and, together with imports, foreign-controlled plants accounted for 70 per cent of the total value of supplied components to Škoda. By 2005, the number of joint ventures had increased to 94 and the number of newly built foreign factories to supply Škoda to 58 (Pavlínek, 2008). Second, the remaining domestic suppliers were pressured to quickly improve the quality of supplied components or face exclusion from the Škoda supplier network. Škoda educated them about new quality standards and its new system of purchasing. Through its quality control processes, Škoda evaluated its domestic suppliers and identified those capable of supplying high-quality components. Based on the terms of the joint venture agreement between Škoda and VW and also on the fact that they were supplying Škoda-specific components, these domestic suppliers were granted a transition period during which they were required to bring the quality of their components to the required level. To continue supplying, domestic suppliers had to undergo a rigorous quality system certification process and pass regular quality audits conducted by Škoda. Domestic companies usually needed three to five years to achieve international quality certification and to meet other Škoda requirements, such as the need to participate in the development of components, which typically required significant investment. Many domestic firms did not succeed (Pavlínek, 2008). Consequently, by 2013 there were only a few domestic suppliers of Škoda remaining in the narrowly defined automotive industry (NACE 29).[4] Most of them supplied simple, low-value-added components. The quality of supplied components increased substantially, as we can see from the following data. In 1993, only 1 per cent of Škoda's suppliers were ranked in the highest category, A, 38 per cent in category B and 61 per cent in the lowest category, C. By 2007, 65.5 per cent of Škoda's suppliers were ranked in category A, 32 per cent were in category B and only 2.5 per cent in category C (interview at Škoda Auto, 1 August 2011).

Developments in the 2000s

VW's takeover of Škoda and its subsequent restructuring proved to be very successful. Škoda's output of passenger cars increased from 172 thousand in 1991 to 962 thousand in 2012 (Figure 22.1). In 2012, Škoda sold 932 thousand vehicles (Škoda Auto, 2013) and

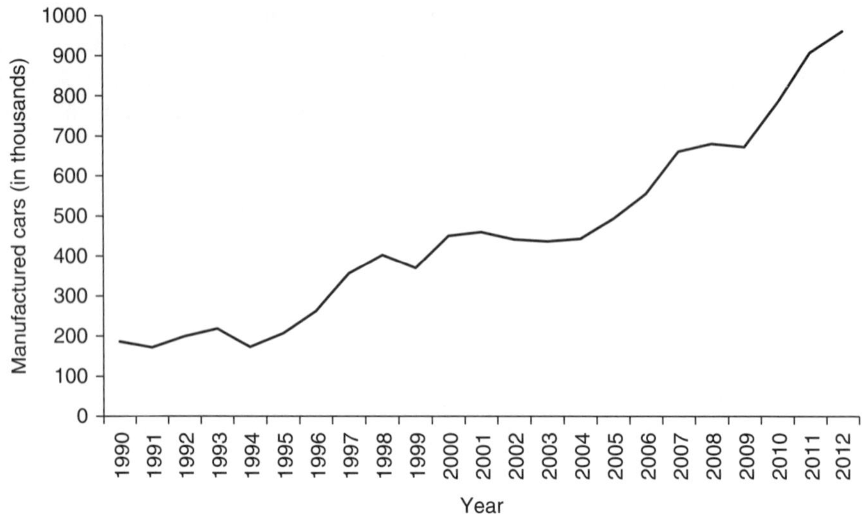

Note: Includes licensed production in China.

Source: Based on data in Škoda Auto Annual Reports 1993–2013.

Figure 22.1 The global production of Škoda passenger cars, 1990–2012

it plans to increase its global sales to 1.5 million vehicles by 2018 (Škoda Auto, 2012). After VW and Audi, Škoda is the third largest brand in the VW Group.

Since 1993, Škoda has had to finance all its investment needs from cash flow, which means that its development has not been financed by VW as is commonly assumed. Its rapid growth was based on the success of Škoda cars in different markets, which convinced VW's headquarters to allow Škoda to develop further, although some of VW's top managers feared that Škoda's success had been achieved partially at the expense of other VW brands. These fears were based on the fact that Škoda's sales growth was increasing at its fastest rate in Western Europe, where more than 50 per cent of Škoda's European output has been sold since 1999. Therefore, there have been attempts to curtail Škoda's relative freedom to develop and the company was asked to pay dividends to VW, which it was previously allowed to use to finance further development. Since 2004, Škoda has been regularly sending dividends to VW's headquarters in Germany; between 2004 and 2012 dividends paid by Škoda to VW amounted to CZK 42 billion (about USD 2.2 billion) (Škoda Auto, 2005–2013). Škoda is not supposed to produce luxury cars and its models should not be as well equipped as those of VW and Audi. The continuing rapid development of Škoda and its ambitious plans to produce 1.5 million vehicles annually by 2018 suggests that VW's headquarters continue to support Škoda's future development and that Škoda should contribute to the goal of the VW Group to become the largest vehicle producer in the world by 2018.

Up until November 1996, Škoda assembled only one model, the Favorit, including its station wagon version called the Forman, and then the Felicia. Since the introduction of the Octavia in 1997, Škoda has gradually added more models, especially in the late 2000s.

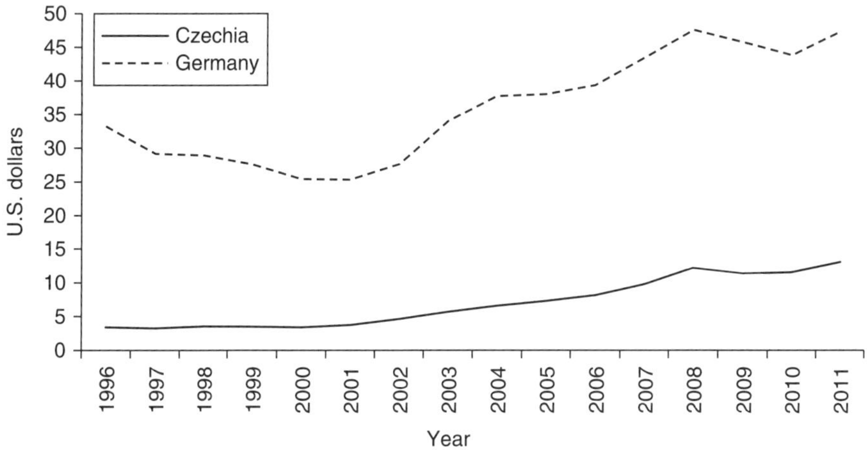

Note: Compensation costs include direct pay, social insurance expenditures and labour-related taxes.

Source: Based on data from U.S. Bureau of Labor Statistics 2012.

Figure 22.2 Hourly compensation costs in manufacturing in Czechia and Germany, 1996–2011

As of 2013, Škoda's model portfolio consisted of seven basic models: the Fabia, Octavia, Superb, Roomster, Yeti, Citigo and the Rapid. In what Škoda calls a 'model offensive' it plans to introduce a new or modernized model every six months in the next several years (Škoda Auto, 2012). Škoda's main competitive advantage continues to be its flexible organization of production, fast pace of work and low-cost skilled labour force. Low wages continue to be Škoda's major advantage although the difference in wages between Škoda and VW decreased from about one-tenth of VW's wages in the early 1990s to more than one-third. Wages continue to account for only 6 per cent of overall production costs at Škoda compared with 17 per cent at VW. In 2012, the average monthly wage of shop floor workers at Škoda was 32,467 CZK (1,300 EUR) (Figure 22.2).[5]

ŠKODA'S ROLE IN RESEARCH AND DEVELOPMENT

In the context of the ECE automotive passenger car assembly, Škoda is unusual (along with the Romanian-based Dacia) because it has important R&D functions. After the takeover by VW, Škoda's Czech managers expected and largely accepted that Škoda's in-house R&D would eventually be transferred to VW's R&D centre in Germany. This was despite the fact that the continuation of Škoda's in-house R&D was one of the conditions set by the Czech government when it was selling Škoda to VW (Pavlínek 2008). However, Škoda's R&D not only survived but it has also been substantially expanded. Škoda employed 584 R&D workers in 1991. By 2012, it employed 1766 R&D workers in a new R&D centre that was opened in 2008 (Pavlínek, 2012; Škoda Auto, 2013).

We can identify two main reasons why R&D continued to grow after the takeover by VW (Pavlínek, 2012). The first reason is related to the fact that Škoda is a distinct brand within the VW Group. Therefore, for marketing reasons, Škoda cars need to have a distinctive look, which is different from VW, Audi and SEAT models. Škoda has been able to maintain responsibility for designing the upper bodies of its cars. These are the visible parts of automobiles that account for about 20 per cent of the finished vehicle. The rest, including the common car platforms and lower bodies for the entire VW Group, is designed in Germany. The major exception is a small three-cylinder 1.2 litre gasoline engine which Škoda developed for the entire VW Group. With the number of Škoda models increasing from one in the early 1990s to seven in 2012, the demand for designers expanded. The second reason why R&D has continued to expand at Škoda is related to lower R&D labour costs in Czechia compared with Germany. This allowed some routine development work, such as computer-aided design operations, to be transferred from Germany to Škoda in the 1990s (CzechInvest, 1997). Overall, Škoda's R&D has the typical attributes of regional automotive R&D centres that lead firms establish at mass production sites or large markets (Pavlínek, 2012).

INTERNATIONALIZATION OF ŠKODA'S SALES AND PRODUCTION NETWORK

In the late 1980s, Škoda sold 72 per cent of its cars on the domestic (Czechoslovak) market and only 28 per cent were exported. The share of the domestic market dropped to 25 per cent in 1991, when domestic sales collapsed due to trade liberalization and 'shock therapy'. 'Shock therapy' was designed to introduce a market economy in former centrally planned economies in the shortest possible time through massive and rapid privatization of state-owned enterprises, domestic price liberalization, limiting governmental intervention and through opening the domestic economy to foreign competition. Škoda exported its cars to 30 countries in 1991. The share of the domestic market continued to decrease in the 1990s and 2000s. In 2012, the domestic market accounted for 6.4 per cent of Škoda's total sales while foreign markets accounted for 93.6 per cent. Since 2008, Škoda automobiles have been sold in more than 100 countries. Since 1991, the territorial structure of exports has changed significantly (Figure 22.3). Western Europe continues to be the most important sales region for Škoda. Throughout the 1990s and early 2000s, the share of Western Europe increased from less than one-third to 56 per cent in 2005. By 2012, it had declined to 38 per cent despite increasing sales. The share of CEE oscillated around one-quarter in the mid-1990s, it peaked at 29 per cent in 2008 and then declined sharply because of the impact of the economic crisis on Eastern Europe. The share of overseas and Asia increased rapidly from 7 per cent in 2006 to 34 per cent in 2012 to become the second most important sales region for Škoda.

The successful internationalization of Škoda's production network has been the main reason behind the rapid growth in output of Škoda automobiles since 2007. Although Škoda assembled its cars abroad from the mid-1990s, it was not until the late 2000s that foreign assembly became really important. Foreign assembly and the continuing internationalization of Škoda's production network, especially in China, Russia and India, are crucial for Škoda's future development (Škoda Auto, 2012). As other car makers, Škoda has increasingly internationalized its production network to overcome trade barriers and comply with governmental policies that control access to large domestic markets.

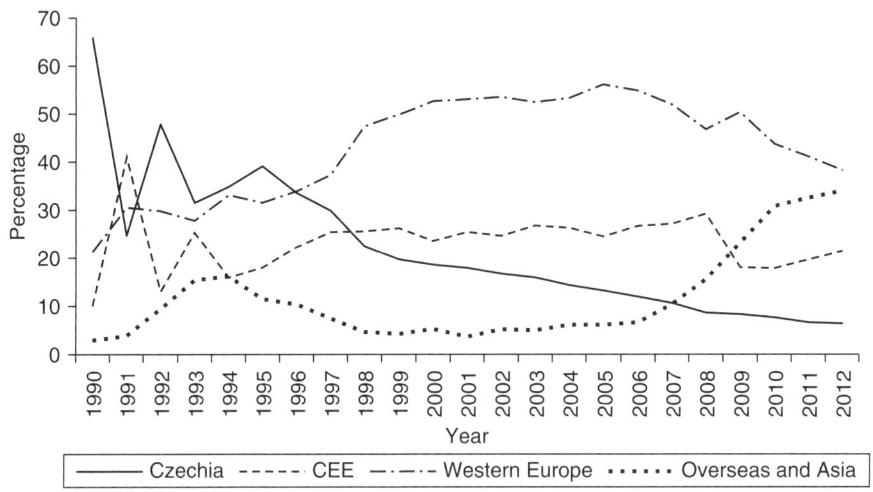

Source: Based on data in Škoda Auto Annual Reports 1993–2013.

Figure 22.3 *Changes in the territorial structure of sales of Škoda passenger cars, 1990–2012*

For foreign assembly, Škoda has used VW's foreign plants (in Poland, China, Bosnia-Herzegovina and India), built new plants together with VW (in Russia) or alone (in India), and also used foreign contract manufacturers (in Ukraine, Kazakhstan and Russia).

Between 1994 and 2002, Škoda assembled its cars in Poland (in Poznań) in VW's factory for commercial vehicles because of a 41 per cent duty imposed on imported Škodas. The production was phased out after the Polish government removed the import tariffs. Between 1998 and 2008, Škoda cars were assembled along with other VW Group vehicles in VW Sarajevo in the town of Vogošća (Bosnia-Herzegovina). The output peaked at 2,720 units in 2008, which was much less than the anticipated annual production of 35,000 vehicles. The production failed to increase because planned exports to the countries of the former Yugoslavia, Turkey and Syria never materialized and the assembly was terminated in December 2008 (see Pavlínek, 2008, for details).

Škoda and VW have used contract manufacturers in Ukraine and Kazakhstan to assemble Škoda, VW and Audi models for the local markets. In Ukraine, the Ukrainian company Eurocar has assembled Škodas under a licence in the town of Solomonovo since December 2001. The output peaked in 2008, when 30 thousand Škodas were assembled, but it decreased by almost 90 per cent during the economic crisis that commenced in 2008 and has only been slowly recovering. In Kazakhstan, the local company Asia Auto has assembled low numbers of Škodas at Ust-Kamenogorsk since 2005 (Table 22.1) (see Pavlínek, 2008 for details).

More importantly, Škoda made consistent efforts to penetrate large markets by local assembly in Russia, India and China in the 1990s and 2000s. By the late 2000s, these three countries became Škoda's three most important foreign production sites, with a combined output of 370 thousand vehicles in 2012 (Škoda Auto, 2013; Table 22.1). China has been the largest market for Škoda since 2010. In 2003, Škoda was allowed to

Table 22.1 Foreign production of Škoda cars, 2005–2012

	2005	2006	2007	2008	2009	2010	2011	2012
Russia	0	0	1109	37,312	24,233	18,053	48,300	*99,100*
India	8710	12,599	12,390	18,342	8986	11,202	15,661	*34,300*
Bosnia and Herzegovina	2267	2170	2579	2720	0	0	0	0
Ukraine	9785	19,013	23,337	30,172	3546	7484	10,909	NA
Kazakhstan	387	1124	1557	281	179	469	NA	NA
China	0	0	38,664	55,920	131,779	197,077	223,030	226,653
Slovakia	0	0	0	18,278	18,666	10,000	1027	36,687
Total	21,149	34,906	79,636	163,025	187,389	244,285	298,927	396,740

Note: Numbers in italic refer to total sales of Škoda vehicles and not to annual production.

Source: Škoda Auto *Annual Reports* 2005–2012.

export only 2,100 of its cars to China because of Chinese quotas and tariffs. Also, VW originally did not want Škoda to enter China because it considered China to be VW's territory. In 2006, only 660 Škodas were sold in China. The situation has changed with China's membership of the World Trade Organization and the abolition of quotas on imported automobiles in 2005. More importantly, VW changed its policy in 2004 and announced that up to 80 thousand Škoda cars would be produced under licence in the existing VW factory in Shanghai. The assembly started in 2007 and the output increased rapidly. In July 2012, a new VW factory was opened at Yizheng which also assembles Škodas. In 2012, 227 thousand Škodas were assembled in China and the company plans to assemble 0.5 million vehicles annually in China by 2018 (Table 22.1; Škoda Auto, 2013).

In Russia, Škoda tried to set up a local assembly of knock-down vehicles from the mid-1990s because high tariff barriers combined with the low value of the ruble made imported Škodas uncompetitive in the Russian market. After three unsuccessful attempts to launch assembly through joint ventures with Russian firms in the 1990s, Škoda and VW decided to build a greenfield factory in Russia. The first plan to open a factory in 2002 failed mainly because the Russian government required 50 per cent local content five years after the start of assembly. An agreement between VW/Škoda and the Russian government was eventually reached and a new factory was opened in Kaluga in October 2007 (see Pavlínek, 2008 for details). The Kaluga factory is a joint venture between VW and Škoda, in which Škoda has a 16.8 per cent stake. In May 2011, VW and the Russian Ministry of Economic Development concluded an additional agreement, according to which the VW Group is mandated to increase its annual production in Russia to 350 thousand cars. This will be achieved by increasing the capacity of the Kaluga plant to 225 thousand cars and by the Russian automaker GAZ assembling 110 thousand VW and Škoda cars per year in Nizhny Novgorod based on the contract manufacturing agreement. The assembly of the Škoda Yeti started in Nizhny Novgorod in October 2011. Škoda's sales in Russia increased to 99,100 cars in 2012, making Russia the third largest national market for Škoda automobiles after China and Germany (Škoda Auto, 2013; Table 22.1).

As in Russia, Škoda has attempted to start local assembly in India since the mid-1990s but it also experienced several failures and delays in setting up production. Attempts to set up a joint venture with an Indian firm in the city of Bangalore failed in the mid-1990s. Then, Škoda's decision to build a greenfield assembly plant was delayed for several years by the changing policies of the Indian government towards foreign investors. Because of these uncertainties and delays, Škoda started to assemble its cars in rented factory space in 2001 and did not open its new factory in Aurangabad until February 2004 (Pavlínek, 2008). Compared with China and Russia, production increased slowly (Table 22.1). In 2009, VW opened a new assembly facility at Pune, where the Škoda Fabia is being assembled along with VW vehicles.

A TIER TWO LEAD FIRM AND STRATEGIC COUPLING

Lead firms are not all the same but differ in a number of attributes despite sharing basic features, such as wielding power over networks of suppliers (e.g. Coe *et al.*, 2008; Dicken, 2011; Gereffi *et al.*, 2005). Škoda is what I call a *Tier Two lead firm* because it has many attributes of automotive lead firms, as defined by the GPN and GVC approaches (Pavlínek 2012; Pavlínek and Janák, 2007). Škoda has built an international production and distribution network, which it manages and controls. It coordinates the vertical network of component suppliers. It has the power to determine who will be included or excluded from this supplier network and has the power to set parameters under which suppliers deliver components to Škoda. It also has its own R&D and testing, although this is limited. Tier Two lead firms, however, differ from Tier One lead firms in the fact that Tier Two lead firms are foreign owned. Foreign ownership means that ultimate decisions and control are not in the hands of Tier Two lead firms but in the hands of their ultimate owners, the Tier One lead firms. In our case, Volkswagen is a Tier One lead firm. Škoda's position as a Tier Two lead firm limits its power and its autonomy. Škoda is completely controlled by VW. Its chairman is German and the personnel director has been the only Czech representative on Škoda's board of directors. This, in turn, limits Škoda's value capture potential through profit repatriation and, consequently, it limits Škoda's potential contribution to economic development in Czechia.

GPN scholars argue that successful regional economic development in the contemporary globalizing economy depends on both internal and external sources of competitiveness (Coe *et al.*, 2004; Yeung, 2009). In order to attract foreign investment, firms and regions in less developed economies need to possess resources that TNCs desire to access because they consider them to be strategically important for their continuing competitiveness. Škoda and VW represent an example of strategic coupling between a domestic firm and foreign TNC. As we have seen, the acquisition of Škoda fitted well with the strategic needs of VW to expand into CEE markets and also to gain access to a low-cost skilled labour force while Škoda desperately needed access to investment capital, technology and managerial expertise.

Despite the limits imposed by Škoda's position in GPN as a Tier Two lead firm, its contribution to the Czech economy has been enormous. As we have seen, Škoda played a critical role in the post-1990 restructuring of the Czech auto supplier industry. Škoda accounts for 7.3 per cent of Czech exports. It directly employs more than 27 thousand workers in its three production complexes in Czechia and thousands more indirectly in its 288 Czech-based suppliers (Škoda Auto, 2012). Škoda accounts for more than 75 per cent of total R&D expenditure in the Czech automotive industry and employs more than 1,700 R&D workers directly in its R&D centre. An additional 425 R&D workers are employed by engineering firms that work in close proximity to Škoda's R&D centre in Mladá Boleslav (Pavlínek, 2012; Škoda Auto, 2012).

CONCLUSION

The case of Škoda Auto is significant for economic geographers for several reasons. First, it illustrates a successful integration of a less developed country firm into a GPN organized and controlled by a core country firm in the automotive industry. Second, Škoda Auto demonstrates an example of successful strategic coupling between the strategic needs of a foreign TNC and the needs of a domestic firm. This particular strategic coupling allowed Škoda to survive, prosper and grow into one of the most successful firms in CEE, which has been beneficial not only for both VW and Škoda but also for the Czech economy as a whole. Third, Škoda Auto suggests that lead firms are very diverse and that foreign-owned firms may behave like typical lead firms. However, foreign ownership means that the ultimate power and control are in the hands of their foreign owner. Ultimately, foreign ownership involves the transfer of value from foreign-owned firms to their owners in the form of dividends. This value was originally generated in foreign locations and its transfer abroad diminishes the regional economic potential of foreign-owned firms in their domestic economy.

There is no doubt that Škoda has experienced successful development since its takeover by VW in 1991. Škoda's current size of output, its product portfolio, profitability, R&D, sales in foreign markets and its international production network have far exceeded the most optimistic expectations at the time of its privatization in 1991. The government decision to privatize Škoda and its privatization policy, including the conditions of privatization and the selection of VW as its original joint venture partner, proved to be correct. Škoda has been successful because of the coupling between its assets and capabilities with the strategic needs of VW and because of its successful integration into global production and distribution networks. Was there any plausible alternative to foreign ownership of Škoda? Given the situation of Škoda in the early 1990s, it was unrealistic to expect that it would survive and prosper in the global automotive industry that is dominated by a relatively small number of very large TNCs. Without a strong foreign partner, Škoda would have most likely followed the fate of Czech truck manufacturers that either ended in bankruptcy and liquidation or never recovered from the production collapse in the early 1990s (Pavlínek, 2008).

The Škoda case study, however, shows that the success of the foreign takeover of Škoda was far from automatic because it required the cooperation and flexibility of both VW and Škoda, including Škoda's managers and Škoda's workers represented by relatively strong labour unions. Still, it is clear that Škoda's future is closely tied to that of VW and its continuing success. As a Tier Two lead firm, its autonomy is restricted and it depends on VW for all strategic decisions. Škoda's value capture capability is limited by foreign ownership. It is important to keep in mind that foreign ownership is not unproblematic and does not automatically lead to success in the automotive industry, as the case of the failed takeover of Polish Fabryką Samochodów Osobowych (FSO) by South Korean Daewoo has demonstrated (Pavlínek, 2006).

Can the success of Škoda be repeated across ECE? Renault, which emulated VW's Škoda strategy with the Romanian Dacia in the 2000s, shows that it is possible. It is unlikely, however, that we will see more Tier Two lead firms emerge in the automotive industry in ECE anytime soon because of overcapacity and also because there are no remaining ECE automotive brands to be acquired by core-based lead firms.

ACKNOWLEDGEMENT

The research and preparation of this chapter were supported by the Czech Science Foundation (Grant Agreement No. 13–16698S).

NOTES

1. CEE refers to the former state socialist countries of Europe, including the former Soviet Union. ECE excludes the territory of the former Soviet Union.
2. Volkswagen, Audi, SEAT, ŠKODA, Bentley, Bugatti, Lamborghini, Porsche, Ducati, Volkswagen Commercial Vehicles, Scania and MAN.
3. These investment plans were scaled down in 1994: the promised investment was lowered to 2.5 billion USD, annual production capacity to 340 thousand vehicles by 2000, and the engine plant construction was cancelled in 1993. In reality, the production capacity of 340 thousand vehicles was reached in 1997 and the engine plant was eventually built in 2001 (Pavlínek, 2008).
4. NACE 29 refers to the manufacture of motor vehicles, trailers and semi-trailers in the statistical classification of economic activities in the European Union (Nomenclature generale des Activites economiques dans les Communautes europeennes).
5. The average Czech monthly wage was 24,514 CZK in the last quarter of 2012.

REFERENCES

Barnes, J. and Kaplinsky, R. (2000), 'Globalization and the death of the local firm? The automobile components sector in south Africa', *Regional Studies* **34**(9): 797–812.

Büsching, G. (1998), 'Výrobní systém Škoda musíme prosadit', *Škoda Mobil* **4**(14): 2.

Coe, N.M., Dicken, P. and Hess, M. (2008), 'Global production networks: realizing the potential', *Journal of Economic Geography* **8**(3): 271–295.

Coe, N.M., Hess, M., Yeung, H.W.C., Dicken P. and Henderson, J. (2004), '"Globalizing" regional development: a global production networks perspective', *Transactions of the Institute of British Geographers* **29**(4): 468–484.

CzechInvest (1997), *Study of Czech Automotive Industry*, Praha: CzechInvest.

Dicken, P. (2011), *Global Shift: Mapping the Changing Contours of the World Economy*, London: Sage.

Dörr, G. and Kessel T. (1996), 'Transformation as a learning process: experiences with joint ventures in the Czech Republic and an example of new modern forms of organizing production', *Emergo: Journal of Transforming Economies and Societies* **3**(2): 44–57.

Dörr, G. and Kessel, T. (1998), 'Transformace jako proces učení: Příklad nové moderní organizace výroby', *Sociologický časopis* **34**(2): 221–233.

Gereffi, G., Humphrey, J. and Sturgeon, T. (2005), 'The governance of global value chains', *Review of International Political Economy* **12**(1): 78–104.

Henderson, J., Dicken, P., Hess, M., Coe, N. and Yeung, H.W.C. (2002), 'Global production networks and the analysis of economic development', *Review of International Political Economy* **9**(3): 436–464.

Humphrey, J. and Memedovic, O. (2003), *The Global Automotive Industry Value Chain: What Prospects for Upgrading by Developing Countries*, Vienna: UNIDO.
Humphrey, J. and Schmitz, H. (2002), 'How does insertion in global value chains affect upgrading in industrial clusters?', *Regional Studies* **36**(9): 1017–1027.
Humphrey, J. and Schmitz, H. (2004), 'Chain governance and upgrading: taking stock', in Schmitz, H. (ed.), *Local Enterprises in the Global Economy: Issues of Governance and Upgrading*, Cheltenham and Northampton, MA: Edward Elgar: 349–381.
Humphrey, J., Lecler, Y. and Salerno, M.S. (2000), *Global Strategies and Local Realities: The Auto Industry in Emerging Markets*, Basingstoke: Palgrave Macmillan.
Kožíšek, P. and Králík, J. (1995), *L&K – Škoda: 1895–1995*, Praha: Motorpress.
Ledgard, J. (2005), 'Škoda leaps to market', *Strategy and Business* (40): 2–12.
Liu, W.D. and Dicken, P. (2006), 'Transnational corporations and "obligated embeddedness": foreign direct investment in China's automobile industry', *Environment and Planning A* **38**(7): 1229–1247.
Liu, W.D. and Yeung, H.W.C. (2008), 'China's dynamic industrial sector: the automobile industry', *Eurasian Geography and Economics* **49**(5): 523–548.
Myant, M. (1999), 'The transformation of Czech enterprises', in Myant, M. (ed.), *Industrial Competitiveness in East-Central Europe*, Cheltenham and Northampton, MA: Edward Elgar: 145–168.
Pavlínek, P. (2002a), 'Domestic privatisation and its effects on industrial enterprises in East-Central Europe: evidence from the Czech motor component industry', *Europe-Asia Studies* **54**(7): 1127–1150.
Pavlínek, P. (2002b), 'Restructuring the Central and Eastern European automobile industry: legacies, trends, and effects of foreign direct investment', *Post-Soviet Geography and Economics* **43**(1): 41–77.
Pavlínek, P. (2002c), 'The role of foreign direct investment in the privatisation and restructuring of the Czech motor industry', *Post-Communist Economies* **14**(3): 359–379.
Pavlínek, P. (2002d), 'Transformation of the Central and East European passenger car industry: selective peripheral integration through foreign direct investment', *Environment and Planning A* **34**(9): 1685–1709.
Pavlínek, P. (2003), 'Transformation of the Czech automotive components industry through foreign direct investment', *Eurasian Geography and Economics* **44**(3): 184–209.
Pavlínek, P. (2006), 'Restructuring of the Polish passenger car industry through foreign direct investment', *Eurasian Geography and Economics* **47**(3): 353–377.
Pavlínek, P. (2008), *A Successful Transformation? Restructuring of the Czech Automobile Industry*, Heidelberg and New York: Physica Verlag.
Pavlínek, P. (2012), 'The internationalization of corporate R&D and the automotive industry R&D of East-Central Europe', *Economic Geography* **88**(3): 279–310.
Pavlínek, P. and Janák, L. (2007), 'Regional restructuring of the Škoda Auto supplier network in the Czech Republic', *European Urban and Regional Studies* **14**(2): 133–155.
Pavlínek, P. and Ženka, J. (2011), 'Upgrading in the automotive industry: firm-level evidence from Central Europe', *Journal of Economic Geography* **11**: 559–586.
Pavlínek, P., B. Domański and R. Guzik (2009), 'Industrial upgrading through foreign direct investment in Central European automotive manufacturing', *European Urban and Regional Studies* **16**(1): 43–63.
Sadler, D. (1999), 'Internationalization and specialization in the European automotive components sector: implications for the hollowing-out thesis', *Regional Studies* **33**(2): 109–119.
Škoda Auto (2005–2013), *Škoda Auto Annual Reports, 2004–2012*, Mladá Boleslav: Škoda Auto.
Sturgeon, T.J. and Lester, R.K. (2004), 'The new global supply base: new challenges for local suppliers in East Asia', in Yusuf, S., Altaf, M.A. and Nabeshima, K. (eds), *Global Production Networking and Technological Change in East Asia*, Washington, D.C.: The World Bank and Oxford University Press: 35–87.
Sturgeon, T., Van Biesebroeck, J. and Gereffi, G. (2008), 'Value chains, networks and clusters: reframing the global automotive industry', *Journal of Economic Geography* **8**(3): 297–321.
U.S. Bureau of Labor Statistics (2012), *International Comparisons of Hourly Compensation Costs in Manufacturing, 1996–2011: Time Series Tables*, U.S. Bureau of Labor Statistics, Division of International Labor Comparisons, 19 December.
Van Biesebroeck, J. and Sturgeon, T.J. (2010), 'Effects of the 2008–09 crisis on the automotive industry in developing countries: a global value chain perspective', in Cattaneo, G.G.O. and Staritz, C. (eds), *Global Value Chains in a Postcrisis World: A Development Perspective*, Washington, D.C.: The World Bank: 209–244.
Yeung, H.W.C. (2009), 'Regional development and the competitive dynamics of global production networks: an East Asian perspective', *Regional Studies* **43**(3): 325–351.

FURTHER READING

Pavlínek, P. (2008), *A Successful Transformation? Restructuring of the Czech Automobile Industry*, Heidelberg and New York: Physica Verlag. (A detailed analysis of the Czech automotive industry, including Škoda.)
Pavlínek, P. (2012), 'The internationalization of corporate R&D and the automotive industry R&D of East-Central Europe', *Economic Geography* **88**(3): 279–310. (An analysis of automotive R&D development in East-Central Europe.)
Pavlínek, P. and Ženka, J. (2011), 'Upgrading in the automotive industry: firm-level evidence from Central Europe', *Journal of Economic Geography* **11**(3): 559–586. (An analysis of upgrading in the Czech automotive industry.)
Sturgeon, T.J., Van Biesebroeck, J. and Gereffi, G. (2008), 'Value chains, networks and clusters: reframing the global automotive industry', *Journal of Economic Geography* **8**: 297–321. (A review of developments in the global automotive industry based on the GVC perspective.)
Van Biesebroeck, J. and Sturgeon, T.J. (2010), 'Effects of the 2008–09 crisis on the automotive industry in developing countries: a global value chain perspective', in Cattaneo, O., Gereffi, G. and Staritz, C. (eds), *Global Value Chains in a Postcrisis World: A Development Perspective*, Washington, D.C.: The World Bank: 209–244. (The GVC perspective on the 2008–2009 crisis and its effects on the automotive industry of less developed countries such as China and India.)

23. Samsung: restructuring, innovation, and global networks*

Sam Ock Park

INTRODUCTION

During the last two decades, global economic spaces have been significantly reshaped by the progress of global megatrends, such as globalization, the knowledge-based economy, and the information society. Manufacturing firms have pursued diverse strategies and interacted with economic spaces to sustain growth under the diverse environments brought about by global megatrends. Innovations of new products and processes have been critical in sustaining competitiveness. Networking with other firms and stakeholders has become important for the accumulation of intangible assets beyond the tangibles which were emphasized in traditional industrial location theories. Innovation is important for both manufacturing and service firms (Bryson and Daniels, 2007). The boundary between services and manufacturing has been considerably blurred by the importance of intangible assets in manufacturing.

The dramatic progress of industrialization and economic development in Korea during the last four decades is embedded in the evolution of large Korean conglomerates. Samsung's growth has also progressed in relation to the overall changes in the Korean economy. In turn, Samsung, as a leading enterprise, has contributed to the development of modern Korea in terms of development of ICT and high tech industries. Accordingly, Samsung can be regarded as both a shaper and a product of the Korean economy (Michell, 2010). Samsung has grown as a leading conglomerate, with total sales of US$247.5 billion and profit of US$18.3 billion in 2011. Samsung has transformed from a low cost manufacturer to a world leader in production, marketing, R&D, and design, ranking ninth in brand value in the world in 2012. The core company of Samsung Group is Samsung Electronics (SEC), with sales of US$149 billion and shares of more than 60 percent of the entire Samsung Group (www.about.samsung.co.kr). SEC has evolved from a subcontractor to a manufacturer and to a leading innovator of electronic products through proper mobilization of a talented team.

The role of intangible assets has been critical in the evolution of the Korean economy and Samsung. Accumulation of intangible assets by firms and by regions has become an important factor for gaining competitiveness. In the processes of a firm's intangible asset accumulation, regions outside the internal organization of the firm become important for acquiring and integrating external knowledge beyond the internal knowledge base of the firm. These processes reflect the firm's interaction with diverse regions and global networks to access region-specific knowledge; as such, the importance of regions as the key driver of innovation is undeniable (Asheim et al., 2011). However, the relationship between corporate system and region in the global economy in the accumulation of intangible assets to increase the competitiveness of both firm and region has not been sufficiently examined.

In considering the evolution of SEC with relation to the development of Korea in the era of globalization and a knowledge-based information society, SEC is an appropriate example to understand contemporary ideas and debates with regard to the manufacturing industry. The purpose of this chapter is to examine Samsung, focusing on SEC, as a case study in terms of corporate evolution and restructuring, brand development, innovation and firm strategy, and global networks. This chapter emphasizes SEC's global network in terms of firm and region relationships in the accumulation of intangible assets. Various materials and firm statistics have been examined, and in-depth interviews with managers, executive directors, and former CEOs have been conducted to clarify the interaction and the accumulation processes of intangible assets.

EVOLUTION OF SAMSUNG: PATH DEPENDENCE, RESTRUCTURING, AND NEW PATH CREATION

Samsung's history can be described in terms of path dependence, restructuring, and new path creation. In broad terms, Samsung Group has evolved from a small commercial firm to a global conglomerate through four stages of development: early group formation stage, diversification of manufacturing industries, reorganization and restructuring, and leading global innovation organization.

Before the launch of SEC in 1969, Samsung was in the early group formation stage. Samsung Group originated from Samsung Sanghoe, a small commercial firm, established in 1938 in Daegu, Korea by the late Byung-Chul Lee, the former chairman of Samsung Group. Before the independence of Korea in 1945, Samsung's business mainly focused on trading, textiles, and food. Samsung expanded its activities to the processing industry, apparel, and life insurance in the 1950s. During the First Five-Year Economic Development Plan launched in 1962, Samsung expanded into more diverse business activities, including a paper mill, hospital, and real estate. In this initial group formation stage, business activities focused on the service sector.

In the late 1960s to the early 1980s, manufacturing became the major business of Samsung. The new stage of the evolution of Samsung Group started with the emergence of Samsung's core business, SEC, established in 1969. In the early industrialization of Korea, the government controlled certain industries by limiting permits by which firms could enter. However, the government encouraged firms to participate in the electronics industry by establishing the Electronics Industry Cooperative Association in 1967 and the Electronics Industry Promotion Law in 1969 (SEC, 2010). Samsung's founding chairman, Lee established a development project at Samsung Moolsan to prepare the establishment of an electronics firm. The establishment of SEC was the result of the founding chairman's vision for the future of Samsung as well as the Korean government's promotion of electronics industrial development. The late founding chairman, Lee concluded that the "electronics industry is perfectly fitted to Korea's economic development stage in all aspects of technology, labor, value added, the prospect of domestic demand and export and so on" (SEC, 2010: 22). Samsung-Sanyo Electronics was established in 1969, and was renamed Samsung Electro-Mechanics in 1975, before merging with Samsung Electronics in 1977. Starting with the electronics industry, Samsung entered a new era of its history. In the 1970s, the Korean government enthusiastically promoted heavy and chemical industrial development, resulting in the emergence of many large firms and conglomerates (*chaebols*) in the heavy and chemical industries (Park, 1991). Samsung Group established diverse firms in the heavy and chemical industries in the late 1970s and early 1980s. Starting with black-and-white TV, Samsung entered the production of diverse household electric appliances, such as color TVs, washing machines, refrigerators, microwave ovens, air conditioners, VCRs, and personal computers. SEC became a major manufacturer in the Korean market and began to export color TVs for the first time in this period. Samsung and all the other large Korean firms benefited from Korea's economic development policy that focused on heavy and chemical industries. Samsung successfully prepared strategic foundations for entering global markets and its future growth with investments in the heavy and chemical industries and the electronics industry. In the early and mid-1980s, Samsung diversified its activities into information technology services, such as system integration, system management, consulting, and network services.

The period from the late 1980s to the late 1990s can be regarded as the period of reorganization and restructuring of Samsung, through its emphasis on technology development and the global market. This period is important for the history of Samsung in two aspects: opening the era of the new chairman, Kun-Hee Lee, son of founding chairman Lee after the latter passed away in 1987; and emphasis on technological development and innovation for competing in a changing world. The new chairman, Kun-Hee Lee, on the one hand continued to focus on talented human resources, following the vision of the founding chairman. On the other hand, he created the vision for leading global products and market through R&D activities and innovation in the field of ICT and electronics. In this aspect, the new chairman manifested leadership with the creation of a new path as well as a path-dependent vision. Samsung Semiconductor and Telecommunications Co. merged with SEC in 1988. Then, SEC selected home appliances, telecommunications, and semiconductors as its core business lines.

New chairman Lee declared a "New Management" initiative, which was the first management innovation that shifted the focus from quantity to quality to enhance competitiveness and to accomplish the new vision of becoming a leading global corporation in the 21st century. The "New Management" initiative integrated Western best practices related to strategy formation, talent management, and compensation into Samsung's existing model (Khanna et al., 2011). "World Best," "World First," and "Worldwide" were emphasized in the new management for its goal of becoming a leading 21st century global corporation (SEC, 2010: 128). Samsung was confronted with tremendous challenges in the high tech sector in the 1990s and invested continuous R&D to overcome such challenges. The title of "world's first" products began to appear in the 1990s.

The other important change in this period was the restructuring of the Samsung Group amid the financial crisis in Korea in 1997. Due to the financial crisis, 16 of the 30 conglomerates in Korea were bankrupted or dissolved, following the government's strong restructuring strategy (SEC, 2010). Samsung faced its most serious crisis since its foundation. However, Samsung had initiated its second management innovation in 1996, one year before the financial crisis. Management innovation in 3P (Product, Process, and Personnel) was emphasized under the motto of independent, speedy, and simple management (SEC, 2010). The strategies of the first and second management innovations were helpful in overcoming the serious crisis in the late 1990s. Despite its management innovation, Samsung was confronted with serious loss of parity exchange in addition to a large loan.[1] Samsung decided to enforce strong restructuring based on the result of the task force team meeting of CEOs. Key points of the restructuring process were the disposal of unnecessary parts of the business that resulted in securing more than US$1 billion, 28 percent reduction in global manpower (23,000 employees), and concentration on the core business by closing down 120 marginal units. This restructuring made certain changes possible, enabling Samsung to open up new opportunities.

In the 21st century, Samsung has moved toward world best by pioneering the digital age. In overcoming the challenges from the Asian financial crisis, it leaped into world best enterprise by finding new business opportunities in the existing group enterprises and in new businesses. The new businesses were unrelated to the existing businesses but fell within Samsung's core capabilities, such as Internet and digital businesses (Michell, 2010). Product and process innovations were continuously emphasized, and new businesses were developed. Strategic alliances for new businesses and new projects, such as health care sectors, as well as green management strategy, have been promoted for continuous restructuring and in search of new opportunities. Samsung has emphasized speed, talents, design, brand power, and soft value of technology in its "creative management," which is related to its line of "creation–innovation–challenge." With strategies for pioneering in the digital age, Samsung exhibited a successful performance during the first decade of the 21st century.

Currently, Samsung has 18 companies in the three major fields of electronics, finance, and heavy and chemical industries. It also has 15 other affiliates that are relatively independent, as shown in Table 23.1. Samsung has managed to hold together its group structure in a synergetic manner, emphasizing fusion and compounding of its affiliates. Michell (2010) pointed out that this complex structure suffered from internal threats of Korean systems of cross-holding as well as external threats caused by systemic risks among Korean financial institutions. However, the strong leadership and fusion of speed and teamwork, inherited from historic ethnic characteristics, are regarded as strengths of its Korean-style management (Kim and Kim, 2013). Samsung's New Management initiative aimed to improve markedly marketing, R&D, and design while retaining core strength in manufacturing, resulting in Samsung's hybrid management system that sought the best of traditional Japanese systems and Western systems (Khanna et al., 2011). More details on Samsung's rise are discussed, focusing on SEC, in the following sections.

INNOVATION AND EVOLUTION OF PRODUCT DEVELOPMENT OF SEC

In understanding the tremendous success of Samsung during the last two decades, SEC should be more closely examined in terms of innovation, as it has been the core of Samsung, with 60 percent of the total sales of Samsung Group. Sales have grown continuously since the late 1990s, even after the Asian financial crisis in 1997 and after the global financial crisis in 2008. Before SEC's innovation is discussed, its global spatial strategy is briefly examined.

SEC has grown remarkably since its establishment in 1969. SEC took about ten years

Table 23.1 Samsung's affiliated companies

Electronics Industries	Samsung Electronics
	Samsung SDI
	Samsung Corning Precision Materials
	Samsung SDS
	Samsung Techwin
	Samsung Mobile Display
	Samsung Digital Imaging
Financial Services	Samsung Life Insurance
	Samsung Fire & Marine Insurance
	Samsung Card
	Samsung Securities
	Samsung Asset Management
	Samsung Venture Investment
Heavy and Chemical Industries	Samsung Heavy Industries
	Samsung Total Petrochemicals
	Samsung Petrochemicals
	Samsung Fine Chemicals
	Samsung BP Chemicals
Other Affiliated Companies	Samsung C&T Corporation
	Samsung Engineering
	Cheil Industries
	Samsung Everland
	The Shilla Hotels & Resorts
	Cheil Worldwide
	S1 Corporation
	Samsung Medical Center
	Samsung Human Resources Development Center
	Samsung Economics Research Institute
	Samsung Lions
	The Ho-Am Foundation
	Samsung Foundation of Culture
	Samsung Welfare Foundation
	Samsung Life Public Welfare Foundation

Source: http://www.samsung.com/us/aboutsamsung/corporateprofile/affiliatedcompanies.html (retrieved January 17 2013).

to join a multinational corporation group. SEC, located in the Capital Region of Korea, started with the production of home appliances, such as black-and-white TVs, refrigerators, and washers, in the early 1970s. It started the production of color TVs in 1977. With the first development of VCR in the nation, SEC began to initiate product development and innovation. Since the 1980s, it has diversified into more high tech-oriented products. SEC has been involved in foreign direct investments since its first investment in the USA in 1978. For ten years, beginning in 1978, SEC invested in foreign subsidiaries in advanced countries, such as the USA, Germany, the UK, Australia, and Canada (Table 23.2). Thus, the 1980s was a key period for SEC's foreign subsidiary investments.

Table 23.2 The evolution of SEC's spatial corporate system through the global networks and localized processes over time

– Foreign local subsidiary (1978–1988): Key period → 1980s: Advanced Countries: USA, Germany, UK, Australia, Canada
– Foreign plant investment (1982–): Key period → 1990s: From Advanced to Industrializing Countries: Portugal, USA, UK; Thailand, Mexico, Malaysia, Brazil, Slovakia, India, Russia
– Foreign R&D Center (1987–2004): Key period → late 1980s and 2000s: From Advanced to Industrializing Countries: USA, Japan, UK, Russia, Israel, China, India, Poland
– Foreign industrial complex (1995; 1996): Key period → 1990s: UK, Mexico

Source: Gathered from SEC internal material, websites, interview surveys, and newspapers (revised from Park 2012).

Since 1982, SEC has also invested in manufacturing plants and sales offices, the key period being in the 1990s. Production investments began from advanced countries, such as the USA and UK, and then extended to industrializing countries, such as Thailand, Malaysia, India, Brazil, and so on (Table 23.2). Foreign R&D investments started from 1987, and the key periods of the investment were the late 1980s and 2000s. These investments also started in advanced countries and extended to industrializing countries. The evolution of the spatial corporate system of SEC suggests that corporate systems successively extend from foreign subsidiaries to manufacturing plants and R&D centers; and from advanced countries to industrializing countries over time.

Innovation has been the most important factor in the successful growth of SEC. Innovation strategy comes from the headquarters of SEC and Samsung Group as a whole, according to the vision and major strategies of Samsung. Several important aspects in the innovation of SEC are discussed as follows.

First, to overcome crises and the highly competitive global market, SEC emphasized R&D activities for product and process innovations. Its R&D investment has been continuously rising, reaching about US$9.2 billion in 2011, which accounted for 6.2 percent of total sales. R&D manpower has also continuously increased, reaching 55,320 in 2011, accounting for a quarter of the total manpower of SEC (Figure 23.1). With intensive R&D investments, SEC has ranked second in the number of patents registered in the USA since 2006.

Second, R&D centers in foreign countries cooperate closely with local institutes to secure regionally strong or specialized technologies. For example, SEC emphasized local R&D cooperation of cloud, next generation web, operating system, and next generation user experience in Silicon Valley; communication, radio frequency/antenna, and servers in Dallas; and security and recognition in Israel. In Bangalore, the thrust is on cooperation for system software, web browsers, and business-to-business solutions; whereas in Russia and Ukraine, SEC focuses on collaboration for algorithms, system software, and hardware enabling. In Beijing, the key area of cooperation is for the Chinese digital TV/communication standard because China controls its own technology standard.

Third, SEC has cultivated strategic alliances with leading global firms, such as IBM,

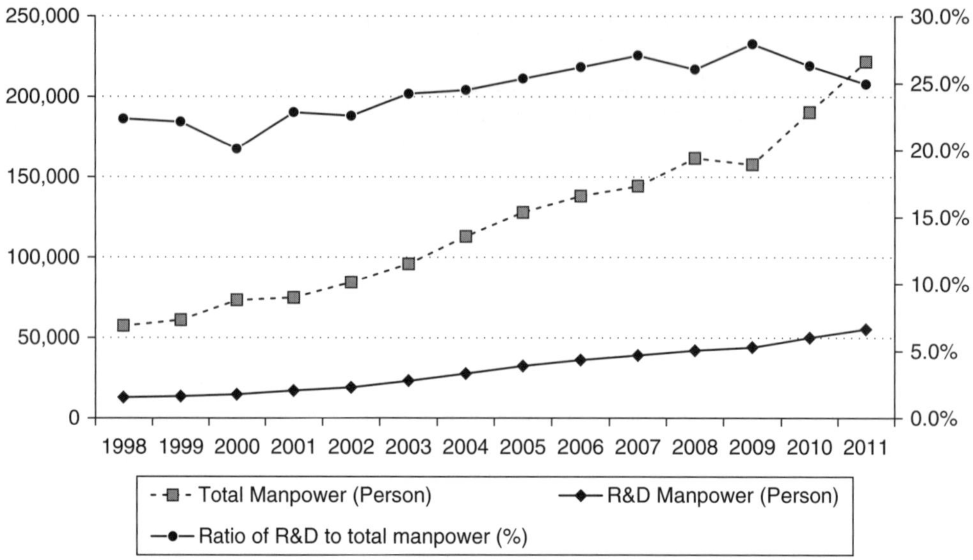

Source: Redrawn from SEC 2010 and SEC 2012b.

Figure 23.1 Growth of R&D manpower and ratio to total employees, SEC

Intel, HP, Sony, Nokia, Discovery, and Charter, for joint technology development and cooperation. Strategic alliances clustered from 2003 to 2007, which coincided with the cluster of the "world's first" product development. Strong drives for global technology alliances with global leading firms surely contributed to the establishment of the global standard of technology and successful innovations. In order to strengthen patent defense, SEC made cross-licensing agreements with global IT firms, such as Toshiba in 2009, Qualcomm, Kodak, International Ventures, Rambus, and Sharp in 2010, and IBM, Microsoft, and Spansion in 2011.

Fourth, SEC expanded transactions in local areas through an open innovation system. In addition to internal R&D activities and cooperation networks with subcontractors, SEC supports diverse activities for small and medium-sized enterprises in terms of manpower, funds, technology, marketing, and so on. SEC has also introduced open sourcing systems for the joint development of parts and facilities with innovative SMEs, despite the latter being excluded as official subcontractors[2] (SEC, 2012a). SEC has also cooperated with universities, R&D institutes, and leading small and medium high tech firms with future technology that complements SEC's technology. R&D centers in foreign countries closely cooperate with leading universities and research institutes in the relevant region. SEC has spearheaded major cooperative activities within local areas, such as investments for start-ups in Silicon Valley, national projects with government research centers and operation of a future technology consulting committee in Dallas, and joint lab establishments with Tsinghai University in Beijing.

Fifth, SEC has also endeavored to foster a creative organizational culture. Recently, a Creative Development Institute has been opened in order to support personnel to actualize their creative ideas into commercial products. SEC personnel with creative ideas in diverse fields, such as product development, business, and operation of organization, can apply for a one-year stint in a task force team in the Creative Development Institute. The first fruit of the Institute was an eyeball-mouse called "eyeCan," through which a user with total paralysis can operate the cursor and computer by moving his/her eyeballs.

With the intensive investment in R&D, global technology alliances, and other internal and external cooperative activities, SEC has evolved as a global leader of product development (Table 23.3). The development of products with the "nation's first" brand started from 1979, with the development of the VCR. Several products were developed "first in the nation" in the early 1980s, and then in 1986 the initiative advanced to the first product development worldwide, the 4mm VTR. In the 1990s, two products were developed for the first time in the firm and two other products as the "first in the world." After the Asian financial crisis, SEC achieved successful results of product development with five "world's first" product developments from 2001 to 2004. From 2005 to 2007, as many as seven "world's first" products were developed, representing a cluster of "world's first" new product development in the early and mid-2000s. After the global financial crisis in 2009, SEC developed four "world's first" products. In 2012, despite the lack of development of "world's first" products, SEC developed several new products.

Continuous evolution of new products has surged in the area of dynamic random access memory (DRAM), which represents the accumulation of technology with continuous integration and cycle of path-dependent and newly created technology (Table 23.3). TV products have also continuously evolved from black-and-white TV, to color TV, to digital TV, and to LED TV. Strong investments in R&D activities, even amid the financial crisis of 1997, and numerous "world's first" developments of new products in the early and mid-2000s made SEC mass produce new products during the global financial crisis in the late 2000s and early 2010s. SEC began to mass produce many products as the first in the world during the global financial crisis, such as 50 Nano 1G DDR2 DRAM in 2008, 40 Nano DDR3 DRAM in 2009, 4G DDR3 DRAM in 2010, 30 Nano DRAM in 2010, DRAM module for Server in 2011, 46" LCD Panel in 2012, and 20 Nano LPDDR2 4Gb mobile DRAM in 2012. SEC has also been the first in the world to introduce several new products to the global market since 2008. Such pioneering mass production and introduction of new products to the global market imply that SEC has significant advantages with premium price in the world market. In 2012, because of the premium price, Samsung recorded a new high of "200 to 20" trillion, which indicated a first record of more than 274 trillion won (US$247.5 billion) in sales and more than 20 trillion won (US$18.3 billion) in profit.

BRANDING SAMSUNG

Before the early 1990s, Samsung did not consider branding as seriously as innovation and production. The majority of Samsung's products sold in foreign countries were with other companies' brands or parts and components inside other firms' products. In the mid-1990s, however, Samsung wanted to build a brand, not only a product, shifting from low end products to premium brand strategy. Samsung's global brand campaign

Table 23.3 Development of major products of SEC

Year	First at Firm Level	"Nation's First" Products	"World's First" Products
1970s		VCR, 1979	4mm VTR, 1986
1980s	Semiconductor 1M SRAM	64K DRAM, 1983	
		256K DRAM, 1983	
		1M DRAM, 1986	
1990s	16M DRAM, 1990		256M DRAM, 1995
	1G DRAM, 1996		1G Alpha Chip, 1999
2001–2004			40"TFT-LCD, 01
	70N Semicon. Process, 02		54"TFT-LCD, 02
			70N 4G NAND Flash, 03
			70N DRAM Process, 04
			60N 8G NAND Flash, 04
2005–2007		70N DRAM, 05	50N 8G NAND Flash, 05
			CTF NAND Tech, 06
			16 Step MCP, 06
			DDI Touch Screen, 06
			1G Mobil DRAM, 06
			65N Receive Chip of DTV, 07
			30N 64G NAND Flash, 06
2008–2012	Green Server Solution, 12	30N 4Gb DDR4 Dram, 12	4G DDR3 DRAM, 09
	1.5Mb STT-MRAM, 12		40N DRAM, 09
	Graphin Semicon, 12		3mm Thick LED TV, 09
	QC Application Process, 12		0.6mm 8 Step Chip, 09
			30N DRAM, 10

Source: SEC 2012a; various newspapers for the year of 2012.

started in 2001, and Samsung's sports marketing initiatives through sponsorship during the winter Olympic Games in Salt Lake City in 2002 gave critical momentum for the successful future brand campaign of Samsung (Michell, 2010). SEC was convinced that increased marketing would benefit sales and profit margins. SEC has strongly supported global brand marketing since the late 1990s. With Eric Kim joining SEC in 1999 as head of responsibility for global marketing, Samsung's investments in marketing activities exceeded US$900 million in 2002. The global campaign has had a successful result.

Before 2000, Samsung's brand was not included in Interbrand's top 100 list. However, in 2000 the brand value of Samsung leaped to US$5.22 billion, ranking 43rd. During the early 2000s, the brand value and rank of Samsung increased significantly. SEC's brand value has again sharply increased since 2010, to US$32.89 billion, ranking 9th in 2012.

The dramatic increase in Samsung's brand value is not solely from the global brand campaign or sports marketing, but is the integrated result of innovation and marketing. The strength of a brand grows as a company introduces new products against competition and as the products grow in sophistication (Michell, 2010). SEC has practiced diverse brand marketing strategies for securing global leadership in the 21st century, as shown in the following (SEC, 2010).

First, SEC has promoted "Market Driven Change" for marketing innovation and has invested about 3–4 percent of total sales in marketing since 2000. Samsung enhanced emotional benefits of digital convergence through campaigns that combined communication, entertainment, and information. It also focused on producing competitive products with the development of leading-edge technology and improvement of design manpower. The key marketing strategies were the construction of brand leadership through enhancing the competitiveness of outstanding digital products, promotion of systematic brand strategy, creation of high value added services from premium products, and promotion of a single master brand that integrated the product brand into the corporate image.

Second, SEC increased brand value through design management. Samsung began to regard design as a primary management factor after the declaration of "New Management" in 1993. Since the announcement of the design revolution in 1996, however, SEC has plunged into its design management system in earnest. SEC recognizes the necessity of the transformation of sensitivity beyond superior function and technology of products to establish a global premium brand, which has led to strategy meetings of major CEOs at the global design core, such as that in Milan in 2005. In that strategy meeting, Samsung announced the "Four Milano Design Strategies," which include construction of original design and User Interface identity, securing excellent design talents, fostering a creative and new organizational culture, and strengthening infrastructure of mold technology. SEC also emphasized the construction of design infrastructure. SEC operates global design research centers in San Francisco, Los Angeles, London, Milan, Tokyo, and Shanghai for market-driven design activities. SEC has produced several hundred designers and operates global design membership in the USA, UK, Italy, China, and Japan.

Third, since 2005 SEC has promoted the second brand strategy for its premium brand and cultural marketing strategy. The focus is on enhancing the premium image of the brand, with the aim of switching the existing "Sensible Brand Buyer" to "High-Life Seeker." The first brand strategy was the campaign "DigitAll," with the slogan "SAMSUNG DIGITall, everyone's invited," promoted from 1999 to May 2005, whereas the second brand strategy was the campaign "Imagine," promoted in June 2005. In the second strategy, the focus has been on customers reestablishing as "High-Life Seekers" with the increase of high value added products. High-Life Seekers are customers who feel that a product and brand speak for their own identity and style and positively seek the mixed value of functional excellence and emotional and aesthetic experience in the product (SEC, 2010). Accordingly, SEC emphasized cultural marketing for premium brand and marketing in gate cities and noted places.

Fourth, SEC has promoted sports marketing to strengthen its brand. Samsung participated as an Olympic partner in the Nagano Winter Olympic Games in 1998, Sydney Olympic Games in 2000, Salt Lake City Winter Olympic Games in 2002, Athens Olympic Games in 2004, and so on. SEC's official sponsorship in the Athens Olympic Games in 2004 and Torino Winter Olympic Games in 2006 contributed to the establishment of Samsung's brand image with Europeans. SEC also promoted football marketing. Samsung contracted official sponsorship with the Asian Football Federation, Chelsea FC of England, and Corinthians FC of Brazil. In addition, SEC has driven diverse sports marketing campaigns to meet with diverse regional characteristics.

SEC'S GLOBAL NETWORK

Along with the evolution of the spatial corporate system, SEC has continuously accumulated intangible assets[3] in the form of knowledge, technology, brand, and so on. SEC shows that formal and informal networks within clusters of local areas, continuous R&D investments, technology cooperation, and strategic alliances, as well as activities such as organizing and participating in temporary clusters, are the major mechanisms of intangible asset accumulation (Park, 2014).

Networks within local areas of foreign countries, which are formed by individual corporate units, are critical in a firm's intangible asset accumulation. In the early stage, larger firms' R&D investments in the innovation cluster of advanced countries have been clearly regarded as activities in search of intangible assets. SEC maintained close local networks to acquire knowledge and technology as well as cooperating with Silicon Valley firms in the earlier stage of R&D investments (Park, 1994, 1996). This asset-seeking behavior resulted in the continuous development of new products, beginning in the 1990s. More recently, SEC has aggressively organized or participated in temporary clusters, such as expo and trade fair events, in order to enhance its innovative networks.

Temporary clusters have made significant impacts on the acquisition of new knowledge and product information and on the establishment of innovative networks (Bathelt and Gluckler, 2011; Rinallo and Golfetto, 2011; Torre, 2008). Trade shows support firms in the development of their marketing capabilities as complementary assets, which are important in ensuring effective innovation and competitive advantage (Rinallo and Golfetto, 2011). Sales network centers are outposts for exploiting marketing bundles of intangible assets in foreign countries. Interactions with other regions and markets within such foreign countries are excellent opportunities for accumulating marketing information and knowledge. Local networks in production centers contribute to savings in transaction costs, and are also helpful in accumulating intangible assets to be transmitted to the R&D centers and headquarters in Korea.

SEC currently operates 206 global bases of diverse functional units, such as regional headquarters, production centers, sales centers, R&D centers, design centers, and others within 75 countries in the world. Among the different functional units, the headquarters, regional headquarters, R&D centers, production plants, and sales offices can be regarded as the major units of the corporate system. In relating the corporate system to regions, cities are classified based on the level of cities within a country, and countries on the level of development following World Bank classification (Park, 2014).

The *headquarters* of SEC should be located in the Capital Region of Korea in order to ensure easy communication with and control over diverse global corporate units, access to the national government, close relations with financial firms, and close networks with leading universities and research labs. In fact, SEC is based in the famous area of Kangnam in Seoul. Despite the emphasis on dispersion of power to individual corporate units, the headquarters of SEC makes major decisions for investments and long-term strategies.

Regional headquarters (RHQs) support SEC affiliates with shared services in finance, tax, legal issues, risk management, and logistics. They plan and introduce local-oriented products through market surveys and deliberation of customer lines. Recently, promotion of sports marketing and conducting local-oriented cultural and corporate social responsibility (CSR) activities, such as "Hope for Children," have also been included in the role of RHQs. RHQs accumulate regional intangible assets through these diverse activities, whereas local firms and institutes accumulate knowledge and technology through interaction with SEC. SEC divides the global market into ten regions, where RHQs are located. The RHQs are all located in the Capital Region or the largest metropolis of a leading country within the divisions of the global market. The location of RHQs clearly matches with the Capital Region or the largest metropolis in the urban hierarchy, which is meaningful given the role of RHQs.

The major roles of *R&D centers* in foreign countries are R&D cooperation with local areas in the field of local superior technologies, promotion of open innovation with local leading universities and institutes, and development of local-oriented products and solutions to respond to the regional technology standard. In the development of local-oriented product development, SEC considers local culture, tradition, and characteristics, generating different strategies by region. For example, SEC emphasizes sensing or asset seeking for technology at Silicon Valley or London, and exploiting and seeking specialized technologies by region in other areas. Accordingly, R&D centers accumulate intangible assets through exploiting and seeking locally specific intangible assets. The locations of R&D centers match with either capital regions or large metropolises, where innovative clusters or regional innovation systems are soundly developed. Each division of global markets has at least one R&D center. Examples of the so-called Capital Regions are London, Beijing, Moscow, and Tel Aviv, where high tech clusters are well developed. Non-capital metropolises are mostly major innovation clusters or high tech clusters of the respective divisions of the global market.

For the location of *production centers*, SEC considers diverse factors such as lead times, transportation and utility infrastructure, labor costs, commercial environment, market merits, and government incentives. Industrializing countries, especially China, have significant advantages for the location of production centers, which is different from the pattern for R&D centers. Only one plant is located in an advanced country, but this is in a low level city. Meanwhile, two plants are located in less industrialized countries, but they are both in the Capital Region. More than 90 percent of the foreign plants are located in industrializing countries, among which about 76 percent are located in larger metropolises. More than 40 percent of the production centers are located in China, whereas the regional divisions of the Middle East and Africa have only one production center each, suggesting that production centers are closely related to industrial clusters as well as to the actual markets. Production centers are mainly located in cities where industrial clusters are well developed in order to make the production networks efficient in local areas. This set-up suggests that production centers consider the savings of transaction costs in production networks of industrial clusters as well as exploiting and seeking intangible assets in relation to the production activities. Due to the progress of the so-called information society, modern production centers now have long-distance networks for transactions, although they are focused on such selected areas as industrial clusters or innovation clusters (Park, 2004).

Sales network centers (SNCs) focus on local marketing and interactions with local areas for developing local-oriented strategies. In some aspects, the role of SNCs is similar to that of RHQs, albeit on a much smaller spatial scale. As such, SNCs are all located in either the Capital Region or large metropolitan areas regardless of the level of the country. SNCs are relatively dispersed throughout the world from advanced to less industrialized countries, which is considerably different from the case with other corporate units. A considerable number of SNCs are located in the less industrialized countries of the Middle East and Africa, Latin America, and CIS and Baltic areas, although these are all in the Capital Region. Different functions are expected from the location of the SNCs based on the hierarchy of urban systems in global economic spaces (Park, 2014). First, SNCs located in advanced countries seek various types of intangible assets in addition to marketing-related knowledge and product information. Second, SNCs in the Capital Region in industrializing countries pursue both asset-seeking and asset-exploiting activities. Third, SNCs in less industrialized countries exploit intangible assets for the future development of the sales market and production activities.

CONCLUSION

The case of SEC reveals that global manufacturing firms are closely related with activities of product and process innovation, marketing, branding, diverse services, and society in the global space economy. During the last four decades SEC has evolved from being a manufacturing firm of low price products to a leading global firm of high value added products. SEC has also transformed from subcontracting manufacturing products to being an innovation flagship for producing world-first new products. Currently, SEC also moves to transform from innovation to integration of innovation, marketing, and culture.

In its evolution, SEC has displayed the processes of path dependence, restructuring, and new path creation. Continuous improvement of technology and upgrading innovation in semiconductor technology can be attributed to the historical inheritance of strong leadership and speedy management. Restructuring and new path creation are related to the adoption of global standards and Western management systems. Successful restructuring during the Asian financial crisis in the late 1990s and recent global financial crisis conferred on SEC resilience against the external crises. Management innovations and restructuring contributed to the creation of the new path of SEC. The processes of path dependence, restructuring, and new path creation are not completely separated, but rather overlap in a continuous complex process. Formerly, production and process innovations were regarded as critical for a firm's competitiveness. However, design innovation, brand marketing, and networking with society are currently regarded as important for a firm's competitiveness, reflecting the need for complex processes and strategies. Through these dynamic processes, SEC evolved into a global technology and market leader.

SEC demonstrates the important role of the CEO and Samsung Group chairman for designing visions and expanding global management. The decision-making power of SEC's headquarters is very important, despite the independent role of RHQs and SNCs in some aspects. The regional dimension is important not only for production networks but also for R&D activities and marketing. Beyond the physical infrastructure and environment, regional tradition and culture are important for deciding the specialty of R&D activities and marketing.

In the evolution of SEC, it promoted diverse strategies for the accumulation of intangible assets. Firms' intangible assets are critical in ensuring competitiveness within a knowledge-based information society. The intangible assets of firms can be increased through the accumulation of knowledge, technology, human capital, brand values, and so on. Such accumulation is possible through formal and informal interactions and networks with suppliers, customers, competitors, governments, and communities (Park, 2014). In the process of intangible asset accumulation and interactions with regions, firms differentiate their strategies with cities through the level of the country's development and status within the urban system. Intangible assets are likewise important in maintaining the competitive advantages of cities and regions. Accumulation of local intangibles is the result of the evolutionary development of the relationships between the corporate and urban systems. The case study also suggests the processes of intangible asset accumulation in terms of asset-seeking and asset-exploiting activities by the interaction of firms and regions (Park, 2014).

Some challenges remain for SEC in the future. In recent years, especially during the period of the global economic downturn due to the global financial crisis, SEC has begun to emphasize corporate partnership and symbiotic development with global society. SEC decided to invest in a production center in Africa in 2012 and participated in diverse activities for education and training programs in Africa in addition to supporting handicapped persons. SEC is moving toward emphasizing CSR. Symbiotic development and CSR cannot be successful without serious consideration of the global society beyond the declaration of responsibility. In a similar vein, SEC has emphasized green management and has tried to apply green management systems throughout its production chains. Considering serious problems brought about by global climate change, SEC should move one step forward to formulate a pro-environmental management strategy in all value chains of economic activities. Recently, SEC has been interested in investments in Africa not only for global network or regional development strategies but also for reconsideration of CSR, which is a positive sign for change. SEC has accumulated immeasurable intangible assets throughout its global networks. From here on out, SEC will be confronted with the challenge of contributing to the accumulation of regional intangible assets in the peripheral areas of the global economy.

NOTES

* This chapter is a revised version of Chapter 8, "Restructuring, innovation, and the global networks of Samsung," in Sam Ock Park (September 2004), *Dynamics of Economic Spaces in the Global Knowledge-Based Economy. Theory and East Asian Cases*, London and New York: Routledge, pp.213–236.

1. At the end of 1997, Samsung's financial structure deteriorated with a loan of 13 trillion won and debt of 6 trillion won.
2. In 2011 there were 651 technology consultations through open sourcing and 23 projects were adopted through peer review.
3. Intangible assets can be categorized with diverse bundles. Anson (2007) categorized the intangible assets with the marketing bundle, IT bundle, technical bundle, and other bundles. In this chapter, for convenience, four bundles are used conceptually to describe the relationship with urban systems.

REFERENCES

Anson, W. (2007), *The Intangible Assets Handbook. Maximizing Value from Intangible Assets*. Chicago, IL: American Bar Association.

Asheim, Bjørn T., Ron Boschma, and Phillip Cooke (2011), "Constructing regional advantage: platform policies based on related variety and differentiated knowledge bases," *Regional Studies*, 45(7), 893–904.

Bathelt, Harald and Johannes Gluckler (2011), *The Relational Economy. Geographies of Knowing and Learning*, Oxford, UK: Oxford University Press.

Bryson, John R. and Peter W. Daniels (2007), *The Handbook of Service Industries*, Cheltenham, UK and Northampton, MA, USA: Edward Elgar Publishing.

Khanna, Tarun, Jaeyong Song, and Kyungmook Lee (2011), "The paradox of Samsung's rise," *Harvard Business Review*, July–August, 1–7.

Kim, Sung Soo and Kyung Jun Kim (2013), "Fusion of nomadic speed with classic leadership . . .'Korean style' management has the hope of future," *Weekly BIZ of Chosun Daily*, January 12–13, C3.

Michell, Tony (2010), *Samsung Electronics and the Struggle for Leadership of the Electronics Industry*, Singapore: Wiley.

Park, Sam Ock (1991), "Government management of industrial change in the Republic of Korea," in P. Rich and G. Linge (eds), *The State and Spatial Management of Industrial Change*, New York, USA: Routledge, pp.74–87.

Park, Sam Ock (1994), "High technology industrial development and formation of new industrial districts: theory and empirical cases," *Journal of Korean Geographical Society*, 29(2), 117–136 (in Korean with English summary).

Park, Sam Ock (1996), "Networks and embeddedness in the dynamic types of new industrial districts," *Progress in Human Geography*, 20(4), 476–493.

Park, Sam Ock (2004), "The impact of business-to-business electronic commerce on the dynamics of metropolitan spaces," *Urban Geography*, 25(4), 289–314.

Park, Sam Ock (2014), Interactions of corporate and urban systems: accumulation of intangible assets. *Urban Geography* (www.tandfonline.com/doi/full/10.1080/02723638.2014.934522).

Rinallo, Diego and Francesca Golfetto (2011), "Exploring the knowledge strategies of temporary cluster organizers: a longitudinal study on the EU fabric industry trade shows (1986–2006)," *Economic Geography*, 87(4), 453–476.

SEC (2010), *Forty Years of Samsung Electronics. The History of Challenge and Creation*, Suwon, Gyunggi, Korea: Samsung Electronics (in Korean).

SEC (2012a), *43rd Business Report, 2011.1.1–2011.12.31*, Seoul: SEC (in Korean).

SEC (2012b), *Global Harmony with People, Society & Environment*, Seoul: SEC (in Korean).

Torre, A. (2008), "On the role played by temporary geographical proximity in knowledge transmission," *Regional Studies*, 42(6), 869–889.

FURTHER READING

Bathelt, Harald and Johannes Gluckler (2011), *The Relational Economy. Geographies of Knowing and Learning*, Oxford, UK: Oxford University Press. (Overall information on the importance of the networked economy in the process of learning and innovation.)

Khanna, Tarun, Jaeyong Song, and Kyungmook Lee (2011), "The paradox of Samsung's rise," *Harvard Business Review*, July–August, 1–7. (Explaining the major factors in the rise of Samsung.)

Michell, Tony (2010), *Samsung Electronics and the Struggle for Leadership of the Electronics Industry*, Singapore: Wiley. (Overall history of Samsung Electronics and the processes of growth.)

Park, Sam Ock (1996), "Networks and embeddedness in the dynamic types of new industrial districts," *Progress in Human Geography*, 20(4), 476–493. (Theoretical framework of networks in the industrial district.)

SEC (2010), *Forty Years of Samsung Electronics. The History of Challenge and Creation*, Suwon, Gyunggi, Korea: Samsung Electronics (in Korean). (Samsung Electronics' history of development and innovation.)

SEC (2012), *Global Harmony with People, Society & Environment*, Seoul: SEC (in Korean). (The firm's responsibility for society and sustainable development.)

http://about.samsung.co.kr/about/achieve.do

http://www.samsung.com/us/aboutsamsung/corporateprofile/affiliatedcompanies.html

http://www.samsung.com/us/aboutsamsung/corporateprofile/history03.html

http://www.samsung.co.kr/samsung/history.do

http://www.samsung.co.kr/samsung/outcome/brand.do

http://www.samsung.co.kr/samsung/business/foreign.do

PART IV

POLICY NARRATIVES IN MANUFACTURING

24. Stability amid industrial change: the geography of U.S. deindustrialization since 1980

Marc Doussard and Greg Schrock

INTRODUCTION

Between 1979 and 1983, the U.S. manufacturing industry shed more than 12 percent of its total workforce, a devastating development that spurred urban economic development innovations across the country and led to the widespread acceptance of the previously controversial neologism "deindustrialization" (Bluestone and Harrison, 1982). The subsequent search for potential job growth in a revamped manufacturing sector has taken many turns, from the prospective identification of stable, new firm networks (Piore and Sabel, 1986; Scott, 1988) to theories of sticky employment in vital urban regions (Markusen, 1996; Storper, 1997) to the embrace of local-serving industries immune from capital flight (Doussard, 2013; Markusen and Schrock, 2009). These developments have produced important questions about economic restructuring in individual regions and industries. But characterizations of manufacturing in the aggregate continue to conceptualize the resurgence of older industries and regions as straightforward results of global dynamics, particularly those seen as likely to sway firms toward "reshoring" previously dispatched industrial activities (Atkins et al., 2011; Friedhoff et al., 2010; Helper et al., 2012), marginalizing important evidence about the spatially, industrially and occupationally uneven character of renewed manufacturing growth.

There is good reason for this. Attention to both the theoretical and empirical sides of industrial restructuring has waned in the past decade and a half. After debating large-scale developments in supply-chain creation, outsourcing and technological change in the production of durable goods (Glasmeier and Leichenko, 1996; Harrison and Kelley, 1993; Harrison, 1994), industrial scholars moved on to regionally situated studies of individual firms, regions and supply chains (Christopherson and Clark, 2007a; Sturgeon, 2002; Whitford, 2005). This intensive engagement with individual production networks has generated significant new ideas about the creation of "sticky" jobs in innovative industries, and the tensions between transnational firms and regional economies, but it has yet to be reconciled with a systematic characterization of the shifting geography of the U.S. manufacturing sector since 1980s deindustrialization.

We address this gap by characterizing the regional dimensions of restructuring within U.S. manufacturing since 1980. To date, accounts of industrial restructuring in the aggregate typically prioritize a single dimension of employment change, such as the regional shift of manufacturing jobs from the Midwest and Northeast to the South and the West (Markusen et al., 1991; Sawers and Tabb, 1984), the growth of science and technology occupations within manufacturing (Chapple et al., 2004; Perry and Watkins, 1977), or the decline and restructuring of individual industries (Christopherson and Clark, 2007b; Markusen, 1986; Weber, 2001). By tracing the industrial and occupational development of U.S. manufacturing over three full business cycles, we trace occupational changes across the U.S., the Rustbelt, the Sunbelt and their nested regions over the course of the 1980s, 1990s and 2000s business cycles. While net manufacturing employment levels and concentrations have changed (albeit not always in the ways assumed by stylized narratives about Rustbelt and Sunbelt), the *types* of work each region performs have remained remarkably stable.

The chapter proceeds in three sections. We begin by detailing the current state of manufacturing research and identifying major developments that would benefit from systematic empirical data. We then detail the unique dataset and methodology used to build a multi-decade, nationwide dataset sensitive to industry, occupation and state. Finally, we detail the restructuring of manufacturing industries across state and region over the course of the 1980s, 1990s and 2000s. Contrary to the commonplace distinction made between a growing Sunbelt and contracting Rustbelt, we find that regional occupational endowments within manufacturing have remained stable as industry employment contracts. Decades after sharp decreases in production spending drew attention to the plight of manufacturing, the Northeast and West remain the most technology-intensive manufacturing regions, while the Midwest and (especially) the South retain lower shares of science, technology and research employees. In periodizing the history of deindustrialization, we identify multiple causal processes, including subcontracting, the development of supply chains, and the growth of distinctive types of industrial districts, that help to explain both the recent history and near future of American manufacturing geographies.

What Happens After Deindustrialization?

The decline in U.S. manufacturing employment now spans more than 30 years (Holzer and Hlavac, 2012). While the received narratives of "early" deindustrialization and the diverging fortunes of the North and South remain familiar to most scholars (Bluestone and Harrison, 1982; Harrison and Bluestone, 1990; Markusen et al., 1991; Tabb, 1984), subsequent geopolitical developments – including the North American Free Trade Agreement and Chinese industrialization – are treated more as shorthand signifiers of job loss than as crucial historical-institutional developments with industry-level policy impacts. Notably, efforts to engage these developments in detail suggest important structural changes to the manufacturing employment that remains in the U.S. For example, Christopherson (2011) found that high-value-added and innovation-intensive industries were likely to be immunized from offshoring pressures. Domestically, the favorable business climates and union avoidance historically advertised by Southern boosters (Harrison, 1984) have migrated to the Midwest and Northeast. Given that Michigan and Indiana have implemented right-to-work laws, and that union density levels in the North and Northeast are generally in line with 1980s union density levels in Texas and the South, the regional political-economic distinctions long thought to determine manufacturing employment growth are at best muddled. On many issues, lingering dichotomies depicting a Rustbelt/Sunbelt divide appear more likely to cloud than illuminate new industry trends and policy developments.

Recent growth in Midwestern manufacturing employment testifies to the obsolescence of older inter-regional distinctions. Between 2010 and 2011, the five metropolitan areas showing the greatest job growth lay in the Rustbelt: Elkhart, Indiana; Detroit; Grand

Rapids, Michigan; Toledo, Ohio; and Dayton, Ohio (Doussard and Schrock 2012). A generation after Youngstown, Pittsburgh, Buffalo and Cleveland stood out as lamented symbols of regional decline, many of those same cities now have favorable manufacturing employment projections (Clark, 2013).

These selected cases of resurgence imply crucial questions about which regions, industries and occupations are likely to flourish, or at the very least contract at a diminished rate. But reading current manufacturing geographies against yesterday's maps only suggests possible pathways of current industrial activity. Because the discussion of manufacturing remains driven by dated stylized facts – the Rustbelt/Sunbelt economy, the search for high-tech occupations and sunrise industries – memory provides a poor basis for policy analysis.

The concept of deindustrialization so often used to describe job loss in the U.S. North has strayed far from its roots. Systematic disinvestment in manufacturing capacity, Bluestone and Harrison (1982: 6) argued, was a nationwide phenomenon that touched North and South alike. But there was no questioning the disproportionate job loss visited upon the Midwest and Northeast – where cities ranging from Chicago to Cleveland to Youngstown to Pittsfield, Massachusetts – lost jobs by the thousand and tens of thousands. Initially, analysts argued that the growth of a Sunbelt spanning the U.S. South and West was driven more by its appeal to high-tech "sunrise" industries than to a declining manufacturing sector (Bluestone and Harrison, 1982; Markusen et al., 1991; Tabb, 1984: 9). As Perry and Watkins, the early chroniclers of the Sunbelt's ascent, pointed out:

> Clearly, this region is more than a low-wage carbon copy of the industrial Northeast and its growth has not been limited solely to those low wage sectors which are looking for a cheaper labor site. Rather, because of its lower pay scales and less stringent labor laws, the South has emerged as a lower wage site for traditionally high wage activities. (Perry and Watkins, 1977: 43)

While the best-known Southern economic development strategies in the 20th century emphasized low factor costs and union avoidance to lure production facilities in textiles, apparel and other deskilled industries, the region's economic growth also carried a marked high-technology component (Clark, 1989). By recruiting federal aerospace and defense facilities and contracts, urban regions as varied as Houston, Dallas, Huntsville, Alabama and Columbus, Georgia benefited from an extensive influx of scientists and engineers, and the subsequent development of dominant firms in innovative manufacturing sectors (Hooks, 2003; Markusen et al., 1991). The substantial historical payoff to this high-tech growth strategy is well documented. Yet evidence of enduring employment strength in Southern manufacturing is difficult to see today. The pace of manufacturing job loss in many Southern states now exceeds that of the North. For example, North Carolina, long known for low union density and the in-migration of well-educated Northerners, lost 30 percent of total manufacturing employment between 2000 and 2010, even as the state added 1.5 million residents; Tennessee, South Carolina, Mississippi and Virginia lost manufacturing jobs at a similar pace.

These trends suggest a sharp disconnect between outcomes and established theory. If the South really captured high-tech and sunrise industries immunized from the cost-based competition that spurred disinvestment across the Rustbelt, the states of the Southeast should not be losing manufacturing jobs at a faster pace than Illinois, Pennsylvania and the nation as a whole. Similarly, the limited Northern manufacturing rebound confounds the supposition that cost-based international competition will steer job losses toward the highest-cost production areas in the U.S.

Switching focus from manufacturing in the aggregate to individual types of manufacturing suggests an answer. Nationwide, job losses have been highest in the least technologically intensive and most cost-competitive industries. Apparel, for example, shed a remarkable 78 percent of total employment between 1980 and 2010. Differences in the intensity of job loss across industries could thus explain the South's losses; the ongoing decline of light-industry employment, for example, may dwarf gains or stability in high-tech industries, just as gains in technologically intensive durable manufacturing jobs in the Great Lakes could offset ongoing job losses in more cost-competitive industries across the Rustbelt.

We propose the alternate explanation of inter-regional stability in occupational structure. Arguments about the spatial distribution of sunrise and sunset industries are at their heart arguments about the types of work and occupations that regions house. Investigating the spatial distribution of management, design and production work in restructuring manufacturing industries is a crucial step in determining the comparative vulnerability and strength of the regions that house them. But the contingent regional distribution of the types of work performed within an industry makes industries themselves an inexact tool for discerning the overall employment structure and opportunities of regions (Morgan and Sayer, 1983). Occupations provide a much clearer lens of analysis, illuminating the types of activities that individual regions perform, and the types of cost and market pressures to which their industries will be susceptible (Markusen, 2004; Thompson and Thompson, 1985).

Deindustrialization in the 2000s

The output of U.S. manufacturing firms grew during the 2000s, but domestic employment contracted at the fastest pace on record. U.S. manufacturing firms shed 20 percent of the total workforce between 2000 and 2010, a pace that far exceeds the 1980s (–7 percent) and 1990s (–8 percent) job losses that spurred policymakers, practitioners and scholars to search for explanations and answers to the demise of millions of mid-wage jobs. The rapid job losses of the 2000s were notable for the fact that they proceeded during boom and bust periods alike. U.S. manufacturers shed more jobs (–11 percent) between 2000 and 2006 than they did during the devastating 2007–2009 recession and subsequent recovery (–9 percent).

In keeping with the past practice of focusing on recessions as transformative moments for industries, firms and workers, current accounts of U.S. manufacturing train their attention on the steep losses and prospective rebound that followed the beginning of recession in December 2007 and the 2008 financial crisis (Christopherson, 2009; Dey et al., 2012; Helper et al., 2012). But the recently concluded recession differs from its predecessors in that it followed a period of steep employment losses. However problematically and unevenly it did so, manufacturing employment had rebounded in the wake of the 1980/1981–1982 and 1990–1991 recessions; subsequent economic contractions punctuated periods of normal industry operation in which output growth and employment growth proceeded hand in hand. Thus, the 2007–2009 recession stands out not just because of the severe disruptions it entailed, but also because of the job-loss expansion it had disrupted.

These ongoing job losses fall to the margins in many accounts of deindustrialization for the simple reason that job loss is no longer novel. But mounting job cuts signify deeper changes in manufacturing activity – in the factors that enable some firms and regions to maintain stability in a shrunken manufacturing core while others do not. The crucial insight here is that contraction is not a one-size-fits-all category. In some cases, a shrinking industry – electronic equipment or automobiles – becomes a progressively larger part of the manufacturing sector, even as it shrinks in the aggregate. And even when industries shed jobs at the same rate, the distribution of job losses across occupational groups differs. Michigan and North Carolina, for example, both lost manufacturing jobs at a rapid pace in the 2000s, but the "rump" of remaining manufacturing employment in Michigan maintains a distinctly higher skill profile.

Finally, this uneven contraction of manufacturing employment leaves each region with a restructured portfolio of manufacturing industries and occupations. Because what regions make, and how they make it, have changed, the next round of manufacturing job losses will operate differently from the one just concluded.

The logical and pragmatic imperative to document the changed landscape of U.S. manufacturing quickly gives way to the practical question of *how* to chart simultaneous changes in manufacturing industries, occupations and geographies. We address this challenge with a unique, time-series dataset that charts net changes in industry and occupational composition for every U.S. state. We provide an updated portrait of U.S. manufacturing that replaces outdated stylized facts with a detailed map of current regional strengths in select industry and occupational groups.

DATA AND METHODS

Our dataset and approach complement richly detailed studies of supply-chain restructuring in individual industries by providing a more systematic analysis of manufacturing. Recoded across decades for consistency in terms of industries and occupations, these data provide the means to track long-range shifts in manufacturing across multiple, simultaneous dimensions. Please see the Methodological Appendix for more detail.

Uneven Contraction: the Changing Contours of Deindustrialization as Job Losses Mount

Analyses of deindustrialization inevitably turn to the regional dimensions of the phenomenon. During the 1970s and 1980s, the growth of Southern and Western states at the expense of the industrial Midwest and Northeast helpfully, albeit imprecisely, encapsulated a broad set of shifts in political power, population and firm-level strategy. As deindustrialization – as measured by net job losses – subsequently deepened, these simple stylized facts began to mislead. By the year 2000, analysts touting the ascent of the Sunbelt at the expense of the Rustbelt had reversed the evident conclusions from the available regional data (Table 24.1).

Ironically, acceptance of the Sunbelt's ascent coincided with the end of manufacturing growth in the region. Manufacturing employment in the Southern states stagnated during the 1980s, and began to shrink (–6 percent) during the 1990s. As U.S. economic output and consumer income expanded during the 1990s, the Midwest, previously

Table 24.1 Manufacturing employment change by state and business cycle

	Manufacturing Employment Change (%)				
	1980s	1990s	2000s	2000–2006	2006–2010
United States	**−7**	**−8**	**−20**	**−11**	**−9**
Midwest	−11	1	−22	−11	−12
Northeast	−21	−22	−24	−16	−9
South	0	−6	−20	−12	−8
West	12	−9	−12	−6	−7
East North Central	−14	0	−25	−13	−14
East South Central	1	−2	−27	−14	−15
Mountain	21	11	−8	−3	−5
Middle Atlantic	−22	−22	−24	−16	−9
New England	−20	−21	−24	−17	−9
Pacific	10	−14	−14	−7	−7
South Atlantic	2	−10	−23	−16	−9
West North Central	−3	5	−13	−7	−6
West South Central	−4	1	−8	−6	−2

written off as a site of potential manufacturing job growth, was the only region of the nation to add jobs. The reality of nationwide job losses during the 2000s points to the epiphenomenal nature of this employment rebound, and the inherent limits of broad generalizations about regions: in cyclical and unstable capitalist economies, stability is the exception to the far messier realities of contingency, restructuring and instability (Storper and Walker, 1989).

While these data trouble the straightforward narratives of the Rustbelt/Sunbelt dichotomy, they also suggest systematic differences across smaller regions embedded within the North and South. Despite the focus on Youngstown, Chicago and the Upper Midwest during early scholarship on deindustrialization, 1980s job losses in the Midwest (–11 percent) were half as intense as those in the Northeast (–22 percent). During the 1990s, the employment trajectories of the Midwest (+1 percent) and Northeast (–22 percent) were the poles between which other regional outcomes could be arrayed. Similarly, Western states fared much better than the historical Confederacy during the period of Sunbelt growth. They expanded manufacturing employment by 12 percent while Southern manufacturing ranks stagnated during the 1980s, declined at a similar rate during the 1990s, and declined at a much slower rate (–12 percent, vs –20 percent) during the extensive job losses of the 2000s. These data indicate that even when it adequately described overall shifts in regional employment, the Sunbelt/Rustbelt dichotomy masked significant divisions within the economies of each region.

Shedding these broad characterizations of regional growth trajectories means losing the underlying explanations on which many policy interventions were built; if the shift of employment to the South and the West does not map directly to growth in areas of the lowest operating costs, a different explanation must be sought. Previous work on emergent high-tech industries and skilled jobs for scientists, designers and technicians in manufacturing suggests an alternate explanation. The analytical container of industries has always provided a limited vehicle for understanding the actual work that firms and regions do (Thompson and Thompson, 1985). While the growth of manufacturing employment in the South and West was tied in the popular imagination to the mobility of runaway shops, their regional manufacturing economies instead benefited from sunrise industries and occupations (Essletzbichler, 2004). While luring runaway shops constituted a notable element of Southern industrial development during the last third of the 20th century, the aggregate impact of firm relocations was dwarfed by the South's development of growing and technologically sophisticated industries tied to its capture of aerospace and defense spending. Seen this way, the capture of Congressionally sanctioned defense spending by politically powerful constituencies explains changing manufacturing geographies better than do models predicated on firms in competitive markets minimizing their factor costs (Fainstein and Fainstein, 1978; Markusen et al., 1991).

This characterization provides a fairly accurate description of the 1980s. But subsequent changes in the employment of scientists, technicians and engineers in manufacturing indicate both changing regional dynamics and the end of net job growth for these once-privileged positions (Figure 24.1). The regional dimensions of high-tech employment change are the most striking. While Illinois and Ohio shed science and technology employment during the decade, numerous other Midwestern states, including Michigan, Minnesota and Wisconsin, added high-tech jobs during the 1980s.

Interestingly the available evidence for a Rustbelt–Sunbelt dichotomy disappeared as soon as the distinction entered the popular lexicon. During the 1990s, science and technology job losses deepened in Ohio, New York, Pennsylvania and other pockets of the industrial Midwest and Northeast – but also in Virginia, Florida, California, Washington, Mississippi, Alabama and Tennessee. This mottled job-loss geography overlaps in substantial part with the Gunbelt (headquartered in Los Angeles, the San Francisco Bay Area, Seattle, St. Louis and Huntsville, Alabama – see Markusen et al. in 1991). But diminished spending at the end of the Cold War cannot explain the entirety of these changes. The Pentagon budget, which grew substantially in the wake of the 2001 terrorist attacks and remained both elevated and stable through the 2007–2009 recession, likewise cannot explain the generalization of science and technology job losses during the 2000s, a decade during which 42 states lost science and technology manufacturing jobs (Figure 24.1). Once the exception to mass job losses, these treasured positions have been reorganized through the same outsourcing practices previously used to relocate production positions (Grossman and Rossi-Hansberg, 2006; Narayanan and Swaminathan, 2007; Srivastava and Theodore, 2005).

Charting the limited but real growth in these highly skilled jobs casts into sharper relief the broader job-change trends of U.S. manufacturing sectors over the past 30 years. The development of occupational strengths in science and technology-related jobs provided regions with pockets of growth amidst seas of decay. While Florida, Mississippi, California, Utah and various other Sunbelt states saw net manufacturing job growth during the 1980s, that growth appears to have been rooted primarily in high-end jobs, as theoreticians would suggest (Figure 24.2). While many states gained manufacturing jobs during that decade, just five gained production jobs. And while popularly remembered for the phenomenon of the runaway shop, 1980s deindustrialization appears to have operated as Barry Bluestone and Bennett Harrison (1982) characterized it: disinvestment and the loss of basic production capacity (Figure 24.2).

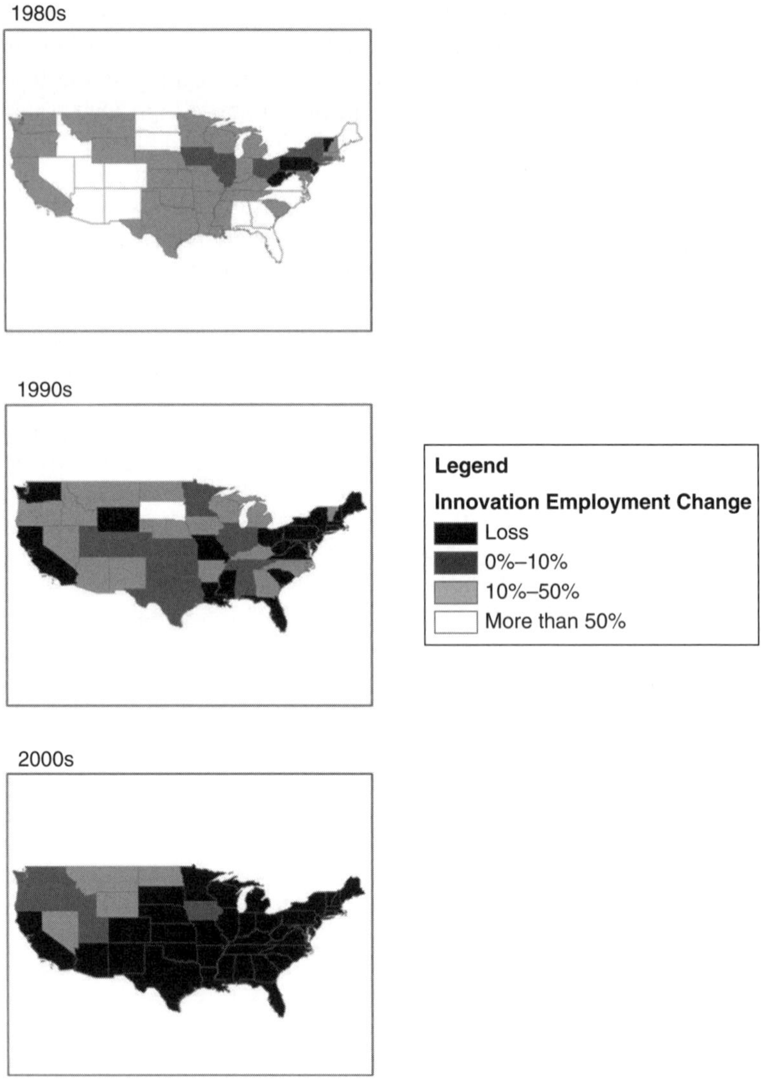

Source: Authors' calculations from U.S. Census Bureau, Public Use Microdata Series and American Community Survey data.

Figure 24.1 Change in manufacturing science and technology employment by state and decade

Subsequent changes in the manufacturing map confirm the divergence between sector-wide and production employment. U.S. manufacturers eliminated manufacturing production jobs at a faster rate in the 1990s than they had in the 1980s (–8 percent vs –7 percent). But more than half of all states added production employment, including Ohio, Michigan, Wisconsin, Minnesota and other centers of Midwest manufacturing

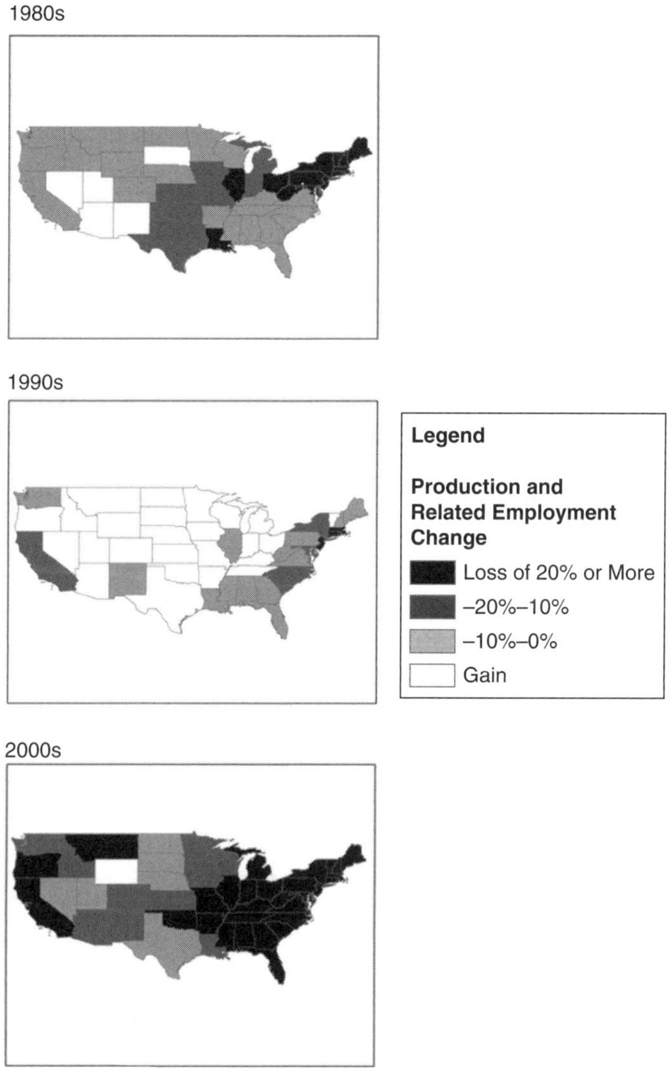

Source: Authors' calculations from U.S. Census Bureau, Public Use Microdata Series and American Community Survey data.

Figure 24.2 Changes in production-related manufacturing employment by state and decade

whose recent job losses had fashioned the basis of the "Rustbelt" neologism. Notably absent in the production recovery were the celebrated winners of the Rustbelt–Sunbelt shift: the Carolinas and California each lost more than 10 percent of production jobs, and total production employment likewise contracted in Florida, Georgia, Alabama, Mississippi and Louisiana. At the precise time that critical geographers had grudgingly acknowledged the victory of the South and Southeast's low factor costs over the more regulated and rigid economies of the Midwest, the Southern states' large concentration of cost-competitive, light manufacturing industries made them net losers in the production rebound (Glasmeier and Leichenko, 1996).

The broad, nationwide demise of manufacturing production jobs during the 2000s indicates the fundamental instability of production jobs in a globalizing economy. Just one state, Wyoming, added production jobs on the decade. By contrast, losses were wide-ranging and deep, typically 20 percent or more. The resulting picture is one of deepening but *structured* deindustrialization. Twenty-two of the 30 major manufacturing industries shrunk during the 2000s (Doussard and Schrock, 2013), but each industry region shrunk in a different way. This suggests that scholars and practitioners would benefit from replacing the contraction/growth dichotomy implicit in most contemporary research with pragmatic efforts to distinguish the various varieties of manufacturing job loss. The research questions and policy prescriptions attached to an atrophied but relatively stable sector such as primary metals fabrication, for example, will differ from those relevant to the continuously contracting and labor-intensive apparel industry. Below we provide basic metrics that distinguish types of deindustrialization.

VARIETIES OF JOB LOSS: INFORMATIVE METRICS FOR SHRINKING MANUFACTURING INDUSTRIES

The basic statistical tool of the location quotient provides a surprisingly strong means of differentiating types of economic contraction. Rather than provide industry-level data that beg the fundamental, policy-relevant question of which industry segments operate in different regions, we supply manufacturing location quotients at the level of states and multi-state regions. These data demonstrate the fundamental data paradox contemporary characterizations of U.S. manufacturing must confront. Despite absorbing the loss of millions of manufacturing jobs over the past 30 years, the Midwest is a bigger part of the manufacturing industry today than at the onset of deindustrialization (Table 24.2).

Strikingly, the regional strength of manufacturing employment in the South and West has changed minimally over multiple business cycles. The Southern states saw their aggregate manufacturing location quotient fall by 1 percentage point from 1980 to 2010. The states of the West saw a slightly larger rise, from 0.78 to 0.86. Meanwhile, the concentration of manufacturing employment in the Rustbelt was polarized. In 1980, Midwestern and Northeastern states had virtually identical location quotients, of 1.16 and 1.13 respectively. Today, the Midwest – the discursive heart of the deindustrialized North – has a manufacturing location quotient of 1.38. The base of U.S. manufacturing jobs has shrunk, but the Midwest represents a more significant portion of that base than at any time in the last 40 years. Meanwhile, the Northeast represents the only region of the U.S. that has shed manufacturing jobs in a linear fashion. Its current-day location quotient of 0.89 is substantially lower than 30 years ago. Nevertheless, manufacturing jobs represent a larger portion of the employment base in the Northeast than in the West, and account for roughly the same share of the regional employment base as does the South's manufacturing sector. The shift of manufacturing employment from North

Table 24.2 Manufacturing and innovation occupation locations by state and year

	Manufacturing Location Quotient				Innovation Occupation Location Quotient			
	1980	1990	2000	2010	1980	1990	2000	2010
United States	**1.00**	**1.00**	**1.00**	**1.00**	**1.00**	**1.00**	**1.00**	**1.00**
Midwest	1.16	1.19	1.31	1.38	0.91	0.87	0.90	0.88
Northeast	1.13	0.99	0.91	0.89	1.13	1.15	1.13	1.12
South	0.92	0.96	0.94	0.91	0.75	0.80	0.82	0.82
West	0.78	0.86	0.83	0.86	1.46	1.40	1.39	1.37
East North Central	1.29	1.30	1.42	1.47	0.89	0.85	0.91	0.90
East South Central	1.12	1.26	1.31	1.25	0.54	0.54	0.56	0.61
Mountain	0.53	0.65	0.64	0.68	1.32	1.45	1.41	1.33
Middle Atlantic	1.08	0.95	0.88	0.86	1.10	1.08	1.03	1.01
New England	1.26	1.10	1.01	0.99	1.21	1.32	1.34	1.35
Pacific	0.87	0.93	0.90	0.95	1.49	1.39	1.38	1.39
South Atlantic	0.93	0.92	0.87	0.80	0.73	0.81	0.84	0.84
West North Central	0.84	0.94	1.06	1.18	0.96	0.91	0.85	0.84
West South Central	0.79	0.84	0.87	0.92	0.96	0.99	1.00	0.94

to South and West represents an important historical and institutional development, but it should not induce scholars to overlook the *growing* dominance of the Midwest in the nation's shrunken manufacturing sector.

Census regions constitute crude containers for regional analysis, of course. Within each of the four basic regions identified in census datasets, divergent trends are evident. The Midwest's growing share of the shrinking manufacturing employment base is most pronounced in Michigan, Wisconsin, Indiana and Ohio, where location quotients rose through each round of job loss. Meanwhile, the manufacturing share of the Illinois economy has fallen slightly, due to its limited participation in the automotive supply chain, and Chicago-area policymakers' embrace of post-manufacturing economic development strategies (Doussard et al., 2009; Rast, 1999). The deindustrialization of the Northeast has been more sweeping and even, as every state displays significant reductions in location quotients from 1980.

Stable Occupational Geographies?

Past research on the occupational dimensions of economic restructuring, and specifically the impact of federally funded technology investment on manufacturing industries, suggests that shifting lenses from industries to occupations greatly alters the assessment of winners and losers in industry restructuring. While other contemporary work (Doussard and Schrock, 2013) identifies the increasing importance of science and technology jobs at the national level and in anchoring an industry to U.S. locations, a review of regional data suggests surprising levels of stability in the science and research component of individual regional economies. Over three business cycles, the regional distribution of "innovation" – science, technology and design – jobs within manufacturing economies has scarcely changed. As early analysts of the Sunbelt suggested, the region has upgraded its share of innovation occupations, rising from a location quotient of 0.75 in 1980 to 0.82 today. This shift can best be characterized as *convergence without equalization*. Despite this significant expansion of the South's regional share of science and technology employment within manufacturing, it remains the least technologically intensive region in the U.S. Equally significant, Southern growth in science and technology jobs appears to have been confined to the 1980s, as the recent stability of the region's innovation location quotient (0.80, 0.82 and 0.82 in 1990, 2000, 2010, respectively) indicates. In 1980, observers of manufacturing could accurately describe the West and Northeast as housing regional manufacturing sectors that specialized in science and technology work relative to other regions. Those observers would still be right today.

These findings underscore a basic paradox of analyses of deindustrialization. While certain periods of industrial restructuring compressed differences between the manufacturing-rich North and the historically less industrialized South, the regions' initial industrial endowments, and the subsequent reversal of industry-location trends that appeared to favor the South, have blunted the net impact of these changes. Thirty years after the onset of deindustrialization, the Midwest represents a growing share of U.S. manufacturing, the Southern share has not changed, and the South has modestly expanded its share of technology-related manufacturing jobs.

CONCLUSION

At the tail end of 1980s restructuring, scholars of manufacturing turned to the mysteries of design-, engineering- and technology-rich regions for suggestions as to which factors, if any, could ensure employment stability in economies suddenly characterized by aggressive firm mobility. The resulting proliferation of theories suggested that the key to making employment "sticky" lay in the combination of research- and design-intensive goods produced through complex networking processes that required constant interaction between skilled professionals. Most memorably embodied in Michael Storper's (1997) identification of "untraded interdependencies" between firms, this scholarship identified a simple reality with important contemporary policy implications: production jobs are more mobile than are the social networks through which high-tech and design work is produced.

This simple distinction helps to explain the durability of the spatial division of manufacturing labor in the U.S. over the past 30-plus years. While Reagan-era defense spending created select science and technology complexes in some Southern and Western metropolitan areas (Huntsville, Alabama, Houston, Dallas, Colorado Springs and the San Francisco Bay area continue to house research networks initially developed decades ago), the creation of new research regions is relatively rare. Thus, Detroit, Chicago and Pittsburgh remain centers of high-technology employment even after decades of deindustrialization, for the straightforward reason that the barriers to moving those complex networks are much greater than the rewards of doing so (Chapple et al., 2004; Treado, 2010). By contrast, production employment facilities can move, expand and contract more freely. While every U.S. region has lost enormous volumes of production employment during the most recent business cycles, the contours of the high-end research complexes on which that production depends are surprisingly resilient. The 1980s shift in science and technology employment to the South was indeed significant – and it was significant because it represents so rare an event that analysts should not expect it to be repeated.

As U.S. manufacturing industries continue to contract and restructure in the current and future business cycles, questions related to this basic observation will become more important. Short-term production employment run-ups continue to occur. They swept the Midwest in the 1990s, and form the basis of currently celebrated job growth in the Great Lakes (see Christopherson, 2011). Given the isolated regional character of job growth, the empirical question of which types of regional manufacturing economies produce these production spikes grows increasingly crucial. Industry-level research conducted during the transformative supply-chain restructuring of the 1990s suggests two possible – and opposing – answers. As firms embraced subcontracting and outsourcing *en masse*, smaller urban regions throughout the U.S. developed unique manufacturing economies ("satellite platforms") based on production capacity alone (Markusen, 1996). Although limited empirical evidence suggests these regions can progressively diversify and add research and development capacity, the question of whether they *have* made that leap remains unaddressed. Answering this crucial question remains essential in determining future geographies of manufacturing job growth.

But decades after the onset of deep job losses, the most important research sites for manufacturing remain older industrial regions themselves. Recent production-job growth in Buffalo, Cleveland and other urban areas in the heavily disinvested Northeast appears to build from a base of lingering research, logistics and sales capacity. There are good reasons to predict that these "innovation" regions will continue to add production capacity as unique sites of high-value-added work in increasingly cost-competitive industries (Clark, 2013). But the list of institutional, market and historical developments capable of restricting that job growth – diminished investment in regional research networks, skills shortages, the successful multinational coordination of previously place-based research processes – is extensive. Our data suggest that the best way to look forward is to look back. Narratives of industrial restructuring identify crucial decisions, pressures and policies driving the employment geographies of individual industries. But those narratives stop in the mid-1990s, a period that predates an especially intensive set period of job losses, offshoring and interfirm restructuring. Identifying the policy opportunities for future manufacturing rests on the ability to answer the basic question of what happened to U.S. manufacturing industries after deindustrialization.

REFERENCES

Atkins, P., P. Blumenthal, A. Edisis, A. Friedhoff, L. Curran, L. Lowry, T. St. Clair, H. Wial and H. Wolman (2011), "Responding to manufacturing job loss: what can economic development policy do?" Brookings Institution, Washington, D.C.

Bluestone, B. and B. Harrison (1982), *The Deindustrialization of America*. New York: Basic Books.

Chapple, K., A. Markusen, G. Schrock, D. Yamamoto and P. Yu (2004), "Gauging metropolitan 'high-tech' and 'I-tech' activity." *Economic Development Quarterly* 18 (1), 10–29.

Christopherson, S. (2009), "Manufacturing: up from the ashes." *Democracy: A Journal of Ideas* 14.

Christopherson, S. (2011), *Riding the New Wave in Manufacturing to More Jobs and a Better Economy*. Research and Policy Brief Series, Vol. 44. Ithaca, NY: Community and Regional Development Institute.

Christopherson, S. and J. Clark (2007a), "Power in firm networks: what it means for regional innovation systems." *Regional Studies* 41 (9), 1223–1236.
Christopherson, S. and J. Clark (2007b), *Remaking Regional Economies*. London: Routledge.
Clark, G.L. (1989), *Unions and Communities Under Siege: American Communities and the Crisis of Organized Labor*. Cambridge, UK: Cambridge University Press.
Clark, J. (2013), *Working Regions: Reconnecting Innovation and Production in the Knowledge Economy*. London: Routledge.
Dey, M., S.N. Houseman and A.E. Polivka (2012), "Manufacturers' outsourcing to staffing services." *ILR Review* 65 (3), 533.
Doussard, M. (2013), *Degraded Work: The Struggle at the Bottom of the Labor Market*. Minneapolis: University of Minnesota Press.
Doussard, M. and G. Schrock (2012), "Post-industrial restructuring? The changing regional landscape of manufacturing in the U.S." *Progressive Planning* Winter, 40–42.
Doussard, M. and G. Schrock (2013), "Dimensions of industrial restructuring: mapping the spatial, occupational and organizational transformation of U.S. manufacturing after deindustrialization." Working paper.
Doussard, M., J. Peck and N. Theodore (2009), "After deindustrialization: uneven growth and economic inequality in 'postindustrial' Chicago." *Economic Geography* 85 (2), 183–207.
Essletzbichler, J. (2004), "The geography of job creation and destruction in the US manufacturing sector, 1967–1997." *Annals of the Association of American Geographers* 94 (3), 602–619.
Fainstein, S.S. and N.I. Fainstein (1978), "National policy and urban development." *Social Problems* 26 (2), 125–146.
Friedhoff, A., H. Wial and H. Wolman (2010), "The consequences of metropolitan manufacturing decline: testing conventional wisdom." Brookings Institution, Metropolitan Policy Program, Washington, D.C.
Glasmeier, A.K. and R.M. Leichenko (1996), "From free market rhetoric to free market reality: the future of the US South in an era of globalization." *International Journal of Urban and Regional Research* 20 (4), 601–615.
Grossman, G.M. and E. Rossi-Hansberg (2006), "The rise of offshoring: it's not wine for cloth anymore." Paper presented at Kansas Fed's Jackson Hole conference for Central Bankers, July 2006. Available at http://www.kc.frb.org/.
Harrison, B. (1984), "Regional restructuring and 'good business climates': the economic transformation of New England Since World War II." In L. Sawers and W.K. Tabb (eds), *Sunbelt/Snowbelt*. New York and Oxford: Oxford University Press, pp. 48–96.
Harrison, B. (1994), *Lean and Mean: The Changing Landscape of Corporate Power in the Age of Flexibility*. New York: Basic Books.
Harrison, B. and B. Bluestone (1990), *The Great U-Turn: Corporate Restructuring and the Polarizing of America*. New York: Basic Books.
Harrison, B. and M.R. Kelley (1993), "Outsourcing and the search for 'flexibility.'" *Work, Employment & Society* 7 (2), 213–235.
Helper, S., T. Krueger and H. Wial (2012), "Locating American manufacturing: trends in the geography of production." Brookings Institution, Washington, D.C.
Holzer, H.J. and M. Hlavac (2012), *A Very Uneven Road: US Labor Markets in the Past 30 Years*. Working Paper #12–09. Ann Arbor, MI: National Poverty Center.
Hooks, G. (2003), "Military and civilian dimensions of America's regional policy, 1972–1994." *Armed Forces & Society* 29 (2), 227–251.
Krippner, G.R. (2005), "The financialization of the American economy." *Socio-Economic Review* 3 (2), 173.
Markusen, A. (1986), "Neither ore, nor coal, nor markets: a policy-oriented view of steel sites in the USA." *Regional Studies* 20 (5), 449–462.
Markusen, A. (1996), "Sticky places in slippery space: a typology of industrial districts." *Economic Geography* 73 (3), 293–313.
Markusen, A. (2004), "Targeting occupations in regional and community economic development." *Journal of the American Planning Association* 70 (3), 253–268.
Markusen, A. and G. Schrock (2009), "Consumption-driven urban development." *Urban Geography* 30 (4), 344–367.
Markusen, A., P. Hall, S. Campbell and S. Deitrick (1991), *The Rise of the Gunbelt: The Military Remapping of Industrial America*. New York and Oxford: Oxford University Press, USA.
Morgan, K. and A. Sayer (1983), *The International Electronics Industry and Regional Development in Britain*. Working Paper no. 34. Sussex, U.K.: University of Sussex School of Urban and Regional Development.
Narayanan, S. and J.M. Swaminathan (2007), *Information Technology Offshoring to India: Pitfalls, Opportunities and Trends*. Working paper. Chapel Hill, NC: Kenan-Flagler School of Business, University of North Carolina.

Perry, D.C. and A.J. Watkins (1977), "People, profit, and the rise of the sunbelt cities." *The Rise of the Sunbelt Cities*. Beverly Hills: Sage Publications.
Piore, M. and C. Sabel (1986), *The Second Industrial Divide: Possibilities for Prosperity*. New York: Basic Books.
Rast, J. (1999), *Remaking Chicago: The Political Origins of Urban Industrial Change*. DeKalb, IL: Northern Illinois University Press.
Ruggles, S., T. Alexander, K. Grenadek, R. Goeken, M.B. Schroeder and M. Sobek (2010), Integrated Public Use Microdata Series: Version 5.0 [Machine-readable database]. Minnesota Population Center [producer and distributor], Minneapolis.
Sawers, L. and W.K. Tabb (eds) (1984), *Sunbelt/Snowbelt*. New York and Oxford: Oxford University Press.
Scopp, T.S. (2003), *The Relationship between the 1990 Census and Census 2000 Industry and Occupation Classification Systems*." Washington, D.C.: US Census Bureau.
Scott, A.J. (1988), "Flexible production systems and regional development: the rise of new industrial spaces in North America and Western Europe." *International Journal of Urban and Regional Research* 12 (2), 171–186.
Srivastava, S. and N. Theodore (2005), "A long jobless recovery: information technology labor markets after the bursting of the high-tech bubble." *WorkingUSA* 8 (3), 315–326.
Storper, M. (1997), *The Regional World: Territorial Development in a Global Economy*. New York and London: Guilford Press.
Storper, M. and R. Walker (1989), *The Capitalist Imperative: Territory, Technology, and Industrial Growth*. New York: Basil Blackwell.
Sturgeon, T.J. (2002), "Modular production networks: a new American model of industrial organization." *Industrial and Corporate Change* 11 (3), 451–496.
Tabb, W.K. (1984), "Urban development and regional restructuring, an overview." In L. Sawers and W.K. Tabb (eds), *Sunbelt/Snowbelt* New York and Oxford: Oxford University Press, pp. 3–15.
Thompson, W. and P. Thompson (1985), "From industries to occupations: rethinking local economic development." *Economic Development Commentary* 9 (3), 12–18.
Treado, C.D. (2010), "Pittsburgh's evolving steel legacy and the steel technology cluster." *Cambridge Journal of Regions, Economy and Society* 3 (1), 105–120.
Weber, R.N. (2001), *Swords Into Dow Shares: Governing the Decline of the Military-Industrial Complex*. Boulder, CO: Westview Press.
Whitford, J. (2005), *The New Old Economy: Networks, Institutions, and the Organizational Transformation of American Manufacturing*. New York: Oxford University Press.

METHODOLOGICAL APPENDIX

For purposes of sample size and presentational clarity, we restrict the tabulation of the data to individual states. Groups of states match neatly to commonplace distinctions about the distribution of manufacturing over various U.S. regions and, as the maps in Figures 24.1 and 24.2 demonstrate, a state-level analysis illustrates fundamental changes in manufacturing geographies in especially clear form. We compiled the data from the Census Bureau's 1980, 1990 and 2000 Public Use Microdata Series files, switching to the American Community Survey (ACS) dataset for 2005–2011.[1] The ACS replaced the "long form" census in 2005, and uses the same basic household sampling and variable structure as the decennial census. This compatibility allowed us to create a continuous dataset. To compensate for the smaller sample size of the ACS (approximately 1 percent of households), we used the Census Bureau's rolling three-year "pooled" samples of data. As a consequence, the data for our final period (2009–2011) are based on observations spread out evenly across 2009, 2010 and 2011.

Methodologically, the main challenge was to develop a consistent coding scheme. The Census Bureau and other U.S. statistical agencies overhauled their industrial and occupational codes between the 1990 and 2000 censuses. In response, we used a detailed correspondence table developed by the Census Bureau (Scopp, 2003) to crosswalk the 497 detailed occupational codes from the old system into 471 occupations based on the current, Standard Occupational Code (SOC) system. To minimize the impact of these inconsistencies and improve sample sizes, we aggregated the data to the level of 22 SOC major occupational groups (e.g. Production, Architecture & Engineering, Management).

The available years for census and ACS data closely match recent business cycles, a fortuitous development that makes the data more policy-relevant. The "double-dip" recession that stands as a central point in the analysis of early U.S. deindustrialization began in January 1980, when our dataset begins.[2] The following business cycle lasted until the middle of 1990, the second year in our dataset. After a mild recession that ended in 1991, the economy began a record-length expansion that lasted until March 2001, a date close to the 2000 census. Business-cycle peaks stopped aligning with the census years afterwards, but the Census Bureau's shift from a decennial long-form census to an annual household survey allows us to map the available data more closely to subsequent punctuations of the business cycle. Our three-year samples of bundled data (2006–2008, 2009–2011) cannot align perfectly with the 2007 entry into, and 2009 exit from, recession, but they closely match longer-run markers of expansion and recession.

NOTES

1. All data were obtained from the Integrated Public Use Microdata Series of the Minnesota Population Center (Ruggles et al., 2010).
2. The recovery from the second "dip" began in November 1982. Please see the National Bureau of Economic Research's Business Cycles Data Page for details, http://nber.org/cycles/cyclesmain.html.

25. Searching for advanced manufacturing in the United Kingdom and United States: definitions, measurement and public policy

Finbarr Livesey

INTRODUCTION

The refocusing of political and economic attention on manufacturing following the global credit crisis and recession of 2008 has led to the re-emergence of an old problem – are our definitions of manufacturing fit for purpose? Do we have definitions that allow us at the national and international level to understand how manufacturing industries are evolving and whether they are declining or thriving? This chapter examines the existing definitions of advanced manufacturing, putting these in the context of how the definition of manufacturing has evolved, and finally discussing current usage of the term in policies developed by the United States and the United Kingdom.

Before the 2008 financial crisis most Western governments had stopped worrying about manufacturing in any real sense, as manufacturing as measured fell as a share of gross domestic product and a precipitous decline in employment occurred over a 20 year period (Pilat et al., 2006). Hand in hand with this lack of focus on manufacturing was a shunning of industrial policy in any explicit form (Livesey, 2012). During this time whether manufacturing was measured or defined appropriately just did not seem to matter – manufacturing was gone, deindustrialisation had occurred and the West's future lay in services, primarily financial services.

Following the financial crisis of 2008 manufacturing appeared almost overnight as a core concern of governments around the world. David Cameron's first major speech after becoming Prime Minister of the United Kingdom in 2010 immediately focused on manufacturing, as he stated that 'Our economy has become more and more unbalanced, with our fortunes hitched to a few industries in one corner of the country, while we let other sectors like manufacturing slide' (Cameron, 2010). Even as a presidential candidate, Barack Obama placed manufacturing at the core of his message on the economy. In April 2008 in a speech in Pittsburgh he commented upon the decline in manufacturing jobs and the factory closures that had occurred during the Bush administration. He cautioned: 'to those who think that the decline in American manufacturing is inevitable; or that manufacturing has no place in a 21st century economy; we say right here and right now that the fight for manufacturing's future is the fight for America's future' (Obama, 2008).

The move from the general to the specific, from manufacturing in the round to advanced or high-technology manufacturing, reflects the role that policymakers either believe or wish such manufacturing to play in the economy. It is this imprinting of role or importance that leads to a diversity of positions on what advanced manufacturing is, creating the difficulty of having one, stable, agreed definition.

But does it matter whether it is possible to define advanced manufacturing? The difficulties that arise from not having a clear definition for advanced manufacturing have impacts for academics and policymakers alike. For example, the UK Commission for Employment and Skills, in its report on advanced manufacturing, admits from the outset that 'Two definitions of the sector have been employed in this report as no single definition of advanced manufacturing is available' (Davis et al., 2012: vi). While the two definitions the report uses are minor variations of one another (essentially adding the standard industrial classification (SIC) code group for research and development (R&D) within business services), it begs the question regarding the comparability of statistics that claim to be representative of advanced manufacturing, especially in an international context.

The following sections review and discuss the foundations for the statistical definition of manufacturing before moving on to considering how advanced manufacturing is extracted from the more general concept of manufacturing. With this understanding, the final sections discuss the re-emergence of manufacturing policy in the United Kingdom and the United States, highlighting what choices have been made when referring to advanced manufacturing. The chapter finishes with some conclusions on the continuing difficulties in defining and measuring advanced manufacturing and provides some recommendations for those attempting to tie down this elusive term.

THE ROOTS OF THE MEASUREMENT OF MANUFACTURING

The key aim in defining advanced manufacturing clearly is to be able to measure and understand changes that are occurring in a specific category of manufacturing. This depends on having a stable definition of manufacturing itself prior to extracting the elements that are considered to be 'advanced'. The roots of the measurement of manufacturing in the economy can be traced back to the 1930s in the United States and the 1940s in the UK, when agencies in both countries realised they could not reconcile statistics collected for different purposes.

The United States formed the Interdepartmental Committee on Industrial Classification in 1937 (Pierce, 1957) while in the UK a similar process was undertaken at the end of the Second World War, leading to the first UK standard being released in 1948 (Beales, 1949). These attempts to quantify the economy have led to the development of SIC codes, which have evolved and now come in a number of different types. For example, the North American Industry Classification System[1] is used to classify industry in the United States, Canada and Mexico, and in the United Kingdom the system of classification is set out in the UK Standard Industrial Classification 2007.[2]

The basis of each of the classifications is an agreed listing of the elements of the economy, with a fundamental separation between manufacturing and services, using a characteristic such as 'activity' or 'final product' to group companies. Manufacturing and services are essentially defined in opposition to one another and '[f]or practical purposes all work is subdivided into two volumes: (1) manufacturing industries and (2) non-manufacturing industries' (Kolesnikoff, 1940). While such a distinction might have been contextually correct 80 years ago, with the rise of outsourcing and offshoring, such a simple definition cannot hold. Nevertheless, this fundamental split continues to be central to the hierarchical structuring of standard industrial classifications and remains part of the general understanding of the role manufacturing plays in regional and national economies. This simple division is reflected in policy formulation, the media and the general public.

Activity-Based Definitions of Manufacturing

Changes in industrial organisation over the past half century have led to a radically different landscape of companies which would consider and refer to themselves as 'manufacturers'. Some of this reorganisation has been a simple case of playing with the boundaries of the firm (Coase, 1937) but reorganisation and restructuring have also challenged the definition of 'manufacturer'. One example of this process is the emergence of hollow firms in which companies remove themselves from production and instead focus on design and distribution. A good example is IBM, which relinquished control of production of its ThinkPad computer range to Lenovo. One response to this has been to redefine manufacturing as a collection of activities, including R&D, design, production and related service provision, and to be looser about which companies qualify as manufacturers (Livesey, 2006). This in many ways is a selective reading of work on value chains, modified to engage with the definitional question of manufacturing rather than the managerial and strategic questions of how to manage manufacturing.

In this approach what counts as manufacturing is much broader, essentially being any company involved in the journey from idea to product, with the actual process or activity of production contained as one set of activities within the broader activities of manufacturing. Some find this approach unappealing as it does not provide a hard boundary between manufacturing and non-manufacturing companies. This revised definition does reflect the complexity of how companies manage their operations in modern manufacturing, with high levels of variation in how vertically integrated companies are in a given sector and across types of manufacturing.

Taking this approach, however, allows for the blurring of the interface between manufacturing and services, rejecting the hard distinction claimed by Kolesnikoff (1940). The literature on servitisation within manufacturing is significant and growing (Neely, 2008; Baines et al., 2009; Visnjic-Kastalli and Van Looy, 2013) and suggests that the definition of manufacturing itself is under threat. Other sectors, specifically those based on digital technologies which have emerged as industries much more recently, are misrepresented by the current standard industrial codes. For example, the UK's digital economy is officially thought to encompass about 120,000 companies, but recent research has indicated that the number of companies is anywhere between two and four times as high (Nathan et al., 2013). The report highlights that '65 years and several revisions later, these SIC codes are no longer adequate. Far from providing an accurate picture of the economy, they only serve to show us how much we don't know. One in ten companies in the UK are now classified vaguely as "other". One in five have no classification at all' (Nathan et al., 2013: 3).

If it were not for the significant costs, and possibly inertia from statistical agencies and companies, our basic framework for measuring and monitoring the economy would have radically changed over the past 10 to 20 years. Nevertheless, current classifications are likely to remain as the foundation for defining the structure of the economy for some time to come, even though such classifications should now be used with considerable caution.

APPROACHES TO DEFINING ADVANCED MANUFACTURING

The difficulties that exist with the general definition of manufacturing raise questions regarding attempts to define the subsection of manufacturing that is advanced. And what is the purpose of developing a definition of advanced manufacturing? This section discusses the various attempts that have been made to define 'advanced manufacturing' and the related terms 'high technology' and 'high value' manufacturing. This section is then followed by an analysis of the application of the concept of 'advanced manufacturing' in policy documents produced by the UK and US governments.

Advanced Manufacturing

The first distinction to be made is whether the identification of advanced manufacturing is based upon advanced manufacturing techniques or companies which produce advanced or complex products, with or without regard for the production technologies that are used in their manufacture (Science and Technology Institute, 2010). The former is a technical discussion on which production technologies are near to the research frontier, having recently emerged from the science base. The latter questions whether the 'advanced' label is about the outputs of these companies or the methods that are used to make the products. This leads to the development of a simple matrix (akin to and modifying the classic Ansoff matrix (Ansoff, 1960)) that contrasts the novelty of the product for a company against the novelty of the production process (Figure 25.1).

Companies that are using existing production processes to produce existing products are unlikely to be involved in advanced manufacturing, but what about the other three

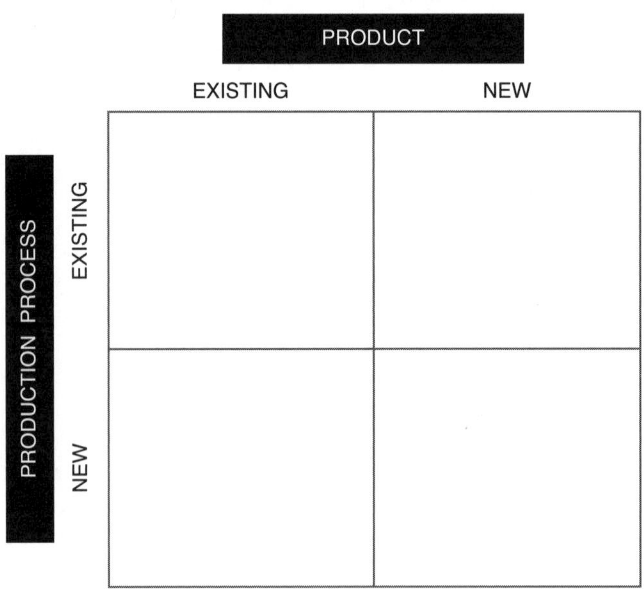

Figure 25.1 Defined advanced manufacturing: contrasting the novelty of product against the novelty of the production process

segments? Can a company only be considered to be 'advanced' if it is using new production techniques? This appears to be the assumption or the implication of a number of the definitions developed to identify and define advanced manufacturing.

These distinctions appear to be at the heart of the broad definition given by the Institute for Defense Analyses (Shipp et al., 2012: 4), which defines advanced manufacturing in the following way: 'Advanced manufacturing improves existing or creates entirely new materials, products, and processes via the use of science, engineering, and information technologies; high precision tools and methods; a high-performance workforce; and innovative business or organizational models.' This would correspond to the bottom of Figure 25.1, with a halo of other aspects including skills and business model added to the technical distinction. A critical reading of this definition would regard it as synonymous with manufacturing, as it does not exclude existing products or indicate that the 'use of science, engineering, and information technology' has to be the latest or most complex. This definition does, however, capture a key element of advanced manufacturing – that it depends on a complex interplay of technology, labour and organisation; it is not uniquely defined by the production technologies that are used to manufacture modern products.

High-Technology Manufacturing

Another term that is associated with advanced manufacturing is 'high-technology manufacturing'. The dominant approach in this vein is that taken by the OECD (OECD, 2011). While there are many characteristics that could define the state of technology used in a given sector, the definition of high-technology manufacturing gathers together those manufacturing sectors that are highest in their intensity of R&D investment in relation to output. The manufacturing sectors which are categorised as high technology are:

- Aircraft and spacecraft
- Pharmaceuticals
- Office, accounting and computing machinery
- Radio, TV and communications equipment
- Medical, precision and optical instruments (OECD, 2011)

This approach is fraught with difficulties. According to Hatzichronoglou (1997: 5):

> [t]he construction of a complete classification of industries according to their technology intensity involves a number of difficulties. The first concerns the criteria for identifying the technology content of an industry. The second concerns the underlying concept. What is a high-technology industry: is it one producing technology, or is it one intensively using technology? A third problem is that there is always some degree of arbitrariness in choosing the cut-off points between the technology classes.

As he goes on to admit, the technology content could be approached in a number of ways but due to data availability R&D intensity was selected as the sole indicator. Given the narrow basis for the OECD definition, this appears to be an unsatisfactory approach to defining advanced or high-technology manufacturing. While it may indicate broad sectors which are investing significantly in R&D, this does not provide any further information on the products produced, the distance of the production technologies to the current production frontier, or how stable or anchored such production is within a national boundary. Overall, it appears to adopt a hyper-pragmatic approach, based on using the best available indicator, without worrying about what the best actual indicator might be to define and measure this type of manufacturing.

High-Value Manufacturing

A very different approach was taken by Livesey (2006) in an attempt to clarify what high-value manufacturing meant in the UK context. This was based on conceptualising manufacturing as a collection of activities, from R&D, through production, and on to service provision. This approach intended to clarify what types of manufacturers would provide high value, either in financial, strategic or social terms, for a variety of stakeholders. The key here is that value is multidimensional and dependent on which stakeholder is asking the question.

The broad types of manufacturers (Figure 25.2) do not align immediately with high and low value – depending on the context and the company there are high-value manufacturing examples in each quadrant. By definition this means that a sectoral approach cannot work; each sector is heterogeneous in terms of value. Reflecting this back into the definitions of advanced or high-technology manufacturing, this may imply that the approaches that use standard industrial codes as their foundation are doomed to failure from the beginning. The SIC approaches trap themselves in a sectoral approach that cannot provide the resolution needed to identify where the companies in question reside within the economy.

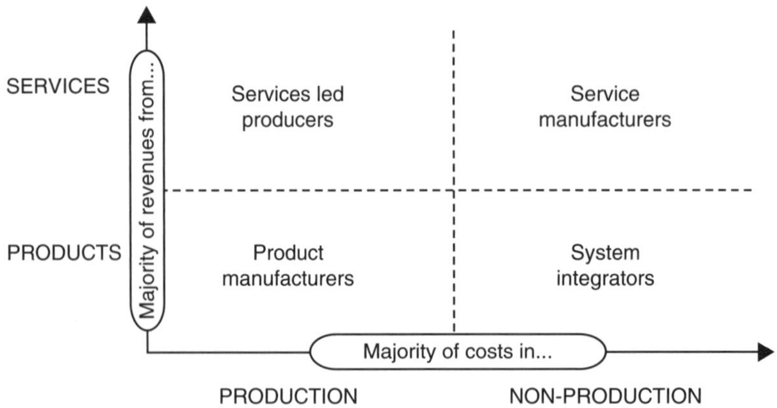

Source: Livesey 2006.

Figure 25.2 Types of manufacturers

ADVANCED MANUFACTURING IN POLICY FORMULATION IN THE US AND UK

Aside from the political use of the term 'advanced manufacturing' within speeches and statements within the US and UK governments, what specific reports and policies have emerged focused on advanced manufacturing? This section highlights these key documents and discusses for both countries whether there is a clear and consistent agenda on advanced manufacturing or whether there is policy confusion.

UK Policy towards Advanced Manufacturing

The term 'advanced manufacturing' emerges in UK policy in 2009 with the publication of a summary note on policies in support of advanced manufacturing. This document adopts the following definition: 'Advanced Manufacturing describes businesses which use a high level of design or scientific skills to produce technologically complex products and processes. Because of the specialised requirements involved, these are usually goods and associated services of high value' (Department for Business, Innovation and Skills (BIS), 2009: 1). It is interesting to note the conjoining of the ideas of complexity and value in this definition, indicating the dual characteristics which are seen to be important. The change of government in 2010 led to a brief hiatus on manufacturing issues, but since coming to power the coalition government has evolved its thinking on manufacturing, moving from a broad call to rebalance the economy to a focus on a set of industrial strategies, including advanced manufacturing. Given the context of the financial crisis it is not completely surprising that, as part of the Growth Review announced by the UK Treasury in November 2010 (BIS, 2010), manufacturing was the first area to receive attention. However, it was not manufacturing broadly – instead it was advanced manufacturing which was the policy target, although not formally defined in the document.

This has been followed by an analysis of the UK's sector performance to support an industrial strategy (as opposed to an industrial policy) (BIS, 2012). The use of standard industrial classification codes is explicitly criticised and, in the analysis, the high-technology categories from the OECD are used to group the elements of manufacturing. There is an unclear link in the analysis, which claims to consider the size and future growth potential of sectors along with the scope for government action. The analysis concludes by stating that sectors matter and that 'Advanced manufacturing is an area of current UK strength, e.g. aerospace and automotive, where rising incomes and environmental drivers are likely to increase demand for more energy efficient products' (BIS, 2012: 34).

In outlining the approach taken towards industrial strategy, Vince Cable, Secretary of State for Business, Innovation and Skills since 2010, walked a fine line between old-fashioned industrial policy ('picking winners') and pure condition-setting interventions. As he noted, 'the truth is that, Britain has for a long while been regarded as an outlier in how purist we are about taking a proactive approach to our economy'.[3] Nine sector strategies have been published following the publication of the statement on industrial strategy, none of which is titled or regarded as being focused on advanced manufacturing. However, following the OECD definitions used in the strategy analysis, it appears that automotive, aerospace, life sciences, offshore wind and possibly agricultural technologies are regarded as forming the advanced manufacturing sector for the UK.

In contrast to the BIS approach the main delivery agency for translational funding in the UK, the Technology Strategy Board (TSB) (which describes itself as the UK's innovation agency) does not have an advanced manufacturing strategy. It has a high-value manufacturing strategy which is based on a matrix of R&D intensity versus relative growth to categorise manufacturing sectors. However, within the strategy the TSB defines high-value manufacturing as 'the application of leading-edge technical knowledge and expertise to the creation of products, production processes, and associated services which have strong potential to bring sustainable growth and high economic value to the UK' (TSB, 2012: 6). At the same time, data on high-value manufacturing exports are quoted in the body of the report from the Office for National Statistics, but the footnote that accompanies this states that 'High value manufacturing is here defined as high and medium-high technology manufacturing as defined by the OECD (SIC codes 20, 21, 25.4, 26, 27, 28, 29, 30.2, 30.3, 30.4, 30.6)' (TSB, 2012: 16). Here the confusion over advanced manufacturing and its various definitions has been written into the text of a delivery agency's strategy, highlighting the persistent problem that results from the absence of a stable and accepted definition of high value or advanced manufacturing.

The core policies which have emerged from the growth review include:

- Funding for Catapult Centres (originally called Technology and Innovation Centres), with the first Catapult established to cover high-value manufacturing.
- The formation of nine Centres for Innovative Manufacturing to undertake research on strategic areas of importance for manufacturing.
- The re-launch of the Manufacturing Advisory Service to provide expert advice to manufacturing companies on productivity and innovation improvements.
- The establishment of an international prize in engineering worth one million pounds (now termed the Queen Elizabeth Prize for Engineering).

Since the growth review the TSB has also founded the Advanced Manufacturing Supply Chain Initiative, which aims to improve the global competitiveness of UK supply chains in advanced manufacturing.[4] The scheme has had four rounds since 2011, with £245 million being allocated through the scheme. It is interesting to note that while this is titled as support for advanced manufacturing, the criteria for the scheme do not exclude or promote any specific type of manufacturing. The stated criteria include collaboration (two or more companies), the support of a prime or tier one company, evidence of support for wider government policy, and good value for money.

Overall the current UK government is very positive towards manufacturing, with advanced or high-value manufacturing receiving attention in name at the very least. However, the support mechanisms appear to be sectorally defined, and while this may be providing useful support broadly to manufacturers it may not be targeting those companies that are considered to be 'advanced'.

US Policy towards Advanced Manufacturing

The Obama administration has broadly supported manufacturing since the 2008 election, for example with significant investments being made to bail out the automotive industry in 2009. The Obama administration's focus, however, on advanced manufacturing began to take definite shape in 2011, when the President's Council of Advisors on Science and Technology (PCAST) released a report entitled *Ensuring American Leadership in Advanced Manufacturing* (PCAST, 2011).

This report makes a general case for the importance of manufacturing and then pivots the analysis on to advanced or sophisticated manufacturing as a type of manufacturing that will not involve competition based on labour costs. This 'advanced manufacturing' is defined as:

> A family of activities that (a) depend on the use and coordination of information, automation, computation, software, sensing, and networking, and/or (b) make use of cutting edge materials and emerging capabilities enabled by the physical and biological sciences, for example nanotechnology, chemistry, and biology. This involves both new ways to manufacture existing products, and especially the manufacture of new products emerging from new advanced technologies. (PCAST, 2011: 9)

Here the link to employment and the aspiration to have manufacturing growth tied to jobs is explicit, providing the foundation for the definition of the type of manufacturing that is thought to meet this challenge.

The 2011 report advocated the establishment of an Advanced Manufacturing Initiative and in June of the same year the Advanced Manufacturing Partnership (AMP) was established with $500 million in funding. The AMP was intended to bring together universities, companies and government to spur the development of new technologies in production, with a particular focus on national security, advanced materials, robotics and the energy efficiency of manufacturing.

The AMP Steering Group produced a new report in July 2012, this time entitled *Capturing Domestic Competitive Advantage in Advanced Manufacturing* (PCAST, 2012), which built on the 2011 report and was consistent in its definition of advanced manufacturing. This advocated the founding of a national manufacturing strategy, based on three themes:

- Enabling innovation
- Securing the talent pipeline
- Improving the business climate

Overall there were 16 recommendations across the three themes, of varying complexity and cost. This may have been too diffuse and possibly led to the initiative lacking some of the focus that it required. One recommendation that has been taken forward is the establishment of a National Network for Manufacturing Innovation (NNMI), proposed as 15 centres funded by a one-time investment of $1 billion. The Institute for Manufacturing Innovation's activities would include but not be limited to:

> Applied research and demonstration projects that reduce the cost and risk of commercializing and manufacturing new technologies or that solve generic manufacturing problems; education and training at all levels; development of innovative methodologies and practices for enhancing the capabilities of and integrating supply chains; and engagement with small and medium-sized manufacturing enterprises (SMEs). (Advanced Manufacturing National Program Office, 2013: 4)

To date a proof of concept centre was established in August 2012: the National Additive Manufacturing Innovation Institute with $30 million in federal funding, and more recently a $200 million investment to fund a competition to found three more centres. The focus for these three centres will be digital manufacturing and design, lightweight and modern metals manufacturing, and clean energy manufacturing.[5]

The political support for advanced manufacturing has been consistent and expressed at the highest level. President Obama focused on manufacturing in the 2012 State of the Union address by stating that:

> We will not go back to an economy weakened by outsourcing, bad debt, and phony financial profits. Tonight, I want to speak about how we move forward, and lay out a blueprint for an economy that's built to last – an economy built on American manufacturing, American energy, skills for American workers, and a renewal of American values.

In the 2013 State of the Union address the references to advanced manufacturing interventions were even more explicit, with direct mention of the NNMI and the launch of the three new centres.

Overall, US policy towards advanced manufacturing appears to be more explicit and potentially more coherent than UK policy. The definition of advanced manufacturing used acknowledges its roots in focusing on manufacturing that will provide domestic growth and opportunities for job creation. Whether a focus on new and emerging technologies provides the best opportunities for growth, both in terms of the overall economy and employment, is an open question.

CONCLUSIONS

The term 'advanced manufacturing' is difficult to define in absolute terms. The meaning depends entirely on the motivations or objectives that influence the person or institution attempting to develop or apply a definition of advanced manufacturing. A number of terms have been used not as synonyms, but in some fashion interchangeably by researchers, policymakers and commentators over the past 10 to 20 years. The terms 'advanced', 'high tech' and 'high value' have all been put forward as candidates to capture the type of manufacturing that is important to a country. This spread of terminology highlights those attempts to capture what is 'advanced' in manufacturing is really a value statement on what role manufacturing might play in a national economy. The high-technology arm of these definitions appears to trace its roots to debates on competitiveness and the response of the United States to the emergence of Japan as a manufacturing power in the 1980s. Advanced manufacturing as a concept or label appeared at the same time that R&D was recognised as being of importance for innovation and growth, through the 1990s. Lastly, the high-value approach appeared in the early 2000s, as manufacturing's place in Western economies was thought to be irrelevant and attempts were made to protect or salvage those elements of manufacturing that were considered to have high systemic importance, for example in supporting a large share of R&D investment.

If the rationale for 'advanced manufacturing' is to highlight that group of companies using the most advanced (i.e. most recent or most dependent on R&D) production technologies then it might be possible to develop an appropriate definition. However, companies would not have a stable place within such a definition and the use of the data would be questionable. Alternatively, if 'advanced manufacturing' is an attempt to measure and analyse manufacturing companies that are most likely to operate in leading economies, or those companies that will have high profit margins or invest the most in R&D, again a definition might be created. Nevertheless, great care would have to be taken to ensure that the term does not elide into other uses of the same term.

Advanced manufacturing as a label or concept is a reflection of what policymakers are attempting to improve or change rather than a pure technological definition. This does not mean that there should be no attempt to define or collect data on advanced manufacturing. The difficulty is in creating a rigorous and robust definition and set of related measures. Governments and statistical agencies must be very clear in defining why they wish to collect such data and what relationship or role they believe advanced manufacturing has to the rest of the economy.

NOTES

1. For full details see http://www.census.gov/eos/www/naics/ (accessed 3 November 2014).
2. For further details see http://www.ons.gov.uk/ons/guide-method/classifications/current-standard-classifications/standard-industrial-classification/index.html (accessed 3 November 2014).
3. Vince Cable speech from 11 September 2012, text available online at https://www.gov.uk/government/speeches/industrial-strategy-cable-outlines-vision-for-future-of-british-industry (accessed 3 November 2014).
4. For further details see https://www.gov.uk/advanced-manufacturing-supply-chain-initiative (accessed 3 November 2014).
5. Details taken from the Advanced Manufacturing Portal website at http://www.manufacturing.gov/nnmi.html (accessed 3 November 2014).

REFERENCES

Advanced Manufacturing National Program Office (2013), *National Network for Manufacturing Innovation: A Preliminary Design*. Executive Office of the President. Washington, DC: National Science and Technology Council.

Ansoff, H. (1960), *Corporate Strategy*. New York: McGraw Hill.

Baines, T., Lightfoot, H.W., Benedettini, O. and Kay, J. (2009), 'The Servitization of Manufacturing: A Review of Literature and Reflection on Future Challenges', *Journal of Manufacturing Technology Management* 20(5): 547–567.

Beales, R.E. (1949), 'Industrial Classification – National and International', *Journal of the Royal Statistical Society. Series A (General)* 112(3): 316–330.

BIS (2009), *Advanced Manufacturing*. London: Department for Business, Innovation and Skills.

BIS (2010), *Growth Review Framework for Advanced Manufacturing*. London: Department for Business, Innovation and Skills.

BIS (2012), *Industrial Strategy: UK Sector Analysis*. London: Department for Business, Innovation and Skills.

Cameron, D. (2010), *Transforming the British Economy: Coalition Strategy for Economic Growth*. London: Prime Minister's Office, 10 Downing Street.

Coase, R.H. (1937), 'The Nature of the Firm', *Economica* 4(16): 386–405.

Davis, C., Hogarth, T. and Gambin, L. (2012), *Sector Skills Insights: Advanced Manufacturing, UK*. London: Commission for Employment and Skills.

Hatzichronoglou, T. (1997), 'Revision of the High-Technology Sector and Product Classification', *OECD Science, Technology and Industry Working Papers*, OECD.

Kolesnikoff, V.S. (1940), 'Standard Classification of Industries in the United States', *Journal of the American Statistical Association* 35(209): 65–73.

Livesey, F. (2006), *Defining High Value Manufacturing*. Institute for Manufacturing. Cambridge: Cambridge University Press.
Livesey, F. (2012), 'Rationales for Industrial Policy Based on Industry Maturity', *Journal of Industry, Competition and Trade* 12(3): 349–363.
Nathan, M., Rosso, A., Gatten, T., Majmudar, P. and Mitchell, A. (2013), *Measuring the UK's Digital Economy with Big Data*. London: National Institute of Economic and Social Research.
Neely, A. (2008), 'Exploring the Financial Consequences of the Servitization of Manufacturing', *Operations Management Research* 1(2): 103–118.
Obama, B. (2008), Speech on manufacturing and trade, Council on Foreign Relations, 14 April, David L. Lawrence Convention Center, Pittsburgh, Pennsylvania.
OECD (2011), *ISIC Rev. 3 Technology Intensity Definition*. Paris: OECD Directorate for Science, Technology and Industry.
PCAST (2011), *Report to the President on Ensuring American Leadership in Advanced Manufacturing*. Washington, DC: Executive Office of the President, President's Council of Advisors on Science and Technology.
PCAST (2012), *Report to the President on Capturing Domestic Competitive Advantage in Advanced Manufacturing*. Washington, DC: Executive Office of the President, President's Council of Advisors on Science and Technology.
Pierce, E. (1957), *History of the Standard Industrial Classification*. Washington, DC: Office of Statistical Standards, U.S. Bureau of the Budget.
Pilat, D. and Cimper, A., Olsen, K. and Webb, C. (2006), 'The Changing Nature of Manufacturing in OECD Economies', *STI Working Papers*, OECD.
Science and Technology Institute (2010), *White Papers on Advanced Manufacturing Questions*. Washington, DC: Science and Technology Institute.
Shipp, S., Gupta, N., Lal, B., Scott, J., Weber, C., Finnin, M., Blake, M., Newsome, S. and Thomas, S. (2012), *Emerging Global Trends in Advanced Manufacturing*. Alexandria, VA: Institute for Defense Analyses.
TSB (2012), *High-Value Manufacturing Strategy*. London: Technology Strategy Board.
Visnjic-Kastalli, I. and Van Looy, B. (2013), 'Servitization: Disentangling the Impact of Service Business Model Innovation on Manufacturing Firm Performance', *Journal of Operations Management* 31: 169–180.

26. National manufacturing policy, local real estate markets and the missing region: prospects for urban industrial development in the U.S.

Laura Wolf-Powers

INTRODUCTION

Manufacturing policy and urban and regional policy coexist uncomfortably in the United States. Although many industrialized nations coordinate federal-level technology and innovation strategies with expressly place-based investments in the commercialization of basic research – and, concomitantly, with support for regional networks of small and medium-sized manufacturers – such mechanisms have not been a prominent feature of U.S. policy (Clark 2010). As a result, attempts to connect economic development strategy with economic geographers' knowledge about place-based innovation often falter. Descriptions and typologies of "learning regions," regional innovation networks and spatially rooted industry clusters are reasonably straightforward. But translating that knowledge into effective, implementable policy for manufacturing development is not straightforward, because there is neither a strong federal government leading regional policy from the top down nor a framework for autonomous regional governance of any sort.[1]

Traditionally de-territorialized, national science and technology policy may be changing. A new manufacturing agenda recently announced by the Administration of second-term President Barack Obama has sparked hope that federal policy will attempt more explicitly to strengthen production networks in ways that draw on the specific strengths of cities and their metropolitan regions. Economic development scholars and practitioners believe that new U.S. national government initiatives in manufacturing contain the seeds of place development strategy, citing the potential to coordinate federal investments with state, regional and local ones, and to maximize the benefits of agglomeration and intra-regional networking (Helper, Krueger and Wial 2012b; Muro and Lee 2013). Many believe that by enfolding local-level assistance to manufacturing SMEs into a cohesive national technology and innovation policy, U.S. officials can seize heretofore missed opportunities to commercialize university-based research, and to diffuse it to and through strong place-based networks of firms (Clark 2012, 2013; Ezell and Atkinson 2011).

Renewed policy and media interest in manufacturing, together with the availability of place-based federal funding through two U.S. federal programs – the National Network of Manufacturing Institutes and the Investing in Manufacturing Communities Partnership, both described later in this chapter – have elevated the efforts of cities and counties pursuing industrial land use and development strategies and stimulated new such efforts (Leigh and Hoelzl 2012). The most appropriate scale for engagement with issues of innovation policy, however, is that of the region, while manufacturing revitalization efforts (due to lack of powerful policy-implementing institutions at the level of the region) are almost inevitably implemented at the level of the municipality.

I argue in this chapter that this gap between scales of action in U.S. policy-making makes the project of manufacturing redevelopment in cities problematic. Because of the "missing region," municipal authorities struggle to meaningfully effect changes in the policy areas most compelling in the national policy discourse. Meanwhile, the national government's main areas of focus – producer networks, commercialization, technology transfer and workforce readiness – become subordinate on the local level to officials' necessary preoccupation with real estate markets and barriers to site development. To illustrate this argument, the chapter foregrounds the city of Newark, New Jersey, a city of 278,000 located ten miles from New York City. In Newark, public and civic leaders intent on rejuvenating a severely disinvested manufacturing base have enlisted the collaboration of national policy actors, aiming to find a place for their city beneath the aegis of U.S. national manufacturing policy initiatives. Newark's case illustrates the fact that while new manufacturing policies have the potential to develop businesses, attract entrepreneurs and create jobs in places that have seen decades of industrial disinvestment, actors at the national and local scales need to acknowledge and resolve their distinct approaches in order for this to happen. National actors must take property development seriously; local ones must think beyond it.

THE U.S. MANUFACTURING AGENDA AND THE MISSING REGION

Before taking up the Newark case, I briefly review recent developments in manufacturing policy at the national level in the United States. While policy rationales and modes of implementation differ markedly under different national regimes, most industrialized nations coordinate national-level technology and innovation strategies (science, technology and innovation policy, or STI) with place-based investments in manufacturing at the regional and local scales (see Ezell and Atkinson 2011; Perry and May 2007). For example, Germany has "a long record of regionalized STI policy" (Koschatzky and Kroll 2007, 1119) through which manufacturing enterprises large and small receive technical support and conduct collaborative product and process research in conjunction with a network of 57 university-linked sector-based institutes.[2] In Japan, where science and technology policy has historically been dominated by the national state, regional bureaus of the national Ministry of Economy, Trade and Industry have played an increasing role in recent years; regional bureaus are implementing regional industry cluster initiatives and playing a larger role in the operation of 182 Kohsetsushi Centers, which assist small and medium-sized firms with research on new products and shop floor process improvement (Kitagawa 2007).

Science and technology policy in the United States has, in contrast, had little territorial component and little relationship to manufacturing.[3] Notwithstanding the Manufacturing Extension Partnership, a U.S. Department of Commerce-funded network of technical assistance providers,[4] the U.S. has in general left industrial development to state governments, where officials pursue it in a context of fragmentation and competition. Ezell and Atkinson contrast the approach of economic development officials in the U.S. with that of officials in Japan, where the national state is a stronger influence:

> Japanese prefectures have the attitude that they cannot co-opt a firm from another prefecture; they can only grow their economy from within through superior technology development, transfer, and commercialization. This is in contrast to the "smokestack chasing" more common in the United States, a "race-to-the-bottom" in which states dangle incentives before businesses to induce them to relocate from one state to another. (2011, 7)

But signs exist that this may be changing. Citing an uptick in manufacturing employment after a decade of steep decline, along with the potential for the "on-shoring" of jobs previously performed overseas, the Obama Administration announced a manufacturing policy agenda in 2012. In July of that year, the Advanced Manufacturing Partnership (part of the President's Commission on Science and Technology) adopted a 16-point strategy that encompasses measures to support innovation, improve the tax and regulatory environment, and invest in training and skills certification, all with the goal of strengthening domestic manufacturing. The Administration's long-term plan is to create a national network of National Manufacturing Innovation Institutes, each embedded in a regional economy.[5]

More significant from an urban policy perspective is the Investing in Manufacturing Communities Partnership (IMCP), announced by the Economic Development Administration in the Department of Commerce in April 2013. Published material on the IMCP expressly promotes alternatives to the typical strategy of attracting footloose firms one at a time via specific incentives (tax abatements, grants, land write-downs). It emphasizes that physical and human infrastructure invested in places – and available to all firms that locate in those places – can offer a return superior to that of conventional firm-oriented subsidy or "smokestack-chasing":

> Through the IMCP, the President is directing Federal agencies to provide coordinated assistance to manufacturing communities through a new partnership that will align Federal economic development resources and help U.S. localities make coordinated, long-term investments in their public goods in partnership with universities and industry. These investments will ultimately help regions become more attractive for manufacturers and supply chains. (U.S. Department of Commerce 2013a).[6]

Thus despite the precarious status of these initiatives at a time when many members of Congress are demanding fiscal austerity,[7] consensus appears to have developed in the federal government around two distinct ideas. The first is that national innovation policy should directly implicate manufacturing. The second is that manufacturing policy should function as place policy, strengthening networks of firms in ways that build on the strengths of individual cities. It is against this backdrop that Newark, New Jersey's current manufacturing strategy has evolved.

CASE: MADE IN NEWARK

The rise and decline of manufacturing in Newark follows a familiar American narrative. As an 18th century colonial town, Newark flourished as a craft manufacturing hub, specializing in carriage-making, malleable iron and especially leather, as well as in leather goods such as boots and shoes, saddles and harnesses. With the advent of steam power and the building of canals and railroads, the city prospered in the mid-19th century, gaining exponentially in population and becoming one of the nation's leading manufacturing cities (Cunningham 1988; Jackson 1985). The city thrived as a diverse industrial center into the mid-20th century: at its early-1950s population peak of around 440,000, Newark had 100,000 people employed in manufacturing, still specializing in leather and iron but making a wide variety of other products as well, including electrical machinery, jewelry and plastics (City of Newark 2012; Cunningham 1988).

As in other U.S. manufacturing cities, however, growth through the mid-1950s masked structural changes that were gradually undermining the city economically and fiscally. Increasingly, Newark's factory workers commuted into the city from outlying suburbs, where aspirant middle-class homebuyers could receive Federal Housing Administration financing (nearly all Newark neighborhoods had been "redlined" by government agencies and real estate lenders).[8] Factories soon followed the largely white industrial working-class population, freed by the growth of trucking from their ties to rail and water transportation. Population growth within the city limits was fueled by African American households whose members had migrated from southern states in search of work just as factories were leaving; geographically constrained by discrimination to just a few, over-crowded neighborhoods and excluded from many industrial jobs, these households were confined to a city that offered low-quality, overpriced housing and diminishing employment prospects. Civil disturbances in the struggling city in July 1967 reflected anger over political and social marginalization, discriminatory housing and employment practices, and a lack of economic opportunity (Mumford 2007). The riots also reinforced the conception that manufacturers were safer in the suburbs, and more industrial firms decentralized.

The deindustrialization that had begun with industry suburbanization soon accelerated as the global division of labor came into being and manufacturing began to be off-shored in the 1970s and 1980s. After a period of precipitous population decline and fiscal stress that lasted through the mid-1990s, a so-called "Newark Renaissance" began under the Mayoralty of Sharpe James (Strom 2002) and continued under Cory Booker, elected in 2006. But this economic and fiscal revival is built on the growth of services, entertainment and real estate – especially residential real estate (Newman 2004). The city's recent population gains (Newark's estimated 2012 population was 278,000, up from 274,000 in 2000, its lowest point in 50 years) rest on large publicly subsidized projects like the New Jersey Performing Arts Center, on the construction and absorption of new office space, on the expansion of higher education functions in the downtown, and on new housing designed to attract middle-class residents. Manufacturing jobs have continued to decline, despite the proximity of a growing seaport and airport (City of Newark 2012; Mistry 2013). While job density is greatest in the port and airport districts (Figure 26.1), that area is characterized by significant unutilized and underutilized space, including 434 vacant parcels.

A "Manufacturing Moment"?

The current state of Newark's industrial sector as a whole (defined as manufacturing, construction and utilities) is summarized in Newark's 2012 Master Plan.[9] This docu-

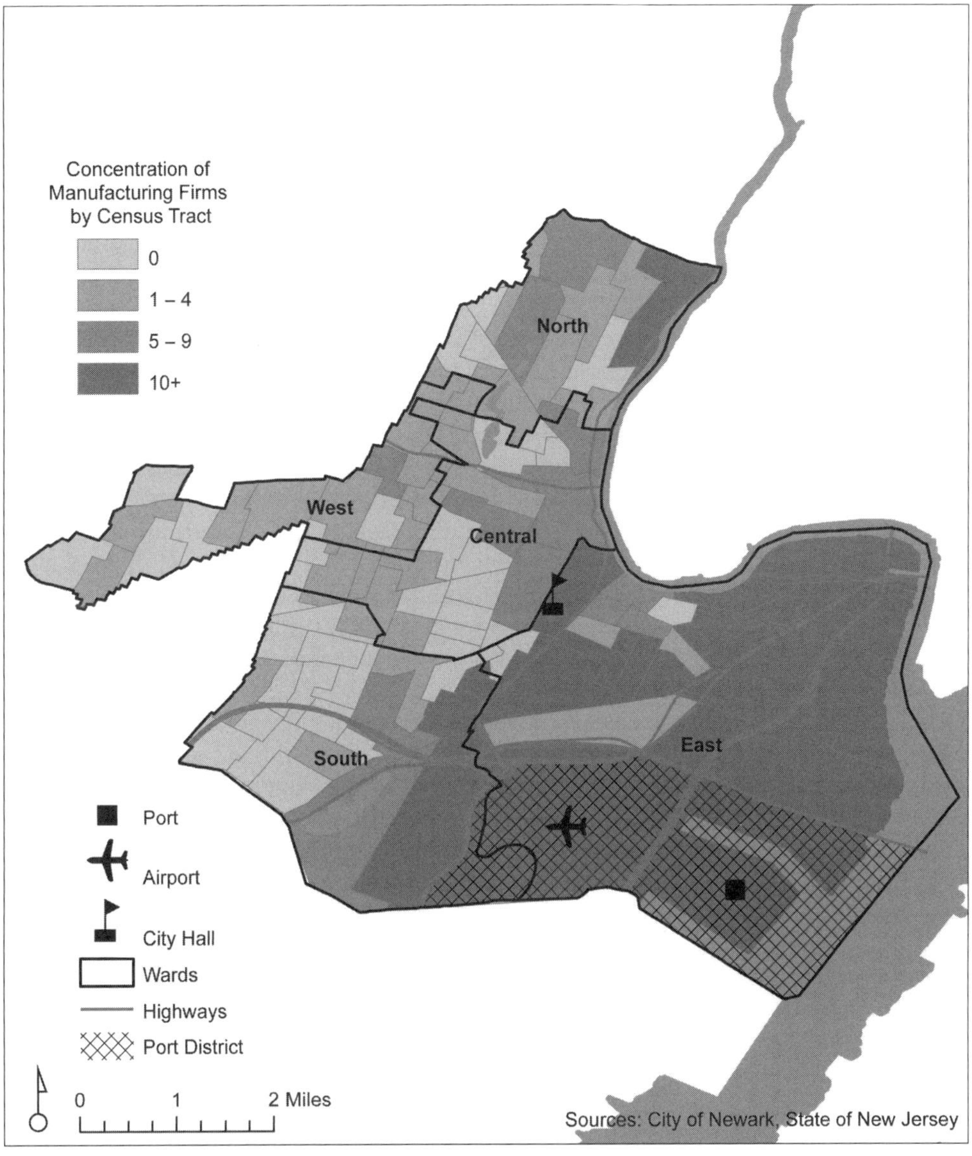

Source: U.S. Census Bureau OnTheMap application.

Figure 26.1 Newark manufacturing firm density

ment, the first comprehensive plan to be developed by the city in two decades, offers insights into shifts in the city's economy and illustrates how these are paralleled in land use patterns. As obsolete industrial buildings and sites have been converted for residential and commercial use, the number of industrial parcels in the city has declined from 1188 in 1998 to 900 in 2010 (Vol. 2, 9). Nevertheless, the plan cites industrial growth as a component of the city's strategy to create 25,000 new jobs by 2025. One explicit goal is to develop more job-dense industrial uses in the port areas; another is to create opportunities for innovation by pairing existing manufacturing businesses and start-up entrepreneurs with technical assistance through area universities (Vol. 1, 19, 248).

A specific look at manufacturing employment and business activity in Newark – as well as a greater sense that manufacturing redevelopment efforts in the city might be part of a broader national trend – comes from the report *Newark's Manufacturing Competitiveness: Findings and Strategies*, published by the Brookings Institution in May 2013 (Mistry 2013). Brookings is a liberal Washington, D.C.-based policy think tank whose Metropolitan Policy Program has had a significant influence on urban and regional policy during the Obama Administration. In 2012, Brookings published two monographs exploring the reversal of employment decline in U.S. manufacturing,[10] tracing the distribution of production activity across metropolitan areas and advocating policies supporting the sector's revitalization (Helper, Krueger and Wial 2012a, 2012b). Brookings researchers in Newark surveyed, interviewed and visited manufacturers as they have done in several other so-called "legacy cities."[11] They also convened leaders from the government and non-profit economic development sectors in the greater Newark area, guiding them toward consensus about the challenges and opportunities inherent in the city's significantly shrunken manufacturing sector. The stakeholders who collaborated on the study also identified priority areas for intervention, and they now intend to work together to implement the study's recommendations.

Newark's Manufacturing Competitiveness: Findings and Strategies begins by detailing Newark's assets and weaknesses as a location for manufacturing activity. On the one hand, the city's location, its airport and seaport, its density and its continued industrial diversity are significant assets. Newark possesses a rich logistics infrastructure that includes Class I, II and III railways, extensive industrial processing, warehousing and distribution capacity, and multi-modal freight and commuter transportation links. Its proximity to consumers (18.9 million of them in the New York/Northern New Jersey Metropolitan Area), to an expanding seaport and to a major international airport positions it to host firms that rely on "next generation" supply chains (Mistry 2013, 23). The expansion of the Panama Canal, along with local infrastructure rehabilitation, promises to double volume at the Port Newark Container Terminal by 2025 (Port Authority of New York and New Jersey 2013; Strunsky 2011).

The report also emphasizes a high concentration of technical talent in New Jersey and a density of educational institutions, some of which are already actively working with industrial firms but with which there is the potential for deeper engagement. Moreover, through an informal network they call "Made in Newark," firms are collaborating to build a joint market identity and promote enthusiasm about and loyalty toward a hometown brand. Twenty-four firms have a presence on the Newark page of the website "Makers Row," which promotes matchmaking between producers, manufacturers sourcing supplies and designers looking to manufacture products (see Schnuer 2013).

Yet Newark's industrial sector remains in decline. Currently, there are 400 manufacturing establishments in the city, employing nearly 10,000 workers and accounting for about 7 percent of the city's jobs. Its still-diverse industrial base includes firms in the food, textiles, metals, printing and furniture sectors. But this base is a shadow of what once existed. The Brookings report emphasizes that there has been little investment by current firms

Table 26.1 Residential locations of people employed in Newark, New Jersey and employment locations of people residing in Newark, New Jersey

	Residency status of Newark workers (134,699) (%)	Job destinations of Newark working residents (81,550) (%)
City of Newark	17	29
Essex County (excluding Newark)	16	14
New York Metropolitan Area (excluding Essex County)	56	52
Outside New York Metropolitan Area	10	5
Total	100	100

Source: U.S. Census Bureau OnTheMap application.

in new product and process development. This is particularly true of the small contract suppliers that prevail in the city. There are few start-up manufacturers testing the viability of new products and technologies, especially in the "advanced manufacturing" area that is the target of most federal policy. Incumbent firms lack the coordination required to optimize supply chain efficiency and integrate environmentally sustainable technologies (Mistry 2013, 28). Despite a highly educated workforce in the region, low educational attainment and skills in much of the city's population make the goal of "Made in Newark by Newark Residents" difficult to reach (Mistry 2013, 33). Meanwhile Newark's population and its job base are mismatched: 71 percent of Newark residents holding jobs work outside the city, while 83 percent of the jobs in the city are held by employees who live elsewhere (see Table 26.1). Labor force participation is low. Employment was 15 percent in 2012 and just 38 percent of households earned enough to be considered economically self-sufficient by the city's definition (City of Newark 2012, 22).

The plan of action articulated in the Brookings document recommends 14 specific interventions to Newark's city government and other regional stakeholders. They include capitalizing on the trend of "near-sourcing" by helping businesses connect to regional market opportunities; leveraging university partnerships to create centers for open innovation and process improvement among small and mid-sized manufacturers; drawing more Newark residents into the manufacturing workforce (and offering employers better-skilled workers) through a variety of recruitment, training and internship programs; raising the city's public profile as a "competitive crossroads for trade" (Mistry 2013, 45–46), a place to site both logistics and production; and redoubling efforts to activate vacant and underutilized industrial parcels, particularly in the port district.

The Brookings study of Newark offers a kind of pivot between the particular situation of Newark and the manufacturing revitalization vision being pursued by the Obama Administration at a national level. Brookings is a national organization whose main agenda in its Metropolitan Policy Program concerns the economic importance of metropolitan regions: urbanized areas defined by labor markets, commuting patterns and business-to-business networks. The purpose of the metropolitan focus is a political one: it provides an identity for metropolitan areas and metropolitan policy in the absence of actual regional governance structures. But when Brookings engages with a local government on an economic development issue of major importance (as it has done here), it is in some sense standing in at the regional scale for a level of governance that exists only as an aspiration in practice. In bridging the city context with the federal one in Newark, the study downplays the real estate challenges that are of the greatest concern to those seeking manufacturing revitalization in the city.

Facts on the Ground

As noted above, *Newark's Manufacturing Competitiveness: Findings and Strategies* connects the condition of and outlook for the sector in the city itself with a national conversation about how to seize a "manufacturing moment." Its strategies – product and process innovation, market expansion, human capital development – are sound. Most of these strategies, however, would be most effective if pursued by a governance entity with a regional scope, operating under the guidance of a national agenda. In contrast, the tools that actors in Newark perceive as being available to effect the trajectory of manufacturing development in the city involve local land use policy and project finance. Officials working with the City of Newark and the affiliated Brick City Development Corporation are responding primarily to problems of how to make parcels suitable for industrial infill development in a city characterized by environmental contamination, multiple industrial landowners (many of whom are absentee) and obsolete buildings. Assembling parcels is difficult in this environment. Owners speculating on future zoning changes quote unrealistic prices for sites the city wants to acquire. Brownfield remediation, even when supported with state grants, adds cost. Unlike in the affordable housing sector, where urban developers are accustomed to projects involving many subsidy sources, gap financing and non-standard building types, industrial brokers and developers are accustomed to a limited range of schematics both financially and with respect to architectural typology (Interview 6).[i]

Viewed in this light, Newark's challenges often seem reducible to spreadsheet math. Space is at a premium in densely built areas with proximity to ports and other transport nodes, and large lots are hard to come by (all but 15 percent of the underutilized parcels in the vicinity of Port Newark-Elizabeth are fewer than five acres in size), so land for sale is more expensive than in outlying areas and may not accommodate the large floor plates required by contemporary producers (Interview 3). Space for rent in Newark, on the other hand, is affordable on average (Table 26.2), but this has the perverse effect of making it difficult to develop structures whose up-front costs can be supported by rent flows. Much of the industrial real estate in the port district of Newark, for instance, is owned by landlords who prefer to rent it to low-end industrial users at $2 to $3 per square foot, earning a profit without investing the capital that would be required to attract Class-A warehousing and manufacturing uses.

Job density figures illustrate the employment impact of this status quo. A comparable port supports 8 to 20 workers per acre of industrial use; the analog in Newark is 5.5 workers per acre. Thus, while Newark has a low industrial vacancy rate, this is large in part because much vacant land and a large number of buildings have not been redeveloped or adapted for the contemporary rental market. What space is available is being rented, but many derelict parcels do not make it into the rental inventory from which

Table 26.2 Newark industrial submarket statistics, year end 2012 (23 submarkets in total)*

Feature	Newark sub-market	In context of Northern New Jersey industrial market (23 submarkets in all)
Number of buildings	1261	7.4% of all buildings
Total rentable built area	44,729,159 square feet	5.5% of total square footage
Vacancy rate	5.7%	Average vacancy rate was 9.0% (Newark's was 2nd lowest)
Year to date net absorption	(57,980) square feet	(Absorption was negative; about 10, or 43%, of the 23 submarkets in the region experienced positive net absorption in 2012)
Year to date deliveries (of new rentable space)	180,000 square feet	18.2% of total new space delivered in region
Square feet under construction	350,000	29.3% of total space under construction in region
Quoted rates (per square foot, average)	$5.35	Average square footage rate was $5.72 (Newark had the 8th least expensive average quoted rate among the 23 submarkets)

Note: *The industrial market in Northern New Jersey encompasses all facilities built for "storing, producing, assembling or distributing product" (CoStar 2013, C). This categorization defies efforts to understand the relative prevalence of warehousing and manufacturing in the city and region's economies.

Source: Adapted by author from CoStar Group 2013.

vacancy rates are calculated. Many owners and even industrial developers do not have a clear understanding of the investments and returns associated with upgrading facilities to contemporary standards for advanced logistics and production (Interview 1, Interview 5).

Moving Forward

Warehouse and distribution facility development is currently the most readily accessible opportunity to effect change, and officials in Newark have made progress in doing so. A July 2012 groundbreaking on a 15 acre site at 60 Lister Avenue in the East Ward (the site of a former Sherwin Williams paint manufacturing facility) marked the first speculative industrial real estate development in the city in several decades (Khavkine, 2012). A second project in the pipeline involves an eight acre site in the port district that is on track to be developed as a mechanized storage facility in anticipation of the Port Newark-Elizabeth expansion. Both of these projects entail environmental remediation, which is being assisted through State of New Jersey brownfield development grants (together with state bonding capacity), and both deals will likely receive equity contributed through the New Jersey Urban Transit Hub Tax Credit (Interview 4).[12]

The prospects for developing or adapting spaces for manufacturing uses are more elusive. Real estate development for contemporary manufacturing may be led by the emergence of the "next generation" supplier firms and innovative start-ups envisioned by the Brookings study. But the experiences of other cities suggest that only where both firms and city governments expressly support the development of land and buildings for manufacturing use, invoking unconventional models of financing and deal-flow, do these firms have a chance of remaining local once they do emerge (Byron and Mistry 2011). The firm most active in advocacy for the manufacturing sector in Newark, UnionWear, is a producer of made-to-order bags, hats and shirts. Mitch Cahn, the company's CEO, says the firm has benefited from the continuous improvement and lean manufacturing training recently embraced by federal policy. He is in Newark, however, because officials there helped him to acquire and renovate a 70,000 square foot building in the North Ward that, while still proximate to an ample labor force, is economically viable for his business in a way that the real estate in Jersey City, his former location, is not. His company is now growing (Interview 2).

Another area of potential for Newark, and one that requires creative real estate strategy to an even greater degree than firms like Cahn's, involves the development or adaptation of spaces for enterprises engaged in small-batch artisanal or DIY technology manufacturing. Enabled by several convergent trends such as anti-establishment entrepreneurialism among young people, open source technology and consumer interest in locally made or sourced products, particularly local food, clothing and furniture, the "Maker Movement" has given rise to demand for urban facilities that combine moderate rents with the potential to share tools and exchange ideas with other producers (see Friedman and Byron 2012). To the extent that start-up manufacturers in this mode access the market connections, technical assistance and university partnerships envisioned in national policy, these resources will be of value. But the fate of such firms in Newark (as well as similar cities) will likely depend more heavily on the availability of adequate spaces at workable rents. In fact, the New York City real estate conditions described by McCormick in this volume (as well of those of Jersey City) could potentially produce a migration of artisanal producers and high-tech start-ups to Newark if real estate can be adapted.

SPEAKING TO CITIES' NEEDS

Practitioners have proposed several concrete strategies for promoting industrial real estate development in problematic urban markets such as Newark's (Byron and Mistry 2011; Pratt Center for Community Development 2013; Urban Manufacturing Alliance 2013). One of these is the branding strategy already being employed by firms participating in the Made in Newark network. Beyond branding, however, lies the ability to affect the real property calculus of manufacturing firms and industrial developers. As Byron and Mistry argue:

> [I]n strong-market cities, because of the speculative nature of real estate, [non-mission-driven] property owners generally perceive that they can receive a higher return if they convert industrial space to other uses. In weak market [i.e. legacy] cities, there is simply inadequate return to attract private investors willing to acquire, renovate, and manage older industrial buildings. (2011, 44)

They conclude that efforts to foster "manufacturing in place" in legacy cities need to involve mission-driven innovation in real estate finance. Most of those who have built and managed affordable urban multi-tenant industrial facilities have, up to this point, been non-profit or mission-driven developers seeking social and community dividends as well as financial return.

Mainstream developers may eventually create a standard, financeable urban multi-tenant industrial product. But for now, urban manufacturing proponents are advocating for a variety of subsidies that can be accessed by mission-driven industrial developers. Similar to the gap financing infrastructure created in the low- and moderate-income housing sector, this might include a recoverable grant or very-low-interest loan program that would provide risk capital to cities or non-profits to finance predevelopment costs on difficult sites (a low-interest loan program might be capitalized by banks as a way of complying with the requirements of the Community Reinvestment Act). It might feature a tax credit designed specifically to provide equity to industrial developers undertaking qualified projects. The rules governing Industrial Revenue Bonds could be revised to allow developers of multi-tenant industrial buildings to borrow in the tax-exempt bond market (Urban Manufacturing Alliance 2013).

Unconventional financing mechanisms need to emerge for many kinds of urban multi-tenant facilities – variously oriented toward artisanal manufacturing, high-tech prototyping or university spinoffs. Some of these facilities might be able to thrive in renovated multi-story buildings[13] while others might require new construction. In any case, they would depart typologically from the one-story greenfield spaces to which most industrial brokers and owners are accustomed. They would also likely be managed by landlords who accept returns on investment lower than would be tolerated by the private market, who encourage tenants to share equipment and ideas, and who flexibly meet the needs of tenants with fluctuating space requirements.

Governance and ownership questions loom large. The participation of non-profit, mission-driven organizations in affordable housing development in the United States has become institutionalized, but the presence of such groups in the industrial market is embryonic. It is unclear whether actors at the city level will create policy conditions under which mission-driven industrial developers survive and thrive (as occurred in the housing sector), and it is unclear what role city-owned land will play in that process. Some advocates suggest that cities, rather than remediating sites to "market-ready" condition and then selling them to developers, should retain them via long-term leases. Another model entails the creation of a distinct, mission-driven entity with a city-appointed board of directors to pursue industrial development goals on city-owned property – either a single property or a portfolio. In the case of the Brooklyn Navy Yard in New York City, an entire complex of buildings, owned by a mission-driven non-profit corporation, has been gradually renovated and repurposed for tenants of varying sizes, with rent proceeds from initial development phases reinvested in additional renovation (Pratt Center for Community Development 2013).

CONCLUSION

Local actors promoting the redevelopment of cities' manufacturing bases construe the problem overly narrowly when they restrict their attention to parcels and property deals. Innovation, university partnerships, market expansion and human capital development – and their manifestations in place-based policy – are also critical to manufacturing revitalization. Thinking beyond property development, creative local economic development officials inevitably will find themselves asking regional and institutional questions. Workforce is of grave importance: while there is a lively controversy about whether those who decry a "skills gap" in manufacturing are overstating the case (see Lowe, this volume), the need to prepare local workforces for contemporary and future manufacturing jobs is undeniable, as is the need to alleviate chronic high unemployment in "legacy cities" like Newark.[14] Workforce policy is an arena where the "missing region" problem is most damaging from the perspective of firms and job-seekers alike (Wolf-Powers 2012). National-level leadership in workforce could make a difference here, overcoming local actors' limited leverage and range of motion and catalyzing new experiments in skill definition and acquisition for the manufacturing labor force.

National manufacturing policy, as it moves away from its de-territorialized past, must take seriously the real estate barriers local officials face as they try to create conditions under which manufacturers will choose central city locations in the coming years. At the same time, local actors must look beyond property issues and respond as fully as they can to national policy priorities with the tools available. A mutual effort to coordinate place-based urban policy with national policy in manufacturing may yield dividends in the form of multi-tenant urban industrial spaces, place-based innovation in which urban density is a driving factor, and workers with the skills to perform production jobs that draw on new innovative capacity. In the long term, the effort could minimize the disconnect between scales of action (national and local) in economic development policymaking, perhaps developing viable regional structures that could take more responsibility for economic policy in the future.

NOTES

1. See Cortwright (2006), Motoyama (2008) and Wolf-Powers (2012) for perspectives on the challenge of operationalizing regional economic analysis, such as the analysis of industry clusters, into policy interventions that have meaningful impact.
2. Filling out the research and commercialization infrastructure that the German state makes available to its manufacturers is a long-standing commitment to production skills training and a strong interface between manufacturing employers and the country's extensive apprenticeship and career preparation systems (Shapira, Youtie and Kay 2011).
3. The exception to this lies with defense industries, where investments made in service of enhanced military capability, especially during the Cold War, wrought huge changes in the nation's industrial geography (Hooks 2002; Markusen, Hall, Campbell and Dietrick 1991).
4. Manufacturing modernization policies were first pursued by individual state governments in the 1980s (Feller 1992, 1997). In the early 1990s officials in the Administration of President Bill Clinton created the Manufacturing Extension Partnership as part of an effort to promote economic development by commercializing defense-bred technologies (Shapira 2008).
5. The National Additive Manufacturing Innovation Institute, a research hub centered on the use of 3-D printing to create products and components, opened in Youngstown, Ohio in August 2012. A competition to choose three other such consortia was announced in May 2013.
6. According to the Economic Development Administration's May 9, 2013 press release for the IMCP, regions and jurisdictions successful in obtaining IMCP grants would be empowered to spend federal funding on "specialized research centers at local universities; business incubators focused on targeted technology sectors; community college programs to train workers in targeted industries; public works projects to upgrade infrastructure or enhance energy efficiency; viable export promotion plans; well-

integrated supply chains; and an engaged community of local government, education, workforce, and business leaders" (U.S. Department of Commerce 2013b).

7. The first of the National Network of Manufacturing Innovation Institutes (the Youngstown facility and three others) and the pilot communities for the Investing in Manufacturing Communities Partnership are both funded from the existing budgets of federal government agencies. To expand these programs, the Obama Administration will need to rely on budget authority from Congress, which at the time of writing seems a slim prospect.
8. Redlining was a practice by which banks and the government agencies that insured mortgages excluded entire neighborhoods – often minority neighborhoods – from consideration for home purchase and improvement loans (see Jackson 1985).
9. Of a total of 150,000 workers in Newark, according to the plan, 18,000 work in the industrial sector and around 8000 in manufacturing. The Brookings study uses data from Moody's Analytics, which uncovers more manufacturing establishments and more jobs than the New Jersey Department of Labor's Quarterly Census of Employment and Wages. Thus Newark's manufacturing sector numbers are slightly higher than those cited in the Master Plan.
10. After a decade of decline, manufacturing employment in the United States began to edge up in 2009. Manufacturing jobs increased by 4.5 percent between December 2009 and March 2013, on par with the growth rate of all nonfarm employment. Given a number of trends, including revived demand for machinery in which U.S.-based manufacturers are specialized and changes in prices and wages that favor "on-shoring" of work previously done in other countries, some analysts assert that this small revival is the beginning of a longer-term trend.
11. "Legacy cities" is a term in use in the U.S. for cities that experienced dramatic job and population decline in the 1970s and 1980s, largely as a result of the loss of industrial jobs, and have not recovered. They are typically characterized by high unemployment, high poverty and fiscal stress (Mallach and Brachman 2013).

i. See Appendix for details of interviews.

12. According to a Newark economic development specialist, requirements for site acquisition, preparation and remediation on a typical industrial site in Newark introduce a 20 to 50 percent financing gap (Interview 5).
13. "Maker spaces" and kindred facilities often are adapted from older, multi-story industrial buildings originally designed for use by single firms. Features rendering these buildings obsolete as sites for standard industrial production, such as small floor plates, large windows and load-bearing columns, become aesthetic assets and create desirable work environments.
14. As Lowe argues, there is also a need for labor market intermediaries to participate in shaping the requirements of and training for those jobs.

REFERENCES

Byron, J. and N. Mistry (2011), *The Federal Role in Supporting Urban Manufacturing*, Washington, D.C.: Brookings Institution and Brooklyn, NY: Pratt Center for Community Development.

City of Newark (2012), *Newark's Master Plan: Our City, Our Future*, Newark, NJ: City of Newark.

Clark, J. (2010), "Coordinating a Conscious Geography: The Role of Research Centers in Multi-scalar Innovation Policy and Economic Development in the US and Canada," *Journal of Technology Transfer*, **35** (5), 460–474.

Clark, J. (2012), "Is There a Progressive Approach to Innovation Policy?" *Progressive Planning*, **Winter** (190), 17–20.

Clark, J. (2013), *Working Regions; Reconnecting Innovation and Production in the Knowledge Economy*, London: Routledge.

Cortwright, J. (2006), *Making Sense of Clusters: Regional Competitiveness and Economic Development*, Washington, DC: Brookings Institution Report, available at: http://www.brookings.edu/research/reports/2006/03/cities-cortright (accessed June 5, 2013).

CoStar Group, Inc. (2013), *The Co-Star Industrial Report: Year End 2012, Northern New Jersey Industrial Market*, Washington, D.C.: CoStar Group, Inc.

Cunningham, J.T. (1988), *Newark: Revised and Expanded Edition*, Newark, NJ: New Jersey Historical Society.

Ezell, S.J. and R.D. Atkinson (2011), "International Benchmarking of Countries' Policies and Programs Supporting SME Manufacturers," available at http://www.nist.gov/mep/ upload/ International-Benchmarking-of-Countries-SME-Support-Programs-and-Policies-2.pdf (accessed June 18, 2013).

Feller, I. (1992), "American State Governments as Models for National Science Policy," *Journal of Policy Analysis and Management*, **11** (2), 288–309.

Feller, I. (1997), "Manufacturing Technology Centers as Components of Regional Technology Infrastructures," *Regional Science and Urban Economics*, **27**, 181–197.

Friedman, A. and J. Byron (2012), "High-Tech, High-Touch, and Manufacturing's Bottom Line," *Innovations*, **7** (3), 83–95.

Helper, S., T. Kruger and H. Wial (2012a), *Why Does Manufacturing Matter? Which Manufacturing Matters?*, Washington, D.C.: Metropolitan Policy Program at Brookings.

Helper, S., T. Krueger and H. Wial (2012b), *Locating American Manufacturing: Trends in the Geography of Production*, Washington, D.C.: Metropolitan Policy Program at Brookings.

Hooks, G. (2002), "Guns and Butter, North and South: The Federal Contribution to Manufacturing Growth, 1940–1990," in P. Scranton (ed.), *The Second Wave: Southern Industrialization, 1940–1970*, Atlanta: Georgia Technological Institute Press, pp. 255–285.

Jackson, K.T. (1985), *Crabgrass Frontier: The Suburbanization of the United States*, New York and London: Oxford University Press.

Khavkine, R. (2012), "Newark Groundbreaking for Facility Expected to Bring in 200 New Jobs", available at http://www.nj.com/news/index.ssf/2012/07/newark_groundbreaking_ for_faci.html (accessed June 17, 2013).

Kitagawa, F. (2007), "The Regionalization of Science and Innovation Governance in Japan?", *Regional Studies*, **41** (8), 1099–1114.

Koschatzky, K. and H. Kroll (2007), "Which Side of the Coin? The Regional Governance of Science and Innovation," *Regional Studies*, **41** (8), 1115–1127.

Leigh, N.G. and N. Hoelzl (2012), "Smart Growth's Blind Side," *Journal of the American Planning Association*, **78** (1), 87–103.

Mallach, A. and L. Brachman (2013), "Regenerating America's Legacy Cities," available at https://www.lincolninst.edu/pubs/dl/2215_1582_Regenerating_Americas_Legacy_ Cities.pdf (accessed June 24, 2013).

Markusen, A., P. Hall, S. Campbell and S. Dietrick (1991), *The Rise of the Gunbelt: The Military Re-mapping of Industrial America*, New York and London: Oxford University Press.

Mistry, N. (2013), "Newark's Manufacturing Competitiveness: Findings and Strategies," available at http://www.brookings.edu/research/reports/2013/05/28-newark-manufacturing-mistry-vey-shearer (accessed June 15, 2013).

Motoyama, Y. (2008), "What Was New About the Cluster Theory? What Could It Answer and What Could It Not Answer?" *Economic Development Quarterly*, **22** (4), 353–363.

Mumford, K. (2007), *Newark: A History of Race, Rights and Riots in America*, New York and London: New York University Press.

Muro, M. and J. Lee (2013), "Revving Up Manufacturing, Region by Region," available at http://www.brookings.edu/blogs/the-avenue/posts/2013/04/11-budget-manufacturing-muro-lee (accessed June 16, 2013).

Newman, K. (2004), "Newark, Decline and Avoidance, Renaissance and Desire: From Disinvestment to Reinvestment," *Annals of the American Academy of Political and Social Sciences*, **594** (July), 34–48.

Perry, B. and T. May (2007), "Governance, Science Policy and Regions: An Introduction," *Regional Studies*, **41** (8), 1039–1050.

Port Authority of New York and New Jersey (2013), "Bayonne Bridge Navigational Clearance Program," available at http://www.panynj.gov/bayonnebridge/ (accessed June 15, 2013).

Pratt Center for Community Development (2013), "Brooklyn Navy Yard: An Analysis of Its Economic Impact," available at http://prattcenter.net/report/brooklyn-navy-yard-analysis-its-economic-impact (accessed June 18, 2013).

Schnuer, J. (2013), "Makers Row Ushers in a New Wave of U.S. Manufacturing," available at http://www.entrepreneur.com/article/226869 (accessed June 24, 2013).

Shapira, P. (2008), "Putting Innovation in Place: Policy Strategies for Industrial Services, Regional Clusters, and Manufacturing SMEs in Japan and the United States," *Prometheus*, **26** (1), 69–87.

Shapira, P., J. Youtie and L. Kay (2011), "Building Capabilities for Innovation in SMEs: Across-Country Comparison of Technology Extension, Policies and Programmes," *International Journal of Innovation and Regional Development*, **3** (3/4), 254–272.

Strom, E. (2002), "Let's Put on a Show! Performing Arts and Urban Revitalization in Newark, NJ," *Journal of Urban Affairs*, **21** (4), 423–435.

Strunsky, S. (2011), "Port Newark Terminal Lease Deal to Double Volume," available at http://www.nj.com/business/index.ssf/2011/06/port_newark_terminal_lease_dea.html (accessed June 16, 2013).

U.S. Department of Commerce, Economic Development Administration, Office of Public Affairs (2013a), "Fact Sheet: The Investing in Manufacturing Communities Partnership," available at http://www.commerce.gov/news/fact-sheets/2013/04/17/fact-sheet-investing-manufacturing-communities-partnership (accessed May 5, 2013).

U.S. Department of Commerce, Economic Development Administration, Office of Public Affairs (2013b), "Obama Administration Seeks Applicants for First Phase of 'Investing in Manufacturing Communities'

Partnership," available at http://www.eda.gov/news/press-releases/2013/05/09/obama_imcp.htm (accessed June 5, 2013).
Urban Manufacturing Alliance (2013), "Policy Brief: Industrial Revenue Bonds," available at http://urbanmfg.org/wp-content/uploads/2013/04/UMA-IRB-Policy-Brief-Final.pdf. (accessed June 7, 2013).
Wolf-Powers, L. (2012), "Human Capital-Centered Regionalism in Economic Development: A Case of Analytics Outpacing Institutions?" *Urban Studies*, **49** (15), 3427–3446.

APPENDIX: INTERVIEWS

Interview	Affiliation	Role	Date	Location
1	Mayor's Office & Dept. of Economic and Housing Development, City of Newark	Fellow and Manufacturing Advisor	May 6, 2013	Newark, NJ
2	UnionWear, Inc.	CEO	May 6, 2013	Newark, NJ
3	Urban Enterprise Zone Office	Administrator	May 17, 2013	Newark, NJ
4	Brick City Development Corporation	Senior Vice-President of Real Estate	May 17, 2013	Newark, NJ
5	Brick City Development Corporation	Director, Real Estate Development and Business Attraction	May 17, 2013	Newark, NJ
6	CoStar Group, Inc.	Commercial Real Estate Manager	June 18, 2013	Washington, DC (by phone)

27. The city and industry: deurbanizing manufacturing in New York City?

Lynn McCormick

INTRODUCTION

In the 1950s, New York City was the largest manufacturing hub in the United States, employing over one million workers in this sector (Liff, 1997). Today, manufacturers in the city employ about 75,000 people, or less than 3 percent of all local workers (Quarterly Census of Employment and Wages (QCEW), 2011). Planners and policymakers face a question of whether to attempt to retain the manufacturing that remains or let it go and foster service sector expansion instead. This mirrors a larger debate in the United States as a whole. Is manufacturing an obsolete activity in a mature economy, one that is more competitive in, and appropriate for, lower-wage and relatively underdeveloped countries? Or, should US workers retain the manufacturing skills that are left and upgrade them with advanced technologies?

Public officials and other stakeholders in New York City are energetically debating these questions today – whether to retain manufacturing or not. They are presenting data and taking political action to support their position on what to do with manufacturing (e.g., see Pratt, 2008; Municipal Art Society (MAS), 2011; Design Trust for Public Spaces for the Council of Fashion Designers of America (DT), 2012 on policy recommendations to strengthen manufacturing; and Armstrong and Lund, 2005, for arguments to rezone industrial areas). This chapter offers a history of these recent events. Although City administrations have been trying to oust manufacturers for many decades – to enable property owners to secure "the highest and best" return from their assets – others in the policymaking environment have spoken up for manufacturers' interests with some success. However, given the hyper-competitive real estate market in this city, manufacturing advocates will have to continue to pressure future administrations to listen to manufacturers' concerns.

After first highlighting what scholars and others say in regard to the future of manufacturing in the United States, I offer some statistics on the manufacturing activity in New York City that remains. Next occurs an accounting of political actors' points of view and tactics to either move manufacturing out of the city or retain it with policy interventions. The last section offers some details as to what challenges New York's urban manufacturers in two sectors – apparel and furniture – face and strategies in which they are engaging to get policymakers' attention. I conclude with lessons that can be drawn from this case.

MANUFACTURING RETENTION? WHY?

The debate on whether to retain manufacturing firms in our reportedly post-industrial era centers on whether one thinks that manufacturing has little importance for advanced economies or can still be competitive in such economies and on the world stage. This debate started in the 1970s, when major manufacturing job losses occurred in many regions of the United States. The country was deindustrializing by sending manufacturing jobs overseas to lower labor cost countries. Several prominent publications in this period initiated a debate on what to do about it, which continues today. For example, Daniel Bell in *The Coming of the Post Industrial Society* (1976) argued that this transition did not matter so much in that we in the United States were substituting higher skill-oriented service sector jobs for the manufacturing ones that we gave up. Others, particularly Stephen Cohen and John Zysman in *Manufacturing Matters* (1987), responded by saying we weren't really losing manufacturing itself, just manufacturing jobs as US firms introduced new technologies (e.g., robots) that replaced workers but also increased manufacturing productivity. Hence, manufacturing still deserved our attention. Others who studied manufacturing in the United States and other advanced economies wrote about industrial clusters of small firms that agglomerated together in a particular region, freely sharing information about the work process and manufacturing challenges, which allowed them, as a group, to innovate faster and compete better in international markets (e.g., Piore and Sabel, 1986; Rosenfeld, 2002). Policies to strengthen local manufacturing clusters in different regions have resulted.

Coming with a different set of concerns, scholars of global cities argue that such places will increasingly only house finance and key service activities to help control global business ventures (Friedmann, 1986; Sassen, 1991, 2005; Hill and Kim, 2000). Multinational corporations operating in a newly integrated global economy, with subsidiaries and production sites in many countries, now require a concentration of key business control and command functions in a concentrated place. New York City is a key prototype. It has international finance and investment companies and related institutions like the stock exchange. It houses major international law, accounting and advertising firms. International media – Bloomberg News, as an example – help drive the information flow that capitalism depends upon, as do other Internet and "dot-com" start-ups. Manufacturing supposedly has no place in such central nodes of business services. It is pushed out in the competition for space that occurs between the various growing business command activities that can pay a higher rent. This results in a situation where the real estate industry, which absorbs the excess investment capital of the rich, controls the urban development process and private property rights reign supreme (Angotti, 2008; Harvey, 2008).

Scholars, activists, progressive economic development practitioners and some manufacturers continue to point out that loss of manufacturing firms and jobs should be avoided since these jobs are often higher-paying than many service sector jobs, unionized, and an opportunity for lower-skilled workers to advance into the middle class. They point out that the future of manufacturing in the United States lies with products that utilize advanced technologies, are design-intensive and offer new innovations needed in global markets; this focus offers manufacturers a competitive "high road" for their operations (Helper, 2008; Hart, Ezell and Atkinson, 2012; National Council for Advanced Manufacturing, n.d.). Manufacturing firms have always been thought of as export-oriented and, therefore, a way for the US, certain cities and neighborhoods to bring in wealth from outside purchasers. The Obama administration has put forth a plan for reenergizing manufacturing that touches on these same themes (Obama, 2012; LoGiurato, 2013).

BACKGROUND: MANUFACTURING IN NEW YORK CITY

The question of what to do with manufacturing has been considered by New York City policymakers since at least the late 1920s, when the Regional Plan Association (RPA) recommended it be moved out of Manhattan to the outer boroughs (Schwartz, 1983; Fishman, 1997). The RPA's first report recognized that manufacturing created much of the wealth of the region, but that it was also still looked on as "blight" and should be moved outward, keeping Manhattan for offices and retail. Later Planning Commissions in the City used Title I, urban renewal funds, to accomplish this displacement in the post-World War period (Schwartz, 1983).

Nevertheless, New York City in the 1950s was the largest manufacturing center in the United States, in terms of employment. It experienced significant manufacturing decline in the following decades, along with other Rustbelt regions, but at a faster pace (Bram and Anderson, 2001). From 1969 to 1999, manufacturing jobs in New York State declined by 52 percent, and in next-door New Jersey by 48 percent – the states with the largest declines in the country. New York City fared even worse during this time period, with a 68 percent loss of manufacturing jobs, showing that a disproportionate share of the state's manufacturing job losses came from the City. Through statistical modeling, Bram and Anderson (2001) attribute about half of the loss in New York State to the outward dispersal of manufacturing companies to other parts of the country, with New York City contributing about 40 percent of the state's decline (see also Drennan, 1991).

Today, manufacturing, at 2.4 percent of all New York City jobs, is among the smallest employing sectors in the local economy (QCEW, 2011). However, it still persists and provides almost 75,000 jobs, with the bulk spread fairly evenly between the boroughs of Manhattan (at 35 percent of all manufacturing jobs), Queens (29 percent) and Brooklyn (26 percent) (QCEW, 2011). The largest manufacturing employer, in terms of sector, is still apparel, as it has been for decades, at 22 percent of all manufacturing jobs in the city (QCEW, 2011). The city's share of apparel jobs is still almost four times greater than the share of apparel jobs in the country (with a location quotient of 3.7). Other large job-producing manufacturing sectors include food, printing, metal and furniture. To what extent are manufacturing job losses and outward migration due to the need to produce in other lower-cost regions? To what extent can dispersal result from active policy that encourages manufacturing removal or displacement?

BI-POLAR POLICY STANCE

The conflictual, or bi-polar, policy stance toward manufacturing in New York City that emerged in the 1920s continues today, even though the City has enacted an explicit industrial policy for probably the first time in its history. Public officials talk frequently about the need to shore up New York City's role as a leader in the global economy. Certain stakeholders (notably the real estate industry) push the City to initiate large development projects so that land values continue rising along with their profits. At the same time, those constrained by these rising land values – including small manufacturers – push for their needs to be heard. This section lays out the tug and pull of those involved in this struggle and the policies, programs and tactics they support and use.

The City's Pro-growth Agenda

Early on, policymakers and scholars debated the importance of manufacturing to the region's economy. The RPA, having decided in the 1920s that industrial activities should be moved out of Manhattan, continued to discuss the issue in its Metropolitan Regional Study that was released in the 1950s. The RPA was able to get the economist and regional scientist, Raymond Vernon, to carry out the study in order to determine the future locational requirements of various economic activities to inform planning by big businesses. The report highlighted the growing financial and business services sectors and concluded that the outmigration of manufacturing was an inevitable and positive change. At the same time, another economist, John Griffin, produced a study of industrial plant location that argued in support of "the 'employment-creating' character of industrial plants" and criticized the loss of 330 such employers between 1947 and 1954 (Schwartz, 1983, p. 252). He identified the expansion of housing into industrial neighborhoods as causing this loss and wanted the City to create industrial districts, reserved exclusively for manufacturing. Vernon's voice won out and became the urban agenda for the Ford Foundation, along with city and regional policymakers of the period (Schwartz, 1983).

Much of recent City policy follows along the lines of Vernon, advocating for explicit municipal support for the command and control functions represented by sectors like finance and real estate. Both Rudy Giuliani, mayor of New York City from 1994 to 2001, and his successor, Michael Bloomberg, have been seen as favoring zoning policies that allow for the expansion of housing, offices and high-end retail over protection for manufacturing (Kotkin, 1998; Barbanel, 2004a; Crean, 2011). Giuliani dealt with the impending outmigration of large banks and other companies to a revitalized Jersey City across the Hudson River by offering sizeable financial incentives to those who threatened to move (Kotkin, 1998; Bagli, 2000). Manufacturers, especially smaller ones, felt their needs were ignored but Joseph Rose, chair of the City Planning Commission, responded by saying: "Our goal is to do nothing that displaces viable manufacturing districts, but . . . we can't tell a landlord what he can do with a commercial lease' (quoted in Kotkin, 1998, p. 1). Rose thought that lost manufacturing jobs could be replaced by new ones in retail, tourism and the "new media" sectors (Kotkin, 1998).

Other actors – outside City Hall – also see the financial and services sectors (versus manufacturing) as the key to the city's future. In the late 1990s, New York City's economy began to boom, which brought attention to the need for more office space. New York State Senator Charles Schumer convened a high-profile task force – the Group of Thirty-Five – to study the problem. It concluded that the city would need to add 60 million new square feet of office space by 2020 to accommodate the projected job growth. Media companies, "dot-coms" and other new technology businesses in particular were flooding

New York City. A New York Media Industry Survey found that 45 percent of all companies in the metro region in that industry were located in New York City. The Schumer report targeted downtown Brooklyn, Long Island City in Queens and the Far West Side of Manhattan – areas that were zoned for manufacturing – as ideal sites for the new offices (Bagli, 2000; Rice, 2000; Trebay, 2000; Wolfson, 2001).

Since 2001, Mayor Bloomberg has proudly pointed to his efforts to rezone much of the city, often industrial land, to accommodate new population growth and commercial economic activities (Barbanel, 2004a). The City Planning Department's website boasts that his administration has undertaken the "largest rezoning agenda since 1961 – now comprising 119 rezonings covering more than 11,000 blocks" to meet the space challenge it foresees for the future (New York City Department of City Planning, 2013, p. 1). In particular, the administration sought to fit in a projected one million new residents by 2030, by adding residential units wherever possible (Roberts, 2006). A recent study of the rezonings concluded most upzonings, to allow for greater density, occurred in lower-income neighborhoods with Latino and Black residents. Wealthier white neighborhoods were more likely to be downzoned to protect their current low-density character. Some contend that the latter were the key voting block for the mayor; hence, the political "haves" got what they wanted (Armstrong, Been, Madar and McDonnell, 2010; Fahim, 2010). In this rezoning environment, manufacturing has always been a weak voice in the city for its interests (Wolf-Powers, 2005).

Advocates for Manufacturing

At the same time that City Hall and other actors seem to be actively working to encourage the displacement of manufacturing from hot real estate markets in the city, advocates for manufacturing – much like Griffin earlier – emerge at times to change the focus of the policy debate. In 1987, for example, when faced with significant loss of garment manufacturers, UNITE, the garment union, persuaded Mayor Koch and the City to enact a protective zoning regulation in Midtown Manhattan for the garment district (Conway and Loker, 1999). Just prior to this, the City joined with garment industry manufacturing trade associations and the union to establish the Garment Industry Development Corporation (GIDC). GIDC has offered employee and managerial training programs, technical/business consulting and marketing services to manufacturers (Conway and Loker, 1999).

More recently, several other stakeholders have strongly advocated for the needs of manufacturers that felt they faced weak City government support. These advocates continue to point out manufacturing's advantages for the city – providing relatively highly paid jobs for lower-educated workers, meeting the input needs of those in more fashionable sectors seen as the key drivers of the local economy (i.e., finance, design and culture, tourism), and offering some protection from downturns in the finance industry by diversifying the economy. They have also argued for relief from manufacturers' key problem – rising rents and dislocation due to upscale real estate developments and gentrification. Some of this effort has been effective.

More specifically, work done by the staff at the Center for an Urban Future, the New York Industrial Retention Network (NYIRN), the MAS and the Pratt Institute Center for Community and Environmental Development (PICCED) resulted in a series of reports arguing for manufacturing land use protection. The Center for an Urban Future, a progressive economic think tank, called for "industrial sanctuaries" for manufacturers that would offer long-term stability in their locations (Bowles, 2000). Discussions between NYIRN, MAS and PICCED staff resulted in a study of the challenges facing manufacturers in eight major industrial neighborhoods in the city. This resulted in recommendations for manufacturing zones that would protect manufacturing property from residential and commercial conversions and guarantee targeted City services and infrastructure investments (PICCED/MAS, 2001). NYIRN has continued to advocate for the needs of, and provide research and technical assistance services to, manufacturers.

A Compromise of Sorts

Most certainly due to this pressure, City Hall enacted a manufacturing retention policy in 2005 that many applaud, but see as insufficient and having weakened over time. This policy drew inspiration, in part, from Chicago's Planned Manufacturing Districts (PMDs) that have provided zoning protections and other incentives to centrally located manufacturing firms since 1988 (Rast, 2005). PMDs allow manufacturers (e.g., printers, prototype or smaller niche companies) to remain in a central location that they require to adequately serve their nearby customers. Mayor Bloomberg installed New York's Industrial Business Zone (IBZ) program in 16 areas throughout the city, with significant concentrations of manufacturing. Similar to Chicago's PMDs, the IBZs restrict land uses in the zones to manufacturing, industrial and other commercial uses, prohibiting residential development and encroachment.

The mayor's industrial policy came out of a consultant study that uncovered similar issues of land use displacement and escalating rents (City of New York, 2005). The study's researchers found that real estate speculation, scarcity and high costs were deterring manufacturers from staying in the city. While some manufacturers (renters) worried about displacement due to the hyper-competitive real estate market, others that owned industrial buildings held their property vacant, hoping to cash in on its increased value should the area eventually be converted. Manufacturers also complained of frequent ticketing by the City for parking, dumpster and traffic violations, which they felt indicated the City's desire for them to move.

The City's industrial policy laid out solutions for each of these complaints. It established an Office of Industrial and Manufacturing Businesses to coordinate policies and programs impacting industrial companies in various other City departments. The IBZs are marketed to industrial businesses by offering tax credits to those relocating into an IBZ. Each IBZ area is served by a nonprofit local development corporation, which contracts with the City to deliver technical assistance services, including access to the NYC Business Solutions program and other City business initiatives. Illegal or inappropriate conversions within IBZs were to be discouraged by enforcing land use restrictions (City of New York, 2005; Failor, 2009).

But, although manufacturing retention advocates approved these measures, they also noted the weakening of the IBZ office over time and feared its elimination after Bloomberg left office. Initially located within the City's Department of Small Business Services, the IBZ office had its own director and staff. Several years later, the director left and a single staff member oversaw the program. Funding to IBZ organizations – those providing business services in each IBZ neighborhood – has also been severely cut since the program's 2005 establishment (Failor, 2009; Crean, 2011).

More recently, the City commissioned another study to determine whether the IBZ program and other industrial initiatives were still warranted (New York City Economic Development Corporation (NYCEDC), 2011). Deciding that they were, the IBZ program was transferred to an Industrial Desk within the NYCEDC, a quasi-public entity in charge of economic development and real estate projects in the city. However, this desk's two staff were hampered in what they could accomplish, given that the Mayor was soon stepping down. Because the IBZ program has no legislative protection, it could easily disappear in the next administration (Failor, 2009; Crean, 2011). By contrast, Chicago's PMDs have legislative approval and appear in the zoning code (Rast, 2005).

In spite of the IBZs, much industrial land has been lost to industry over the years. NYIRN estimates that 23.4 million square feet of industrial space was lost to manufacturers and other industrial companies from 2001 to 2008 due to approved property zoning variances for residential uses near or in industrial neighborhoods (NYIRN, 2004). At the same time, the city experienced a 42 percent decline in manufacturing jobs – much worse than that for the country as a whole (at a 25 percent decline) (Crean, 2011). Besides the contribution of global competition to manufacturing job losses, NYIRN staff contended that other contributing factors included the amount of illegal conversions occurring inside the IBZ zones proper and legal conversions to other commercial uses. Illegal conversions to residential uses occur because the City does not always enforce existing zoning regulations. Other commercial companies can also legally rent or develop space inside the IBZs. This has resulted in industrial tenant displacement, as big box stores (e.g., Home Depot), hotels, bowling alleys, nightclubs, supermarkets, office condos and other commercial uses have moved in because they are able to pay for higher land prices (Failor, 2009; NYIRN, 2009; Crean, 2011). In addition, some criticize the City as giving variances too freely to developers in industrial zoned areas outside the IBZs to convert to other uses (e.g., residential or commercial). This acts to price out industrial tenants over time (Hevisi, 2001; Barbanel, 2004b).

While manufacturers, wholesalers and other industrial companies wait to see what will happen after a new mayor takes office in 2014, PICCED has been building alliances with other similar groups in urban areas throughout the country. Along with SFMade, a nonprofit formed in 2010 to help develop local manufacturing capacity in San Francisco (SFMade, 2012), PICCED organized the Urban Manufacturing Alliance in 2012 to bring together other organizations and stakeholders focused on strengthening urban manufacturing throughout the country. These stakeholders – including, for example, the Allentown Economic Development Corporation, the Illinois Manufacturing Extension Center, the Boston Redevelopment Authority and others – are developing a platform of common challenges that urban manufacturers face and best practices for solving them (Pratt, 2012). Part of the Alliance stakeholders' agenda is to move beyond local governmental lobbying and engage Federal agencies for support. In a joint study with the Brookings Institution, PICCED researchers describe the unique characteristics of urban manufacturers today and recommend Federal actions to support them. Urban manufacturers are more likely very small, produce custom or low-volume items that can command a higher-than-average price, and require proximity to designers and buyers with whom they discuss product design changes frequently (Mistry and Byron, 2011).

SPECIFIC SECTORAL RESPONSES

How has this tug-of-war affected specific manufacturing sectors? I profile the furniture and apparel industries next.[1] To what extent do manufacturers in these sectors need to reside within New York City? Is City policy helping to strengthen their operations? Or, is the City stimulating their exodus by encouraging large real estate developments and the expansion of finance, banking and other service sectors?

The Garment Industry

The fight for space for apparel manufacturers has been a longstanding issue. The 1987 protective garment manufacturing district is still standing. However, there are many stakeholders who want to eliminate it. Although the garment manufacturing industry is only a fraction of its former size, the manufacturers that remain – those that sew clothing, cut the cloth and design the patterns – are critical to TV's *Project Runway* and the fashion industry that it represents.

Fashion is something Mayor Bloomberg has wholly supported during his administration (NYCEDC, 2013). The fashion industry in all of New York City employed 173,000 workers in 2010, including all of the industry's component businesses and manufacturing (DT, 2012, p. 16). But City policy has been undermining it by pushing apparel producers out of Manhattan. Although the protective zoning in the garment district prohibits building conversions on side streets to uses other than manufacturing, some argue that the City never enforced this, allowing conversions to happen anyway (Rice, 2000). After many years of trying to attract these other tenants, local property owners and the Fashion Center Business Improvement District (FCBID) that represents them petitioned the City in 2007 to end these zoning protections so they can rent to higher-paying commercial tenants (FCBID, 2013a; FCBID, 2013b).

Afterward, the City's Economic Development Corporation presented a new proposal to allow other uses in the district and approached the Council of Fashion Designers of America (CFDA) to secure its support for the rezoning (Rice, 2000; FCBID, 2013a). This, however, catalyzed a campaign in opposition. The CFDA wants to retain manufacturers in the district. It started a Save the Garment Center organizing campaign and the 'Made in Midtown' study, with the research assistance of a nonprofit, the Design Trust for Public Spaces. This project has documented the extent of manufacturers remaining in the district and the disastrous economic effects that wholesale conversion would produce. Garment designers argue that they, retailers and others at the end of the fashion production chain critically depend on having manufacturers (sewers, cutters, patternmakers) nearby. The newer, start-up designers are especially dependent on the fashion-related cluster.

As Nanette Lepore, a designer, explains on the Made in Midtown website:

> I couldn't have started my business if it wasn't for the New York City Garment District . . . If you start out like I did, out of my apartment, I could sell 20 of something. I would go see the buyer at Barneys every month and show her a new dress, and she'd order 50. And I could take it to a factory around the corner.

Today, although she has a much larger firm, she says she makes 80 percent of her products within a ten-block radius from her office in the garment district (Madeinmidtown, n.d.).

The City tried to broker a solution that would put all of the apparel manufacturers in one building, operated by a nonprofit organization (Bagli, 2009; FCBID, 2013a). But this project would protect space for, at most, half the manufacturers currently operating in the garment district. The CFDA and the Design Trust are recommending another space solution in a recent report of area redevelopment options (DT, 2012). They propose that the City use existing incentives in the zoning code, as have been successfully used to stop theater displacement in another Midtown neighborhood, to encourage property owners to voluntary retain and improve space over the long run for manufacturing and related uses. Property owners doing this would be able to sell development rights they are foregoing to other property owners in the district that can build higher elsewhere in Midtown.

A basic purpose of the Design Trust and CDFA's report, and others before it, is also to educate a wider audience about the longstanding and complex business ecology that the apparel industry has built over many decades in the Midtown area (Crean, 2003; MAS, 2011; DT, 2012). It comprises a cluster of fashion-related businesses, all of which need each of the components there to be successful. These include clothing "designers, schools [e.g., the Fashion Institute of Technology], trade magazines and newspapers, fabric and button suppliers, department stores and boutiques [i.e., buyers of designers' garments], shoppers [who are drawn to the area for high-end shopping], and garment manufacturers" (DT, 2012, p. 3). Because the cluster in Midtown consists of all elements of this wider industry, it acts as an incubator for emerging designers. It also encourages innovation as each part of the production process can work together for greater efficiency and quality – and lower costs.

Quick turnarounds give designers test labs of a sort, with the final product often shaped by the manufacturers. "When you're working on a higher end collection," says designer Shelley Steffee, sewers and patternmakers "are almost like designers themselves." They help solve design problems, figure out ways to operationally reduce costs in ramping up to larger-scale production, and even introduce new techniques. "Sometimes in problem solving, innovation happens . . . at every level," says designer Yeohlee Teng (DT, 2012, p. 23).

In order to retain this cluster, the CDFA and the Design Trust are requesting that the City and property owners rethink what a mixed-use neighborhood can be like. They believe that with the proper incentives to retain apparel manufacturing and related activities, some additional buildings can convert to other uses so that the needs of all stakeholders in the area can be met. To date, there has been no more movement on this space issue by the City.

Furniture

The furniture manufacturing industry in New York City is also heavily design-oriented and organized into clusters of small shops in the outer boroughs, mostly Brooklyn and Queens.[2] They produce customized or low volumes of products, often sold to interior designers and mostly for local consumption. As one manufacturer put it: "I'm not a typical manufacturer . . . I'm an artisanal studio furniture maker." Another said: "I specialize in prototypes for mass production – one of a kind, custom work, or for other designers and architects. We also fabricate artworks . . . the art world has been doing pretty well throughout the recession."

Some of these shops' products may be exported – one shop-owner who was interviewed sold 90 percent of his work out of state even though it is custom work. But often furniture makers' largest customers are local – those who are remodeling expensive homes, store interiors, high-end restaurants or other properties. One shop owner said: "decorators and designers hire us to build the furniture they want. A designer is redecorating an entire home for a client. What we do is usually a pretty small part of the entire job. Usually the owner will gut rehab the apartment . . . put in new everything, lighting, fabrics. . . ."

Furniture-manufacturing companies distinguish themselves between those that make tables and chairs versus those that make cabinets and other furniture installations. Cabinetmakers are in demand, especially by those who live in the older and smaller residences in Manhattan and Brooklyn, which can only house furniture of specific dimensions. Whereas cabinetmakers used to work for a wider clientele up until a couple of decades ago, since the wide availability of standard cabinetry from Home Depot, IKEA and other major retailers, these companies now mostly produce for the upscale market. These include wealthy homeowners as well as companies wanting a specially designed and built executive office, reception area or other unique installation.

Those who make tables and chairs may design and produce a distinctive line of furniture. A few of these may operate established factories; many others operate small woodshops wherein they make artisan furniture that may be displayed in museum collections as often as in the homes of wealthy Manhattanites. Only the larger, more established companies (employing maybe 20 workers) also bear the expense of operating centrally located showrooms near the interior design hub of companies in Midtown Manhattan. Others find customers through the Internet but most often they find customers by word of mouth through doing business with various interior designers or contractors.

When asked about their need to stay in the City, manufacturers often cited the need to be close to their customers or close to others in the design-related professions for their supply needs and ideas. For example, one company owner in Queens who makes custom cabinetry and other installations for high-end boutique stores, such as Ann Taylor, responded:

> Why am I in Long Island City [in Queens]? Because I am five minutes away. I can be at a client's place in 15 to 30 minutes. I like to go and see the space and talk to them about what they want. A lot of what we do is customer service. I like to go out and look at an installation when it's being installed [to make sure there aren't any problems].

Another owner, the one with mostly out-of-town sales, said he could move upstate where the rents are lower but would have trouble getting skilled workers. Another disadvantage of being upstate, he said: "lack of inspiration. I have to be somewhere where there's constant inspiration. I'm constantly looking at things . . . I get my inspiration from shopping and from magazines. For inspiration, it's good to be in New York."

Not only do the manufacturers want to be near to their customers but also to other manufacturers or suppliers they use in the production of their product. Clusters offer economic advantages for furniture makers too. One manufacturer in Brooklyn rents an entire floor of a loft building but sublets to other related businesses that he often uses in his work (e.g., a sander and wood finisher). He also uses other companies in the wider neighborhood on a frequent basis for the specialization he does not do in-house (e.g., one company he uses is especially good at wood inlays). Another interviewee, a restorer of antique furniture and maker of reproductions, had to move his workshop from Manhattan to Long Island City when his landlord raised his rent to force his relocation. Even though still relatively close to Manhattan, he was not near his cluster. He enticed others with whom he does regular business (e.g., an upholsterer, a metal findings manufacturer) to relocate to his new neighborhood. They did, but now face new displacement pressures since the City encouraged nearby office development when it developed a plan for and rezoned the area in 2001 (Kinetz, 2001; Wolf-Powers, 2005). He has joined with others nearby who have manufacturing shops or related industrial businesses to form an artisans' guild that would include the custom-build manufacturers along with their customers who are designers, architects and high-end retailers (Artisans' Guild of America, 2012). Although the group has only just incorporated and held two events, it follows a similar logic to what the CDFA did by organizing the whole industry of manufacturers, designers, retailers and buyers into an organization that could lobby for the necessity of space for all components of the cluster, including manufacturing, and not just the "sexy" end-user sectors of design, fashion and culture.

CONCLUSION

This story of New York City's land use fight over what to do with manufacturing in a supposedly post-industrial era offers several important lessons for policymakers and activists. First, although there has been a significant decline in manufacturing activity in cities, some still remains. Urban manufacturing looks and acts much differently than what is commonly thought of as manufacturing. Urban manufacturers are very small, make one or a few products, and can operate successfully in old (or new) multistory buildings. Their high-value products are often sold to local customers, rather than exported outside the region.

Second, small urban manufacturers depend on the proximity of suppliers and customers, with whom they confer frequently about product design and specifications. This proximity promotes innovations in product design, production processing and marketing strategies. The need for proximity results in a complex cluster of businesses and other educational and support institutions that improve the cluster's international competitiveness. Mayors of large, global cities eagerly promote business services, tourism, design and cultural sectors to attract international investors and visitors. But they are largely unaware of the dependence of these "sexy" downstream industries on small manufacturers and others that are being displaced because of their hot real estate markets.

Third, because space is tight in large, global cities, urban manufacturers and related uses compete with other businesses that may be international in scope and can afford to pay higher prices for space. This competition is encouraged by real estate developers and investors to increase their wealth at the expense of many urban space users – including residents and manufacturers. This dilemma requires some way to restrict the free operation of the real estate market to balance the needs of different groups of space users. Because restraint of private property rights is difficult to obtain by policymakers and activists in the current neoliberal environment, in New York City it has required constant organizing, political action and lobbying to block the destruction of its manufacturing clusters.

NOTES

1. Data and information regarding the manufacturing debate in New York City were gathered from more than two dozen stakeholder interviews, public data sources, published reports, histories and news articles. Because interviews were conducted to allow respondents to remain anonymous, information gathered from them is not explicitly cited.
2. Students interviewed furniture makers in Brooklyn as part of a graduate urban planning seminar on manufacturing retention (McCormick, Borrero, Imperatrice and Manglavil, 2012).

REFERENCES

Angotti, T. (2008), *New York for Sale: Community Planning Confronts Global Real Estate*, Cambridge, MA, USA: MIT Press.

Armstrong, A., V. Been, J. Madar and S. McDonnell (2010, March), "How Have Recent Rezonings Affected the City's Ability to Grow?" New York, NY, USA: Furman Center for Real Estate & Urban Policy, New York University, available at http://furmancenter.org/files/publications/Rezonings_Furman_Center_Policy_Brief_March_2010.pdf (accessed November 20, 2012).

Armstrong, R. and T. Lund (2005), "Up from the Ruins: Why Rezoning NYC's Manufacturing Areas for Housing Makes Sense," Manhattan Institute for Policy Research, Rethinking Development Report, No. 2, available at http://www.manhattan-institute.org/html/rdr_02.htm (accessed November 3, 2014).

Artisans Guild of America (2012), AGA website, available at http://www.artisansguildofamerica.org/ (accessed November 15, 2012).

Bagli, C.V. (2000, September 19), "The Crowded City: Office Shortage Imperils Growth in New York City," *The New York Times*, available at http://www.nytimes.com/2000/09/19/nyregion/the-crowded-city-office-shortage-imperils-growth-in-new-york-city.html?pagewanted=all&src=pm (accessed November 15, 2012).

Bagli, C.V. (2009, August 19), "New York Seeks to Consolidate its Garment District," *The New York Times*, available at http://www.nytimes.com/2009/08/20/nyregion/20garment.html?pagewanted=all&_r=0 (accessed November 15, 2012).

Barbanel, J. (2004a, January 18), "Remaking or Preserving the City's Face," *The New York Times*, available at http://www.nytimes.com/2004/01/18/realestate/remaking-or-preserving-the-city-s-face.html?pagewanted=all&src=pm (accessed November 15, 2012).

Barbanel, J. (2004b, April 12), "A Zoning Board's Quiet Work has Neighbors Making Noise," *The New York Times*, available at http://www.nytimes.com/2004/04/12/nyregion/a-zoning-board-s-quiet-work-has-neighbors-making-noise.html?pagewanted=all&src=pm (accessed November 15, 2012).

Bell, D. (1976), *The Coming of the Post Industrial Society: A Venture in Social Forecasting*, New York, NY, USA: Basic Books.

Bowles, J. (2000, February), "The Empire Has No Clothes," New York, NY, USA: Center for an Urban Future, available at http://nycfuture.org/research/publications/the-empire-has-no-clothes (accessed January 5, 2013).

Bram, J. and M. Anderson (2001), "Declining Manufacturing Employment in the New York–New Jersey Region: 1969–99," *Federal Reserve Bank of New York Current Issues in Economics and Finance* **7** (1), 1–6.

City of New York (2005, January), "New York City Industrial Policy: Protecting and Growing New York City's Industrial Job Base," available at http://www.nycedc.com/system/files/files/industry/New%20York%20City%20Industrial%20Policy%20%282005%29.pdf (accessed January 15, 2013).

Cohen, S.S. and J. Zysman (1987), *Manufacturing Matters: The Myth of the Post Industrial Economy*, New York, NY, USA: Basic Books.

Conway, M. and S. Loker (1999), *The Garment Industry Development Corporation: A Case Study of a Sectoral Employment Development Approach*, Washington, DC, USA: The Aspen Institute.

Crean, S. (2003, August), "NYC's Garment Industry: A New Look?" New York, NY, USA: Fiscal Policy Institute, available at http://www.fiscalpolicy.org/NYC%20sGarmentIndustry.pdf (accessed January 5, 2013).

Crean, S. (2011, January 3), "Did City's Industrial Policy Manufacture Defeat?" *City Limits: City Conversations*, available at http://www.citylimits.org/conversations/125/industrial-policy-in-new-york (accessed November 15, 2012).

Drennan, M.P. (1991), "The Decline and Rise of the New York Economy," in J.H. Mollenkopf and M. Castells (eds), *Dual City: Restructuring New York*, New York, NY, USA: Russell Sage Foundation, pp. 25–41.

DT (2012, October), "Making Midtown: A New Vision for a 21st Century Garment District in New York

City," New York, NY, USA: Design Trust for Public Spaces, available at http:// www.designtrust.org/projects/project_11makingmidtown.html (accessed January 15, 2013).
Fahim, K. (2010, March 21), "Despite Much Rezoning, Scant Change in Residential Capacity," *The New York Times*, available at http://www.nytimes.com/2010/03/22/nyregion/22zoning.html (accessed November 15, 2012).
Failor, T. (2009, December 8), "Combining Land Use Tools and Community Based Technical Assistance for Industrial Retention: Observed Outcomes of the 2005 NYC Industrial Policy," Mimeo, University of North Carolina – Chapel Hill, Department of City and Regional Planning, NC, USA.
FCBID (2013a), "Zoning Overview," available at http://www.fashioncenter.com/real-estate/zoning-overview (accessed November 15, 2012).
FCBID (2013b), "BID History," available at http://www.fashioncenter.com/about/who-we-are/bid-history (accessed January 5, 2013).
Fishman, R. (1997), "The Regional Plan and the Transformation of the Industrial Metropolis," in D. Ward and O. Zunz (eds), *The Landscape of Modernity: Essays on New York City, 1900–1940*, New York, NY, USA: Russell Sage Foundation, pp. 106–125.
Friedmann, J. (1986), "The World City Hypothesis," *Development and Change* **17** (1), 69–83.
Hart, D.M., S.J. Ezell and R.D. Atkinson (2012, December), "Why America Needs a National Network for Manufacturing Innovation," The Information Technology & Innovation Foundation, available at http://www2.itif.org/2012-national-network-manufacturing-innovation.pdf (accessed January 2, 2013).
Harvey, David (2008, September–October), "The Right to the City," *New Left Review* **53**, 23–40.
Helper, S. (2008), "Renewing U.S. Manufacturing: Promoting a High-Road Strategy: Economic Policy Institute Briefing Paper #212," available at http://www.epi.org/publication/bp212/ (accessed December 15, 2011).
Hevisi, D. (2001, July 8), "When New Homes Drive Out Old Jobs," *The New York Times*, available at http://www.nytimes.com/2001/07/08/realestate/when-new-homes-drive-out-old-jobs.html?pagewanted=all&src=pm (accessed January 15, 2013).
Hill, R.C. and J.W. Kim (2000), "Global Cities and Developmental States: New York, Tokyo and Seoul," *Urban Studies* **37** (12), 2167–2195.
Kinetz, E. (2001, April 29), "Neighborhood Report: Long Island City; Making the Citicorp Tower Less Lonesome," *The New York Times*, available at http://nytimes.com/2001/04/29/nyregion/neighborhood-report-long-island-city-making-the-citicorp-building-less-lonesome.html (accessed December 5, 2013).
Kotkin, J. (1998, September 13), "Viewpoint: Little Industries Mean a Lot to Manhattan", *The New York Times*, available at http://www.nytimes.com/1998/09/13/business/viewpoint-little-industries-mean-a-lot-to-manhattan.html?pagewanted=all&src=pm (accessed November 15, 2012).
Liff, B. (1997, December 22), "City's Zoning Out on Manufacturing," *Daily News, Suburban*, p. 4.
LoGiurato, B. (2013, February 13), "Obama: Here's my 4-Step Plan to Make America a 'Magnet for Jobs," *Business Insider*, available at http://www.businessinsider.com/obama-manufacturing-jobs-plan-2013-2 (accessed February 14, 2013).
Madeinmidtown (n.d.), Made in Midtown website, available at http://www.madeinmidtown.org (accessed November 10, 2011).
MAS (2011, October), "Fashioning the Future: NYC's Garment District," available at http: www.mas.org/urbanplanning/garment-district/ (accessed January 15, 2012).
McCormick, L., E. Borrero, S. Imperatrice and R. Manglavil (2012, January 1), "A Role for Manufacturing in the Real Estate Capital of the World? Furniture and Apparel in New York City," *Progressive Planning Magazine* Winter (190), 27–30.
Mistry, N. and J. Byron (2011, April), "The Federal Role in Supporting Urban Manufacturing," Washington, DC, USA: Brookings Institute, available at http://www.brookings.edu/~/media/Research/Files/Papers/2011/4/urban%20manufacturing%20mistry%20byron/04_urban_manufacturing_mistry_byron.PDF (accessed December 10, 2012).
National Council for Advanced Manufacturing (n.d.) "About Us," available at http://www.nacfam.org/Home/About Us/tabid/58/Default.aspx (accessed January 2, 2013).
New York City Department of City Planning (2013), "Neighborhoods Count: Celebrating DCP Rezonings," available at http://www.nyc.gov/html/dcp/html/rezonings/index.shtml (accessed January 5, 2013).
NYCEDC (2011, June 7), "The New Industrial NYC," available at http://www.nycedc.com/industry/industrial (accessed January 5, 2012).
NYCEDC (2013), "Why NYC? Fashion. NYC. 2020," available at http://www.nycedc.com/industry/fashion (accessed July 7, 2013).
NYIRN (2004, May), "Illegal Residential Conversions in the East Williamsburg In-Place Industrial Park," available at http://prattcenter.net/sites/default/files/publications/illegal_residential_conversions.pdf (accessed January 5, 2013).
NYIRN (2009, October 28), "Commercial Uses Invading Mayor's Industrial Business Zones," available at

http://prattcenter.net/sites/default/files/publications/commercial%20uses%20invading%20IBZs%20final.pdf (accessed January 5, 2012).
Obama, B. (2012), "The New Economic Patriotism: A Plan for Jobs and Middle Class Security," available at https://www.documentcloud.org/documents/557639-jobs-plan-booklet-from-the-obama-2012-campaign.html (accessed February 12, 2013).
PICCED/MAS (2001), "Making it in New York: The Manufacturing Land Use and Zoning Initiative, Vol. 1: Report," available at http://www.nylpi.org/images/FE/chain234siteType8/site203/client/EJ%20-%20Making%20it%20in%20NY%20-%20PICCED%20Study.pdf (accessed November 15, 2012).
Piore, M. and C. Sabel (1986), *The Second Industrial Divide: Possibilities for Prosperity*, New York, NY, USA: Basic Books.
Pratt (Pratt Center for Community Development) (2008, August 8), "Protecting New York's Threatened Manufacturing Space," available at http://prattcenter.net/research/protecting-new-yorks-threatened-manufacturing-space (accessed November 3, 2014).
Pratt (Pratt Center for Community Development) (2012, June 12), "Urban Manufacturing Alliance," available at http://prattcenter.ent/news/urban=manufacturing-alliance (accessed January 5, 2013).
QCEW (Quarterly Census of Employment and Wages) (2011), United States Government, Department of Commerce, USA, available at http://www.census.gov (accessed November 15, 2012).
Rast, J. (2005, November), "Curbing Industrial Decline or Thwarting Redevelopment? An Evaluation of Chicago's Clybourn Corridor; Good Island, and Elston Corridor Planned Manufacturing Districts," Center for Economic Development, University of Wisconsin-Milwaukee, WI, USA, available at http://www4.uwm.edu/ced/publications/pmdstudy1.pdf (accessed November 15, 2012).
Rice, A. (2000, March 13), "Goodbye, Garmentos: Rag Trade Surrenders to Dot-Com Riches," *New York Observer*, available at http://observer.com/2000/03/goodbye-garments-rag-trade-surrenders-to-dotcom-riches/ (accessed December 19, 2012).
Roberts, S. (2006, February 19), "By 2035, Planners See a Million New Stories in the Crowded City," *The New York Times*, available at http://www.nytimes.com/2006/02/19/nyregions/19population.html?pagewanted=all&_r=0 (accessed January 15, 2013).
Rosenfeld, S. (2002), *A Governor's Guide to Cluster-Based Economic Development*, Washington, DC, USA: National Governors' Association.
Sassen, S. (1991), *The Global City: New York, London, Tokyo*, Princeton, NJ, USA: Princeton University Press.
Sassen, S. (2005), "The Global City: Introducing a Concept," *The Brown Journal of World Affairs* **11** (2), 27–43.
Schwartz, J. (1983), *The New York Approach: Robert Moses, Urban Liberals, and Redevelopment of the Inner City*, Columbus, OH, USA: The Ohio State University Press.
SFMade (2012), "About SFMade," available at http://www.sfmade.org/about-us/about-sfmade/ (accessed January 15, 2013).
Trebay, G. (2000, November 14), "Dot-Coms Intrude in the Land of Needle and Thread," *The New York Times*, available at http://www.nytimes.com/2000/11/14/style/dot-coms-intrude-in-the-land-of-needle-and-thread.html?pagewanted=all&src=pm (accessed November 15, 2012).
Wolf-Powers, L. (2005), "Up-Zoning New York City's Mixed-Use Neighborhoods: Property-Led Economic Development and the Anatomy of a Planning Dilemma," *Journal of Planning Education and Research* 24, 379–393.
Wolfson, J. (2001, June 25), "Space Wars," *Gotham Gazette*, available http://www.gothamgazette.com/iotw/spacewars/ (accessed January 15, 2013).

28. Manufacturing in the knowledge economy: innovation in low-tech industries

Teis Hansen and Lars Winther

INTRODUCTION

Recent studies stress the relevance of a broad conceptualization of the knowledge economy which goes beyond the strong, current policy focus on high-tech industries. Today, low-tech industries continue to have a significant role in European manufacturing in terms of employment, value added and export. However, the character and activities of these industries are profoundly changing as they become increasingly knowledge intensive. Thus, innovation processes and knowledge production in manufacturing are much more complex than suggested by the classic division into high-, medium- and low-tech industries. This calls for a rethinking of manufacturing's position in contemporary capitalism and a redefinition of the central categories based on research and development (R&D) intensity that dominate the debate on the knowledge economy.

Our use of the term 'the knowledge economy' refers to an economy where acquisition, creation and utilization of knowledge are the key factors explaining competitive advantage of firms. The increasing speed of knowledge diffusion on a global scale implies that strongholds and favourable positions only exist temporarily. Firms must constantly innovate to maintain profitability. Importantly, this is the case in all types of industries. Thus, this conception of the knowledge economy does not emphasize specific industries, for instance high-tech industries or knowledge-intensive business services, nor a science-based view on innovation. Neither does it emphasize specific types of knowledge, such as scientific knowledge. Rather, it highlights that the economy in general is becoming more knowledge intensive.

This chapter is organized as follows. We start by introducing the taxonomy of high- and low-tech industries and its rather one-dimensional understanding of innovation. We introduce industrial knowledge bases as an alternative framework for analysing innovation in low-tech industries. The empirical part of the chapter accounts for the importance of low-tech industries in Europe, analyses manufacturing investments in capital goods and human capital, and relates the investment patterns to low-tech innovation. The final section concludes and considers implications for policymakers.

HIGH- AND LOW-TECH INDUSTRIES

The relatively low R&D investment levels in Europe compared with the USA have often been stressed as a main explanation for the former's lower GDP growth and productivity during the 1990s (Sapir et al. 2003; European Commission 2004). The underlying assumption is that the less research-intensive parts of the economy are unable to withstand competition from low-wage countries with increasingly skilled labour forces due to the high European wage levels (Dicken 2011). Therefore, continuous investments in R&D leading to radical innovations are prioritized to achieve long-term economic prosperity. Associated with this is an emphasis on high-tech industries defined by their R&D intensity and subsequently on the large city regions where these industries are concentrated. Thus, non-R&D-intensive industries are receiving decreasing attention from policymakers and this will eventually also have significant effects on regions focused mainly on low-tech industries. An important way in which this science-based view of innovation influences policymaking is through industrial classifications based on a linear conception of innovation processes. The most widespread and influential classification system is that developed by the OECD and Eurostat, which is based on the R&D intensity of different industries. Outcomes of R&D are characterized by significant elements of novelty and scientific or technological progress (OECD 2002), and the R&D intensity is given by the ratio of R&D expenditures to the output value of the sector. Four categories are used (Smith 2005):

1. High-tech industries: R&D intensity above 5 per cent.
2. Medium high-tech industries: R&D intensity between 3 and 5 per cent.
3. Medium low-tech industries: R&D intensity between 1 and 3 per cent.
4. Low-tech industries: R&D intensity below 1 per cent.

Table 28.1 below shows the resulting classification by the OECD of manufacturing industries that we use to illustrate the problems of using this taxonomy to understand competitiveness of manufacturing industries. The attractiveness of this taxonomy in a policy context is closely related to its simplicity, which makes it very precise and easy to measure (Jacobson and Heanue 2005; Godin 2006), in contrast to other, more elaborate alternatives. For instance, the taxonomy by Pavitt (1984) overcomes this limited view

Table 28.1 Manufacturing industries classified according to R&D intensity (NACE Rev. 1.1 codes in brackets)

High-tech	Medium high-tech
Pharmaceuticals (24.4)	Chemicals – excl. pharmaceuticals (24 excl. 24.4)
Computers, office machinery (30)	Non-electrical machinery (29) (2.5)
Electronics-communications (32)	Electrical machinery (31)
Scientific instruments (33)	Motor vehicles (34)
Aerospace (35.3)	Other transport equipment (35.2 + 35.4 + 35.5)
Medium low-tech	**Low-tech**
Coke, refined petroleum products and nuclear fuel (23)	Wood, paper products, printing and publishing (20–22)
Rubber and plastic products (25)	Food, beverages and tobacco (15–16)
Non-metallic mineral products (26)	Textile and clothing (17–19)
Basic metals (27)	Other manufacturing and recycling (36–37)
Fabricated metal products (28)	

Source: OECD 2004.

of innovation, as it considers employment of skilled employees, learning-by-doing and learning-by-using. Firms can therefore be classified as high-tech even though the actual R&D intensity is relatively low. Similarly, Bar-El and Felsenstein (1989) consider that an industry's technological intensity depends on the percentage of academic and skilled labour, the technological intensity of capital (investments in sophisticated machinery or processes) and the technological intensity of the product (R&D intensity). Finally, Laestadius et al. (2005) include similar measures to Pavitt (1984) and Bar-El and Felsenstein (1989), and furthermore stress the importance of a number of other aspects, such as organizing capacity and the ability to design solutions through synthesizing different fields of knowledge. They argue that such measures are necessary to describe the full innovativeness of firms and sectors, and that any attempt that relies on just one indicator is reductionism; a complex issue such as innovation needs different indicators, which cannot be compiled into one. However, even the more complex measures remain within the framework of defining high- and low-tech industries. We suggest a different perspective focusing on the use of different types of industrial knowledge bases.

KNOWLEDGE BASES AND LOW-TECH INNOVATION

Thus, another way to distinguish innovations is to consider their relation to current levels of technology (Fagerberg 2005). A constant improvement of a product or a production process is considered an incremental innovation, as it is not fundamentally different from well-known technologies in the field. A radical innovation, in contrast, is completely different from the current products or processes available on the market. Whether the firms in an industry primarily create incremental or radical innovations depends fundamentally on the characteristics of the industry. In general, this can be termed the 'industrial knowledge base' (Asheim and Gertler 2005; Asheim 2007). Asheim and Gertler (2005) distinguish between two main knowledge bases. A key feature of the analytical knowledge base is the use of scientific methods and modelling. Codified knowledge is a major input as well as a major output of the knowledge creation process, and links to universities and research institutions are therefore crucial to the innovativeness of firms with analytical knowledge bases. Innovations are often radical, contrary to the predominantly incremental innovations in industries with synthetic knowledge bases. The firms in these industries primarily utilize existing knowledge in new product and process developments, and tacit knowledge acquired through learning-by-doing and learning-by-using is therefore central to innovation processes. While links to knowledge institutions can be of importance, especially within the field of applied research, up- and down-stream relations are particularly significant, as innovation projects often revolve around identifying engineering-based solutions to technical problems.

The insight that industries rely on different knowledge bases is important, as it opens up an alternative understanding of knowledge creation and innovation processes to the dominant science-based view and the taxonomy of high- and low-tech industries. Recently, interest has increased in the types of capabilities and innovation processes that allow low-tech industries to maintain competitiveness (e.g. Bender and Laestadius 2005; Arundel et al. 2008; Hirsch-Kreinsen 2008; Kirner et al. 2009; Petrou and Daskalopoulou 2009; Hansen 2010). While, on the one hand, some low-tech firms increasingly move into technology-intensive product categories (Robertson and Patel 2005; Mendonça 2009), the majority of low-tech firms, on the other hand, rely on a predominantly synthetic knowledge base that allows them to develop customer-specific solutions and continuously improve products and production processes. Low-tech firms utilize transformative capabilities to convert global knowledge, which is available worldwide, to knowledge that is specific to a certain context, and configurational capabilities to synthesize novelties by organizing knowledge, artefacts and actors in new ways through collaborating, (re-)designing and tapping into various types of knowledge pools (Bender and Laestadius 2005). The importance of such capabilities is underlined by recent research which shows that these factors explain firm innovativeness better than R&D investments in all types of manufacturing industries – not just low-tech (Hervas-Oliver et al. 2011).

Thus, there is an increasing recognition that the importance of science-based knowledge production has been exaggerated, and that it is necessary to specifically understand knowledge production in low-tech industries. However, despite this development, there is still very limited knowledge on the investments of low-tech industries in human capital and capital goods. As noted by Smith (2002), such investments are of central importance to the innovative capacity of firms, and the size of these expenditures is considerably larger than R&D investments across industries. In the subsequent sections, we analyse these issues and relate them to low-tech innovation processes; but first, we account for the economic importance of low-tech industries in high-wage countries.

ECONOMIC IMPORTANCE OF LOW-TECH INDUSTRIES

The relevance of analyses of low-tech innovation patterns is evident considering the continuing importance of low-tech industries in Western countries. Non-R&D-intensive industries maintain a high and stable share of manufacturing value added and employment (Kaloudis et al. 2005; Hansen and Winther 2011). Figure 28.1 illustrates this point. Even though low-tech employment has decreased over the years, these industries continue to account for a larger share of manufacturing employment than the other types of manufacturing. Further, the medium low-tech industries' share of employment has in fact increased over the period from 25.1 per cent in 1995 to a maximum of 27.0 per cent in 2007, which is the latest year with available data. The share constituted by high-tech manufacturing remained stable throughout the period, varying only between 8.3 per cent and 8.8 per cent. This enduring economic importance of non-R&D-intensive industries is also reflected in developments of labour productivity over time. The growth rates of labour productivity in the four different types of industries have been very similar since the millennium in European manufacturing (see Hansen and Winther 2011).

This persistence of non-R&D-intensive industries in high-wage countries is remarkable considering the policy emphasis on R&D and high-tech industries. The priority given in Europe to research and high technology is evident through EUREKA, an intergovernmental R&D programme, as well as the EU-wide support programmes which focus on promoting R&D activities and neglect non-research-intensive innovations (Hirsch-Kreinsen 2008). An important factor explaining the continuing competitiveness of these low-tech industries is investment patterns. While low-tech industries – by

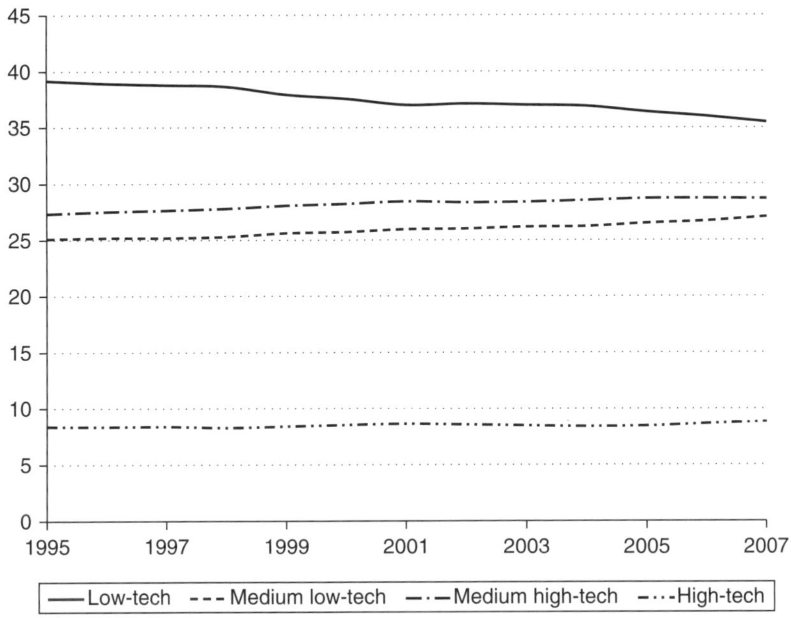

Note: Based on figures for nine European countries with available data throughout the period: Austria, Belgium, Denmark, Finland, France, Germany, Italy, Spain and Sweden.

Source: Own calculation, data from OECD's STAN database.

Figure 28.1 Shares of manufacturing employment (%)

definition – have low R&D investment levels, they may nevertheless maintain competitiveness through investments in capital goods and human capital.

INVESTMENTS IN CAPITAL GOODS AND HUMAN CAPITAL

The development of investment patterns in Danish manufacturing highlights the possibility of a route to sustained manufacturing competitiveness which does not primarily depend on R&D investments. The changing composition of Danish manufacturing over time is quite similar to that described above for a larger sample of countries, even if the share of low-tech manufacturing employment is relatively large in Denmark (38.5 per cent in 2007 compared with an average of 35.5 per cent for all nine countries; see also Hansen 2010). Figure 28.2 presents the development in capital intensity (the capital/labour ratio) for Danish manufacturing industries in the period 1990–2008. Figure 28.2A includes all types of capital investment – from buildings and machines to software and art. The graph shows that the high-tech industry has the largest growth over the period, but the capital intensity of the low-tech industry increases to nearly the same extent.

Examining the development of capital investments solely in machinery (Figure 28.2B)

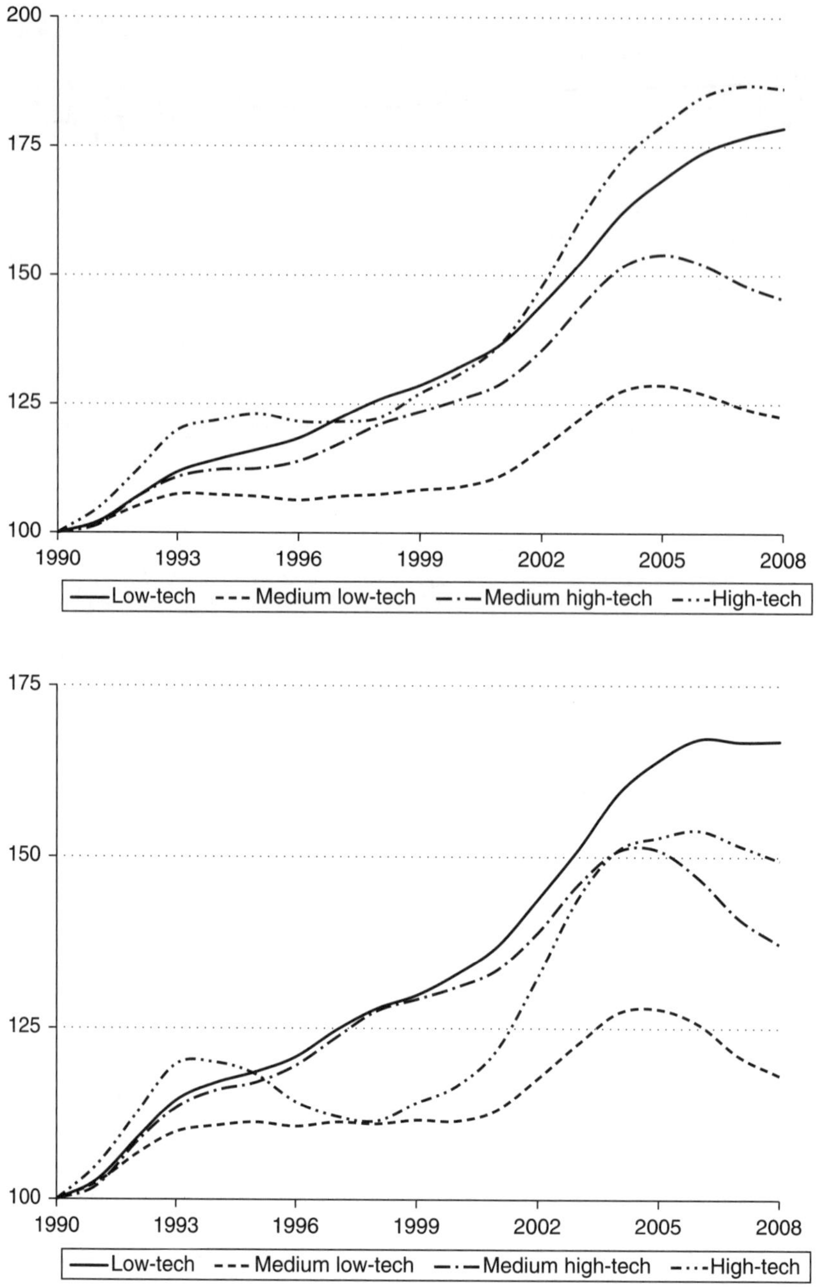

Note: To reduce fluctuations, values are calculated based on three-year averages of investments in the preceding, the actual and the following year. Values for 1990 and 2008 are two-year averages.

Source: Own calculation, data from Statistics Denmark.

Figure 28.2 Capital intensity over time in Danish manufacturing (1990 = Index 100)

Table 28.2 Growth in the share of highly skilled labour (1993 = Index 100) (brackets indicate absolute shares)

	1993		1996		2002		2004		2006	
Low-tech	100	*(1.7)*	130	*(2.3)*	194	*(3.4)*	214	*(3.7)*	258	*(4.5)*
Medium low-tech	100	*(1.3)*	121	*(1.5)*	150	*(1.9)*	180	*(2.3)*	196	*(2.5)*
Medium high-tech	100	*(3.5)*	120	*(4.2)*	174	*(6.1)*	212	*(7.4)*	239	*(8.4)*
High-tech	100	*(5.9)*	119	*(7.0)*	127	*(7.4)*	151	*(8.9)*	186	*(11.0)*
Manufacturing total	100	*(2.4)*	125	*(3.0)*	174	*(4.1)*	204	*(4.9)*	238	*(5.6)*

Note: Highly skilled labour defined as employees with formal education equalling bachelor's degree, master's degree or PhDs, equivalent to ISCED categories 5A and 6 (UNESCO 1997).

Source: Own calculation, data from Statistics Denmark, Hansen et al. 2013.

presents a different picture. Low-tech industries have by far increased machinery investments the most. Further, it is worth noticing that the low-tech industries maintain a stable investment level into the global recession, while the three other types of industries decrease their investments in machinery towards the end of the period.

Capital – and in particular machinery – investments are not the only types of investments of increasing importance to low-tech manufacturing. While many studies point to the role of highly skilled labour for the development of high-tech industries (e.g. Zucker et al. 1998; Murphy and Siedschlag 2013), it is often overlooked that low-tech industries are increasingly investing in human capital. In fact, in a Danish context, the growth in the low-tech industries' use of highly skilled labour is significantly higher than the growth in the three other types of industries (Hansen et al. 2014). No significant distinctions are found between medium low-, medium high- and high-tech industries, emphasizing that human capital is of increasing relevance in low-tech manufacturing in particular. Still, the low-tech industry has a considerably lower percentage of highly skilled labour than the medium high- and high-tech industries (see Table 28.2). However, it should be noted that the share of highly skilled labour is consistently higher in the low-tech industry than in the medium low-tech industry, underlining that there is not a one-to-one relation between industries' R&D investment levels and their use of human capital.

Qualitative studies of innovation processes in Danish low- and medium low-tech industries confirm that investments in capital goods and human capital are crucial for the competitiveness of these manufacturing firms (Hansen 2010; Hansen and Winther 2011). Naturally, investments in machinery are the most labour-saving capital investments, as they often aim at increasing automation. Such investments may be of significant importance for the competitiveness of low-tech firms as they decrease the share of firm expenditures constituted by labour costs. While reduced labour costs are by far the most important benefit of such investments, consistent levels of quality, increasing production capacity and optimal use of materials are additional advantages of the growing emphasis on machinery investments in low-tech firms (Groth et al. 2012).

The use of increasingly advanced machinery has considerable consequences for the skill requirements of employees in low-tech firms. First, while low-tech industries have traditionally been major employers of unskilled labour, the accelerating substitution of machinery for labour implies that this is gradually less the case. Second, firm managers of low-tech firms note that the processes of selecting new machinery are increasingly challenging and time demanding due to both the complexity of the machines and the number of available models on the market. At the same time, making the right decision when acquiring new equipment is also very important due to the substantial financial costs associated with the purchase of many types of sophisticated machinery. Consequently, these processes are highly prioritized by managers of low-tech firms, who therefore hire employees with considerable technological knowledge which can ensure that the right decisions are made. Third, implementation, operation and maintenance of advanced machinery require increasingly higher skill levels of employees. In summary, the increasing use of sophisticated machinery is a main driver of the significant upgrading of skills in low-tech manufacturing.

In addition to the close relation between the use of advanced machinery and investments in human capital, highly skilled labour is also of growing importance simply due to the increasing sophistication of product development in low-tech firms. One important role of highly skilled labour in low-tech firms is to facilitate collaborations with high-tech partners. Increasing diversification of low-tech firms into more research-intensive fields (Robertson and Patel 2005; Mendonça 2009) is associated with a high frequency of joint innovation projects between high- and low-tech firms where partners bring together analytical and synthetic knowledge (Hansen and Winther 2011). These collaborations are beneficial for both sides as firms relying on either analytical or synthetic knowledge tend to be less innovative than firms that combine the two knowledge bases in their development processes (Jensen et al. 2007), and the projects provide access to competencies and knowledge which are otherwise difficult to acquire (synthetic knowledge for high-tech firms, analytical knowledge for low-tech firms). In these partnerships, low-tech firms typically contribute with knowledge on materials, production processes and engineering techniques. To enter the collaboration, the high-tech firms will often require that the low-tech firms have highly skilled employees working in development positions to ease communication in the project and ensure that work routines are not too dissimilar between the two firms (Hansen 2014). In this way, highly skilled workers are a necessary condition for successful collaborations between high- and low-tech firms.

An example of the benefits of such collaborations is the partnership between a manufacturer of plastic products and pharmaceutical firms (Hansen and Winther 2011). The lack of plastic-specific knowledge among many pharmaceutical firms offers a possibility for the plastics producer to get involved in collaborative innovation processes on both drug delivery devices and production facilities, most notably clean rooms. Over the last decade, the firm has engaged in a considerable number of such development projects, thus developing a strongly specialized expertise on plastic moulding technology with relevance for the pharmaceutical industry. These collaborations lead to results – such as reduction of production time by 50 per cent – which are beyond the reach of the firms individually. The plastics firm invests heavily in facilities, education and training that underpin the specialization of the firm, but it is still necessary to collaborate with firms in the pharmaceuticals sector, which are at the forefront of R&D, to achieve successful innovations in these product categories. This exemplifies the mutual benefits to high- and low-tech firms of these collaborations with regard to execution time, knowledge build-up and market access.

Summing up, this section emphasizes that it is a mistake to equate the lack of R&D intensity in low-tech industries with a lack of knowledge intensity. Increasing investments in advanced production equipment and highly skilled labour are fundamental to the competitiveness of these industries in high-wage countries. However, the changing character of low-tech manufacturing is also resulting in some worrying prospects concerning the position of unskilled labour in an increasingly advanced manufacturing industry. As argued by Christopherson and Clark (2007), while not all industries have a great need for employees with PhD degrees, medium-skilled labour is increasingly becoming a minimum requirement across different types of industries. In this way, the developments that underpin the continuing competitiveness of Western low-tech firms are associated with an increasing employment problem for unskilled labour.

CONCLUSIONS

In this chapter we have argued that low- and medium low-tech industries maintain significant economic importance in high-wage countries, but that the character and activities of these industries are changing profoundly. Contrary to the assumptions of much academic and policy work, low- and medium low-tech manufacturing industries continue to be present in Western countries. While the shares of manufacturing employment and value added constituted by low-tech industries have decreased slightly over recent decades, this is not the case for medium low-tech industries. Consequently, non-R&D-intensive industries continue to make up the clear majority of both manufacturing employment and value added in Europe.

This fact stresses that low-tech firms are not passively waiting to be outcompeted by firms from low-cost countries. Instead, they are actively pursuing strategies to maintain competitiveness and increase the value added of products and processes. The innovation strategies of non-R&D-intensive firms are founded on inherited capabilities that provide a platform for incremental improvements of products and production processes as well as the customization of products. A strong synthetic knowledge base allows firms to, for example, change from producing large batches of standardized products to specializing in prototype fabrication.

However, it is equally important to recognize that increasing investments in machinery and human capital are fundamental to the competitiveness of low-tech industries. Low- and medium low-tech firms are in general not particularly threatened by increasing global price competition. Of course, some firms will maintain a focus on cost competition, and not intensify investments in skills and equipment. While the future viability of these firms is highly uncertain, the majority use a variety of means to stay competitive, including investments in sophisticated machinery resulting in reduced labour costs, and increasing employment of highly skilled labour. Thus, framing the conclusion in a positive tone, there is clearly significant potential in advancing the position of low- and medium low-tech manufacturing industries in high-wage countries, to which policymakers should pay attention. This involves, for instance, assisting firms in the increasingly difficult selection process of acquiring sophisticated machinery, or introducing policies towards attraction and retention of highly skilled labour specifically in low-tech industries. In fact, following the argument of Porter (1990), a selective factor disadvantage concerning wages can stimulate a transition towards innovative practices that may secure a long-term strong position of Western European low-tech industries.

Based on this analysis, a final question to consider is the relevance of the low-tech/high-tech taxonomy. Certainly, the R&D intensity of industries is an important characteristic, but the emphasis in policymaking on high-tech industries suggests that this taxonomy does more harm than good. There is a need for a broader conception of manufacturing in the knowledge economy, which goes beyond R&D and radical innovations. Thus, more nuanced taxonomies (e.g. Pavitt 1984; Bar-El and Felsenstein 1989; Laestadius et al. 2005; Asheim 2007) that pay attention to, among other things, human capital levels and capital investments have more to offer. Rather than attempting to rank industries according to their knowledge intensity, there is a need for understanding and underlining inter-industrial differences in the types of critical knowledge and the ways this knowledge is sourced. Today, all manufacturing industries in high-wage regions, such as Western Europe, are knowledge intensive, and this should be reflected in the labels we apply to them. While the institutionalization of the low-tech/high-tech taxonomy in policy circles makes it unlikely that it will be abandoned any time soon, this chapter emphasizes that there is an alternative to this narrow understanding of the knowledge economy.

REFERENCES

Arundel, Anthony, Catalina Bordoy and Minna Kanerva (2008), 'Neglected innovators: How do innovative firms that do not perform R&D innovate?', INNO-Metrics Thematic Paper.

Asheim, Bjørn T. (2007), 'Differentiated knowledge bases and varieties of regional innovation systems', *Innovation: The European Journal of Social Science Research*, 20 (3), 223–241.

Asheim, Bjørn T. and Meric S. Gertler (2005), 'The geography of innovation: Regional innovation systems', in Jan Fagerberg, David C. Mowery and Richard R. Nelson (eds), *The Oxford Handbook of Innovation*, Oxford: Oxford University Press, pp. 291–317.

Bar-El, Raphael and Daniel Felsenstein (1989), 'Technological profile and industrial-structure – Implications for the development of sophisticated industry in peripheral areas', *Regional Studies*, 23 (3), 253–266.

Bender, Gerd and Staffan Laestadius (2005), 'Non-science based innovativeness: On capabilities relevant to generate profitable novelty', *Perspectives on Economic, Political and Social Integration*, Special Edition XI (1–2), 124–170.

Christopherson, Susan and Jennifer Clark (2007), *Remaking Regional Economies: Power, Labor and Firm Strategies in the Knowledge Economy*, London: Routledge.

Dicken, Peter (2011), *Global Shift: Mapping the Changing Contours of the World Economy*, 6th edition, London: Sage.

European Commission (2004), *Facing the Challenge: The Lisbon Strategy for Growth and Employment*, Brussels: European Commission.

Fagerberg, Jan (2005), 'Innovation: A guide to the literature', in Jan Fagerberg, David C. Mowery and Richard R. Nelson (eds), *The Oxford Handbook of Innovation*, Oxford: Oxford University Press, pp. 1–26.

Godin, Benoît (2006), 'The linear model of innovation – The historical construction of an analytical framework', *Science, Technology & Human Values*, 31 (6), 639–667.

Groth, Niels B., Lars Winther, Heikki Eskelinen, Timo Lautanen, Pasi Saukkonen, Olaf Foss, Frants Gundersen, Mats Johansson and Marcus Adolphson (2012), *Urban Patterns of Growth: The Geography of Suppliers of High-Tech Companies in the Nordic Countries*, Copenhagen: Nordic Council of Ministers.

Hansen, Teis (2010), 'The Danish fabricated metal industry: A competitive medium-low-tech industry in a high-wage country', *Danish Journal of Geography*, 110 (1), 65–80.

Hansen, Teis (2014), 'Juggling with proximity and distance: Collaborative innovation projects in the Danish cleantech industry', *Economic Geography*, 90, 375–402.

Hansen, Teis and Lars Winther (2011), 'Innovation, regional development and relations between high- and low-tech industries', *European Urban and Regional Studies*, 18 (3), 321–339.

Hansen, Teis, Lars Winther and Ronnie F. Hansen (2014), 'Human capital in low-tech manufacturing: the geography of the knowledge economy in Denmark', *European Planning Studies*, 22, 1693–1710.
Hervas-Oliver, Jose-Luis, Jose Albors-Garrigos and Ignácio Gil-Pechuan (2011), 'Making sense of innovation by R&D and non-R&D innovators in low technology contexts: A forgotten lesson for policymakers', *Technovation*, 31 (9), 427–446.
Hirsch-Kreinsen, Hartmut (2008), '"Low-tech" innovations', *Industry & Innovation*, 15 (1), 19–43.
Jacobson, David and Kevin Heanue (2005), 'Policy conclusions and recommendations', *Perspectives on Economic, Political and Social Integration*, Special Edition XI (1–2), 359–416.
Jensen, Morten B., Björn Johnson, Edward Lorenz and Bengt-Åke Lundvall (2007), 'Forms of knowledge and modes of innovation', *Research Policy*, 36 (5), 680–693.
Kaloudis, Aris, Tore Sandven and Keith Smith (2005), 'Structural change, growth and innovation: The roles of medium and low tech industries, 1980–2002', *Perspectives on Economic, Political and Social Integration*, Special Edition XI (1–2), 49–74.
Kirner, Eva, Steffen Kinkel and Angela Jaeger (2009), 'Innovation paths and the innovation performance of low-technology firms – an empirical analysis of German industry', *Research Policy*, 38 (3), 447–458.
Laestadius, Staffan, Trond E. Pedersen and Tore Sandven (2005), 'Towards a new understanding of innovativeness – and of innovation based indicators', *Perspectives on Economic, Political and Social Integration*, Special Edition XI (1–2), 75–121.
Mendonça, Sandro (2009), 'Brave old world: Accounting for "high-tech" knowledge in "low-tech" industries', *Research Policy*, 38 (3), 470–482.
Murphy, Gavin and Iulia Siedschlag (2013), 'Human capital and growth of information and communication technology-intensive industries: Empirical evidence from open economies', *Regional Studies*, 47, 1403–1424.
OECD (2002), *The Measurement of Scientific and Technological Activities – Proposed Standard Practice for Surveys of Research and Experimental Development: Frascati Manual*, Paris: OECD.
OECD (2004), 'High-tech trade, employment and value added in high-tech industries and knowledge-intensive services', *Statistics in Focus*, 2.
Pavitt, Keith (1984), 'Sectoral patterns of technical change: Towards a taxonomy and a theory', *Research Policy*, 13 (6), 343–373.
Petrou, Anastasia and Irene Daskalopoulou (2009), 'Innovation and small firms' growth prospects: Relational proximity and knowledge dynamics in a low-tech industry', *European Planning Studies*, 17 (11), 1591–1604.
Porter, Michael E. (1990), *The Competitive Advantage of Nations*, New York: The Free Press.
Robertson, Paul L. and P. R. Patel (2005), 'New wine in old bottles: Technological diffusion in developed economies', *Perspectives on Economic, Political and Social Integration*, Special Edition XI (1–2), 271–304.
Sapir, André, Philippe Aghion, Giuseppe Bertola, Martin Hellwig, Jean Pisani-Ferry, Dariusz Rosati, José Viñals and Helen Wallace (2003), *An Agenda for a Growing Europe: The Sapir Report*, Brussels: European Commission.
Smith, Keith (2002), 'What is the "Knowledge Economy"? Knowledge intensity and distributed knowledge bases', UNU-MERIT Working Paper Series, #2002-6.
Smith, Keith (2005), 'Measuring innovation', in Jan Fagerberg, David C. Mowery and Richard R. Nelson (eds), *The Oxford Handbook of Innovation*, Oxford: Oxford University Press, pp. 148–177.
UNESCO (1997), *International Standard Classification of Education (ISCED)*, Paris: UNESCO.
Zucker, Lynne G., Michael R. Darby and Marilynn B. Brewer (1998), 'Intellectual human capital and the birth of U.S. biotechnology enterprises', *American Economic Review*, 88 (1), 290–306.

FURTHER READING

Asheim, Bjørn T. (2007), 'Differentiated knowledge bases and varieties of regional innovation systems', *Innovation: The European Journal of Social Science Research*, 20 (3), 223–241. (A detailed description of knowledge bases.)
Bender, Gerd and Staffan Laestadius (2005), 'Non-science based innovativeness: On capabilities relevant to generate profitable novelty', *Perspectives on Economic, Political and Social Integration*, Special Edition XI (1–2), 124–170. (An examination of the capabilities that allow low-tech firms to maintain competitiveness.)
Fagerberg, Jan, David C. Mowery and Bart Verspagen (2009),'The evolution of Norway's national innovation system', *Science and Public Policy*, 36 (6), 431–444. (An analysis of the non-R&D-intensive Norwegian economy.)
Hansen, Teis and Lars Winther (2011), 'Innovation, regional development and relations between high- and low-tech industries', *European Urban and Regional Studies*, 18 (3), 321–339. (Emphasizing the interdependencies between high- and low-tech industries.)

Hirsch-Kreinsen, Hartmut (2008), '"Low-tech" innovations', *Industry & Innovation*, 15 (1), 19–43. (On different innovation strategies in low-tech firms.)

Jensen, Morten B., Björn Johnson, Edward Lorenz and Bengt-Åke Lundvall (2007), 'Forms of knowledge and modes of innovation', *Research Policy*, 36 (5), 680–693. (Highlighting the positive impact of combining different innovation modes.)

Laestadius, Staffan, Trond E. Pedersen and Tore Sandven (2005), 'Towards a new understanding of innovativeness – and of innovation based indicators', *Perspectives on Economic, Political and Social Integration*, Special Edition XI (1–2), 75–121. (Presenting an alternative to the high-/low-tech taxonomy.)

Pavitt, Keith (1984), 'Sectoral patterns of technical change: Towards a taxonomy and a theory', *Research Policy*, 13 (6), 343–373. (The seminal study on different sources of technological change.)

Sapir, André, Philippe Aghion, Giuseppe Bertola, Martin Hellwig, Jean Pisani-Ferry, Dariusz Rosati, José Viñals and Helen Wallace (2003), *An Agenda for a Growing Europe: The Sapir Report*, Brussels: European Commission. (A central report for the last decade's EU policymaking.)

Smith, Keith (2002), 'What is the "Knowledge Economy"? Knowledge intensity and distributed knowledge bases', UNU-MERIT Working Paper Series, #2002–6. (Questioning the science-driven conception of the knowledge economy and highlighting the role of capital investments.)

29. Crafting a comeback: cultivating an innovative ecosystem in mature regions

Maryann Feldman and Lauren Lanahan

1. INTRODUCTION

Older industrial cities that once served as the backbone of the 20th century economy are struggling to redefine themselves. The fortunes of these regions have followed the fortunes of the entrepreneurial firms that grew into large corporations within their boundaries and provided stability and prosperity for the local economy for almost a century. As these large corporations lost ground, either squandering their competitive advantages or being overwhelmed by the forces of globalization, older manufacturing cities such as Pittsburgh, Cleveland, Akron and Youngstown, among others in the US and internationally, have struggled to redevelop their economies. Entrepreneurship – defined as new firm formation and growth – and innovation – defined as the creation of value in an economy – emerge as the most viable strategies for the economic future of older cities. Yet how this transformation might happen and what types of public policies are needed remain open questions. While entrepreneurship is frequently held out as a policy panacea, we observe many cases where regions, despite considerable dedicated public resources, have not been able to restructure or even remain resilient.

There is growing attention in the literature placed on the specific process of regional restructuring and how the potential of entrepreneurial firms is influenced by the array of public programs provided (Bristow 2010; Clark 2013). An assumption is that regions are defined by the product life cycle of industries (Markusen 1985; Markusen 1986). Renewal is predicated on the logic that once one product cycle matures and declines, another new product cycle will take its place, *Q.E.D.* Yet in reality transitions are far from straightforward and entail significant human costs. The skills of one generation are not easily adapted and transferable to the needs of new industry in the regional economy.

Although much has been written about social and community resilience, little has been said about how government policy can aid the longer-term development of a renewed and vibrant regional industrial base. The literature has not been able to provide compelling explanations or policy prescriptions. Little is known about what types of government policies best ameliorate negative social impacts while promoting innovation and entrepreneurship. Unless scholars address these questions, policy makers will be left to hope that their investments will yield the desired results with little knowledge to guide them. Public programs take on new importance in mature industrial regions that are restructuring or trying to craft an economic comeback.

This chapter provides an exploratory case study of one key mature regional economy in the United States – the cities of Pittsburgh, Cleveland, Akron and Youngstown, which have banded together to define themselves as the *Tech Belt* – a continuous geographic region with similar industrial structure, and the need to secure their economic future.

Located in the adjacent states of Pennsylvania and Ohio, these cities have more in common with each other than with other cities in their respective states. Pittsburgh is very different from Philadelphia, the largest city in Pennsylvania, which is proximate to New York City. Cleveland, Akron and Youngstown in northeastern Ohio, once the largest and most politically powerful cities, have been eclipsed by the more southern cities of Columbus and Cincinnati. Moving beyond the brand of the RustBelt, which stood as a prominent symbol for the region, the *Tech Belt* initiative emerged as a response to assist the region in transitioning to a technology- and knowledge-based economy. Rather than evaluate a specific program, our objective is to consider how innovative firms actually use the array of local[1] and national programs available to them and to consider how these programs contribute to an innovative ecosystem.

We find that an ecosystem of innovation and entrepreneurship is emerging but fragile in this older industrial economy. Restructuring efforts are complicated by multiple jurisdictions that define their loyalties to competing sports teams. Innovative companies are surviving and trying to grow with the limited resources available. Institutions such as universities and local economic development agencies understand the role they need to play in developing a thriving regional innovation cluster. They lack resources and are limited in what they can accomplish alone yet find it difficult to coordinate to work together. Local innovation programs connect well with entrepreneurs but are limited in scale, and are unable to address the range of perceived needs of innovative firms. The full complement of local actors – businesses and universities, state economic development agencies, community colleges, venture capitalists and commercial bankers – depend on current national innovation policy funds to pay for, or augment, their own efforts.

However, we find that coordination with national programs is problematic: the levels of government do not reinforce one another to form a cohesive infrastructure but instead appear to operate in silos. National programs are relatively better funded than state and local programs and do yield benefits for participating firms, but significant information problems exist as entrepreneurs do not believe they qualify for the national programs targeted toward them or find the application process to be a barrier to their participation. In addition, national programs appear disconnected from one another and entrepreneurs are confused about to whom they might apply. Despite their limitations, national innovation programs remain important to regional industrial restructuring (Clark, 2010).

The next section considers restructuring in the older industrial region, drawing on the context of the Tech Belt region. In section 3, we review the national and local public resources to support innovation among firms. We then present the survey that we conducted of innovative firms in the Tech Belt region in section 4. Section 5 details the results from the survey. Last, we offer reflective conclusions and an assessment of the region's potential to develop as a new high tech cluster; we discuss implications for high tech-based economic development policies in mature industrial regions.

2. THE PROBLEMS OF REGIONAL RESTRUCTURING WITH DIMINISHED CAPACITY

The decline of older manufacturing industries and the associated problems of the places that contained them is well documented (Bluestone and Harrison 1982; Rowthorn and

Ramaswamy 1997; Whitford 2005). The fortunes of industries and regions are inextricably tied and arguably follow the logic of the product life cycle. Places benefit as an industry develops and skilled labor and knowledge spillovers predominate (Audretsch and Feldman 1996; Breschi and Malerba 2001). Resources become tailored to the dominant industrial activity and even subsidiary industries are so connected in local value chains that the growth period offers great opportunity. As the industry matures, growth slows down. When the dominant industries decline the region suffers and stagnates.

There are a number of predictions for what might potentially happen in the region at this point. One argument is that the region will be able to successfully make a transition to a new industry and the cycle of growth will begin again. Glaeser (2005) argues that Boston has been able to successfully remake its economy at least four times since its colonial history through entrepreneurial efforts and due to the presence of skilled capital. The logic is that economic stagnation occurred during the mature stages of the industrial life cycle, which lowered costs that then allowed experimentation and new industries to form. Seattle, which suffered the loss of shipping, timber and manufacturing during the 1970s, provides another example of successful restructuring.

Another contrasting option is that the regional economy will continue to decline, and never regain its former vibrancy or stature (Safford 2004). The outcome may be a significantly smaller city, with a typically older and poorer population – those individuals who are unable or unwilling to relocate. For the individuals that make the declining region their home this outcome is difficult to accept. Moreover, older cities often have massive infrastructure investments and cultural amenities. While these could be written off as sunk costs, the costs of a resulting migration to both the migrants and the places where they relocate are considerable. Destination cities need to levy taxes to build new schools and infrastructure while existing infrastructure in declining places is underutilized. We can question if this relocation of people is efficient national policy.

It has proven difficult to attract relocating firms to declining regions. Entrepreneurism and the growth of new industrial sectors offer the seemingly best strategy. Yet, there is considerable uncertainty about how to best promote entrepreneurship and the transition of the economy. The potential to build on existing expertise declines with time as essential skills atrophy and decline. A contemporary academic observer finds herself waiting for the outcome – for the end of the story – to understand if, and how, regions will be able to redefine themselves.

For the Tech Belt region, our case study of interest, we document the loss of capacity in an attempt to understand the prospects of restructuring. Beginning in the 1980s, despite each city's loss, we trace the commonalities that create their sustained cohesion and help to define the region. Only after understanding their histories and current characteristics can we then begin to speculate how this older industrial region will redefine itself and move forward. In this section, we detail and discuss the implications of the region's industrial decline and the loss of Fortune 500 companies; the current innovative activity, as measured by patents; and the population redistribution from the cities to the suburbs.

The question of how to define a region has been the focus of a long debate (Meyer 1963). While regions have been recognized as an appropriate unit of analysis, there is still debate about the appropriate spatial scale. Regions may be defined as supranational entities that encompass and supplant nation states (Ohmae 1995), or sub-national entities as large as blocks of states, as in US census regions, or as small as the environs of cities, such as Silicon Valley in California. Supplanted by the modern term "clusters," there is still no accepted empirical method to define a region. There is general agreement that we know a region exists when we are confronted with its history, the empirical details of the economy, the mechanics of the social relationships and common conventions in the region.

2.1 A Similar Industrial History

The area from Pittsburgh in western Pennsylvania (PA) to Cleveland in eastern Ohio (OH) has long been defined as a region due to a common industrial history. Economic geographer Richard Hartshorne (1936) describes the economic importance of this region and notes that beginning in the 1870s dominance in heavy industry and manufacturing defined a unified regional economy. Their rich industrial history and the complementary nature of their manufacturing efforts unite these cities with a sense of a shared historical experience of being a locus of industrial innovation from a prior era. Similar to other older industrial regions, there was once great opportunity and large fortunes were made. Each city in this region has its own story: they are united by a rise to industrial prominence followed by a period of decline.

Pittsburgh once dominated world steel production until a lack of investment and failure to innovate in both product and processes amid increased competition led to massive plant closings and layoffs in the early 1980s (Hoerr 1988). Cleveland had a more diversified economy and was best known for its precision production of component parts for a range of other industries. The industrial decline of Cleveland has been more gradual than in Pittsburgh, punctuated by many efforts to restructure alongside a common belief that recovery was within reach. Akron was synonymous with tires due to the agglomeration of Goodyear, B.F. Goodrich, Firestone and General Tire and Rubber (Buenstorf and Klepper 2009). The demise of these firms is attributable to their failure to innovate, despite the fact that historically they had the market locked up (Sull 1999; Sutherland 2006). Competition was redefined by superior performance – most notably by Michelin Tires in France.

Despite their differences, we argue that these cities collectively define a region. In the midst of their separate declines, these cities have moved in concert by the strength of their regional collaborations. Although this area spans the distance of two US states and multiple jurisdictions, a degree of similarity and complementariness unites them and makes them more alike than the other cities within their respective states. As emphasized throughout our interviews with local entrepreneurs and university officials from the area, there was an expression that they were part of a greater, integrated region that spanned the two states. Moreover, the venture capitalists that we spoke with highlighted that those who are interested in investing in this area see it as a region. Visits to Pittsburgh are often followed by stops to Akron or Cleveland and vice versa. Not only do those who comprise the area view it as a region, those from outside the area tend to define it that way as well.

While the region once specialized in specific industries, the high location quotients from the region's golden age are gone. The regional economy now more closely reflects national patterns in both employment and industrial output. The lack of any defined comparative advantage signals that a region has little to build upon or that perhaps any emerging strength is too fragile to be identified.

Examining the presence of corporate headquarters in the region provides another means to assess the loss of economic capacity. The loss of these large firms is greater than simply the associated loss of employment and makes restructuring efforts more difficult as there are fewer corporate resources. In addition to the obvious employment benefits, headquarters serve as anchor tenants for the regional economy, increasing the flow of people and ideas in the regional economy. Many other firms develop as suppliers to headquarter firms, which further supports business services (Gray, Golob and Markusen 1996). Large corporations also increase civic capital through their philanthropic and community efforts, providing funding for arts and cultural amenities that increase the attractiveness of cities.

To examine the location of large corporate headquarters for the region we rely on a list provided by Fortune Magazine that ranks firms based on the dollar amount of profits. The Fortune 500 list was first compiled in 1955; however, we trace the distribution of Fortune 500 firms in Pittsburgh, Cleveland, Akron and Youngstown over three consecutive 22-year intervals – 1965, 1987 and 2009.[2] These years were selected to illustrate the distribution of Fortune firms in the region during their more prosperous years, before the regional economic downturns in the 1970s and 1980s, up through to the present. We examine the most current year for which data were available, 2009, and include 1987 as a midpoint. Table 29.1 provides a listing of the corporate headquarters located in this region and their rankings at the three different time intervals.

Since 1965, the geographical distribution of Fortune 500 headquarters in the United States has shifted away from the Northeast and Midwest.[3] The Pittsburgh, Cleveland, Akron and Youngstown region, however, has undergone a significant and disproportionately severe loss. In 1965, the region had 42 firms ranked in the Fortune 500, with 13 in the Fortune 100. By 1987, the number of headquarters had declined to 28, with 8 firms in the Fortune 100. By 2009, there were 11 firms in the Fortune 500 and none in the Fortune 100. The industrial focus also changed. Headquartered firms in 1965 and 1987 were industrial- and manufacturing-based firms while by 2009 services had become more prevalent. Industrial firms that remained in the region lost standing; for example, Goodyear was 20th in 1965, 35th in 1987 and 141st in 2009. While Goodyear has remained in the region, many headquarters relocated out of the region due to their own set of reasons. As an example, BF Goodrich started as an entrepreneurial firm in Akron in 1870. However, in 1988 it left the tire industry and also left Akron. The firm changed its focus to making components for the aviation industry and moved its headquarters to Charlotte, North Carolina. Not only has the number of Fortune 500 firms headquartered in this region decreased, but those remaining have also dropped in ranking.

For the states of Pennsylvania and Ohio, the number of Fortune 500 firms decreased from 48 to 25 and 42 to 23 from 1965 to 2009, respectively. However, the cities in our region of interest lost 75 percent of their Fortune 500 firm rankings, while other cities in the same states fared better. While the number of Fortune 500 ranked companies in Cleveland dropped from 16 to 4 between 1965 and 2009, its Ohio counterpart, Cincinnati, actually increased the number of Fortune 500 firms over this time frame from 4 to 6 firms. Columbus, Ohio also increased in the number of Fortune 500 firms from zero in 1965 to 4 headquarters in 2009. As for Pennsylvania, Pittsburgh's decrease in the

Table 29.1 List of Fortune 500 firms located in the region, by city

Region								
1965			1987			2009		
Firm	Rank	City	Firm	Rank	City	Firm	Rank	City
Gulf Oil	10	Pittsburgh	BP America	22	Cleveland	Goodyear	141	Akron
Westinghouse Electric	18	Pittsburgh	USX	23	Pittsburgh	FirstEnergy	179	Akron
Goodyear	20	Akron	Rockwell Automation	27	Pittsburgh	PPG Industries	190	Pittsburgh
Firestone	31	Akron	Westinghouse Electric	33	Pittsburgh	Easton	194	Cleveland
Republic Steel	41	Cleveland	Goodyear	35	Akron	United States Steel	211	Pittsburgh
Aluminum Co. of	53	Pittsburgh	Alco	51	Pittsburgh	Parker Hannifin	230	Cleveland
America			PPG Industries	78	Pittsburgh	H.J. Heinz	233	Pittsburgh
National Steel	55	Pittsburgh	H.J. Heinz	88	Pittsburgh	Sherwin Williams	319	Cleveland
Jones & Laughlin Steel	62	Pittsburgh	Bayer Corporation	103	Pittsburgh	Key Corp	356	Cleveland
Goodrich	64	Akron	Easton	110	Cleveland	Wesco International	448	Pittsburgh
General Tire & Rubber	69	Akron	PNC Financial Services	123	Pittsburgh			
Pittsburgh Plate & Glass	73	Pittsburgh	Group					
Youngstown Sheet &	85	Youngstown	Goodrich	180	Akron			
Tube			Sherwin Williams	185	Cleveland			
Eaton, Yale, & Towne	100	Cleveland	Parker Hannifin	203	Cleveland			
TRW	107	Cleveland	Koppers	224	Pittsburgh			
White Motor	109	Cleveland	Reliance Electric	245	Cleveland			
Standard Oil (OH)	116	Cleveland	Allegheny International	250	Pittsburgh			
Heinz	130	Pittsburgh	American Greetings	298	Cleveland			
Rockwell-Standard	133	Pittsburgh	Cyclops Industries	303	Pittsburgh			
Koppers	182	Pittsburgh	Wheeling Pittsburgh	326	Pittsburgh			
Sherwin Williams	197	Cleveland	Aristech Chemical	344	Pittsburgh			
Allegheny Ludlum Steel	205	Pittsburgh	Allegheny Technologies	350	Pittsburgh			

Consolidation Coat	217	Pittsburgh	Ferro	394	Cleveland
Crucible Steel	220	Pittsburgh	Joy Technologies	403	Pittsburgh
Gildden	227	Cleveland	Sealy	429	Cleveland
Midland-Ross	234	Cleveland	Robertson-Ceco	451	Pittsburgh
Porter (HK)	246	Pittsburgh	Cleveland-Cliffs	496	Cleveland
Westinghouse Air Brake	252	Pittsburgh	Standard Products	497	Cleveland
			A. Schulman	498	Akron
Addressograph-Multigraph	256	Cleveland			
Diamond Alklai	309	Cleveland			
Hupp	334	Cleveland			
Rockwell Manufacturing	335	Pittsburgh			
Blaw-Knox	339	Pittsburgh			
Pittsburgh Steel	340	Pittsburgh			
Joy Manufacturing	391	Pittsburgh			
Hanna Mining	396	Cleveland			
Copperweld Steel	426	Pittsburgh			
Harris Intertype	442	Cleveland			
Harbison-Walker Refractories	452	Pittsburgh			
Roberston (H.H.)	453	Pittsburgh			
Clevite	480	Cleveland			
Warner & Swasey	484	Cleveland			
Island Creek Coal	500	Cleveland			

number of Fortune 500 companies was more similar to Cleveland while Philadelphia was relatively stable. This trend suggests that the Pittsburgh, Cleveland, Akron and Youngstown region shared a similar degree of loss compared with other cities in their respective states, while the balance of corporate power relocated to other cities in the state. This loss of corporate headquarters makes restructuring more difficult.

2.2 Population Redistribution

All of the cities in this region have suffered population decline that accompanied industrial decline. However, the trend toward suburbanization has further complicated restructuring efforts. Compared with national trends, the city/suburban distribution for this region is considerably skewed toward the suburbs. Data from the US Department of Housing and Urban Development report that the 2003 national ratio of suburban residents in a metropolitan region was approximately 62 percent, while the average ratio of suburban residents in our region of interest for the same year was 75.5 percent (State of the Cities Data System 2010). As a result of this uneven population distribution, the geographic spread of the region now covers some 220 cities, towns and villages. Suburbanization has united the region as the boundaries of Pittsburgh, Cleveland, Akron and Youngstown are now more blended. Nevertheless, local governance for this region and its unbalanced population distribution is complicated by the fact that the political systems in place were designed for a larger population size.

While regional population has declined, political jurisdictions have not merged or consolidated. As suburbanization has redistributed the population to new locations, additional schools and infrastructure are required. This has resulted in great inefficiencies, redundancy in service provisions and higher taxes. Leaving aside the additional demands of an aging population, these jurisdictional inefficiencies further inhibit the governmental action needed to aid in restructuring. Despite this population redistribution, however, inventive activity is still concentrated in the cities.

2.3 Inventive Activity

Inventive activity is still strong in the region.[4] From January 2006 to May 2010, a total of 8,367 US patents were granted or filed as applications, with over 1,000 unique organizations in the region listed as assignees for the inventions that required intellectual property protection. The distribution has a long tale, with some large established firms in the region applying for several hundred patents over this three-year five-month time frame. Table 29.2 provides a listing of the top patent holders for Pittsburgh, Cleveland, Akron and Youngstown. A small number of established large firms – including Goodyear and the Eaton Corporation – hold a disproportionately large number of patents. Goodyear, for example, accounted for 944 patents, or 11 percent of the region's patents. In addition, universities and other large research organizations located within these cities figure prominently. The University of Pittsburgh, Carnegie Mellon, the Cleveland Clinic and Case Western Reserve are among the most prolific patenting organizations. This reflects the academic stature of these organizations; however, we must remember that university intellectual property can only increase commercial activity through licensing or start-up activity.

Table 29.2 The ten most active inventing organizations

Akron		Cleveland		Pittsburgh		Youngstown	
Assignee/Firm	Patents/ Applications	Assignee/Firm	Patents/ Applications	Assignee/Firm	Patents/ Applications	Assignee/Firm	Patents/ Applications
Goodyear	944	Eaton Corp.	792	University of Pittsburgh	195	R & L Marketing & Sales, Inc.	11
University of Akron	86	PPG Industries	497	Alcoa Inc.	181	Exal Corporation	8
Theken Spine, LLC	17	Electrolux Home Products	243	Chem Image Corp.	154	Gasser Chair Company, Inc.	6
Apteryx, Inc.	16	Parker Hannifin	201	Carnegie Mellon	119	Maui Toys, Inc.	2
Lockheed Martin	16	The Cleveland Clinic Foundation	118	Bayer Material Science LLC.	76	George A. Mitchell Company	1
Advanced Elastomer System, L.P.	14	Case Western Reserve University	109	Westinghouse Electric Co. LLC	48	King Tool & Manufacturing Company, Inc.	1
Ecology Coating, Inc.	14	Ferro Corporation	108	Vocollect Healthcare Systems, Inc.	35	Next Energy Wave, Inc.	1
OrthoHelix Surgical Designs, Inc.	12	Sherwin Williams	80	Dynamics Inc.	32	Renal Clean Technology	1
GOJO Industries, Inc.	8	Lubrizol Advanced Materials, Inc.	58	Jenmar Corporation	31	The Ink Factory, Inc.	1
Akron Special Machinery; Khyber Technologies Corporation; Spinematrix, Inc., T. K. M. Unlimited, Inc.	6	The L.D. Kichler Co.	50	Emerson Process Management Power & Water Solutions, Inc.	27	Turning Technologies, LLC	1

The majority of patents are attributed to small and medium sized firms: 90 percent of the firms hold 12 or fewer patents. What is more, 53 percent of the organizations had filed exactly one patent. This may be indicative of new start-up activity – a question that warrants follow-up investigation.

Data from the Delphion Patent database show that the large majority of the most recent patent activity is taking place within the city boundaries, leaving only a portion of the activity in the region's connected suburbs. Between January 2006 and May 2010, 94 percent of the patent activity in the Akron metropolitan area took place within Akron's city limits, 80 percent within Cleveland's city limits, 98 percent within Pittsburgh's city limits and 35 percent within Youngstown's city limits. Despite the inconsistency with Youngstown's patent activity compared with the other three cities, the large majority of patent activity throughout the region was most vibrant within the city limits, with the weighted average of the four cities approximating 86 percent of all patent activity. Even with the suburbanization of the population, inventive activity is still geographically concentrated in the older industrial core cities.

The inventive capacity of the region holds promise for future economic growth. A typical story is that an individual will decide to become an entrepreneur to capitalize on an idea from their former employment. The new firm they form receives business advice from a mix of public and private programs, and may be introduced to a public financing program for start-up capital or an angel investor. The US firm may apply for a Small Business Innovation and Research (SBIR) grant to aid in developing the technology, or it may need to negotiate patents and licenses with the university. Technological success may lead the firm to graduate and qualify for an Advanced Technology Grant. It may also require the transition of the firm from public to private venture capital. In navigating these programs and services, the firm encounters a variety of organizations with distinctive cultures, procedures and levels of awareness of one another. What can often be one of the largest obstacles for a firm during this process, however, is navigation through this array of national and local programs supporting various steps of this entrepreneurial endeavor.

3. PUBLIC RESOURCES FOR RESTRUCTURING

In an ideal world, entrepreneurs with good ideas would be able to develop those ideas into new products and processes that would generate good paying jobs and bring wealth and prosperity to their local economy. However, in reality entrepreneurs generally face obstacles that comprise the infamous "valleys of death" (Auerswald and Branscomb 2003). In moving an idea through the initial stages of basic research to development and then ultimately to commercialization, challenges including insufficient early stage funds and market barriers often thwart this progression. These shortcomings demand the need for aligned public policy efforts to provide resources for entrepreneurs engaged in technological innovation. There has been a great deal of experimentation in the organizational arrangements that implement the national and local programs. Firms as a result have been able to tap into the services and resources across these agencies; university, national and industry laboratories; and business and engineering consultants from the public, private and non-profit sectors. In an effort to describe this public effort, we briefly detail the most prominent national and local public innovation programs accessible in the Pittsburgh, Cleveland, Akron and Youngstown region. We address each set of programs in turn.

3.1 National Innovation Programs

Over the past 30 years, largely in response to a crisis of competitiveness in key industries, US R&D investment has become dedicated to stimulating industrial technological innovation and development. At the national level there has been a good deal of political controversy associated with these programs, drawing charges of "picking winners" and engaging in "corporate welfare." However, both Democratic and Republican administrations have supported a number of recent initiatives targeted to innovation and entrepreneurship. The most prominent of these national programs supporting firm-level innovation are the SBIR and Small Business Technology Transfer (STTR) programs.

While the US Small Business Administration oversees the central administration of these two programs, 11 different executive-branch departments and agencies are involved with either one or both of these programs.[5] The Small Business Administration program awards competitive funds to firms through three Phases. In addition, US federal mission agencies that include the National Science Foundation (NSF), Department of Commerce (DoC), Department of Labor (DoL) and Department of Energy (DoE) administer complementary programs that provide financial grants, training and resources for early stage firm development. The NSF typically grants financial support to competitive firms, while the National Institute of Standards and Technology program through the DoC, the Workforce in Regional Economic Development program through the DoL, and the Energy Innovation Hubs through the DoE assist firms through means other than grants. This national support ranges from knowledge transfer and training programs to firm management skills. The most prevalent programs are outlined in Table 29.3.

3.2 Local Innovation Program for the Tech Belt Region

In addition to the national government's efforts, US states have dramatically increased their support for science and technology over the same time period, with the majority of funds dedicated to stimulating industrial interactions with local university research. There is little political controversy associated with such programs at the local level (Plosila 1986). In fact, states tend to demand an explicit link between science and technology research and economic development in both program design and evaluations. The most salient evaluation metric to state political leaders is job growth and retention, which has taken on new immediacy in the current economy.

In the Pittsburgh, Cleveland, Akron and Youngstown region, a number of agencies provide resources on the local level. Nevertheless, we found that these local agencies were cognizant of each other's activity and supportive of each other's overarching missions to support local innovation and economic development. Only in 2012 did the Tech Belt region receive public support to establish the National Additive Manufacturing Innovation Institute.[6] Up until this pilot was nationally funded, public efforts to support this entire region were fragmented, although we did not find an agency that catered to the entire Tech Belt region.

In Pittsburgh, Innovation Works and the Pittsburgh Life Sciences Greenhouse were

Table 29.3 US National innovation programs

National Agency	National Programs[1]	Budget	Other Services
Small Business Administration	Small Business Innovation Research (SBIR)	$250,000 (Phase I grant) $2,000,000 (Phase II grant)	Aim to stimulate technological innovation; to use small business to meet national research needs; to broaden participation; and to increase private sector commercialization
	Small Business Technology Transfer (STTR)	$250,000 (Phase I grant) $2,000,000 (Phase II grant)	Expansion of public–private sector partnership to include joint venture opportunities for small business and non-profits
National Science Foundation	Partnerships for Innovation (PFI)	$600,000 (three-year grant)	Foster connections between public and private organizations to spur innovation
	Advanced Technology Education Program (ATE)	$200,000 (three-year grant)	Help to facilitate the education of high technology technicians via two-year colleges
	Industry–University Cooperative Research Partnership Program (I/UCRP)	$10,000 (seed grants) $70,000 (initial five-year annual award)	Assist in transfer of university basic research to industry
Department of Commerce	Manufacturing Extension Program (MEP)	$125 million proposed FY 2010 budget	Network of 59 centers in 393 locations in every state and in Puerto Rico to help small businesses embrace process improvements and growth strategies to increase competitiveness
	Technology Innovation Programs (TIP)	$69 million FY 2009 budget	Created by America COMPETES Act of 2007 to accelerate high risk, high reward, private sector research
Department of Labor	Workforce in Regional Economic Development, Employment & Training Administration (ETA)	$326 million budget since 2005	Focus on building a skilled workforces
	Community-Based Job Training Program	$125 million annual budget	Develop partnerships between the workforce system and community college–vocational education networks
Department of Energy	Energy Innovation Hubs	$280 million proposed budget for FY2010	Support cross-disciplinary R&D transforming energy technologies into commercial products

Note:[1] This selection of programs is by no means exhaustive – there are many other programs in many other agencies, such as the Rural Business Enterprise Grants Program in the Department of Agriculture, the Enterprise Zones Initiatives in the Department of Housing and Urban Development, and the Brownfield Redevelopment Programs in the Environmental Protection Agency, among others. While all of these contribute in some way or another to the development of innovation clusters, the programs outlined in this table are of the greatest size, impact and prominence.

Table 29.4 Local public sector innovation organizations in the Tech Belt region

Local Organization	Mission Statement	2010 Operating Budget (approx. estimations)
BioEnterprise (OH)	To make Northeast Ohio a nationally recognized center of bioscience commercialization by creating, attracting and accelerating the growth of high potential bioscience businesses	$4,500,000
GSP Consulting (PA)	To enable the growth of organizations, companies, communities and economies through a suite of innovative government, business and economic development consulting services	$6,000,000
Innovation Works (PA)	To create a vibrant technology-based economy in southwestern Pennsylvania by providing investment and business assistance to high growth companies	$9,000,000
JumpStart (OH)	A non-profit organization creating economic transformation in Northeast Ohio by providing resources to entrepreneurs leading high potential, early-stage companies	$10,662,500
NorTech (OH)	A catalyst for expanding Northeast Ohio's technology industries, is leading an effort to develop regional innovation clusters that will spur job creation, capital attraction and long-term positive economic impact	$2,500,000
Ohio Third Frontier (OH)	Unprecedented commitment to create new technology-based products, companies, industries and jobs. Its strategic intent is to create an "innovation ecosystem" that supports the efficient and seamless transition of great ideas from the laboratory to the marketplace	$145,000,000
Pittsburgh Life Sciences Greenhouse (PA)	To create, nurture and help establish a globally dominant life sciences industry in western Pennsylvania	NA
University of Akron Research Foundation (OH)	Organized and operated exclusively for the benefit of the University of Akron, and such other public Ohio institutions of higher education, regional hospitals, regional municipalities, and regional economic development and innovation support organizations, that qualify as charities	$4,500,000

among the larger agencies. In Akron, the non-profit University of Akron Research Foundation supported early stage firm activity, while JumpStart, BioEnterprises, NorTech and GSP Consulting supported activity in Cleveland. The Ohio Third Frontier was the largest of all these programs, providing assistance not only to the Ohio part of our region of interest, but for the entire state of Ohio as well. Table 29.4 lists these local innovation organizations and details their mission statement and annual level of funding.

With the array of national and local programs available to firms in the region, the important question that remains is: how well, if at all, do these organizations complement one another to succeed in their mission of supporting innovative firms through the early stages of development? It is not only important to examine whether national and local programs provide sufficient flexibility for the firms to use these services effectively, but it is even more critical to assess the extent to which firms are even aware of these programs. Many high tech firms by-pass this array of services or have only limited interactions en route toward establishing a product or business despite these organizations' efforts to provide support along the way. In light of this trend, we seek to understand the level of interactions between these national programs, local public organizations and firms. What are the demands of innovative firms, and how adequately do public resources support their needs and the needs of the region?

To do this we turn to the survey, which asks technology-driven firms in the Tech Belt region these very questions. While this analysis only serves as a case study, and thus the implications of these results are not explicitly generalizable, this analysis serves as an initial step toward understanding the relationship between these public organizations and technology-driven firms as they both seek to promote economic development.

4. DATA AND METHODOLOGY

In the summer of 2010, we conducted a survey of firms in the Tech Belt region. The overarching objective of the survey was to study the efficacy of national and local innovation programs and intergovernmental cooperation for the Pittsburgh, Cleveland, Akron and Youngstown regional economy. We began our research by conducting a series of interviews with 25 individuals in the region representing a mix of public and private enterprises engaged in innovation. Our interviews included representatives from university technology transfer offices, technology incubators, state-funded technology-based economic development organizations, start-ups, small and medium sized companies, angel investors and locally focused venture capitalists. Based on what we learned, we designed a survey that was sent to the universe of technology-driven firms in the region. The survey comprised 38 questions asking each firm to identify recent national and local public support, to comment on the outcomes from the public funding, to note challenges faced both by the firm and regional economy, to assess the level of intergovernmental cooperation between the varying levels of government, and to provide general information on the organizational characteristics of the firm.

Our challenge was to compile a comprehensive census of technology-driven small and medium sized firms. This information is essential to understanding the state of a regional economy, yet is elusive to find (Reedy 2011). There was no central repository of firms for any individual city, much less one comprising a region bridging multiple cities and states. In addition, while such a list would likely be out of date as soon as it was available – as new firms are started and others failed – such data are the only way to study regional dynamics.

Given these limitations, we drew from a number of sources to identify a robust list of innovative firms in technology-based industries within this region. We searched for firms that complied with the industrial and geographic restrictions for the scope of the study; in addition, the time frame specified for this search spanned from January 1, 2006 to May 20, 2010 to include the most recent activity. As an initial step, representatives from a number of regional economic development organizations[7] provided lists of their client firms or firms with which they had some relationship or contact. To increase our sample frame and mitigate any self-selection issues, we additionally used information from the following three sources – patent applications,[8] firms that had received funding from a list of national research programs, and a list of firms from the Capital IQ database[9] – to define the sampling frame.

Given that patent applications provide one means to understand inventive activity, we compiled a list of US patent assignees located in the regional area.[10] For national programs, we pulled information from the following: SBIR and STTR awardees; Technology Innovation Program- and Manufacturing Extension Partnership-affiliated firms;[11] NSF Partnerships for Innovation, Advanced Technological Education, and Industry and University Cooperative Research Program awardees; and National Institute of Health research grant recipients. Lastly, we utilized the Capital IQ financial transaction database to pull a list of firms in the region which had received a financial investment (Feldman and Zoller 2011).

Altogether, after eliminating duplicates, our list contained 4,195 firms. An email survey was sent to the CEO (or equivalent). Immediately, however, approximately 850 emails (roughly 20 percent) bounced back due to invalid or inactive email addresses. The relatively large number reflects the turbulent and inherent nature of start-ups and entrepreneurial activity. After sending two electronic requests to the firm CEOs, 269 replied, yielding an 8.04 percent response rate. The lack of information about the universe of firms makes it difficult to assess the representativeness of our survey respondents. Following Armstrong and Overton (1977), we analyzed the responses of the sample based on when they responded. We find no evidence of systematic differences between the first 75 percent and the last 25 percent of the responses. We also find that the firms that were on the client lists of the local development agencies had a relatively low response rate (less than 10 percent of the client lists of any agencies responded; this statistic, however, mirrored the response rate from the larger sample). This indicated that our results are not overly influenced by selection bias.

Table 29.5 presents the industrial make-up for the firms sampled. Using the Standard Industrial Classification codes to classify the firms in the sample, the two most prominent industries are as follows: 23 percent self-identified as part of the information technology industry and 14 percent indicated health care.

In our effort to survey the active innovative, technology-driven firms in the region, besides classifying the firms by industry, we found that the firms fell into three distinctive categories. The first consists of old, large established companies that started between 1870 and the end of World War II in 1945. This group includes the Fortune 500 firms or spin-off companies; over the past 50 years they have served as the industrial backbone for the region. The second group comprises older mid-sized firms that started between 1946 and 1999. On average these firms had 85 employees and exhibited competitive levels of patent and licensing activity compared with the other firms. The third group of firms consists of newer start-ups that launched after 2000. With less experience and limited capacity, these firms qualified for the largest portion of local and national programs.

Table 29.5 Distribution of firms in region by Standard Classification Code

Industry	Distribution (%)
Aerospace and defense	7.8
Automobiles, motorcycles and components	3.7
Bioscience	5.9
Capital goods	6.8
Commercial and commercial services	6.5
Consumer goods and services	3.7
Energy	7.8
Finance	0.6
Health care	14.3
Information technology goods and services	23.3
Materials	6.8
Telecommunication services	1.2
Transportation services	0.6
Utilities	1.8
Not elsewhere classified	9.0

With the old, large established firms generally in strong standing given their history and capacity, and the new start-up firms supported by a plethora of public programs, we found the mid-sized, mature firms to be an underappreciated and underserved innovative resource in the economy. While most of the public programs are designed to assist new firms through their initial stages of development and commercialization, these mid-sized firms participate less in government programs compared with their younger counterparts, given qualification restrictions. If the revitalization of a depressed economy is a central goal for a region, then perhaps public programs designed to support more experienced firms with demonstrated levels of production and operation would be a viable alternative for stimulating economic development.

5. RESULTS FROM THE SURVEY

5.1 National and Local Public Funding Activity

To gauge the degree of public funding in the region, we asked firms to indicate their type and level of activity between 2007 and 2010. Of those firms which responded, 37 percent applied for or received some form of national support over the three-year period, while 44 percent of firms indicated that they received local support over the same time frame. This difference suggests a greater level of activity between the firms and local programs as compared with the national programs.

When asked if they would like to work with the various public programs in the future, 34 percent indicated that they would be interested in working with national programs in the future, while 40 percent expressed interest in working with the local programs. These responses stand in contrast to the 5 percent and 6 percent which indicated that they would not want to work with national and local programs, respectively. These results not only indicate a pronounced interest in working with the public programs, but again these respondents have suggested a greater level of activity and interest in local programs compared with the national ones.

We also found that a portion of firms were not only aware of local funding programs in their neighboring states, but that they were interested in the prospect of working with those regional funding sources as well. Of those firms which responded to this question, 24 percent of those from PA expressed interest working with the OH organizations, while 26 percent of the OH respondents expressed interest in PA organizations. This demonstrates the firms' interest in the greater region, which supports the claim that this area does in fact consider itself a region.

5.2 Outcomes from Public Funding

In addition to estimating the frequency of activity and interest in working with the various public programs, we asked firms to provide information on the outcomes that resulted from their recent support. For both the national and local programs, the three most prominent benefits were new or significantly improved: goods, services, and methods of manufacturing or producing goods or services. Table 29.6 lists the distribution of responses for both sources – national and local.

As is illustrated by their responses, what differed most was not the order of their answers, but rather the degree to which the source of funding affected the firms. While the relative level of activity between firms and local programs exceeded the activity with national programs, these survey results indicate the outcomes gained from national support outweighed those from local sources. Proportionally more firms reported higher levels of benefits from national programs than local programs. This trend remained consistent in the results requiring firms to specify additional benefits. The most prominent additional benefits from national funding were as follows: (i) gained experience that was not elsewhere available (40 percent); (ii) helped secure additional local government funding (38 percent); (iii) helped secure additional private funding (32 percent); and (iv) provided help with business development (32 percent). As for additional benefits gained

Table 29.6 Outcomes of US national and local funding

Outcome of funding	National (%)	Local (%)
New or significantly improved goods	54.4	24.4
New or significantly improved services	39.1	17.7
New or significantly improved methods of manufacturing or producing goods or services	45.7	10.9
New or significantly improved logistics, delivery or distribution methods of inputs	2.2	4.2
New or significantly improved support for processes, including maintenance systems or operations for purchasing, accounting or computing	8.7	5.0

Note: 'Local' includes state and local programs.

from local programs, the most prominent were as follows: (i) provided industrial contacts (19 percent); (ii) helped secure additional private funding (19 percent); (iii) helped secure additional local government funding (17 percent); and (iv) provided help with business development (17 percent).

The discrepancy in the difference between the outcomes gained by the two sources is perhaps indicative the relative size of national grants compared with local resources. Nevertheless, these results suggest that these various sources together play a critical role in supporting firms in the area. It appears as though the firms in the area demand support from both national and local programs. What remains to be determined, however, is how well these various levels of public support work together to complement one another. The results from the survey on intergovernmental cooperation are discussed in the next sub-section.

We included a question designed by Edwin Mansfield (1992) to assess the contribution of government funding by asking, "if your organization did not receive this funding, what percentage of current products would not have been developed or would have taken longer to develop?" Regarding national funding, survey respondents indicated that 40 percent of their current products would not have been developed, and 50 percent of their products would have been developed but would have taken longer. As for local funding, 46 percent of their current products would not have been developed, and 43 percent of their products would have been developed but would have taken longer. Despite the similarities with these responses, the subtle difference between them suggests that local funding may serve a more central role in product development. Nevertheless, given the limited scale of this survey, we hesitate to draw generalizations from this difference.

On a final note, we asked firms to comment on their experience with the one-time national stimulus funding administered through the 2009 American Recovery and Reinvestment Act (ARRA). Of the respondents, 35 percent of the firms applied for ARRA funding. Of those which applied, 57 percent were denied funding and 17 percent had applications pending, leaving 26 percent as grant recipients. The results from this question were not very positive. Given the severity of the economic climate in 2009 when the Act was passed, this nominal amount of activity suggests that either the firms were not aware of the program, or the administrative burdens to apply for funding outweighed the prospective benefits of receiving public support. While the results above suggest notable activity and interest between firms and public resources, the limited amount of activity with the ARRA funds suggests the firms' hesitation or limited knowledge of this program.

5.3 Intergovernmental Cooperation

The potential benefits gained from greater cooperation between local governments and US federal agencies are notable. One of the goals motivating this interest is to explore the opportunities for partnership between the local governments and national agencies in the development and implementation of innovation policies. This is an explicit focus of the America COMPETES Act and studies commissioned by the National Governor's Association and the State–Federal Partnership task force. Each of these studies has called for greater recognition of the ability to bridge the gap between nationally supported research, local technology development programs and the commercialization of technology.

Table 29.7 Firm survey responses regarding efficacy of intergovernmental cooperation

Intergovernmental cooperation	Agree (%)	Disagree (%)
Local economic development programs provide good information about national R&D programs	26.5	41.4
Local economic development programs provide matching funds for national programs	12.6	41.8
Local economic development programs help with grant applications for national programs	26.3	39.3
Local programs work with national programs	16.9	33.1

With respect to intergovernmental cooperation, we asked the firms a series of questions to gauge how well they perceived that local governments did the following: (i) provide information about national R&D programs; (ii) provide matching funds for national programs; (iii) assist with grant applications for national programs; and (iv) work with national programs. Despite the level of subjectivity inherent in these questions, the results consistently suggested that firms felt intergovernmental cooperation could be improved.

As illustrated by the results in Table 29.7, with every question asked, more firms disagreed with the statement than agreed. The general trends of these results suggest an underlying skepticism among the firms regarding how well the different levels of government cooperate with one another. Taken together, these results suggest a fundamental flaw in the US national government's traditional approach to economic development and local government's capabilities of tapping into those national programs. National innovation programs are commonly deployed around the country without any regard for the cluster-based development objectives of local innovation programs. Given that firms depend on support from both national and local agencies, streamlining the various programs is imperative for maximizing the benefit to firms.

5.4 Challenges Facing the Firms and the Region

In addition to gauging the level of activity, benefits and perceived intergovernmental cooperation between the various public programs, we asked firms a series of questions regarding their current challenges and strengths. Regarding the most prevalent challenges, of those firms which responded to this question, 73 percent indicated that lack of expansion capital was either a moderate or significant problem. This response was followed by 62 percent which specified that lack of short-term funding and health care costs posed moderate or significant problems. These shortcomings appeared to be consistent among all the firms in the sample, regardless of their age or location within the region.

As for problems faced by the region, the firms offered similar responses, noting that lack of access to seed-stage and early-stage venture capital (65 percent), limited working capital (64 percent) and lack of expansion capital (63 percent) were among the most prevalent challenges. These results suggest a demand among the firms for public support throughout various stages of firm development.

Turning to the firms' strengths, 54 percent indicated that access to vendors with expertise matched to the firm's industry sector or growth stage was not a problem; 50 percent felt that legal expertise was easily accessible; and 48 percent found the university expertise for their industry/technology to be a notable strength. As for the region's strengths, the responses listed above likewise were reported as the most prevalent.

The results suggest that the region has the infrastructure for sustained economic growth; what is lacking, however, is the short-term financing needed for supporting these firms during their initial stages of growth alongside the expansion capital needed to grow over the long term. These regional strengths and weaknesses are critical to understanding how local and national programs can be better deployed to help the region's economy grow.

5.5 Industry Cluster Coordinator

We asked the firms a series of questions regarding their thoughts on an industry cluster coordinator. The large majority of those which responded indicated they would like to work with an industry cluster coordinator if one were available (84 percent). As for the administration of the position, 47 percent felt that the coordinator should be a regional, multistate function rather than a national (10 percent), state (10 percent), local (19 percent) or private (19 percent) function. Over half of the respondents (52 percent) indicated that public and private sources should jointly fund the coordinator position, but that it should be privately run. Creating a position to administer this task is a notable and desirable aim; what the results from this survey suggest is that a large portion would welcome a cluster effort to promote bottom-up competitive advantages for the region.

6. REFLECTIVE CONCLUSIONS

We examine a region that spans at least two US states, multiple counties and cities, and smaller political jurisdictions. This region is united by a common industrial history that influences the current industrial composition; however, efforts to move forward are complicated by this jurisdictional patchwork. While there are always dangers of generalizations based on case studies, this chapter provides insights about the detailed nature of the use of government programs seeking to support the industrial restructuring of firms in a mature region. This study further demonstrates the utility of surveying innovative firms to learn about industrial restructuring and the extent to which firms consider themselves to be part of a region.

Even as firms view themselves as part of a larger region and are aware and interested in programs in the adjacent state, their specific location in a state within the region dictates what programs they are eligible to participate in and receive benefits from. This seems to be inefficient at best and duplicative. Theory would dictate that there would be gains if government treated the region as a unified whole. While the outside world views the firms in the region similarly, they struggle to define institutions of collaborative governance and their efforts are inhibited by jurisdictional boundaries and rigidities. A regional policy approach would benefit these firms and also allow for programmatic learning.

We find that in the eastern Midwest of the US an ecosystem of innovation and entrepreneurship is emerging but fragile, requiring the sustained efforts of local and national agencies working together to help firms survive and thrive. The problem is, we also found that local innovation programs that connect well with entrepreneurs are limited in scale, and the handoff with national programs can be problematic at best. Within this one region we find companies that have worked well with the limited resources available to them. In addition, universities and local economic development agencies thoroughly understand the role they need to play in developing a thriving regional innovation cluster, but lack resources and are limited in what they alone can accomplish. Amid the current stumbling economic recovery, our intention was to understand how regional economies stitch themselves together from the bottom up – what makes them tick and what they need to grow and thrive.

It is crucial for policymakers to consider a variety of measures to help businesses compete and expand their workforces. Alas, national innovation policy aimed at boosting the competitiveness of the US economy through investments in science and technology commercialization are grounded in 20th century economic development strategies that no longer match the pressures of the 21st century global economy. At the regional level, however, many businesses and universities, local economic development agencies and community colleges, venture capitalists and commercial bankers all depend on current national innovation policy funds to pay for or complement their own efforts to boost economic development. Despite the clear limitations in the scale of existing national programs, they remain important to economic competitiveness. Understanding the efficacy of these national innovation programs is the first step toward understanding how to improve or replace them.

NOTES

1. Throughout this chapter, *local* programs refer to public programs at the state level or smaller; this includes sub-state and city-level initiatives.
2. The data for the Fortune 500 companies in Pittsburgh, Cleveland, Akron and Youngstown were retrieved from the Fortune 500 magazine. The 1965, 1987 and 2009 Fortune 500 lists were printed in Fortune Magazine in the May 1966 issue, April 25, 1988 issue, and May 3, 2010 issue, respectively.
3. Between 1965 and 2009, the number of Fortune 500 companies in Texas and California increased significantly from 8 to 57 and 30 to 56, respectively.
4. Data from the US Patent and Trademark Office (USPTO) provide patent activity by Metropolitan Statistical Area from 1990 to 1999 (source: http://www.uspto.gov/). Over two consecutive four-year intervals – 1992 to 1995 and 1996 to 1999 – the aggregated number of utility patents for the Pittsburgh, Cleveland, Akron and Youngstown MSAs were 6,252 and 6,753, respectively. Compared with the patent activity for the region between January 2006 and May 2010, this provides evidence that the area is increasing the level of patent activity. It is important to note, however, that the time frame and geographical areas used in the USPTO report and this research analysis are similar but not explicitly comparable. For this study, we collected data over a three and a half-year interval and expanded our count to include a greater part of the Tech Belt region than merely the four individual MSAs. Despite these discrepancies, these data suggest that patent activity has been increasing for the region.
5. The 11 participating US federal departments and agencies in the SBIR program are the Department of Agriculture, Department of Commerce, Department of Defense, Department of Education, Department of Energy, Department of Health and Human Services, Department of Homeland Security, Department of Transportation, Environmental Protection Agency, National Aeronautics and Space Administration, and National Science Foundation. The departments that participate in the STTR program are the Department of Defense, Department of Energy, Department of Health and Human Services, National Aeronautics and Space Administration, and National Science Foundation.
6. Source: http://namii.org/, accessed July 2010.

7. These regional partners included representatives from JumpStart, NorTech and BioEnterprises in Ohio and Innovation Works and the Pittsburgh Life Sciences Greenhouse in Pennsylvania.
8. Patent data were pulled from the Delphion database.
9. We utilized the Capital IQ financial transactions database to pull a list of firms and dealmakers in the regional area. The result was an additional 1,300 firms added to the sample.
10. From the responses that we received, more than half of our survey respondents had applied for US patents starting in 2007. Firms that did not apply for patents were younger or working on software, research or professional services that do not require patent protection. This indicates that using patent data provides coverage of about half of the population of innovative firms in the regional economy.
11. Specifically those associated with Catalyst Connection, the WV Manufacturing Extension Program, and the Manufacturing Advocacy and Growth Network.

BIBLIOGRAPHY

Armstrong, J.S. and T.S. Overton (1977) "Estimating Nonresponse Bias in Mail Surveys," *Journal of Marketing Research*, **14**, 396–402.

Audretsch, D.B. and M.P. Feldman (1996) "R&D Spillovers and the Geography of Innovation and Production," *American Economic Review*, **86**(3), 630–640.

Auerswald, P.E. and L.M. Branscomb (2003) "Valleys of Death and Darwinian Seas: Financing the Invention to Innovation Transition in the United States," *Journal of Technology Transfer*, **28**, 227–239.

Bluestone, B. and B. Harrison (1982) *The Deindustrialization of America*. New York: Basic Books.

Breschi, S. and F. Malerba (2001) "The Geography of Innovation and Economic Clustering: Some Introductory Notes," *ICC*, **10**(4), 817–833.

Bristow, G. (2010) "Resilient Regions: Replacing Regional Competitiveness," *Cambridge Journal of Regions, Economy and Society*, **3**(1), 153–167.

Buenstorf, G. and S. Klepper (2009) "Heritage and Agglomeration: The Akron Tyre Cluster Revisited," *Economic Journal*, **199**(537) 705–733.

Chinitz, B. (1961) "Contrasts in Agglomeration: New York and Pittsburgh," *American Economic Review*, **51**(2), 279–289.

Clark, J. (2010) "Coordinating a Conscious Geography: The Role of Research Centers in Multi-scalar Innovation Policy and Economic Development in the US and Canada," *Journal of Technology Transfer*, **35**(5), 460–474.

Clark, J. (2013) *Working Regions: Reconnecting Innovation and Production in the Knowledge Economy*. Oxford: Routledge.

Clark, J., H. Huang and J.P. Walsh (2010) "A Typology of 'Innovation Districts': What it Means for Regional Resilience," *Cambridge Journal of Regions, Economy and Society*, **3**(1), 121–137.

Feldman, M.P. and L. Lanahan (2010) "Silos of Small Beer: A Case Study of the Efficacy of Federal Innovation Programs in a Key Midwest Regional Economy," *Science Progress*, September.

Feldman, M.P. and T. Zoller (2011) "Dealmakers in Place: Social Capital in Regional Entrepreneurial Economies," *Regional Studies*, **46**(1), 23–37.

Glaeser, E.L. (2005) "Reinventing Boston: 1630–2003," *Journal of Economic Geography*, **5**, 119–153.

Gray, M., E. Golob and A. Markusen (1996) "Big Firms, Long Arms: A Portrait of a 'Hub and Spoke' Industrial District in the Seattle Region," *Regional Studies*, **30**(7), 651–666.

Hartshorne, R. (1936) "A New Map of the Manufacturing Belt of North America," *Economic Geography*, **12**(1), 45–53.

Hoerr, J.P. (1988) *And the Wolf Finally Came: The Decline of the American Steel Industry*. Pittsburgh, PA: University of Pittsburgh Press.

Lorant, S. (1999) *Pittsburgh, the Story of an American City*, 5th edn. New York: Derrydale Press.

Mansfield, E. (1992) "Academic Research and Industrial Innovation: A Further Note," *Research Policy*, **21**(3), 295–296.

Markusen, A. (1985) *Profit Cycles, Oligopoly, and Regional Development*. Cambridge, MA: MIT Press.

Markusen, A. (1986) *High Tech America: The What, How, Where, and Why of the Sunrise Industries*. Boston: Allen & Unwin.

Markusen, A. (1996) "Sticky Places in Slippery Space: A Typology of Industrial Districts," *Economic Geography*, **72**(2), 293–313.

Markusen, A. (2003) "Fuzzy Concepts, Scanty Evidence, Policy Distance: The Case for Rigour and Policy Relevance in Critical Regional Studies," *Regional Studies*, **37**(6–7), 701–718.

Martin, R. and P. Sunley (2003) "Deconstructing Clusters: Chaotic Concept or Policy Panacea?" *Journal of Economic Geography*, **3**(1), 5–35.
Meyer, J. (1963) "Regional Economics: A Survey," *American Economic Review*, **1**(1), 19–54.
Miller, C. and R.A. Wheeler (1997) *Cleveland: A Concise History, 1796–1996*, 2nd edn. Bloomington, IN: Indiana University Press.
Noyelle, T. and T. Stanback (1984) *The Economic Transformation of American Cities*. Totawa, NJ: Rowman & Allenheld.
OECD (2007) *Competitive Regional Clusters: National Policy Approaches*. Paris: OECD.
Ohmae, K. (1995) *The End of the Nation State: The Rise of Regional Economies*. New York: Free Press.
O'Neill, B. (2009) *The Paris of the Appalachia: Pittsburgh in the Twenty-First Century*. Pittsburgh, PA: Carnegie Mellon University Press.
OPTI (2000) *A Technology Assessment and Forecast Report*. Alexandria, VA: US Patent and Trademark Office.
Plosila, W. (1986) "Government Initiatives to Encourage Technological Innovation: The Ben Franklin Partnership Initiative," in *Entrepreneurship and Technology: World Experiences and Policies*, edited by W.S. Brown and R. Rothwell, 83–90. Harlow: Longman.
Reedy, E.J. (2011) "How the Statistical System Can Help Create Jobs," AMSTATNEWS, February 1, 2011. http://magazine.amstat.org/blog/2011/02/01/statviewfeb11/.
Rosenfeld, S. (1997) "Bringing Business Clusters into the Mainstream of Economic Development," *European Planning Studies*, **5**(1), 3–23.
Rowthorn, B. and R. Ramaswamy (1997) *Deindustrialization: Causes and Implications*. Washington, DC: International Monetary Fund.
Safford, S. (2004) "Searching for Silicon Valley in the Rust Belt: The Evolution of Knowledge Networks in Akron and Rochester," *MIT IPC Local Innovation Systems* Working Paper no. 04–002.
State of the Cities Data System (2010) Department of Housing and Urban Development: Policy Development and Research Information Service, Information retrieved at http://socds.huduser.org/Census/Census_Home.html, accessed July 2010.
Storper, M. (1997) *The Regional World: Territorial Development in a Global Economy*. New York: Guilford Press.
Sull, D. (1999) "The Dynamics of Standing Still: Firestone Tire & Rubber and the Radial Revolution," *Business History Review*, **73**(3), 430–464.
Sutherland, J. (2006) "The Ideas Interview: Tom Kelley," *The Guardian*, August 22, available at http://www.theguardian.com/technology/2006/aug/22/news.ideas, accessed July 2010.
Tittle, D. (1992) *Rebuilding Cleveland: The Cleveland Foundation and Its Evolving Urban Strategy*. Columbus, OH: Ohio State University Press.
Whitford, J. (2005) *The New Old Economy: Networks, Institutions, and the Organizational Transformation of American Manufacturing*. New York: Oxford University Press.

30. From skill mismatch to reinterpretation: challenges and solutions for manufacturing worker retention and recruitment

Nichola J. Lowe

> The ultimate resource of an industrial economy is its people.
> (Dertouzos et al., *Made in America*, 1989: 21)

> Skill builds by moving irregularly, and sometimes by taking detours.
> (Sennett, *The Craftsman*, 2008: 238)

Manufacturing jobs are coveted, and for good reason. On average they provide significantly better wages and benefits compared with most service sector jobs that share a similar labor pool. Yet changes to educational requirements for new and incumbent workers potentially threaten job access for moderately educated workers that have long enjoyed a middle class existence as result of their employment in manufacturing. This emergent labor market challenge motivates a more critical examination of the way that skill, and more specifically changing perceptions of skill, might affect manufacturing worker recruitment, retention and mobility. While this shifting educational requirement is especially pronounced in the United States, related pressures have emerged in labor markets in Europe and Asia, and thus indicate an opportunity for international comparison and policy reflection.

An educational uptick in the U.S. manufacturing workforce obviously represents an opportunity to enhance job quality and worker status. Still, it is also emblematic of a profound and prolonged institutional transformation that has the potential to produce the opposite effect for many workers in this sector. The post-World War II employment model supported quasi-lifetime employment and created pathways for upward mobility for moderately educated workers that were supported by company-specific training systems (Doeringer and Piore 1971; Hall 1984; Osterman 2000;). Job mobility in U.S. manufacturing is less certain today and often reflects the adoption of a truncated employment model which divides the workforce along educational and professional lines and results in "deskilling" and job churning at lower occupational levels with more minimal educational requirements (Davidson 2012a; Fligstein and Shin 2004; Kalleberg 2011).

The conventional policy response is to push more job seekers to secure academic degrees from colleges or universities. But is this the best policy option and how realistic is it to expect incumbent workers, especially older workers, in manufacturing to invest time and resources in higher education? As an alternative, is there maneuvering room to shape employment practices in ways that reestablish earlier norms of company-supported pathways for career development and occupational mobility? And, if so, what might this look like and how might this be achieved?

This chapter explores these questions by looking at sector-based experiments in workforce development in manufacturing that seek to promote job access for less educated workers and job seekers by influencing hiring and training practices of U.S.-based manufacturers. Common to these sector initiatives is a recognition that skill, as a concept, is inherently ambiguous and as such can be subject to reinterpretation. Treating skill in this way opens up the possibility for mediated interventions that encourage manufacturing companies to recognize greater degrees of freedom when it comes to accessing and assessing skill. External educational and training institutions, including community colleges and vocational training centers, support mediated practices of skill interpretation by creating an institutional scaffolding or exoskeleton that sector initiatives draw on to reinforce skill development practices and priorities internal to firms. This support role especially valued by resource-constrained and smaller-sized manufacturers that face greater constraints when it comes to internalizing worker training. But skill reinterpretation is fundamentally about getting employers to look beyond these supports and accept greater responsibility for upskilling through ongoing investments in their incumbent workforce and the development of internal pathways for career advancement. Recasting the labor market challenge from skills mismatch to skills reinterpretation has implications for sustained job access in manufacturing, especially for moderately educated workers. It also creates a framework for labor market interventions that are potentially more enduring and certainly more inclusionary than the more standard educational-fix approach.

Still, the labor market reach of this sectoral approach is uneven and lacks consistency. Most sectoral interventions are limited to a select group of cities and industries. As a consequence, this labor market reproduction system is ad hoc and geographically continent. Renewed Federal policy interest in U.S. manufacturing creates an opportunity to extend mediated interventions capable of dynamically certifying skill content and that encourage more predicable hiring and compensation decisions based on skill.

SKILL AMBIGUITY AS INSTITUTIONAL RESOURCE

Standard recommendations to boost the educational attainment of manufacturing workers rest on a widely held belief that a more formally educated workforce will encourage job creation in manufacturing and better use of enhanced worker knowledge for introducing new and better manufactured products. Accordingly, without an educational uptick, it is presumed manufacturing job growth will stall, reflecting what is commonly described as an intensifying "skills mismatch" or skills gap.

This perspective is especially prevalent in the United States. Concern about a looming "skills gap" is visible in numerous, recently published media and scholarly accounts about the future prospects of U.S. manufacturing. Business consulting reports give additional strength to this argument. An oft-cited 2011 survey of U.S. manufacturers conducted by the Manufacturing Institute in partnership with Deloitte Consulting estimated that close to 70 percent of firms faced "moderate to severe shortage" of qualified labor. The results of this survey are often used to explain why, despite a high national unemployment rate, 600,000 U.S. manufacturing jobs remain unfilled. The survey also notes that over 50 percent of U.S.-based manufacturing employers anticipate facing an intensifying skills shortage in coming years.

Fear of a manufacturing skills gap is certainly not unique to the 2010–2012 post-recession recovery. In fact, this has been a recurrent theme in recent decades, with intensified pressure to upskill coinciding with periods of economic downturn and when U.S. policy and economic analysts are most concerned with the problem of higher than normal rates of unemployment (Katz and Autor 1999; Levy and Murnane 1992). But as national discourse around industry skill requirements heats up, so have counterclaims that challenge fears of anticipated skill shortages.

In the 1980s and throughout the 1990s, U.S. labor economists called into question the widely held assumption that the introduction of computer technologies significantly altered industry skill requirements and with it undermined job and wage security for incumbent manufacturing workers (Howell 1994; Osterman 2000). This phenomenon – commonly referred to as skill-biased technical change – was believed to be a major contributor to rising income inequality in the 1980s, reflecting a seemingly common sense argument that those lacking computer knowledge would experience wage stagnation in relation to those with more experience with personal and micro-computing. But this claim was soon refuted by labor market data that demonstrated widespread wage stagnation, even in industries that experienced a significant increase in educational attainment and during periods with little technological upgrading (Card and DiNardo 2002).

In response to evidence of a "skills myth," labor market analysts searched for alternative explanations for rising income inequality that were less dependent on technological transformation and focused instead on changes to the institutional structure of the labor market that undermined job tenure, income security and occupational mobility (Fligstein and Shin 2004; Levy and Temin 2007; Osterman 2000; Osterman et al., 2007). While institutionalists often incorporated a discussion of skill development, it was folded into a larger statement about a weakening of internal structures for supporting on-the-job training and upskilling and what this might imply for worker displacement and disenfranchisement.

Today's renewed interest in the role of upskilling in economic recovery has led to similarly powerful counterclaims that unfilled jobs in manufacturing are not the result of a skills mismatch, but rather reflect below market wages that make it difficult for companies to attract and retain qualified candidates (Boston Consulting Group 2012; Cappelli 2011, 2012). As one analyst put it, "trying to hire high-skilled workers at rock-bottom rates is not a skills gap" (Davidson 2012b). The logical conclusion, then, is for companies to increase the wages they offer successful job applicants.

But what this recurring debate around skill obscures is the notion that skill, as a concept, is inherently slippery and ambiguous (Attewell 1990; Handel 2003, 2005). Proposing to simply end this debate by pushing companies to raise wages may not be sufficient, for it too fails to get at a deeper set of issues that affect how and whether worker skill gets recognized and valued in the first place.

To some extent, obscurity around skill has been attributed to research design and specifically to differences in how analysts represent skill to those they turn to for its assessment (Sandberg 2000). As one example, surveys of employers – a common data source for assessing industry skill requirements – rarely generate consistent lists for which types of skill are most valued within an industry and, related to this, which may be in short supply (Handel 2005). Michael Handel, in his in-depth review of employer surveys, found strong evidence for this. While some surveys emphasize the need for cognitive skills, especially in the areas of basic math and reading comprehension, others stress instead a set of work habits and interpersonal and communication skills. This inconsistency supports recommendations for improved survey design (Handel 2005).

But sources of ambiguity are not simply the result of a misspecification. They also stem from the "interpretative" nature of skill that often makes it difficult, if not impossible, to reduce skill into an objective and quantifiable metric (Sandberg 2005; Sennett 2008). Theories of "situated knowledge" help reinforce this point, by highlighting skill as context dependent and therefore impossible to fully separate from or understand outside the social relations and work environments in which it is developed and deployed (Lave and Wenger 1991; Lave, Murtaugh and de la Rocha 1984; Rogoff and Lave 1999). As Paul Attewell points out, "the terms grounded and situated imply that features of the context in which the work is done play a very important role in how the work is done, such a large role that it is meaningless to talk of a particular skill outside of the situations and practices in which it is used" (Attewell 1990: 425). In other words, skill is not a set of fixed capabilities that workers carry with them to all work settings and environments. Rather, skill is "situated" in the work environment itself and results from a dynamic and ongoing interplay between worker(s) and their work.

This line of reasoning is used to explain why, for example, individuals have been known to underperform in testing environments that differ significantly from the actual setting in which the skill sets being tested are developed or applied (Rogoff and Lave 1999). Additionally, context-dependent perspectives help to underscore sources of "tacit" skill that remain hidden or obscure, at times even to those possessing that skill (Attewell 1990; Doeringer and Piore 1971; Iskander and Lowe 2010). And this also helps us better understand sources of skill that are collectively produced, rather than individually controlled – in other words, that stem not from or through individual action but rather from combined knowledge that is created in a group setting or team-based work environment (Brown and Duguid 2002).

But equally, the idea that skill is context dependent implies that processes of skill formation can take place outside a formal classroom setting and are likely to reflect what happens within the workplace itself and as a result of on-the-job learning opportunities and exchanges. This, in turn, underscores the need to uncouple formal, academic education from skill development in order to recognize competencies that are produced through a variety of channels and experiences, only some of which depend on a traditional educational format (Iskander and Lowe 2010; Rose 2005).

Noting this, scholars have adopted a number of innovative methodologies for trying to capture skill that is context dependent and tacit in nature. In most cases, the approach is to utilize in-depth ethnologies, including participation observation and in-depth interviewing, in order to document skill-in-action and engage workers in an interpretative research process that helps broaden understandings of skill and identify hidden sources of talent (Iskander, Lowe and Riordan 2010; Piore 2006; Sandberg 2000). Side stepping a longer discussion about the value of these methodological approaches for now, it is useful to consider what these improved methods also imply for skills identification and skill interpretation more generally. Examining skill at the worksite, and through in-depth observation and discussion with those active in skill development, deployment and assessment, suggests there is room for collective reinterpretation of skill. That is to say, for multiple actors to engage in a negotiated process over what is meant by skill, who is presumed to possess that skill, and what structures and supports might be needed for workers to demonstrate and defend skill.

As this implies, the ambiguous and slippery nature of skill is more than just a complicating factor for skill assessment. It also represents a potential "resource" for agency, advocacy and intervention. We see evidence of the transformative power of ambiguity in recent studies of product and process innovation. As Lester and Piore note, "ambiguity is the critical resource out of which new ideas emerge" (Lester and Piore 2004: 54). And as they also note, interpretation "plays in the space of ambiguity" by establishing an interactive process that is "open-ended, and part of an ongoing conversation" (Lester and Piore 2004: 53). Lester and Piore have applied the concepts of interpretation and ambiguity to a study of industrial engineers and designers to gain insight into processes that lead to both radical and sustained forms of innovation. But given the ambiguous nature of skill across all occupations and classes of work, these concepts are also applicable to studies of skill on the shop-floor.

Skill interpretation, in this context, might involve an ongoing dialog with employers that helps to challenge deeply held assumptions and perceptions of shop-floor skill that potentially undermine or hinder job access and job security for moderately educated workers. Following from this, heightened public awareness of and interest in industry upskilling provides the basis for opening that conversation with employers. With this possibility in mind, I turn next to a brief historical overview of the role that labor market institutions have played in reinterpreting sources of shop-floor skill in manufacturing and their role in motivating emergent institutional supports.

LABOR MARKET INSTITUTIONS AND THE DEFENSE OF SKILL

U.S. labor market institutions, most notably labor unions, have long positioned themselves as defenders of worker skill. But how they do so has differed considerably across industry lines and has also changed with time. In the U.S. construction industry, for example, the defense of worker skill was formalized through industry-sanctioned apprenticeships that labor unions have tightly controlled (Palladino 2005). While the vast majority of new worker training received through construction apprenticeship programs occurred at the jobsite and often involved informal learning processes, there was considerable specificity with regard to required hours of training and how phases of training, once completed, would be compensated and rewarded. Union-friendly construction companies supported this formalized training structure by assigning skilled journeymen to the mentoring process and allowing a specified number of apprentices to work each site.

Until quite recently, the role of U.S. labor unions in demonstrating and defending worker skill in manufacturing has been less direct. Formal apprenticeship programs in U.S. manufacturing remain rare and, when they exist, are limited to a handful of specialized occupations, such as tool-and-die maker or airplane mechanic. Reflecting this, union wages in manufacturing tend not to be associated with "mastery" of skill, as is the case with construction, but rather have reflected years of industry work (Doeringer and Piore 1971). While one could argue this temporal form of seniority has functioned as a de facto signifier of skill, historic accounts of labor relations in manufacturing suggest a potentially different interpretation. Beyond a handful of specialized occupations that have been formally recognized as "skilled crafts," most manufacturing jobs have been defined instead as "semi-skilled," regardless of the seniority level of those filling these positions (Carnevale, Leila and Metzer 1988; Slichter, Healy and Livernash 1960). This perspective continues even today, most notably through descriptions of manufacturing jobs as "middle-skilled."

The view that manufacturing requires less skill relative to other manual trades appears to be unique to the United States. In Western Europe, where labor unions have long represented manufacturing workers, there is considerably less differentiation. In Germany, for example, apprenticeships are not only commonplace in both manufacturing and construction, but even today are structured and revised according to government-sanctioned educational and skill standards (Streeck 1989; Thelen 2004). In Britain too, labor unions have long promoted manufacturing workers as highly skilled.

So why have manufacturing skill requirements been portrayed so differently in the United States? Are these jobs substantively that different from their counterparts in Europe and, therefore, require less worker intellect and skill? Some comparative accounts of manufacturing imply this might be the case (Gertler 2004). Others suggest instead a more complicated institutional story and one that is not simply about identifiable differences in the organization of production or in how U.S. workers relate to technology, but rather is an outgrowth of a long-standing struggle over the control of skill and its determinants (Thelen 2004; Weir 1993).

History is replete with examples of U.S. employers and managers treating their manufacturing workforce as less skilled than they actually are. Some accounts portray managers as woefully ignorant of the existing knowledge and skill that manufacturing workers bring to their jobs and further develop as they encounter complex problems at work (Juravich 1988). Tom Juravich's (1988) rich ethnographic study of an electrical wire-making plant in New England not only illustrates this point, but also helps to highlight the way that managerial ignorance can manifest itself in subsequent decisions that undermine channels through which workers share knowledge and collectively solve problems.

Other accounts are less benign. Some labor scholars argue instead that U.S. managers intentionally "deskilled" tasks by applying principles of scientific management, a.k.a. Taylorism, and introducing a detailed division of labor that lowered skill requirements for most manufacturing jobs (Braverman 1974). Doing so allowed managers to retain control over the labor process by creating an easily replenished manufacturing labor pool and with it made credible their threats to replace workers that resisted this structure or attempted to collectively organize against it.

Still, claims of purposeful deskilling by U.S. managers are complicated by conflicting accounts that show their equally strong desire to engage worker intellect and ingenuity, including at the shop-floor level (Lazonick 2004; Osterman 2000). Assumptions of widespread deskilling are further undermined by detailed studies of contradictory shop-floor practices that manufacturing workers themselves created and willingly reproduced.

In the 1960s, Doeringer and Piore, in their seminal book on internal labor markets, not only identified informal processes through which manufacturing workers develop skills on-the-job but, paradoxically, their simultaneous role in concealing such practices from their supervisors. "In some cases they [workers] also have a strong incentive to hide what they do from management who might use this knowledge to extract greater output from them or to correct a loose incentive rate" (Doeringer and Piore 1971: 20). Building from this, Michael Burawoy's book *Manufacturing Consent* painted a rather troubling portrait of a complex power play between manufacturing workers and managers over skill and work responsibilities (Burawoy 1982). And while Burawoy ultimately concluded that manufacturing workers were pawns in a rigged and unfairly structured game, he nonetheless reinforced the point that skill was contested and open to interpretation.

Despite this, U.S. labor unions – at least through the late 1980s – made few attempts to alter the perception that U.S. manufacturing jobs were semi-skilled, not highly skilled. Some of this reflects their own legacy in specifying and reinforcing these differences in the early 20th century, when union jobs were initially categorized as either "craft" or "industrial," with the former treated as highly skilled and the latter as semi-skilled (Thelen 2004). Labor unions fought among themselves over these classifications and what they implied for labor market jurisdictions and control. But long after this internal conflict subsided and craft and industrial unions united through the merger of the American Federation of Labor and Congress of Industrial Organizations, there remained little need for "industrial" unions in the United States to explicitly challenge entrenched definitions of skill as long as manufacturing job growth remained robust and wage-setting processes remained in place to reward workers for years of industry service.

This all changed in the late 1980s, when deindustrialization and intensifying threats from global competition ultimately pushed U.S. labor unions to make a stronger case that worker skill was a central contributor to productivity gains in U.S. manufacturing. Front and center in this effect by industrial unions to wrest control over and better defend the skill of manufacturing workers was a novel American experiment (though admittedly one that drew heavy inspiration from Japan) called the high performance work organization.

SKILLS AND TRAINING IN THE HIGH PERFORMANCE EXPERIMENT

In its idealized form, the high performance work organization involved a range of company-supported human resource practices that were designed to increase the participation of rank-and-file workers in product and process innovation (Appelbaum and Batt 1994). These practices included the use of more flexible work routines which allowed workers to move more freely across traditional job categories and, in the process, engage with, learn from and combine multiple tasks. As this implies, job rotation was a key element, as were open channels for communication and collective problem solving that crossed the traditional labor–management divide.

U.S. manufacturing firms – especially those in automotive manufacturing – were early adopters of the high performance model (Osterman 1993). The United Auto Workers (UAW), the main labor union representing workers in the U.S. automotive industry, was an especially eager participant and, as such, was prepared to relax traditional systems of seniority and rules of job allocation in order to support a more fluid work environment. The UAW worked closely with General Motors to introduce high performance models at Saturn's production facilities in Tennessee and at the New United

Motor Manufacturing, Inc. (NUMMI) plant, a joint venture with Toyota, in Fremont, California (Adler and Cole 1993; Kochan and Rubinstein 2000). They also worked with Japanese automotive manufacturers, especially Mazda, to incorporate high performance elements at U.S.-based assembly plants (Shaiken 1993).

In light of this commitment, labor scholars initially rejoiced the high performance experiment, viewing it as a new internal structure for engaging and valuing worker skill and ingenuity, and linking it to improvements in job quality (Appelbaum 2000; Kochan and Osterman 1994). Further research however revealed potential limits to high performance strategies and noted especially the tendency for partial adoption within manufacturing and in other sectors as well (Knauss 1998; Osterman 1993, 2000). While a handful of U.S. manufacturing facilities, namely GM's Saturn and NUMMI, adhered closely to the model's core principles and tenets, the vast majority of firms that classified themselves as high performance chose only select elements of the model. And the procedures that did get implemented often had the perverse effect of increasing worker stress and turnover and foreclosing channels for skill development that high performance work systems were meant to support. This led to claims by some labor market scholars that high performance firms were also adopting "low road" labor practices, in turn undermining job quality and job security.

Interestingly, labor union involvement was a determining factor in whether or not a firm adopted a more worker-friendly version of high performance (Osterman 2000; Shaiken 1993). Comparative studies in manufacturing noted that high performance companies with strong union representation were also more likely than those without it to prioritize investments in worker training and take steps to ensure shop-floor workers were given sufficient time and resources to learn, master and apply newly acquired skills and knowledge. Without this institutional check, employers tended to underinvest in worker training and, by extension, undermine the ability of their labor force to leverage skills for ongoing productivity gains and innovation. In this regard, unions formalized the link between worker skills, productivity and job quality, and ensured gains from high performance strategies were shared among employers and employees alike. They also helped to make visible the skills contribution of shop-floor workers and defended formal structures and training investments that would ensure manufacturers continued to value and leverage this contribution.

EMERGENT INSTITUTIONAL APPROACHES

The U.S. manufacturing workforce has experienced a significant decline in union representation in recent decades, dropping from a high of around 40 percent of the manufacturing workforce in the early 1970s to less than 10 percent today. The declining role of U.S. labor unions threatens to undermine progress in forging a stronger connection between worker skill and productivity gains, and, subsequently, for pushing U.S.-based manufacturers to make a stronger commitment to worker training and career development. But emergent institutional actors have stepped in to help fill this void and, in the process, provide an additional vehicle for moving strategies of skill reinterpretation to the next level. One recent effort is the Digital Badges initiative sponsored by the Mozilla and MacArthur Foundations that seeks to certify skills developed in informal and non-traditional training environments, including those acquired outside a formal classroom setting. In 2012, 30 U.S-based organizations were awarded funds to develop digital badges, with the goal of helping job seekers, and especially youth, communicate less visible skills and qualifications to prospective employers.

The next step is convincing prospective employers to recognize non-traditional qualifications. In this regard, sector-based initiatives are proving to be an especially effective institutional actor. Defined by the National Network of Sector Partners as "regional, industry-focused approaches that improve access to good jobs and/or increase job quality in ways that strengthen an industry's workforce," sector initiatives emerged in response to evidence of declining job quality and, more specifically, in an effort to remove barriers to rewarding job opportunities and career advancement for less educated job seekers. In the manufacturing sector, over 200 initiatives have been cataloged thus far and represent a subset of workforce development-oriented non-profits, unions, community colleges and Workforce Investment Boards, regional labor market entities established under the U.S. Workforce Investment Act of 1998.

What differentiates these initiatives from other forms of labor market intervention is their ability to use specialized industry knowledge to mediate the relationship between job seeker and employer, and in ways that also influence industry norms and practices around hiring, wage setting and occupational mobility. Often this involves intentional actions designed to reveal and defend "hidden" or less visible sources of worker talent that may be obscured by low levels of formal education.

While a comprehensive review of all manufacturing sector initiatives is beyond the scope of this chapter, it is useful to consider how their practices contribute to skill reinterpretation. Table 30.1 provides a list of five sector-initiatives that have helped to establish an institutional standard with respect to manufacturing worker skill. All five were initially created in response to perceived skill shortages on the part of manufacturing employers in their regional economy. But each has used its labor market position to engage U.S. manufacturers in a negotiated process over skill, and more specifically with the goal of expanding employment and advancement opportunities for individuals that lack academic degrees from colleges or universities.

First and foremost, these initiatives work closely with manufacturing employers to influence which skills and qualifications are recognized during the hiring process. Key here is their ability to help match job seekers with employers through pre-employment screening and assessment services. In that capacity, however, they do not simply act as agents of employers, as a private staffing agency might. Rather, they mediate the hiring process to support less educated job seekers that might otherwise be excluded from quality employment opportunities.

As part of this effort, they encourage employers to stop making hiring and compensation decisions on the basis of how an applicant looks "on paper." Why? Because standard application formats, including the ever-popular single-page resume, tend to reinforce an educational bias and can obscure sources of tacit knowledge which may be hard to demonstrate and defend in writing. Sector initiatives instead work with manufacturers to create evidence-based skill assessments, which they often develop in close partnership with incumbent workers within the company. As subject area experts, these technical workers provide detailed information on skill requirements for specific tasks and at multiple occupational levels. This knowledge, which is not always in reach for human

Table 30.1 Examples of manufacturing sector initiatives in the United States

	Date Started	Location	Type	Manufacturing Focus
Wisconsin Regional Training Partnership	1992	Milwaukee	Jointly union- and industry-driven, grant funded	Manufacturing, various
Lancaster County Workforce Investment Board (WIB)	1999	Lancaster County, PA (west of Philadelphia)	Workforce Investment Board	Food processing, metals
BioNetwork	2004	North Carolina statewide	Community-college driven non-profit	Biomanufacturing, biofuels, food processing
Manufacturing Works	2005	Chicago	Workforce Investment Act (1998) community affiliate, driven by mayor's office	Food processing, metals & machinery
Partners for a Competitive Workforce	2011	Greater Cincinnati (including Indiana & Kentucky)	Umbrella organization for regional workforce programs; non-profit with some grants, some WIB funds	Auto suppliers, aviation, metal fabrication

resource managers, is helpful in clarifying skill requirements. But more importantly, it is also used to expand job access by encouraging employers to conceptualize bundles of attributes that can reflect a variety of work experiences and backgrounds.

But even with this broader understanding of skill, there is the possibility that specific skill gaps will emerge for some job applicants. Sector initiatives are mindful of this and are quick to point out their desire to provide employers with quality candidates that will contribute value to the company. As one manufacturing sector-initiative director described it, our goal is to help employers "figure out who is a good candidate to invest in who will work for a long time." As such, they differentiate themselves from more traditional social service providers that provide job placement support to a pre-selected group of candidates. In contrast, sector initiatives focus on establishing a long-term match between a job seeker and employer, which can make it more difficult for them to place all prospective applicants. This is why partnerships with community organizations and other support agencies prove crucial to the work of sector initiatives, as they provide a social safety net for those not placed and also help identify and address more persistent barriers to employment.

Still, by bundling worker attributes, sector initiatives are in a powerful position to advocate on behalf of job seekers that have most, but not all, preferred qualifications. Sometimes this involves convincing employers to rank order attributes, thereby creating greater maneuvering room for applicants with some gaps lower down that weighted list. In other cases, sector initiatives step in to provide pre-employment training to address more prevalent skills gaps within an applicant pool. Typically this is done in partnership with external training specialists, such as community colleges or non-profit vocational training programs.

In addition to shaping hiring decisions, manufacturing sector initiatives also seek to influence how employers engage with and treat their incumbent workforce. Important here are interventions that encourage manufacturers to better harness worker knowledge when introducing improved production practices and techniques – in other words, turning to incumbent workers when introducing and developing process innovations. According to one sector initiative director,

> the most valuable asset that any of these manufacturing firms have is the workers who are handling the product . . . so we start off with that premise that investing in workers and involving workers in some of the decision-making process is the wisest thing to do . . . and sometimes with a little bit of help they [the employers] listen to not only what their workers need but also take value in what their workers can offer from a floor perspective.

Sector initiatives first show evidence of underutilization of worker knowledge, including structural barriers that limit involvement by shop-floor workers in decision-making processes. They then intervene to remove these barriers, but also take steps to formalize processes for augmenting and defending shop-floor knowledge. For more established manufacturing firms, sector initiatives have created interim apprenticeships that essentially act as skills equalizers for incumbent workers by codifying and augmenting skills learned earlier in their careers. But they also intervene to establish linkages to external institutions, including community colleges, that can support upskilling and related certification. As this suggests, sector initiatives do find ways to incorporate formal educational programs and credentials. In doing so, however, they avoid blindly pushing a college or university degree. Rather, they use their labor market position to create a flexible training and credentialing resource that helps to prepare industry newcomers, but equally ensures the hidden talents of the incumbent workforce become more visible and valued.

What enables these sector initiatives to influence employment practices is their ability to embed themselves within the local business community and establish ongoing relationships with business owners and managers. Admittedly, this has not been the case for all sector initiatives, and directors of some initiatives, including those in manufacturing, report difficulty in establishing and maintaining strong employer relationships. In this regard, the five initiatives reviewed here – and numerous others modeled after them – stand apart for their ability to maintain deep connections with industry that enable them to keep open a conversation around skill, and during times when definitions and perception of skill may shift and evolve.

A POLICY CROSSROADS

The United States is at a policy crossroads with respect to manufacturing. Federal and state industrial policy, while not quite set in the United States, has the potential to influence manufacturing job creation, including where those jobs are located and how accessible they are to less educated job seekers. But it could also undermine job access as well. Key Federal agencies that influence U.S. manufacturing policy direction and discourse seem all too willing to take employer statements about their skills gaps at face value. In response, these agencies narrowly focus on bolstering external educational supports and getting more individuals through these systems – the popular policy mantra being "college for all." In the process, less attention is given to interventions that might influence internal human resource strategies and have the goal of expanding options for workplace learning and related structures for occupational mobility.

If we start instead with the assumption that employers are inclined to overstate their skill requirements and in ways that unnecessarily conflate skill with academic degrees, we may find ourselves in a position to push for a more comprehensive institutional solution. Efforts that are designed to shape and reshape employer perceptions of and practices around skill may result in more encompassing strategies for skills identification and development. But equally, they could prove more effective in stabilizing manufacturing employment and protecting established pathways into the middle class.

Sector initiatives provide one institutional model for influencing manufacturing employer thinking and decision making, even down to the shop-floor level. Their interventions extend earlier efforts by labor unions to mediate the relationship between worker and supervisor in an attempt to forge a stronger connection between productivity gains and the application and augmentation of worker skill. But more can be done to extend practices of skill reinterpretation and widen their labor market reach.

Several options exist for bolstering sector initiatives, including securing additional and more stable funding sources that can replace those recently eliminated through a budget appropriation bill affecting the Federal Workforce Investment Act. Beyond this, there is the possibility to strengthen partnerships between sector initiatives and Federal agencies. Especially relevant is the federally funded Manufacturing Extension Partnership (MEP), a 25-year old initiative housed under the National Institute of Standards and Technology. As part of its recent strategy overhaul, MEP has committed to extending workforce development services to small and medium-size manufacturing firms through its vast network of state and local technical assistance centers and a new platform called SmartTalent. Traditionally the MEP network has concentrated its resources on helping firms introduce new technologies and coupling that investment with lean manufacturing principles. The decision to add workforce to the mix reflects its desire to have greater influence on manufacturing employment growth. But equally, it reflects MEP's recognition that worker skill is an essential resource for industry innovation and upgrading – something labor unions recognized much earlier with their support of the high performance experiment in manufacturing.

Still, for a partnership between sector initiatives and the MEP to flourish, it is also important to recognize inherent differences in orientation across these networks, with the former approaching workforce development from a worker-centered perspective and the latter as part of an integrated business solution. At first glance, these differences could create intractable sources of conflict that could undermine opportunities for cooperation. Under a closer light and given evidence of some initial partnering within specific localities, we instead see an opportunity where ambiguity can again function as resource for change – in this case, a resource for institutional innovation and reinterpretation.

REFERENCES

Adler, P. and R. Cole. 1993. Designed for learning: A tale of two auto plants. *MIT Sloan Management Review* 34:85.
Appelbaum, E. 2000. *Manufacturing advantage: Why high-performance work systems pay off*. Ithaca: Cornell ILR Press.
Appelbaum, E. and R. Batt. 1994. *The new American workplace: Transforming work systems in the United States*. Ithaca: Cornell University Press.
Attewell, P. 1990. What is skill? *Work and Occupations* 17:422–448.
Boston Consulting Group. 2012. *Skills gap in U.S. manufacturing is less pervasive than many believe*. Boston: Boston Consulting Group.
Braverman, H. 1974. *Labor and monopoly capital: The degradation of work in the twentieth century*. New York: Monthly Review Press.
Brown, J.S. and P. Duguid. 2002. *The social life of information*. Cambridge: Harvard Business Press.
Burawoy, M. 1982. *Manufacturing consent: Changes in the labor process under monopoly capitalism*. Chicago: University of Chicago Press.
Cappelli, P. 2011. Why companies aren't getting the employees they need. *The Wall Street Journal*, October 24.
Cappelli, P. 2012. *Why good people can't get jobs*. Philadelphia: Wharton Digital Press.
Card, D. and J.E. DiNardo. 2002. *Skill biased technological change and rising wage inequality: Some problems and puzzles*. Cambridge: National Bureau of Economic Research.
Carnevale, A.P., G. Leila and A. Metzer. 1988. *Workplace basics: The skills employers want*. San Francisco: Jossey-Bass Publishers.
Davidson, A. 2012a. Making it in America. *The Atlantic*, January/February, available at http://www.theatlantic.com/magazine/archive/2012/01/making-it-in-america/308844/3/.
Davidson, A. 2012b. Skills don't pay the bills. *New York Times*, November 20, available at http://www.nytimes.com/2012/11/25/magazine/skills-dont-pay-the-bills.html?pagewanted=all.
Dertouzos, M.L., R.K. Lester and R.M. Solow. 1989. *Made in America: Regaining the productive edge*. Cambridge, MA: The MIT Press.
Doeringer, P.B. and M.J. Piore. 1971. *Internal labor markets and manpower analysis*. New York: ME Sharpe Inc.
Fligstein, N. and T.J. Shin. 2004. The shareholder value society: A review of the changes in working conditions and inequality in the United States, 1976 to 2000. In *Social inequality*, 401–432. New York: Russell Sage Foundation.
Gertler, M. 2004. *Manufacturing culture: The institutional geography of industrial practice*. Oxford: Oxford University Press.
Hall, R.E. 1984. *The importance of lifetime jobs in the US economy*. Cambridge: National Bureau of Economic Research.
Handel, M.J. 2003. Skills mismatch in the labor market. *Annual Review of Sociology* 29:135–165.
Handel, M.J. 2005. *Worker skills and job requirements: Is there a mismatch?* Washington, D.C.: Economic Policy Institute.
Howell, D. 1994. The skills myth. *The American Prospect* 18:81–90.
Iskander, N. and N. Lowe. 2010. Hidden talent: Tacit skill formation and labor market incorporation of Latino immigrants in the United States. *Journal of Planning Education and Research* 30:132–146.
Iskander, N., N. Lowe and C. Riordan. 2010. The rise and fall of a micro-learning region: Mexican immigrants and construction in center-south Philadelphia. *Environment and Planning A* 42:1595–1612.
Juravich, T. 1988. *Chaos on the shop floor*. Philadelphia: Temple University Press.
Kalleberg, A.L. 2011. *Good jobs, bad jobs: The rise of polarized and precarious employment systems in the United States, 1970s to 2000s*. New York: Russell Sage Foundation.
Katz, L.F. and D.H. Autor. 1999. Changes in the wage structure and earnings inequality. In *Handbook of labor economics*, eds C.A. Orley and C. David, 1463–1555. Amsterdam: Elsevier.
Knauss, J. 1998. Modular mass production: High performance on the low road. *Politics & Society* 26:273.
Kochan, T. and S. Rubinstein. 2000. Toward a stakeholder theory of the firm: The Saturn Partnership. *Organization Science* 11:367–386.

Kochan, T.A. and P. Osterman. 1994. *The mutual gains enterprise: Forging a winning partnership among labor, management, and government*. Cambridge: Harvard Business Press.
Lave, J., M. Murtaugh and O. de la Rocha. 1984. The dialectic of arithmetic in grocery shopping. In *Everyday cognition: Development in social context*, eds B. Rogoff and J. Lave, 67–94. Cambridge: Harvard University Press.
Lave, J. and E. Wenger. 1991. *Situated learning: Legitimate peripheral participation*. Cambridge: Cambridge University Press.
Lazonick, W. 2004. Corporate restructuring. In *The Oxford handbook of work and organization*, eds S. Ackroyd, R. Batt, P. Thompson and P. Tolbert, 577–601. Oxford: Oxford University Press.
Lester, R. and M. Piore. 2004. *Innovation: The missing dimension*. Cambridge: Harvard University Press.
Levy, F. and R.J. Murnane. 1992. U.S. earnings levels and earnings inequality: A review of recent trends and proposed explanations. *Journal of Economic Literature* 30:1333–1381.
Levy, F. and P. Temin. 2007. *Inequality and institutions in 20th century America*. Cambridge: National Bureau of Economic Research.
Osterman, P. 1993. How common is workplace-transformation and who adopts it? *Industrial and Labor Relations Review* 47:173.
Osterman, P. 2000. *Securing prosperity: The American labor market: How it has changed and what to do about it*. Princeton: Princeton University Press.
Osterman, P., T.A. Kochan, R.M. Locke and M.J. Piore. 2007. *Working in America: A blueprint for the new labor market*. Cambridge: MIT Press.
Palladino, G. 2005. *Skilled hands, strong spirits: A century of building trades history*. Ithaca: Cornell University Press.
Piore, M.J. 2006. Qualitative research: Does it fit in economics? *European Management Review* 3:17–23.
Rogoff, B. and J. Lave. 1999. *Everyday cognition: Development in a social context*. Cambridge: Harvard University Press.
Rose, M. 2005. *The mind at work: Valuing the intelligence of the American worker*. New York: Penguin Group USA.
Sandberg, J. 2000. Understanding human competence at work: An interpretative approach. *Academy of Management Journal* 43:9–25.
Sandberg, J. 2005. How do we justify knowledge produced within interpretive approaches? *Organizational Research Methods* 8:41–68.
Sennett, R. 2008. *The craftsman*. New Haven: Yale University Press.
Shaiken, H. 1993. Beyond lean production. *Stanford Law and Policy Review* 5:41.
Slichter, S.H., J.J. Healy and E.R. Livernash. 1960. *The impact of collective bargaining on management*. Washington, D.C.: Brookings Institution.
Streeck, W. 1989. Skills and the limits of neo-liberalism: The enterprise of the future as a place of learning. *Work, Employment & Society* 3:89–104.
Thelen, K. 2004. *How institutions evolve: The political economy of skills in Germany, Britain, the United States, and Japan*. Cambridge: Cambridge University Press.
Weir, M. 1993. *Politics and jobs: The boundaries of employment policy in the United States*. Princeton: Princeton University Press.

FURTHER READING

Appelbaum, E. 2000. *Manufacturing advantage: Why high-performance work systems pay off*. Ithaca: Cornell ILR press.
Berger, S. 2005. *How we compete: What companies around the world are doing to make it in today's global economy*. New York: Doubleday.
Burawoy, M. 1982. *Manufacturing consent: Changes in the labor process under monopoly capitalism*. Chicago: University of Chicago Press.
Cappelli, P. 2012. *Why good people can't get jobs*. Philadelphia: Wharton Digital Press.
Doeringer, P.B. and M.J. Piore. 1971. *Internal labor markets and manpower analysis*. New York: ME Sharpe Inc.
Gertler, M. 2004. *Manufacturing culture: The institutional geography of industrial practice*. Oxford: Oxford University Press.
Handel, M.J. 2005. *Worker skills and job requirements: Is there a mismatch?* Washington, D.C.: Economic Policy Institute.
Juravich, T. 1988. *Chaos on the shop floor*. Philadelphia: Temple University Press.

Kochan, T.A. and P. Osterman. 1994. *The mutual gains enterprise: Forging a winning partnership among labor, management, and government*. Cambridge: Harvard Business Press.
Osterman, P. 2000. *Securing prosperity: The American labor market: How it has changed and what to do about it.* Princeton: Princeton University Press.
Rogoff, B. and J. Lave. 1999. *Everyday cognition: Development in a social context*. Cambridge: Harvard University Press.
Thelen, K. 2004. *How institutions evolve: The political economy of skills in Germany, Britain, the United States, and Japan*. Cambridge: Cambridge University Press.
Weir, M. 1993. *Politics and jobs: The boundaries of employment policy in the United States*. Princeton: Princeton University Press.

PART V

CONCLUSION

Conclusion: regeneration economies: manufacturing as the next industrial revolution

Jennifer Clark, John R. Bryson and Vida Vanchan

Manufacturing continues to matter and its study must form a central part of the social and engineering sciences. Manufacturing involves all countries – from high-cost developed market economies to emerging economies. Manufacturing may be a labor- or capital-intensive production process in which labor has been replaced by machines. In most cases, the manufacture of a product involves elements that are labor and capital intensive. The manufacture of the simplest product requires knowledge, skill and expertise, and often relies on a complex geographically distributed production system. There are also localized agglomeration economies embedded within distributed systems of production. The economy is complex and requires many different tasks to create products (Bryson and Rusten, 2010); tasks include manufacturing activities, services, finance, logistics, retailing, education and training, and research and development. There is a danger in segmenting an economy into discrete packages. Thus, a focus just on financial services may isolate finance from retailing or manufacturing. Similarly, a focus on a narrow definition of manufacturing may ignore the role logistics, finance and retailing play in the production of products. In this collection, the focus has been on manufacturing, but it has also highlighted the importance of many other economic sectors and activities for the manufacture of products.

Throughout the book, contributors have illustrated in case after case how industries and firms are adapting to new technologies and new markets. In the aggregate, these chapters present a narrative about industrial restructuring and innovation in the world economy that is dynamic and design driven. The authors provide a nuanced account of how firms and industries modify existing processes to produce products for established and emerging markets. The scope of this volume allows a rare opportunity to consider transformations in production systems and processes across sectors, technologies and industries. In other words, the manufacturing described in this volume is not a linear project but rather a cumulative exercise. The manufacturing firms and industries in the 21st century world economy iterate not from a blank slate but from a complex set of initial conditions.

The empirical evidence in this book includes contributions from a range of interdisciplinary scholars – from engineering to policy to economic geography. This evidence underscores the reality that manufacturing not only matters in the world economy but that industrialized economies are participants rather than bystanders in this process. National and regional actors – firms, institutions, intermediaries and governments – have significant agency in these manufacturing stories. That agency includes how and where things are made. The Maker Movement in particular highlights a shift in the understanding of agency in the production process (described in Wolf-Powers' Chapter 26). But it should be noted that this serves as a parallel story to those of small, established manufacturing companies which have been making conscious and deliberate choices about how and where they manufacture and how they provide product-related services.

The underlying debate among policymakers over the past several decades has not actually been about the ability to make strategic policy choices around manufacturing capacity but about the desire to do so (Clark, 2012; Feller, 1997; Helper, 2009; Zysman and Tyson, 1983). Since the 1990s, advanced economies have often debated whether they could retain manufacturing. But at the center of this debate was also a question about whether industrialized countries wanted to maintain broad production capacities in the new global economy or whether a service-oriented economic policy model would produce better outcomes in a world economy where wealth creation appeared concentrated in financial services and other high-value business services like marketing, advertising and intellectual property (Bryson, Daniels, and Warf, 2004; Corbridge, Martin, and Thrift, 1994; Reich, 1992).

Since the "Great Recession" of 2008, the voices expressing a conventional wisdom that advanced economies can sustain economic shocks without a strong production base have all but been silenced. This U-turn in macroeconomic thinking has returned policymakers to some persistent questions about how to understand the relationship between technological innovations and existing production processes (Amin, 1994; Bluestone and Harrison, 1982; Bluestone, Harrison, and Clayton-Matthews, 1986; Bryson and Rusten, 2010; Harrison and Bluestone, 1988; Noyelle and Stanback, 1983; Piore and Sabel, 1984). The research from the early 1980s highlights – again and again – how manufacturing processes adapt to new technologies incrementally and unevenly across time and place.

The adoption of new technologies is highly contingent. Technological adaptation depends on industry, sector, location, capital availability, labor skills and costs, policy, and individual firm strategies. Again, this nuanced, dynamic, and non-linear experience of how new manufacturing has emerged in the 21st century economy points to the need for continued research into industries and sectors. It also means that generalizations are hard to justify about how manufacturing is changing and why, and the role key factors play in this transformation: industry, sector, location, capital availability, labor skills and costs, policy, the blurring of manufacturing and services, and firm strategies.

Throughout this book, authors describe the effects of these variables on particular processes, industries, and individual firms. The final section engages in a discussion of how policies at the national, regional, and local scale shift the landscape for manufacturing firms and alter their calculus when it comes to labor reproduction and other factor inputs such as land and energy. The authors also highlight the increasingly important role of institutional intermediaries in mediating supply chains and accessing factor inputs like labor, land, and technology. Some of these policy frameworks are clearly relicts of an earlier manufacturing era. Others are surprisingly innovative efforts to coordinate across institutions and scales of governance.

Across cases, what stands out is the richness of the stories of each industry and process. Again, this work points to the need for empirical work that focuses on understanding the ongoing transformation of manufacturing production systems. In an age of "big data" and the move in policy to focus on databases and comparable proxies, what we find in the analysis of manufacturing industries is that fieldwork is critical. Big data can perhaps describe the past but it cannot clearly predict the future. In the manufacturing sector, projections based on past performance and choices plot a line that reflects a reality understood to be already antiquated. In fact, one could argue that the inability to understand the rapidity and the nature of the changes in manufacturing has led to many of the misapprehensions made by policymakers, business leaders, and politicians over the past quarter century.

TECHNOLOGICAL CHANGE AND PRODUCTION PROCESSES: SUSTAINABLE MANUFACTURING AS AN ECONOMIC CONSTRUCT

Throughout the last quarter century, many academics and policy analysts emphasized the promise of emerging technologies as the target for export-oriented strategies (Cooke, 2007; Robinson, Rip, and Mangematin, 2007). Policy discussions highlighted product innovations in industries such as software and biotechnology and the importance of research and development in industrial policy (Tassey, 2010). Simultaneously, many engineers pointed to innovations in materials and processes like lightweight composites, additive manufacturing, and digital fabricating (including computer-aided design or CAD) as the future of advanced manufacturing (Bryson and Rusten, 2008). References to "high-tech manufacturing" by policymakers and economic developers often referred to the product rather than the process behind its manufacture (Livesey, Chapter 25).

There is a tension between investing (private and public sector) in product rather than process innovation, but these two strategies vary in their relationship to existing industries. Product innovation tends to privilege start-ups firms, technology transfer, and entrepreneurship. It emphasizes disruptive technologies, and new systems and institutions. Alternatively, a focus on process innovation emphasizes how new materials, methods, and modes of production alter and adapt the practices of incumbent firms and industries. The emphasis in process innovation is on adaption and "technology uptake" within existing systems rather than disruptive shifts in organizational models and markets.

Interestingly, it is the process innovations enabled by digital fabrication including 3D printing, e-retailing, and computer-aided design that are behind the emergence of small-scale manufacturing seen in the Maker Movement. These process innovations allow a new generation of manufacturing entrepreneurs to make products that are familiar (textiles, furniture, household goods, consumer products) but to use these niche and project-based production methods rather than mass production techniques. It is in this way that the shift in technology-enabled small-scale design plus manufacturing process innovations is moving a segment of manufacturing in the US towards a "flexible specialization 2.0" model (Christopherson and Storper, 1989; Clark, 2013; Shaiken, Herzenberg, and Kuhn, 1986).

During the period preceding the 2008 recession, business and policy commentators pointed to the potential of new products far more frequently than new processes (Lohr, 2010). This was the case even when new technologies were built from an existing industrial base. For example, biotechnology was rarely cast as an evolution or merger of the underlying manufacturing industries of medical devices or pharmaceuticals (Clark, 2013). Instead, advocates and investors promoted biotechnology as a market that was both new and novel. Similarly, nanotechnology was often cast as a new site of investment rather than an evolution of existing processes (Bozeman, Laredo, and Mangematin, 2007). Although referred to as "emerging technologies," advocates rarely identified where these technologies emerged from. An underlying production or manufacturing sector was not often emphasized.

This is, in part, simply a commercial bias towards new income streams from new markets (Kaplan, 2011). A bias that is expected rather than exceptional. What is surprising is how broad the shift away from thinking about and investing in process and material innovations became. It was not until the 2008 recession that a rethinking of the potential for process innovations to increase efficiency (economic and environmental) really gained a foothold in policy discussions. Additive and digital manufacturing have become an important part of this discussion about the future of manufacturing for emerging firms, incumbent firms, and even large firms. For small and emerging firms the goal is to enter these new markets, and for large or incumbent firms, the goal is to maintain market share in the face of both disruptive and incremental technology-driven change. What becomes evident in reading the cases and commentaries in this book is the clear case for a broad consideration of the systems that enable sustained manufacturing capacity. As a matter of industrial ecosystems, the challenge is increasingly one related to technology adoption into an incumbent system rather than the emergence of a wholly new system and the receding of an older, seemingly antiquated one.

EMERGING PROCESSES: THE IMPLICATIONS OF DIGITAL AND ADDITIVE MANUFACTURING

The history of most manufacturing is based around subtractive methods in which material is removed by cutting, drilling, and other forms of machining. Subtractive manufacturing has been supplemented by technological developments that have given rise to additive manufacturing techniques, rapid prototyping or 3D printing (Kumar and Kruth, 2010; Wohlers, 2001; Yan and Gu, 1996). Additive manufacturing is a process in which a three-dimensional solid object is produced by laying down successive layers of material (liquid, powder, or sheet materials) controlled by a digital file. The digital file can be produced by 3D CAD model data, Computed Tomography and Magnetic Resonance Imaging scan data, and data created from 3D digitizing systems. The development of additive manufacturing means that it is now possible for components and products to be developed using CAD and printed using a 3D printer. The cost of 3D printers is declining and very small firms are now able to afford to purchase small 3D printers. Developments in additive manufacturing have the potential to transform manufacturing and to produce a new geography. In theory, additive manufacturing could eliminate the subcomponent section of the supply chain. Thus the emphasis would shift from cost competition for the subcomponents to pre-production elements, including 1) product design, 2) process design (software), and 3) materials (especially quality and purity). Indeed, this would result in a very different production geography.

Of course, the key to understanding why and how additive manufacturing matters to manufacturing is an awareness of when and why an incumbent firm shifts its existing production process to "uptake" this new process technology and to simultaneously understand why and how an emerging firm can use this technology to enter new markets or become competitive in existing markets. Additive manufacturing makes it possible to produce lightweight structures that are extremely strong and optimize the relationship between material content and performance. As Ward notes in Chapter 7, all this could lead to the development of entirely new businesses and business models. For existing sectors such as aerospace (see Drury's Chapter 19), the introduction of these additive manufacturing processes enhances the extensive investments in lightweight materials. And, of course, the implications for energy consumption (see Brown and Kim's Chapter 9) and thus logistics (see Hall's Chapter 4) are not trivial. For industries where the final product often involves precision assembly like optics and photonics (see Clark's Chapter 16) this technology could allow end producers to fabricate on-site rather than source subcomponents through a distributed supply chain.

Additive manufacturing may prove to be the most important technological development in manufacturing since the introduction of CAD. Developments in additive manufacturing mean that individuals and small firms have the capability to develop, design, and print new products in-house. For small producers and the emerging Maker Movement, which highlights customizable and craft-based production, small-scale 3D printers make it possible for an individual to purchase a product as a digital file to then print at home. Many collaborative "maker spaces" already included communally use small, affordable "maker bots" for prototyping.[1] And, small-scale producers can modify and adapt products for each individual customer. In effect, additive manufacturing has the capability to fulfill the potential of flexible specialization (Amin, 1994).

This technological development makes it extremely simple to customize products and to alter the relationship between manufacturing and consumers. Technically minded individuals will be able to develop products that might be tested by companies that specialize in selling products that can be printed at home or by a provider of 3D printing services. This means that warehouse inventory levels will be significantly reduced as more products can be printed close to the point of consumption. There are important sustainability and climate change implications, as additive manufacturing has the potential to localize and customize the production of many manufactured products. This would lead to the development of a much more distributed geography of production, but with centers of design and development. Indeed, additive manufacturing makes urban manufacturing and shared spaces for small-scale production, "maker spaces," increasingly viable regeneration and land use strategies. In effect, places that retain integrated design and production capacities will benefit, especially those with experience absorbing process innovations.

It is important not to over-emphasize the potential impact that 3D printing might have on manufacturing. Currently, 3D printers are used in craft-based industries, fashion, and in the design and development of prototypes. They also make an important contribution to the production of parts for older machines. Nevertheless, compared with mass production based on subtractive technologies 3D printing is a slow and less efficient technology. It is energy intensive and there are problems with the finish of products that have been 3D printed. This technology continues to improve, but it is likely that it will remain an additional technology to support established mass production processes.

LABOR MARKETS AND SKILL IN THE WORLD ECONOMY

Although technology tends to capture the headlines, skill remains critical to firm strategies and a crucial aspect of policy design and implementation at the regional and national scales. Ultimately, the uptake of new technology is a skill question. Manufacturing firms substitute capital for labor when it is economically rational to do so. That rationale is calculated both within the firm and across the global supply chain. In practical terms, the promise of additive manufacturing depends not only on its own cost curve but on that for its immediate and existing substitutes.

The debate about reshoring, or "repatriation," of manufacturing is therefore a question not just of technology. It is also a question of the cost, quality, and risk associated with production alternatives like the labor-intensive manufacturing processes that drove distributed supply-chains into lower-cost economies in the first place. Cowell and Provo's Chapter 5 underscores the complexity of the reshoring debate. Hall's Chapter 4 on logistics reinforces the role of transportation costs. But fundamentally, the transitions in manufacturing are increasingly trade-offs between types of labor and capital configurations in which risk and quality are as important as cost. Weller's Chapter 2 reminds readers of the often forgotten issues embedded in how firms approach labor market reproduction and the assignment of responsibility for skill upgrading to the firm, individual, or various scales of government (cities, states/provinces, and national governments).

One factor driving the rapid changes in the available sets of labor and capital configurations is demographic transition in China. Lower-cost Chinese manufacturing labor has been a key driver in the outsourcing opportunities available to manufacturing firms in the US, UK, and other industrialized countries. This holds true for large producers as well as smaller manufacturers. Doussard and Schrock's analysis in Chapter 24 of the deindustrialization in the US in the period preceding the Great Recession describes the impact of this offshoring on manufacturing capacity.

The demographic transition in China is likely to set up a push factor to the reshoring trend that complements the pull factor of higher-technology production processes in industrialized countries. The demographic problem is also combined with inflation. Production costs are rising and manufacturing is already beginning to relocate from China. Combined with the development of China's internal consumer market, this suggests that the flow of manufactured goods from China will slow down and production will have to take place elsewhere (Daniels et al., 2011). Potentially there will be a manufacturing renaissance in countries like the UK focused on the design and production of high-value added products but also low-value products using efficient production systems (Bryson and Taylor, 2010; Bryson, Taylor, and Cooper, 2008).

Demographic issues are exacerbated by the complexities of China's own social policy challenges. The adoption in the 1970s of a one-child policy was instrumental in reducing the country's historic high birth rates. Four decades on, the consequences of this policy are now beginning to exert profound demographic effects nationally (Chen, 2009; Daniels et al., 2011). A rapidly ageing population, greater incidence of chronic diseases and disability, improved longevity, and higher standards of living among an emerging middle class are increasingly regarded as threats to the future well-being of Chinese society. The country already has about 108 million elderly (people aged 65 and over), or over one-fifth of the world's elderly population, and this number is expected to triple by 2050. In other words, China will become an "old" society before its economic development is ready to underwrite the fundamental economic and political challenges posed by such demographic transition. China's demographic time bomb is ticking and should explode in 10–15 years' time (Chen, 2009; Daniels et al., 2011; Linge, 1998; Zhang and Goza, 2006).

The transition in the Chinese labor market shifts the production cost calculus for manufacturing firms (Zhang and Goza, 2006). If exceptionally low cost labor is not readily available, then technological solutions become more viable from a cost perspective. And, higher-tech production processes generally require higher-skilled labor. Unfortunately, the erosion of demand for manufacturing skills in countries like the US and UK for the last quarter century has led to atrophying in the workforce development system (Doussard, 2013; Lowe, 2007; Osterman, 1999; Tilly, 1996). One key policy question going forward is whether and to what extent advanced industrialized economies will take seriously the issue of skill development. In all likelihood, a sustained retooling of the existing institutional frameworks will be necessary. Innovative new approaches to skill definition and delivery by regional institutional intermediaries capable of collaborating between workers, firms, educational institutions, and governments are essential to this process (Lowe, Chapter 30).

CONCLUSIONS: INSTITUTIONAL RETHINKING AND POLICY INNOVATION

In a world economy often described as "flat," it is the uneven distribution of assets and factor conditions between places that often stands out. And here, policy is a key determinant. Increasingly, policy variation shapes regional variation in costs, risks, and capacities. Firms make production decisions based on a set of "knowable" variables. A key conclusion from this collection is that manufacturing motives, incentives, and choices are not a set of mysteries. The lack of commitment by policy actors to invest in learning about manufacturing processes and thus adapting policy models to support and sustain localized manufacturing is a fundamental problem.

Firms and industry intermediaries are very clear about the production challenges and opportunities facing industrialized countries. Although many small US- or UK-based manufacturers are unaware of the policy dimensions influencing demographic transition in China, they do know that production costs are increasing rapidly for outsourced production. They also know that niche, customized, fast fashion products, craft, and batch product modes favor small-scale domestic manufacturing and dis-incentivize offshoring. Policies that support this mode of production would directly support domestic manufacturing as a whole.

The empirical challenge for researchers involves understanding the landscape of small manufacturers – their composition and their variation. What is increasingly evident in the case study literature is that the manufacturing landscape in both the US and UK includes a set of small and medium-sized firms that have survived the erosion of manufacturing over the past 30 years in addition to a set of emerging firms. These incumbent firms include the ACME Whistle Company in Birmingham, UK, which is among several manufacturers in and around the historic Jewellery Quarter which continue to produce specialized products and compete on quality rather than price (Bryson and Rusten, 2010). In the US, one example is WH Bagshaw in Nashua, New Hampshire, the oldest pin manufacturer in the country, manufacturing pins, pin assemblies, and wire since 1840. The national narratives about manufacturing have all but asserted that these firms do not and cannot exist as a category. And yet, they do. They are competitive and innovative. In these specific cases, the firms have taken different approaches to innovation over time. ACME Whistles has innovated through the development of new product lines and by expanding into new, niche markets like orchestral whistles and birdcalls. WH Bagshaw has been a process innovator building on quality and technology.[2]

In addition to these established small manufacturers, there are new manufacturing firms emerging to meet a growing market for small-scale, design-driven consumer goods like textiles, personal accessories, recreational equipment, and household items. These consumer end products require supply chains for everything from rivets to leather to ceramics. These suppliers must be geographically proximate to meet the scale and specifications of the end producers, which necessarily locate near relevant consumer markets to respond to trends, taste, and fashion (Bryson and Rusten, 2010; MacPherson and Vanchan, 2010). This re-emergence of demand for craft-based consumer end products has instigated a new round of flexible specialization. The challenge, however, is to reconnect and re-create these supply networks after decades of seemingly de-territorialized production systems. Online retail portals like ETSY enable small-scale producers to reach wider markets. And ETSY has begun to partner with other actors to connect its successful producers with the supplier networks they need to sustain and expand production. And here is where regionally embedded yet scalable intermediaries play a critical role in information sharing and coordination.

Information gaps and asymmetries are the subject of increasing study in the workforce development literature. There is a growing consensus among firms, workers, and educators that the skill delivery system is not working efficiently. Innovations in this area include information technology firms deploying beta models to codify and transfer worker knowledge across industry and technology contexts. The most notable of these is Mozilla and the MacArthur Foundation's "Open Badges" project.[3] The project works to establish transferable mechanisms for aggregating skills acquired from formal and informal training and present that array of skills as a legible "badge" digitally attached to a worker. The goal is to facilitate job matching for firms and workers and career planning for individuals.

Recent forms of supply-chain intermediaries closely associated with reshoring and the Maker Movement are intermediaries brokering partnerships, providing advocacy for producers, and acting as an information clearinghouse. Some of these intermediaries bridge the space between traditional production systems, new firm entrants, and labor market intermediaries at the regional scale. A leading example is Nortech in Northeast Ohio. Nortech is a regional not-for-profit enterprise that coordinates technology-based economic development efforts across sectors with an emphasis on small and medium-sized firms.

These coordination functions between regional economic development actors have been increasing over the last two decades. The governance and policy models remain in flux and are often updated and revamped (as they are in the UK currently through the Local Enterprise Partnership scheme). Less attention has been paid to an emerging set of firm network intermediaries that support localized manufacturing by brokering information, making connections, and advocating on behalf of a geographically distinct membership (Rantisi, 2002). Among this group of supply-chain or firm network intermediaries are SF Made based in San Francisco, California, Maker's Row based in Brooklyn, New York, and ADX Portland based in Portland, Oregon.

SF Made is a not-for-profit organization that builds the San Francisco area's economic base by supporting local manufacturing through information dissemination, advocacy, marketing, technical assistance, and training.[4] Maker's Row is a for-profit enterprise that provides information on manufacturers to designers and other manufacturers in order to bring together supply chains.[5] ADX Portland is a variation on the maker's space model that combines instruction (both for adults and children), access to space and machinery through memberships and space rental, and fabrication project management to form an incubator for incumbent and emerging small manufacturers to expand the feasibility of smaller (and incremental) scales of production.[6] These supply-chain intermediaries connected to the Maker Movement have a particularly powerful, if subtle, effect of recasting manufacturing as a practice of working with others rather than working for others. This is consistent with a "flexible specialization 2.0" approach to production in which a design-driven production model blurs the lines between development and deployment or innovation and production (Clark, 2013). In all these cases, the emerging model is one that provides a geographically distinct set of producers with more information about how to do what they do in partnership with producers, designers, and innovators in their immediate community. Essentially these intermediaries are building tightly bounded networks of production practice to localize and embed manufacturing capacities.

Localized manufacturing specializations are not disconnected from the national economy but rather are engines that drive distinct nodes in an organizationally complex and geographically distributed supply chain. Agglomeration economies – geographically embedded technology/industry specializations generating scale economies – do not exist independently of national and global production systems. Instead, these regional economies constitute an integrated system of specialized design and production sites in a distributed network of interlocking supplier and customer firm relationships. Increasingly policymakers are recognizing that the geographic distribution and organization of innovation and manufacturing matters. And, it matters a lot.

The research in this book identifies the potential for sustained growth in established sectors where design and innovation play a key role in production processes not just end products. Industrial geographers have tracked the spatial implications of shifts in the organization of production processes in response to process innovations for some time and made similar points (Christopherson and Clark, 2007; Clark, 2013; Glasmeier, 2000; MacPherson and Vanchan, 2010; Rutherford and Holmes, 2008; Stone, 1973). However, the policy impacts of this research have lagged. In May 2012 a Brookings Institution report entitled *Locating American Manufacturing* pulled together a body of empirical research on the geographic distribution of manufacturing in the US and made the argument for policies that take spatial organization into account – not just to leverage politicians or interest groups but in the design and implementation of manufacturing policies (Helper, Krueger, and Wial, 2012).

And this is a key point. Although consideration of the nuances of industrial geography is not unheard of in policy circles, it is generally applied to leveraging political interests instead of designing policies and programs. This distracts from the more empirical economic realities surrounding how and why the spatial organization of economic activities is arranged – and rearranged – in "the new economy." Recent debates in the US about the viability and sustainability of the American auto industry illustrate these points. In the US, automotive manufacturing is associated with a single city: Detroit. The automotive industry is also associated with a single set of end products: cars and trucks. In fact, the automotive industry is far broader in terms of geographic distribution and product diversity. The "automotive industry," and particularly its supplier networks, is interwoven across a range of civilian and military manufacturing applications. The automotive industry is far more geographically distributed and diverse in scope than the popular discourse would indicate. Further, policies that support the automotive industry tend to support manufacturing across the economy because the industry is woven into the national industrial ecosystem. In other words, the criticism that targeting the automotive industry is government "picking winners and losers" does not match the empirical reality (Helper et al., 2011).

To reinvigorate a resilient manufacturing sector, a renewed focus on small and medium-sized enterprises (SMEs) is necessary. This is not because small firms are the job generators of the new economy. It is because SMEs are critical for functional supply chains capable of absorbing innovations. Manufacturing policy should be concerned with pushing technology down the supply chain, not simply transferring innovations to high-tech start-ups. The implication is that this involves modifying existing intermediaries to provide technical assistance to SMEs on innovative production processes as well as designing and prototyping innovative products. These innovative processes include energy efficiency, life-cycle product design, and adoption of better, greener, safer materials. Intermediaries should focus on international standards and certifications, environmental, labor, corporate codes of conduct, systems and logistics, that provide suppliers with increased credibility with end producers in a global supply chain. The good news is that with additional empirical research, like that included in this volume, there is a real opportunity to build on the work of innovative intermediaries. Using that base, advanced industrialized countries have an opportunity to bring production back into localized networks of small producers which can support sustainable regional economies and promote urban regeneration.

NOTES

1. "Maker bots" are small-scale 3D printers. They are often seen in architecture and industrial design offices but are increasingly becoming part of the small-scale and craft production of the Maker Movement. See Maker's Row: http://makersrow.com. Retrieved November 25, 2013.
2. See Makers Row. (2013). How It Works. Retrieved November 25, 2013, from http://makersrow.com/how-it-works.
3. See Mozilla. (2013). Open Badges. Retrieved November 25, 2013, from http://openbadges.org/about/.
4. See SF Made. (2013). About SF Made. Retrieved November 25, 2013, from http://www.sfmade.org/about-us/about-sfmade/.
5. See Makers Row. (2013). How It Works. Retrieved November 25, 2013, from http://makersrow.com/how-it-works.
6. See ADX Portland. (2014). About ADX. Retrieved May, 2014, from http://www.adxportland.com.

REFERENCES

ADX Portland (2014). About ADX. Retrieved May, 2014, from http://www.adxportland.com.

Amin, A. (1994). *Post-Fordism: A Reader*. Oxford; Cambridge, MA: Blackwell.

Bluestone, B. and Harrison, B. (1982). *The Deindustrialization of America: Plant Closings, Community Abandonment, and the Dismantling of Basic Industry*. New York: Basic Books.

Bluestone, B., Harrison, B., and Clayton-Matthews, A. (1986). Structure vs Cycle in U.S. Manufacturing Job Growth. *Industrial Relations*, *25*(2), 101.

Bozeman, B., Laredo, P., and Mangematin, V. (2007). Understanding the Emergence and Deployment of "Nano" S&T. *Research Policy*, *36*(6), 807.

Bryson, J.R. and Rusten, G. (2008). Transnational Corporations and Spatial Divisions of "Service" Expertise as a Competitive Strategy: The Example of 3M and Boeing. *Service Industries Journal*, *28*(3), 307–323.

Bryson, J.R. and Rusten, G. (2010). *Design Economies and the Changing World Economy*. Abingdon, Oxon; New York, NY: Routledge.

Bryson, J.R. and Taylor, M. (2010). Competitiveness by Design and Inimitability through Service: Understanding the Dynamics of Firm-Based Competition in the West Midlands Jewellery and Lock Industries. *Service Industries Journal*, *30*(4), 583–596.

Bryson, J.R., Daniels, P.W., and Warf, B. (2004). *Service Worlds: People, Organisations, Technologies*. London: Routledge.

Bryson, J.R., Taylor, M., and Cooper, R. (2008). Competing by Design, Specialization and Customization: Manufacturing Locks in the West Midlands (UK). *Geografiska Annaler: Series B, Human Geography*, *90*(2), 173–186.

Chen, S. (2009). Aging with Chinese Characteristics: A Public Policy Perspective. *Ageing International*, *34*(1), 172–188.

Christopherson, S. and Clark, J. (2007). *Remaking Regional Economies: Power, Labor, and Firm Strategies in the Knowledge Economy* (1st ed.). New York: Routledge.

Christopherson, S. and Storper, M. (1989). The Effects of Flexible Specialization on Industrial Politics. *Industrial & Labor Relations Review*, *42*(3), 331.

Clark, J. (2012). Is There a Progressive Approach to Innovation Policy? *Progressive Planning*, (Winter), 17–21.

Clark, J. (2013). *Working Regions: Reconnecting Innovation and Production in the Knowledge Economy*. London: Routledge.

Cooke, P. (2007). How Benchmarking Can Lever Cluster Competitiveness. *International Journal of Technology Management*, *38*(3), 292.

Corbridge, S., Martin, R.L., and Thrift, N.J. (1994). *Money, Power and Space*. Oxford; Cambridge: Blackwell.

Daniels, P.W., Rubalcaba, L., Stare, M., and Bryson, J.R. (2011). How Many Europes? Varieties of Capitalism, Divergence and Convergence and the Transformation of the European Services Landscape. *Tidschrift voor Economische en Sociale Geografie*, *102*(2), 146–161.

Doussard, M. (2013). *Degraded Work: The Struggle at the Bottom of the Labor Market*. Minneapolis, MN: The University of Minnesota Press.

Feller, I. (1997). Manufacturing Technology Centers as Components of Regional Technology Infrastructures. *Regional Science and Urban Economics*, *27*(2), 181.

Glasmeier, A. (2000). *Manufacturing Time: Global Competition in the Watch Industry, 1795–2000*. New York: The Guilford Press.

Harrison, B. and Bluestone, B. (1988). *The Great U-turn: Corporate Restructuring and the Polarizing of America*. New York: Basic Books.

Helper, S. (2009). The High Road for U.S. Manufacturing. *Issues in Science and Technology*, *25*, 39.

Helper, S., Krueger, T., and Wial, H. (2012). *Locating American Manufacturing: Trends in the Geography of Production* (T.M.P. Program, Trans.). Washington, DC: The Brookings Institution.

Helper, S., Park, K.W., Kuan, J., Krueger, T., Warofka, A., Zhu, J., Eisenmenger, W., and Peshek, B. (2011). The US Auto Supply Chain at a Crossroads: Implications of an Industry in Transformation. *Driving Change: Greening the Automotive Workforce*. Cleveland: Case Western Reserve University.

Kaplan, T. (2011, September 27). High-Tech Companies to Invest $4 Billion in New York State, Cuomo Says. *The New York Times*.

Kumar, S. and Kruth, J.P. (2010). Composites by Rapid Prototyping Technology. *Materials and Design*, *31*, 850–856.

Linge, G. (1998). An Overview: Towards Sustainable Development? In G. Linge (ed.), *China's New Spatial Economy: Heading Towards 2020*. Oxford: Oxford University Press.

Lohr, S. (2010, February 18). Turning Patents Into "Invention Capital". *The New York Times*.

Lowe, N. (2007). Job Creation and the Knowledge Economy: Lessons from North Carolina's Life Science Manufacturing Initiative. *Economic Development Quarterly*, *21*(4), 339–353.

MacPherson, A.D. and Vanchan, V. (2010). Locational Patterns and Competitive Characteristics of Industrial Design Firms in the United States. In G. Rusten and J.R. Bryson (eds), *Industrial Design, Competition and Globalization*. Houndmills: Palgrave Macmillan.

Makers Row (2013). How It Works. Retrieved November 25, 2013, from http://makersrow.com/how-it-works.

Mozilla (2013). Open Badges. Retrieved Novemeber 25, 2013, from http://openbadges.org/about/.

Myhrvold, N. (2010). The Big Idea: Funding Eureka! *Harvard Business Review*, *88*(3), 40–50.

Noyelle, T.J. and Stanback, T.M. (1983). *The Economic Transformation of American Cities*. Totowa, NJ: Rowman & Allanheld.

Osterman, P. (1999). *Securing Prosperity: The American Labor Market: How It Has Changed and What To Do About It*. Princeton, NJ: Princeton University Press.

Piore, M.J. and Sabel, C.F. (1984). *The Second Industrial Divide: Possibilities for Prosperity*. New York: Basic Books.

Rantisi, N.M. (2002). The Competitive Foundations of Localized Learning and Innovation: The Case of Women's Garment Production in New York City. *Economic Geography*, *78*(4), 441.

Reich, R.B. (1992). *The Work of Nations: Preparing Ourselves for 21st-Century Capitalism*. New York: Vintage Books.

Robinson, D.K.R., Rip, A., and Mangematin, V. (2007). Technological Agglomeration and the Emergence of Clusters and Networks in Nanotechnology. *Research Policy*, *36*(6), 871.

Rutherford, T. and Holmes, J. (2008). "The Flea on the Tail of the Dog": Power in Global Production Networks and the Restructuring of Canadian Automotive Clusters. *Journal of Economic Geography*, doi: 10.1093/jeg/lbn014.

SF Made (2013). About SF Made. Retrieved November 25, 2013, from http://www.sfmade.org/about-us/about-sfmade/.

Shaiken, H., Herzenberg, S., and Kuhn, S. (1986). The Work Process Under More Flexible Production. *Industrial Relations*, *25*(2), 167.

Stone, K. (1973). The Origins of Job Structures in the Steel Industry. In M.R. Richard Edwards and David Gordon (eds), *Labor Market Segmentation*. Toronto: D.C. Heath and Company.

Tassey, G. (2010). Rationales and Mechanisms for Revitalizing US Manufacturing R&D Strategies. *Journal of Technology Transfer*, *35*, 283–333.

Tilly, C. (1996). *Half a Job: Bad and Good Part-Time Jobs in a Changing Labor Market*. Philadelphia: Temple University Press.

Wohlers, T. (2001). *Wohlers Report: Rapid Prototyping and Tooling*. State of the Industry Annual Worldwide Progress Report. Fort Collins, CO: Wohlers Associates.

Yan, X. and Gu, P. (1996). A Review of Rapid Prototyping Technologies and Systems. *Computer-Aided Design*, *28*(4), 307–318.

Zhang, Y. and Goza, F.W. (2006). Who Will Care for the Elderly in China? A Review of the Problems Caused by China's One-Child Policy and their Potential Solutions. *Journal of Aging Studies*, *20*(2), 151–164.

Zysman, J. and Tyson, L.D.A. (1983). *American Industry in International Competition: Government Policies and Corporate Strategies*. Ithaca: Cornell University Press.

Index